COMPUTER ORIENTED CIRCUIT DESIGN

COMPUTER APPLICATIONS IN ELECTRICAL ENGINEERING SERIES

Franklin F. Kuo, *editor*

Davis *Computer Data Displays*
Jensen & Lieberman *IBM Circuit Analysis Program: Techniques and Applications*
Kuo & Magnuson *Computer Oriented Circuit Design*

PRENTICE-HALL INTERNATIONAL, INC., *London*
PRENTICE-HALL OF AUSTRALIA, PTY. LTD., *Sydney*
PRENTICE-HALL OF CANADA, LTD., *Toronto*
PRENTICE-HALL OF INDIA (PRIVATE) LTD., *New Delhi*
PRENTICE-HALL OF JAPAN, INC., *Tokyo*

COMPUTER ORIENTED CIRCUIT DESIGN

Edited By

Franklin F. Kuo

Professor of Electrical Engineering
University of Hawaii

Waldo G. Magnuson

Electronic Engineering Department
Lawrence Radiation Laboratory

Prentice-Hall, Inc.
Englewood Cliffs, N.J.

Current Printing:

1 0 9 8 7 6 5 4 3 2 1

13–165860–3

Library of Congress Catalog Card Number 72-100083

Printed in the United States of America

PREFACE

The digital computer has become a highly useful tool in engineering design. With the advent of time-shared computers and on-line graphical input-output devices, the computer has become a real-time partner in the creative processes of the successive stages of engineering design. This book concerns a number of different aspects of design of electronic circuits using the digital computer. What is computer-aided design? To us it means programs for performing (a) circuit analysis, (b) graphical input and output, and (c) printed or integrated circuit layout. The ingredients that go into computer aided design programs for electronic circuits are: network topology, state space techniques, numerical analysis, modeling of solid state devices, optimization techniques, graphic data processing, and automated circuit layout. This book covers all of these areas in sufficient depth to enable the reader to gain familiarity with the methods used so that he could develop special purpose programs for his own use or so he could use some of the existing programs described.

The first chapter by Kuo and Magnuson is a broad survey of the field of computer-aided analysis and design of electronic circuits. The coverage emphasizes breadth rather than depth. The next three chapters present the basic ingredients of computer-based methods of network analysis. Chapter 2 by Dawson covers the topological aspects of circuit analysis in terms of both the connectivity and the algebraic properties associated with linear graphs. In the following chapter, Branin presents methods for the computer generation and solution of circuit equations based upon state space concepts. Many of these techniques have been developed by Branin himself. Chapter 4, by Chua, deals with procedures of nonlinear circuit analysis and compares different algorithms for nonlinear analysis by computer. The next two chapters, both by Temes, are perhaps the most *design* rather than *analysis* oriented chapters in the book. Chapter 5 is concerned with the use of optimization techniques in circuit design. These techniques may eventually lead to *completely automated* circuit design by computer, wherein no decisions are left to the engineer in the design process. The second Temes chapter treats insertion loss filter design by computer. This is an area which has been almost completely taken over by the computer. Whereas in pre-computer days, an elaborate insertion loss filter design could have required days to complete, now, using one of the many elaborate and sophisticated filter design programs, the same design requires only a few minutes of computer time. Moreover, the computer design is probably more efficient and accurate than the human design. Chapter 7 by Daniel concerns a vital ingredient in all electronic circuit analysis programs—that of accurately simulating the behavior of a semiconductor device, such as a transistor. Daniel's treatment is complete in that both small and large signal device models are covered. In Chapter 8, Magnuson and Walsh critically examine a number of analysis programs from the standpoint of the user and compares their available options, features, advantages, and drawbacks. This chapter is particularly recommended for the *user* rather than the *programmer.*

At this point, a natural dividing line in the organization of the book occurs. The first 8 chapters deal with circuit analysis and design *per se.* In the last 5 chapters, the concentration is upon *computer graphics* as applied to circuit analysis and design. The first two graphics chapters are general and serve to put the whole area in perspective. The final three chapters concern specific applications of computer graphics to circuit design. Chapter 9 by Cralle and Michael, and Chapter 10 by Michael and Bertran are concerned respectively with the hardware and software requirements of computer graphics. In Chapter 11, Magnuson, Kuo and Walsh apply the methodologies discussed in the previous two chapters to the graphic input system GINA: *G*raphical *I*nput for *N*etwork *A*nalysis. The final two chapters deal with computer aids to the layout of printed and integrated circuits. Chapter 12, by Zane and Wilber, deals with the algorithms and internal organization of the program, PUZZLE, for the layout of printed circuit boards. In Chapter 13, Hazlett and Beemiller discuss methods and procedures for applying computer techniques for the automatic generation of film photomasks for LSI (large scale integration) integrated circuits.

Reference is made to related material at the end of each individual chapter. These bibliographies are ordered according to the appearance of the entries in the text.

The influence on the authors of their many colleagues is gratefully acknowledged. The editors wish to acknowledge the help and advice provided by A. J. Stripeika and E. H. Hulse of the Lawrence Radiation Laboratory, Livermore, California. In addition, we thank Rita Bonivert and Marie Garibaldi for their assistance in the preparation of the manuscript.

Honolulu, Hawaii F. F. Kuo
Livermore, California W. G. Magnuson, Jr.

CONTENTS

COMPUTER ORIENTED CIRCUIT DESIGN

CHAPTER 1

COMPUTER AIDED DESIGN OF ELECTRONIC CIRCUITS †

F. F. KUO, University of Hawaii
W. G. MAGNUSON, Jr., Lawrence Radiation Laboratory

INTRODUCTION

As engineering systems become increasingly complicated, the designer must rely on more sophisticated design techniques. This book attempts to present those modern day design techniques for electronic circuits which make use of the digital computer as the central design tool. What is design? To us, design is iterative analysis until certain specifications are met. The design process may be characterized by a

---

† Parts of this chapter have been extracted from the following articles:

D. F. Dawson, F. F. Kuo, and W. G. Magnuson, Jr., "Computer aided design of electronic circuits - a user's viewpoint," *Proc. IEEE*, vol. 55, pp. 1946-1954, November, 1967.

W. G. Magnuson, Jr., "Network characterization for efficient computer time-domain circuit analysis," *10th Midwest Symp. on Circuit Theory (Conf. Record)*, pp. I-4-1 to I-4-10, 1967.

simple block diagram (Fig. 1.1). From available technology, design techniques, and experience, the design engineer formulates a possible design to meet the specifications. The design must then be analyzed to find out if the proposed design is reasonable. For the case of electronic circuit design the next step is to construct a breadboard model and test the design. Feedback occurs at the analysis and experimental stages (and possibly even includes the specifications) until the final design is achieved.

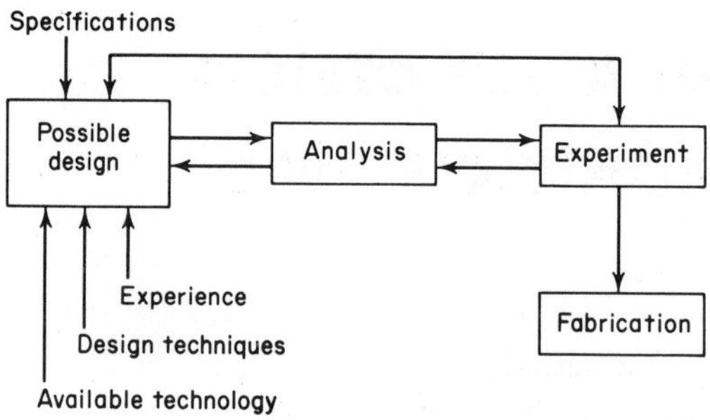

FIG. 1.1. The design process.

In recent years, engineers have turned with ever increasing attention to the use of computers in electronic circuit design. The central tools at their disposal are electronic circuit simulator programs which perform dc, ac, transient, and in some cases, sensitivity, and statistical analyses. These programs are used in the design process by performing repeated analyses of a network. Most of these programs include equation compilers - that is, the programs generate the necessary network equations from a node-to-node and element-type description of the network. Generally, the circuit description input is highly user-oriented, and often allows considerable freedom in the description of numerical values, order of input, numbering of nodes, etc. Some examples of user-oriented input language analysis programs are: NET-1 [1], SCEPTRE [2], CIRCUS [3], CALAHAN† [4], ECAP [5], CIRCAL [6], and AEDNET [7]. In this chapter we will examine the common features of some of these programs and discuss their use for circuit design.

First we briefly discuss transistor modeling; although the emphasis is on transistor equivalent circuits, the results may easily be extended to include other active devices. Following that are short discussions of

† The "Linear Network Analysis Digital Computer Program" by D. A. Calahan is often referred to as CALAHAN.

general network analysis in terms of a state-variable approach and of the computer solution procedure. Only the salient features are presented for analysis and the solution procedure. In the next few sections we examine the requirements on the operating environment for effective use of the analysis programs in circuit design. In particular, we discuss the implications of time-sharing upon circuit design and the impact of using graphic consoles for on-line design. We next examine the features of some representative circuit layout programs. Finally, we discuss some requirements for a *design supervisory program*, whereby an opaque interface is created between the engineer and the computer. The engineer would not need to know the capabilities, input requirements, and output options of any of the individual programs that he might use in the course of the design. He would merely place the schematic on the graphic console. The design supervisor then asks a series of questions, and selects the appropriate programs for use in the design.

1.1 TRANSISTOR MODELING

To obtain meaningful results from a network analysis, passive and active linear and nonlinear devices must be represented to an acceptable degree of approximation. Furthermore, the model chosen should be simple as to present to the design engineer a readily interpretable equivalent circuit, yet be accurate in describing the physical device to predict the circuit performance.

Basically, here, we are interested in modeling transistors for large-signal transient analysis and in particular, characterizing transistors for efficient analysis by computer programs. Although there are many large-signal models that may be used, three have received much attention in the literature:

1. Ebers-Moll large-signal model [8],
2. Beaufoy-Sparkes charge-control model [9],
3. Linvill lumped model [10].

All three predict equally well the large-signal behavior of transistors but differ in their representation of the physical processes. Generally, transient circuit analysis computer programs are developed to predict the network response in terms of element values and externally measurable device parameters. The programs are circuit design (not device design) oriented; therefore the object is not to describe the physical processes of device operation, but to compute the circuit response.

Consequently, the Ebers-Moll and the charge-control models are generally more useful for computer analysis programs.

Several considerations must be taken into account when representing an active device by an equivalent circuit:

1. *Operating region.* If the device performs only over a small range in voltages and currents, the parameters in the model may be regarded as approximately linear. Otherwise the device model may require nonlinear parameter representation.

2. *Frequency range.* When the analysis is over a limited frequency range, additional simplification may be made in representing the device.

3. *Accuracy.* Generally, in any analysis, the degree of accuracy required will greatly affect the complexity of the equivalent circuit used in the modeling.

4. *Measurability of parameters.* Not only must numerical values for the circuit elements in the model be determined, but their dependency on electrical and physical conditions must be determined as well.

5. *Efficiency in use.* For a general time domain analysis, computer solution times are highly dependent upon the equivalent circuit model.

An especially useful large-signal equivalent circuit for the transistor is that proposed by Ebers and Moll [8]. Figure 1.2 depicts this model with its extension to include dynamic behavior. In small-signal analyses, the signal levels are sufficiently small so that all elements can be assumed to be constant. This situation is altered drastically for the large-signal dynamic model. In this instance, the elements are explicit functions of the network voltages and currents. For example, the base-emitter capacitance for a transistor has the form

$$C_{be} = C_{te} + C_{de} \equiv \frac{C_0}{(V_z - V_{be})^n} + \frac{q}{kT\omega_\alpha} \frac{I_{es}}{1 - \alpha_n \alpha_i} e^{\frac{qV_{be}}{kT}} \tag{1.1}$$

That is, the capacitance has essentially two components: transition and diffusion capacitance – both nonlinear with junction voltage. On specification sheets, C_{te} is generally given by a plot of capacitance vs reverse bias voltage.

Generally, the emitter and collector static characteristics are represented by curves (current vs voltage) on device specification sheets; these curves may be used directly or, alternatively, the parameters for the analytic representation may be found from the curves. In the same manner, the extrinsic base resistance changes with current level; to represent the

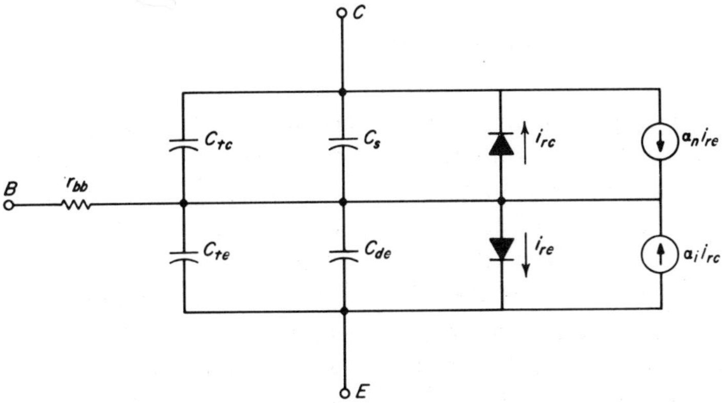

$$i_{re} = I_{es}\left(e^{qV_{BE}/kT} - 1\right), \quad i_{rc} = I_{cs}\left(e^{qV_{BC}/kT} - 1\right)$$

$$C_{te} = \frac{C_0}{(V_z - V_{BE})^n}, \qquad C_{tc} = \frac{C_0}{(V_z - V_{BC})^n}$$

$$C_{de} = \frac{q}{kT} \cdot \frac{i_{re}}{2\Pi f_a}, \qquad C_s = \frac{q}{kT} \cdot T_s \cdot i_{rc}$$

FIG. 1.2. Ebers and Moll large-signal transistor equivalent circuit.

transient network response accurately, we must represent it as a nonlinear element in the analysis. A similar requirement might be necessary for the collector saturation resistance.

As pointed out above, there are a number of ways to describe the electrical behavior of an element. In terms of network analysis by computer programs, however, these different representations can be grouped into four classifications:

1. A numerical constant,
2. A table,
3. An analytic expression,
4. Or a combination of the above.

In performing an analysis, not all four element representations will be equivalent in terms of program efficiency. For example, a table-lookup requires more time than extracting a single constant value for an unknown. Furthermore, evaluating an analytic expression may require many lengthy program operations compared to a simple table-lookup procedure.

1.2 GENERAL NETWORK ANALYSIS

The development of a general network analysis computer program must be based on the framework of a mathematical description of the general loop and node system of equations [11, 12]. Otherwise the program will be reduced to solving special cases and would consequently be of limited usefulness. Basically, the formulation of a program can be divided into two phases: Obtaining a set of simultaneous ordinary differential equations that describe a general network, and establishing a numerical procedure to solve the network equations. This section considers these two phases and relates the details of a general analysis to the problem of excessive solution times.

The starting point in formulating the necessary equations in a general network analysis using state variables is Kirchhoff's current and voltage laws. These laws, expressed in matrix notation, are

$$\mathbf{A}\mathbf{i}_{\mathrm{b}}(t) = 0 \tag{1.2}$$

$$\mathbf{B}\mathbf{v}_{\mathrm{b}}(t) = 0, \tag{1.3}$$

where $\mathbf{A}$ is the *reduced incidence matrix* and $\mathbf{B}$ the *fundamental-loop matrix*.

Kirchhoff's laws are concerned only with the branch voltages and currents and with the topology of the network currents. They do not specify the nature of the branch elements themselves (resistor, capacitor, inductor, sources) or the linearity of the branch element.

Physically, each branch of an electrical network may consist of three distinct electrical devices: (1) a passive element; (2) an ideal voltage source in series with the passive element; and (3) an ideal current source in parallel with the passive element.

The relative position and orientations of these electrical elements for a general rth branch are shown in Fig. 1.3; if we use the notation of that figure as definitions for voltages and currents, then Kirchhoff's equations become

$$\mathbf{AI} - \mathbf{AJ} = 0 \tag{1.4}$$

and

$$\mathbf{BV} - \mathbf{BE} = 0. \tag{1.5}$$

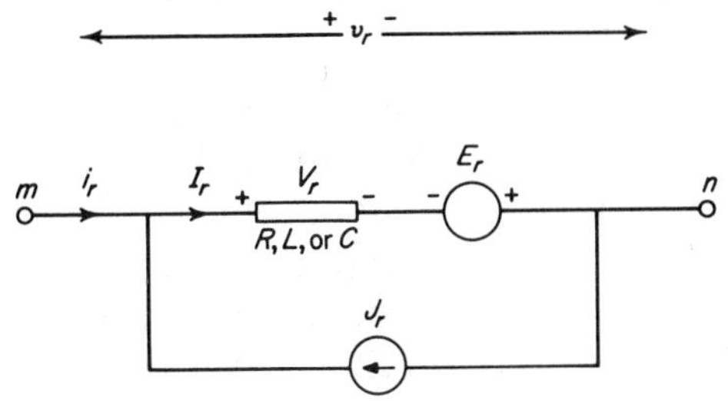

FIG. 1.3. A general network branch.

Again from the notation of Fig. 1.3, a generalization of Ohm's Law can be represented as

$$\mathbf{E} + \mathbf{v} = \mathbf{Z}(\mathbf{J} + \mathbf{i}) \tag{1.6}$$

and

$$\mathbf{J} + \mathbf{i} = \mathbf{Y}(\mathbf{E} + \mathbf{v}), \tag{1.7}$$

where the **Z** and **Y** matrices consist of complex numbers in the most general case, and are mutually inverse, that is, $\mathbf{YZ} = \mathbf{ZY} = \mathbf{U}$, where **U** is a unit matrix.

Ohm's Law leads to a system of ordinary integro-differential equations with constant coefficients and is generally solved by the Laplace transform method. The normal procedure for solving a system of simultaneous ordinary integro-differential equations numerically would be first to express the system as a higher-order set of differential equations, removing the integrals through differentiation. Then the system of higher-order differential equations is reduced to a set of first-order equations by making a simple change of variable.

An alternative development consists of choosing a set of network variables that leads automatically to a set of first-order differential equations [13, 14, 15]. The minimal set of branch currents and voltages whose instantaneous values are sufficient to determine completely the instantaneous state of the network is called a *complete set of dynamically independent network variables* or a *state variable set*. The process used to arrive at such a set depends upon forming the branch equations in a particular manner. This form involves classifying the network branches into the following six groups:

Class	*Description*
α	Capacitive links
β	Resistive links
γ	Inductive links
δ	Capacitive branches
ϵ	Resistive branches
ζ	Inductive branches

This branch classification is used to partition the network vectors and matrices into link-contained (subscript L) and branch-contained (subscript T) vectors and matrices. For example, the source vector becomes

$$\mathbf{E} = \begin{bmatrix} \mathbf{E}_L \\ \mathbf{E}_T \end{bmatrix}. \tag{1.8}$$

Each of the subvectors is then further partitioned according to the six classes; i.e.,

$$\mathbf{E}_L = \begin{bmatrix} \mathbf{E}_\alpha \\ \mathbf{E}_\beta \\ \mathbf{E}_\gamma \end{bmatrix} \qquad \mathbf{E}_T = \begin{bmatrix} \mathbf{E}_\delta \\ \mathbf{E}_\epsilon \\ \mathbf{E}_\zeta \end{bmatrix}. \tag{1.9}$$

It turns out that the voltages across capacitors and the currents through inductors form a minimal set of state variables for electrical networks. This pair, $\mathbf{I}_\gamma$ and $\mathbf{V}_\delta$, can be used to completely describe the state of a network. Then $\mathbf{I}_\gamma$ and $\mathbf{V}_\delta$, as dependent variables, can describe the network behavior by a system of first-order differential equations.

In the same way, the incidence matrix **A** and the fundamental-loop matrix **B** may be partitioned as

$$\mathbf{A} = [\mathbf{A}_L\, \mathbf{A}_T] \tag{1.10}$$

and

$$\mathbf{B} = [\mathbf{U}\mathbf{B}_T], \tag{1.11}$$

where, again, **U** is the unit matrix.

Besides the **A** and **B** matrices, there is an additional matrix of interest in the following development – the fundamental cut-set matrix **C**. It may be constructed from either the **A** or **B** matrices and is defined as the system of cut-sets in which each cut-set includes exactly one branch of a tree of the graph. The cut-set provides a completely general method to choose an appropriate set of network equations, i.e., a sufficient number of independent equations.

The fundamental cut-set matrix equations are analogous to the fundamental loops. In fact, the rows of the incidence matrix are expressible as linear combinations of rows of **C** (Ref. 12, p. 96). Since each fundamental cut-set contains a tree branch that is in no other cut-set, the Kirchhoff current law equations within **C** are linearly independent. This fact makes the cut-set matrix more attractive to use than the incidence matrix because linearly independent equations are ensured, whereas they are not in the reduced incidence matrix.

From the fundamental cut-set matrix, Kirchhoff's current law expression becomes

$$\mathbf{CI} - \mathbf{CJ} = 0. \tag{1.12}$$

If the columns of the **A**, **B**, and **C** matrices are ordered according to branches and links as previously, then the fundamental cut-set matrix can be partitioned as

$$\mathbf{C} = [\mathbf{C}_L\,\mathbf{U}], \tag{1.13}$$

and the following relations result (Ref. 12, p. 98)

$$\mathbf{C} = \mathbf{A}_T^{-1}\,\mathbf{A} \text{ and } \mathbf{C}_L = -\mathbf{B}_T' = \mathbf{A}_T^{-1}\mathbf{A}_L\,. \tag{1.14}$$

From (1.14), (1.13) becomes

$$\mathbf{C} = [-\mathbf{B}_T'\,\mathbf{U}]. \tag{1.15}$$

Finally we can construct a combined matrix equation that describes the network equations. To do this, we use Ohm's Law, Kirchhoff's voltage law, and Kirchhoff's current law in terms of the fundamental cut-set matrix. This single matrix equation becomes

$$\begin{bmatrix} \mathbf{O} & \mathbf{B} \\ \mathbf{C} & \mathbf{O} \\ \mathbf{Z}_{RL} & -\mathbf{Y}_C \end{bmatrix} \begin{bmatrix} \mathbf{I} \\ \mathbf{V} \end{bmatrix} = \begin{bmatrix} \mathbf{BE} \\ \mathbf{CJ} \\ \mathbf{O} \end{bmatrix} \tag{1.16}$$

This equation may be written in expanded form (Fig. 1.4) using the six classes for network branches given earlier.

$$
\begin{array}{l}
\text{Kirchhoff voltage laws} \left\{ \begin{array}{l} \\ \\ \\ \end{array} \right. \\
\text{Kirchhoff current laws} \left\{ \begin{array}{l} \\ \\ \\ \end{array} \right. \\
\text{Ohm's law equations} \left\{ \begin{array}{l} \\ \\ \\ \\ \\ \\ \end{array} \right.
\end{array}
\left[\begin{array}{cccccccccccc}
\cdot & \cdot & \cdot & \cdot & \cdot & \cdot & 1 & 0 & 0 & B_{\alpha\delta} & 0 & 0 \\
\cdot & \cdot & \cdot & \cdot & \cdot & \cdot & 0 & 1 & 0 & B_{\beta\delta} & B_{\beta\epsilon} & 0 \\
\cdot & \cdot & \cdot & \cdot & \cdot & \cdot & 0 & 0 & 1 & B_{\gamma\delta} & B_{\gamma\epsilon} & B_{\gamma\zeta} \\
-B'_{\alpha\delta} & -B'_{\beta\delta} & B'_{\gamma\delta} & 1 & 0 & 0 & \cdot & \cdot & \cdot & \cdot & \cdot & \cdot \\
0 & -B'_{\beta\epsilon} & -B'_{\gamma\epsilon} & 0 & 1 & 0 & \cdot & \cdot & \cdot & \cdot & \cdot & \cdot \\
0 & 0 & -B'_{\gamma\zeta} & 0 & 0 & 1 & \cdot & \cdot & \cdot & \cdot & \cdot & \cdot \\
1 & \cdot & \cdot & \cdot & \cdot & \cdot & -C_{\alpha}\frac{d}{dt} & \cdot & \cdot & \cdot & \cdot & \cdot \\
\cdot & R_{\beta\beta} & \cdot & \cdot & \cdot & \cdot & \cdot & -1 & \cdot & \cdot & \cdot & \cdot \\
\cdot & \cdot & L_{\gamma\gamma}\frac{d}{dt} & \cdot & \cdot & L_{\gamma\zeta}\frac{d}{dt} & \cdot & \cdot & -1 & \cdot & \cdot & \cdot \\
\cdot & \cdot & \cdot & 1 & \cdot & \cdot & \cdot & \cdot & \cdot & C_{\delta}\frac{d}{dt} & \cdot & \cdot \\
\cdot & \cdot & \cdot & \cdot & R_{\sigma\sigma} & \cdot & \cdot & \cdot & \cdot & \cdot & -1 & \cdot \\
\cdot & \cdot & L_{\zeta\gamma}\frac{d}{dt} & \cdot & \cdot & L_{\zeta\zeta}\frac{d}{dt} & \cdot & \cdot & \cdot & \cdot & \cdot & -1
\end{array} \right]
\left[\begin{array}{c} I_{\alpha} \\ I_{\beta} \\ I_{\gamma} \\ I_{\delta} \\ I_{\epsilon} \\ I_{\zeta} \\ V_{\alpha} \\ V_{\beta} \\ V_{\gamma} \\ V_{\delta} \\ V_{\epsilon} \\ V_{\zeta} \end{array} \right]
=
\left[\begin{array}{c} BE \\ BE \\ BE \\ CJ \\ CJ \\ CJ \\ 0 \\ 0 \\ 0 \\ 0 \\ 0 \\ 0 \end{array} \right]
$$

FIG. 1.4. Combined matrix equation describing Ohm's and Kirchhoff's laws in terms of the fundamental cut-set matrix.

Starting with the network equations of (1.16) (or Fig. 1.4), we eliminate variables until a dynamically independent set is obtained. This process depends on forming the branch equations in a certain form, which involves classifying the network branches into the six classes stated earlier. This produces twelve classes for the branch voltages and currents. The elimination process consists of expressing ten of these classes algebraically in terms of the remaining two classes.

The branch equations used in the elimination process are formed from a particular tree construction [15, 16]. Tree-branch elements are chosen in this order: capacitors, resistors, and inductors. Then, in carrying out the elimination process, the variables

$$\mathbf{I}_\alpha, \mathbf{I}_\beta, \mathbf{I}_\delta, \mathbf{I}_\epsilon, \mathbf{I}_\zeta, \mathbf{V}_\alpha, \mathbf{V}_\beta, \mathbf{V}_\gamma, \mathbf{V}_\epsilon, \mathbf{V}_\zeta$$

are expressed in terms of the state variables

$$\mathbf{I}_\gamma \text{ and } \mathbf{V}_\delta,$$

together with the various generators and their differentials.

If one first starts with Kirchhoff's laws in terms of the six voltage and current classes, Ohm's Law can be used to solve directly for the resistor voltages and currents. These can then be used to solve for the inductor and capacitor voltages in currents. More will be said of this in Chapter 3.

The elimination process leading to the equations used in the final solution process is different from, but entirely equivalent to, Bryant's method [14]. With but minor differences, the equations summarized in this section are similar to those used by the SCEPTRE circuit analysis program in solving for the time response of a network [2].

1.3 SOLUTION PROCEDURE

The set of first-order differential equations established from topological circuit considerations and the elimination procedure can be evaluated at each value of time if the state variables are known at each time. The normal procedure for obtaining a time-history of the circuit response is to start with time equal to zero and evaluate the network initial conditions to establish values for the state variables.

Solution of the network equations is then accomplished through matrix operations that produce state variable derivatives, which must be numerically integrated to produce new state variables for the evaluation at the next time step. That is, the state variable **S** at time $i + 1$ is evaluated from

$$\mathbf{S}_{(i+1)} = \mathbf{S}_{(i)} + \Delta t \dot{\mathbf{S}}_{(i)}, \tag{1.17}$$

where Δt is the time step between evaluations of the network equations. After the computed derivatives of the state variables are numerically integrated, the variable elements and sources are updated and the system of network equations is evaluated at the next time step. The value of each element voltage and current is available at each time step as output. This solution procedure is shown schematically in Fig. 1.5. A repetitive evaluation of the transient solution and subsequent integration of the state variable derivatives produces the time-history of the network response.

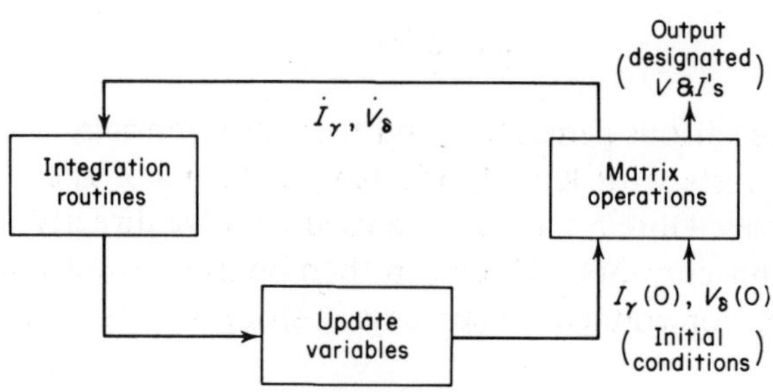

FIG. 1.5. Transient solution procedure.

The solution of the network equations, among other things, depends on integrating the differential equations describing the state variables at each time step. The rate at which the solution progresses depends on the method of performing this integration. There are two basic categories of methods available for this process:

1. One-step methods that do not depend on the past history of the solution. The most widely used one-step method is the *Runge-Kutta*.

2. Multistep methods in which the next solution point is obtained from previous solutions. Most methods of this type are called *predictor-corrector*.

Basically, the process can be described as follows: We take the starting point, compute the slope, and move in that direction for a short distance, and select the next point. From this new starting point the slope is again evaluated, and so on. Control of the errors between the true solution and the computed solution is not achieved by this method; so, generally, a slightly different procedure is followed.

In general, the solution to the first-order differential equation

$$\frac{dx}{dt} = f(x, t) \tag{1.18}$$

effectively replaces the result of truncating a Taylor-series expansion of the form

$$x_{n+1} = x_n + h\dot{x}_n + \frac{h^2}{2!}\ddot{x}_n + \frac{h^3}{3!}\dddot{x}_n + \cdots \tag{1.19}$$

by an approximation in which x_{n+1} is calculated from a formula of the type (Ref. 16, p. 233)

$$x_{n+1} = x_{n-p} + h(a_{-1}\dot{x}_{n-1} + a_0\dot{x}_n + a_1\dot{x}_{n-1} + \cdots + a_r\dot{x}_{n-r}). \tag{1.20}$$

In the special case where $f(x, t) = Ax$, where A is a constant, the preceding equation takes the form

$$(1 - a_{-1}Ah)x_{n+1} = x_{n-p} + Ah(a_0x_n + a_1x_{n-1} + \cdots + a_rx_{n-r}), \tag{1.21}$$

where at least one of the coefficients of a_r is nonzero. The general solution to this equation is (Ref. 16, p. 203)

$$x_n = c_0p_0^n + c_1p_1^n + \cdots + c_rp_r^n\ , \tag{1.22}$$

where the c coefficients are determined by the initial values of x_0, x_1, ... x_{r-1} and p_1, p_2 ... p_r are their roots of the characteristic equation obtained by substituting the particular solution $x_n = p^nx_0$ into (1.21).

The principal solution of (1.22) will be generated by one of the roots and the other $r - 1$ roots represent parasitic solutions. These parasitic solutions result because the order of the difference equation exceeds the order of the approximated differential equation by r. If any one of the parasitic roots is greater than unity, the corresponding term in (1.22) will increase without limits as n increases (Ref. 16, p. 205). This situation, called *numerical instability*, arises when the integration interval h is too large.

A pair of predictor-corrector formulas used frequently for numerical integration employs *Milne's method* and is predictor:

$$x_{n+1} = x_{n-3} + \frac{4h}{3}(2\dot{x}_n - \dot{x}_{n-1} + 2\dot{x}_{n-2}) + \frac{14}{45}h^5 x^{(5)}(\mu) \qquad (1.23)$$

corrector:

$$x_{n+1} = x_{n-1} + \frac{h}{3}(\dot{x}_{n+1} + 4\dot{x}_n + \dot{x}_{n-1}) - \frac{h^5}{90}x^{(5)}(\mu) \qquad (1.24)$$

where the values of μ lie between the largest and smallest argument involved in the formulas and, in general, are not equal.

To avoid numerical instability from the above integration formulas, and if (1.18) is rewritten as $\dot{x} = -ax + b$, then the integration step h must satisfy the relationship $ha \leqslant 0.83$. In network equations a will be a matrix, and, consequently, the above restriction on the integration step size requires that h be less than the smallest natural time constant of the network [17]. Although these roots contribute least to the solution, they force the numerical integration to proceed at a rate dependent on them and consistent with the allowable error in the solution.

Then, in designing a system for solving the set of differential equations, it is desirable to use as large a value for h as possible. Besides the formulas for integrating each step, it is necessary, to prevent numerical instability, to have formulas for increasing or decreasing the integration interval and criteria for when to do so. These criteria are provided by comparing the predicted and corrected values. These formulas are given in Hamming (Ref. 18, p. 208).

In contrast to the predictor-corrector methods, the Runga-Kutta methods do not use information from previously calculated points and consequently are self-starting and, in fact, are generally utilized in starting the methods that do require previous solution points.

In both predictor-corrector and Runga-Kutta integration methods automatic step-size control is possible. Generally, predictor-corrector methods require two derivative evaluations per time step as against four for the Runga-Kutta method; but Warten [19] has shown that, from stability considerations, the extra derivative evaluations per single time step for the Runga-Kutta method do not appear as serious a drawback as might first appear. Warten's method estimates the local truncation error in the Taylor-series expansion of the solution and then, using the error, prescribes smaller steps during the transient region and larger steps in the steady state portion. This allows the solution to progress at a rate that depends on the numerical integration errors and greatly speeds the solution process.

Most integration methods have been linear in the sense the value of the function at a point is expressed as a linear combination of function values and derivative values at this point and of previous points. A modification of the linear multipoint methods uses an exponential term in the integration routine, which in most cases, provides better stability properties so that a larger integration step can be taken. Both Pope [20] and Fowler and Warten [21] have developed exponential integration methods which have been used in general purpose network analysis programs (CIRCUS and SCEPTRE are examples). The exponential method significantly reduces the computing time for certain large systems of ordinary differential equations in which there is a large spread of time constants. Chapter 8 discusses this problem in further detail.

1.4 ON-LINE DESIGN PHILOSOPHIES

A primary factor that has limited widespread use of circuit analysis programs for design is the time delay involved between analyzing successive designs. Before the day of the time-shared interactive computer†, circuit analyses were generally run on the *batch-processing* mode. A time delay of an hour to perhaps 24 hours could be expected between submitting a job and receiving printout of results. If the designer relied exclusively on circuit analysis programs to test successive trial designs, the design process could require several weeks. Whereas, if the designer bread-boarded the circuit in the laboratory and measured the results of parameter adjustments on an oscilloscope, the design might take only a few days to complete.

With the advent of the time shared computer, we are now able to design circuits by computer in the same manner as in the laboratory. A time-shared computer system such as the OCTOPUS system of Lawrence Radiation Laboratory, Livermore, shown in Fig. 1.6, has many remote input-output terminals tied, on-line*, through transmission

---

† A computing system whereby a number of users can concurrently execute programs with which the users may interact during execution, and may be generally assured of some minimum time (called a time slice) for program execution.

* A generic reference to a device which can be directly controlled, by a computing system to which it is electrically connected.

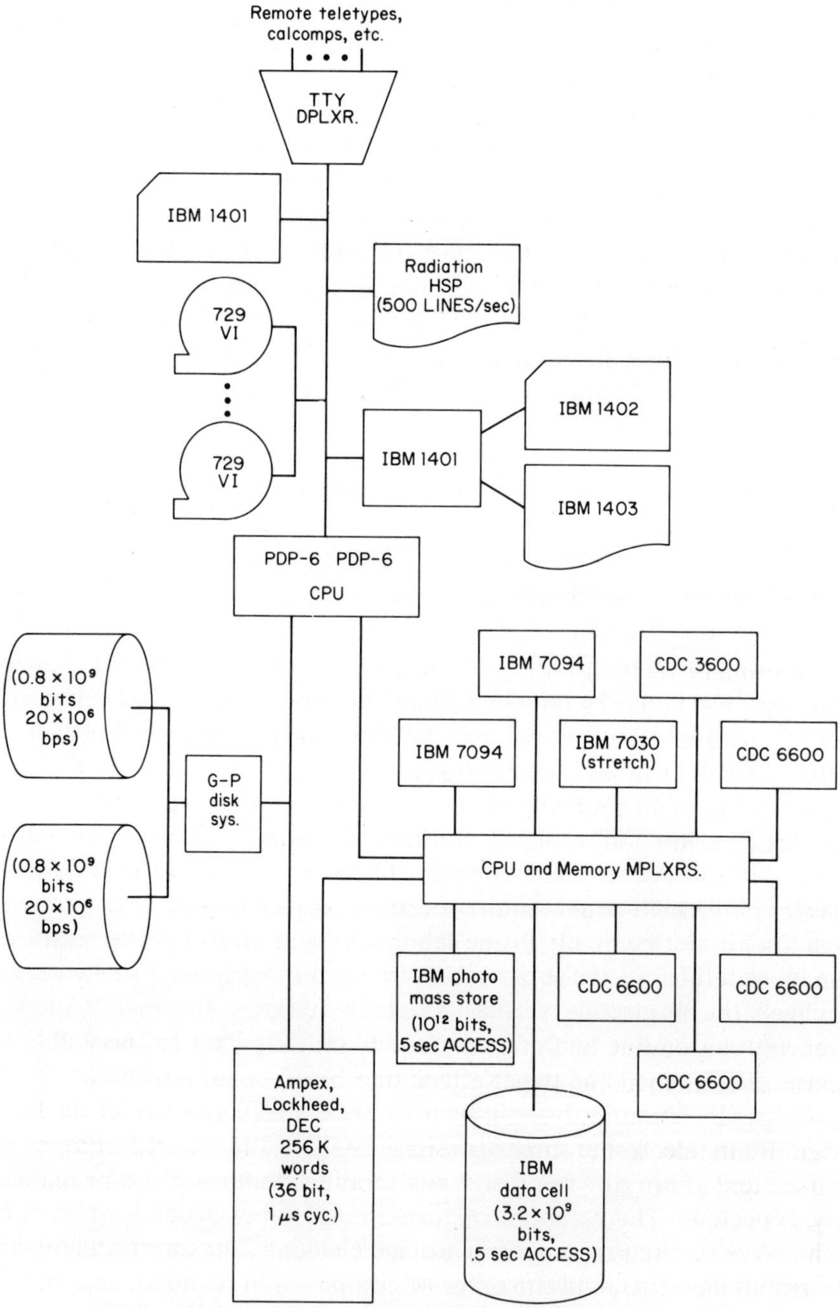

FIG. 1.6. Simplified schematic of the LRL OCTOPUS time-share computer system. For the sake of clarity, several systems (remote video displays, DD-80 displays, on-line data collecting network, CDC 160-A, multiplexing units, etc.) are not shown.

channels to the central processor control. The processor allots a given number of time slices to each user and serves each user in sequence (according to priorities assigned by a *scheduling program*). The main reason that a time-shared computer is effective is that the central processor computes thousands of times faster than the user's reaction time. When a user sends a command to the CPU, it seems only a "moment" before he gets the results. (The exact duration of the "moment" will be discussed below.) Therefore, the turn-around time for a time-shared computer is measured in seconds or minutes rather than the hours which batch processing requires.

A. On-Line Design

The impact of on-line computer systems on computer-aided electronic circuit analysis is rather recent, and we can expect the influence to become much more pronounced in the near future. We should not expect, however, that all computer systems will go predominantly time-sharing, or that circuit analysis will be only done on-line. On the contrary, batch-processing will be used for those problems where an intimate man-machine interaction is not required, or longer turn-around times are unimportant. Time-sharing and batch-processing are not mutually exclusive, and can and will be done concurrently on the same computer system.

The on-line use of computers is very valuable in a number of applications. For example, the debugging of programs is a popular use of the system. Information retrieval is of growing importance. We are concerned with electronic circuit design and here, on-line computer utilization is as useful as in the above examples. It is much faster, for example, to check the magnitude versus frequency response for even a simple filter with an on-line analysis program than it is either to check the response with a pencil and paper attack or a bread-board approach.

Good turn-around time is essential to the effective use of analysis programs in electronic circuit design. With a time-shared computer, turn-around times of several seconds to a few minutes could be reasonably expected. The variation of turn-around times depends on factors such as size of circuit, number of storage elements, the time constants of the circuit elements, and the range of computation required, and on the characteristics of the time-share system being utilized. For large circuits and programs, time-sharing is perhaps not the most efficient procedure to employ. There are several reasons for this statement. First, programs which require substantial memory capacity take much longer to transfer between core (fast) memory and disk, drum, or tape (slow memory). If the size of time slices allotted to each program being run is small, much of the computing time is used up in rolling in and out of core memory. Second, for a large circuit, suppose each analysis takes five or ten minutes execution time, on-line. The user at the remote console can do very little except wait. In terms of human factors, the 5 to 10 minute pause between each output can be intolerably long.

B. Program Design for Time Sharing

The design of a network analysis program for use in an on-line environment involves many considerations. The first consideration will, of course, be the characteristics and capabilities of the system on which the program will be operating. When the system is small (has rather limited memory storage available to the user) several special considerations must be taken into account. An example of a network analysis program developed for a time-share system (General Electric 265 computer) with rather restrictive memory available (approximately 5300 words) is the LINGO program [22]. However, when the system memory available is large, different considerations come into play. These considerations involve size of programs, size of problems, and remote console response times.

When the exchange between the person at the console and the program is designed to be continuous, then the response time or waiting time should be fairly short (a few seconds). This requires then, either small programs (so they will not be locked out of memory) or for large programs, few console users. At Lawrence Radiation Laboratory good response times are obtained with programs of up to a core size of approximately 40,000 to 45,000 (octal). The CALAHAN program is 45,600 and ECAP (dc section only) is 35,600 (octal) in length, so both have good response times.

Usually, general network analysis programs tend to be large and consequently response times will suffer. Instead of a high degree of exchange between program and user, it is then better to create an input file, reference the analysis program to this file, and let the program place results in an output file which the user can access at any time. In this way, fast turn-around times can be achieved but yet the engineer is not tied to the console. If he wishes output as it is produced, this option can be easily placed into the program.

An advantage of creating a file of input data is that the file can be proof-read before starting the analysis for transmission errors, errors in typing, format errors, etc. Also, the input is retained and can be slightly altered for a subsequent analysis after the results are inspected. Two advantages of collecting the output on a disk file are that it is faster than teletypewriter output and the disk file can be edited.

Another very important consideration in the design of a program for on-line circuit analysis is the inclusion of adequate error diagnostics. When the user incorrectly specifies input he should be told (politely, of course) in a understandable and non-fatal way. When the program or system generates an error or encounters numerical difficulties, the program should also indicate these in a way that can be interpreted by the user without an understanding of the program or system.

It is not always clear, when a program is being developed for on-line use, what is the best way to implement program-user interaction. At Lawrence Radiation Laboratory we seem to be going in the direction of having a high interaction for small programs (linear analysis programs, filter design programs, etc.,) and toward using input/output disk files for larger programs where the response times are degraded.

1.5 THE ROLE OF COMPUTER GRAPHICS IN CIRCUIT DESIGN

In order to render on-line analysis feasible, an effective communication language must be established between man and machine. In circuit design the natural language of the engineer is the schematic. Therefore, the communication language between man and machine should be graphical. Such communication is possible with an input-output device known as a *graphics console*. A number of graphics consoles are available commercially [23,24]. These consoles usually consist of a short persistence cathode-ray tube (CRT) with local memory, an input device known as a *light pen*, and a teletypewriter. The graphics displayed on the CRT are software generated from local memory. The memory also serves to "refresh" the display periodically to reduce flicker effects. Using the graphics console, a man sitting at the CRT display could enter data (such as a circuit diagram) directly on the CRT face by using a light pen. The schematic which the designer enters on the CRT is actually stored as data in the local memory. These data are then converted into an internal form which is acceptable by the circuit analysis program residing in the central computer. Under the multiprogramming mode, the translated graphic input data are sent to the central computer. At the

connecting the components If the design is to be an integrated circuit, programs exist which determine the sizes and shapes of the mask patterns for the capacitors, diodes, resistors, and transistors.

The final two chapters in this book are devoted to circuit layout by computer. Chapter 12, by Zane and Wilber, deals with printed circuit layout by means of the program, PUZZLE. Under PUZZLE, component placement is fixed by the input data, and the program determines the routing of the interconnections. In Chapter 13, Hazlett discusses basic techniques for automatically generating masks for integrated circuits by computer through the CAMP program. We will not attempt to summarize the methods described in the last two chapters, but will let the reader go on to the chapters themselves.

What we can do here is to describe some programs which are not mentioned in the book. The two programs which are typical examples of the power of the computer in circuit layout are ACCEL [26], for printed circuit board layout, and CADIC [27], an integrated circuit layout program. The central features of ACCEL are: (1) an *automatic component layout* algorithm by a unique linear force-field analogy, and (2) an *automatic routing procedure* which combines topographic simulation and minimum connection path algorithms [28, 29]. The outputs of ACCEL are:

(a) a schematic of the circuit
(b) a parts list
(c) an assembly drawing
(d) a hole drilling list
(e) an enlarged mask of the printed circuit
(f) a jumper listing

ACCEL is probably the most sophisticated and most automatic of the layout programs to date. It is more elaborate compared to PUZZLE in that it performs component layout as well as routing interconnections. The automatic routing feature could be a mixed blessing however. ACCEL either finds the complete solution for the routing or considers the problem unsolvable. There is no provision for interjecting human intuition in the procedures employed by ACCEL. This is because ACCEL was designed to be, and still is, an off-line program.

On the other hand, CADIC (which stands for Computer Aided Design of Integrated Circuits) is a program that is meant to be used on-line using an IBM 2250 graphics console. The primary difference between

CADIC and ACCEL are: (1) CADIC is meant primarily for integrated circuit layout whereas ACCEL is intended for printed circuit boards upon which elements are wired. Using CADIC, all components are realized in integrated form, with masks for the resistors, capacitors, etc. being determined by the program itself. On an initial display, all components are represented in true relative size and shape. CADIC then lets the engineer manipulate, on-line, the component placement until an efficient layout is achieved. Next the components are interconnected either by on-line manual routing, or by means of automatic maze solving procedures based upon Lee's algorithm [28]. It should be noted that ACCEL also uses Lee's algorithm for routing. When the layout is completed, CADIC generates final mask patterns for production. Using programs such as CADIC, the cost of building integrated circuits is reduced considerably over the purely manual photo-etching processes. Individually tailored integrated circuits are thus rendered (almost) economically feasible.

We should note here that the CADIC design philosophy differs from that employed in large scale integration (LSI). With CADIC, the circuit is transposed into integrated form in its entirety. In the LSI scheme, many different logic circuits are grown in multiples on a single monolithic substrate and the problem is to interconnect the individual circuits (cells) to meet various design specifications. The generation of interconnections between cells involves the design of photomasks consisting of multiple layers of metal and insulation. In Chapter 13, Hazlett discusses the automatic generation of film photomasks for interconnecting logical cells. Another system which also provides a computer aid for interfacing layout programs directly with the LSI manufacturing process, is the Texas Instrument Company's effort called the *Multilevel Interconnect Generator*, or MIG system [30].

Finally it should be pointed out that all the programs described in this section were originally meant to be stand-along programs. The input to these programs is a circuit schematic and the output is a printed or integrated circuit mask. A great advantage could be gained if the data structure for the circuit schematic could be made identical to the data structure of the circuit description from an analysis program such as SCEPTRE. Once this is done, a designer could go from the analysis stage directly to the layout program without going through an intermediate I-O process. We will describe this interface problem in greater detail in a later section.

1.7 PROGRAM REQUIREMENTS FOR COMPUTER AIDED DESIGN

In this section we will examine certain pragmatic requirements which must be satisfied before an engineer can effectively design circuits by computer. The requirements are given in terms of (a) Availability (b) Language and (c) Monitor System Compatibility.

A. Availability, Documentation and Testing of Programs

There has been a great deal of publicity in recent months concerning the capabilities of various circuit analysis programs. Often disregarded in the publicity are the vitally important questions of documentation and availability. No matter how good a program is, it is useless unless both requirements are met.

Availability. After reading the publicity announcements, the engineer is shocked when he discovers that many of the programs are either non-existent, non-working or not available due to proprietary rights. Since the basic theory is straightforward and the actual programming represents most of the effort involved, a paper describing a new program and omitting the program itself is of limited value. Unless the program is available, journals should not allow papers extolling the merits of the program. A point involved here is that unless the program is available to other readers there is no check on the work of the author. This is at variance with the open lines of science inquiry.

Documentation. A number of good electronic circuit analysis programs such as PREDICT [31] and CIRCUS [3] were originally developed to test radiation effects on transistor circuits. It became apparent that these programs were applicable to general circuit analysis work. Even though the programs themselves were excellent, they suffered from inadequate publicity and documentation, especially in the case of CIRCUS. From a user's standpoint, an out-of-house program must be completely documented before it can be used. Since the formation of the CADAR Council, the CADAR Newsletter provides an excellent channel of communications between potential users and program developers.

Testing. After a program becomes operational it still must be tested before it is turned over for general design use. However, even with exhaustive testing, esoteric bugs still remain. A large program is never completely debugged, and program modifications to correct errors are necessary from time-to-time. Occasionally other errors arise, such as a magnetic tape becoming defective, and must be corrected. Although a program may be advertised as "general" there are restrictions both listed and unlisted in the user's manuals. Program checking involves familiarity with the program to recognize where and what the program restrictions and limitations are. Finally, the program must be checked to ensure the answers it produces are trustworthy. This is difficult to do – one method consists of analyzing a network with entirely different programs and comparing results.

B. Language

Most computers have a FORTRAN capability. Programs such as CIRCUS, SCEPTRE and ECAP which are written in FORTRAN can be readily adapted to most machines. On the other hand, Katzenelson's nonlinear network program, AEDNET, which was written in the AED-O language at Massachusetts Institute of Technology, cannot be used on other computers without an extensive and tedious translation process. If the program is in FORTRAN, then it may be possible to make major improvements with very little effort. Although symbolic language programming is more flexible, more opportunity for errors exist. In addition, program debugging is easier in FORTRAN than in symbolic language programming. Some of the programs (such as NET-1, and portions of PREDICT) were written in a machine assembly language such as FAP, limiting, without extensive translation, the use of the program to an IBM 7090-7094 computer. Many of the programs require a computer similar in capability to an IBM 7094 for effective utilization. Even if the program can be adapted to the local computer, this adaptation is non-trivial and could require several months of programming time.

C. Monitor and Memory Requirements

For ease of use, a circuit analysis program should be adaptable to the user's computer monitor system. This allows the program to be run by batch-processing, or in an increasing number of cases, by time-sharing. A program, such as NET-1, which does not run under any monitor requires that the user control the program from the operator's console. In a large computer facility such as Livermore's, scheduling often requires that a NET-1 user run his program at inconvenient times. It should be said, however, that adapting a program to run on an alien monitor is a most tedious task. Another source of difficulties in getting outside-originated programs to run is compiler differences. The errors produced in this case may be obvious and easily remedied or they may be very obscure, and then are usually difficult to locate. Compiler length is likewise a source of difficulties. When moving a program from one computer to another memory overflows can occur due to a difference in compiler lengths.

Program memory requirements are often a source of problems. Quite often a user selects a program without first determining whether the program fits into his computer's memory. There is some confusion as to the memory size requirements of a given program due to the variation between versions of the same program and also due to the ambiguity in specifications being given in both word and byte size.

Although the memory size for a specific computer is generally the same at different installations, the operating system resident in core will, more often than not, occupy a different amount of memory. It seems that programmers, although recognizing this problem, ignore it. Notable examples of this is the PREDICT program which, although chained, did not allow any extra room for the possibility of other operating systems being larger than the one under which PREDICT was developed. The SCEPTRE program is a second example. SCEPTRE is a second generation of PREDICT and again, the programmers failed to allow for operating systems with even slightly larger memory requirements than the one the program was developed under.

1.8 DESIGN SUPERVISORY PROGRAM

In a recent issue of the Proceedings of the IEEE, a panel report was given on computer-aided design.† In the report, J. F. Kaiser gives such a succinct summary of the applications on computer-aided design of electronic circuits that we will quote his remarks here in some length.

"The computer is presently being applied to many different phases of the design problem, for example:

(1) Modeling and automated testing of basic devices to determine their specific parameter values
(2) specification of device designs to meet specifc performance requirements

---

† "Panel discussion on computer-aided design" *Proc. IEEE*, vol. 55, November 1967, pp. 1775-1784. See particularly J. F. Kaiser's comments on pp. 1778-1779.

(3) determination of how the devices are to be connected together in a circuit to perform the desired function
(4) evaluation of the performance of the designed system by simulation and analysis
(5) determination of parts lists, wiring layouts, mask patterns, and other manufacturing information
(6) determination of optimum grouping of circuits prior to integration
(7) generation of suitable test, maintenance, and fault detection procedures for the constructed systems.

At present the computer-aided design programs solving these individual problems and similar ones are separate entities. What is required now is serious thought given to the design of data structures these individual programs work with, so that linking of programs together either directly or by real-time interaction can be readily achieved. The linking together of programs must be much more than just patching together the individual pieces. Instead it must be done on a sound theoretical basis taking into account the detailed methods used in each part and changing these where necessary to give a more effective overall design program."

One of the major difficulties in the use of circuit analysis programs is that every program has a different input language for describing the topology and the element types and values of the circuit. There are two ways to handle the interface problem between the graphics I-O console and a large number of analysis programs. The first requires that the non-display data associated with the graphics input be tailored for each individual program. The second method involves leaving the graphic input system unchanged, and devising a common interface language for every analysis program in use. The common interface language seems inherently to be a more efficient approach, since this would enable an engineer to use a program on a batch-processing mode, also, without learning the details of the input language for the particular program. This common input language is the first step in establishing a *Design Supervisory Program*. Such a program would create an opaque interface between the engineer and the computer. The engineer would not need to know the capabilities, input requirements, and output options of any of the individual programs that he might use in the course of the design. He would merely place the schematic on the graphics console. The Supervisory Program then asks a series of questions, and then selects the appropriate programs for use in the analysis. After a number of iterations, if the responses fail to converge quickly to meet the given specifications, a light button called OPTIMIZE could be picked. Control then passes on to an optimization program using a method such as the Fletcher-Powell [32]. The engineer, again, does not need to know the internal details of the optimization program. At any time, the engineer could ask for return of control to the graphics console and proceed with his trial-and-error design method. The process involving cut-and-try and optimization is repeated over and over until the responses meet the original design specification. The final circuit is then constructed by using printed circuit layout programs such as ACCEL and PUZZLE. As graphic software evolves, physical layout of the circuit could also be done, on-line on the graphic console. We believe that the Design Supervisory Program would be the final link in making on-line, computer-aided design a practical reality for all engineers.

REFERENCES

1. A. F. Malmberg, F. L. Cornwell, and F. N. Hofer, "NET-1 network analysis program," Los Alamos Scientific Laboratory, Los Alamos, N. M., Rept. LA-3119, August 1964, 7090/94 version.

2. H. W. Mathers, S. R. Sedore, and J. R. Sents, "Automated digital computer program for determining responses of electronic circuits to transient nuclear radiation (SCEPTRE)," vol. I, IBM Space Guidance Center, Owego, N. Y. IBM File 66-928-611, February 1967.

3. L. D. Milliman, W. A. Massena, and R. H. Dickhaut, "CIRCUS – A digital computer program for transient analysis of electronic circuits – User's guide," The Boeing Company, Seattle, Wash., Harry Diamond Laboratories Rept. AD-346-1, January 1967.

4. D. A. Calahan, "Linear network analysis and realization digital computer programs, and instruction manual," University of Illinois Bulletin, vol 62, no. 58, February 1965.

5. "1620 Electronic Circuit Analysis Program [ECAP] [1620-EE-02X] User's manual," IBM Application Program File H20-0170-1, 1965.

6. M. L. Dertouzos, "CIRCAL: On-line circuit design," *Proc. IEEE*, vol. 55, pp. 637-654, May 1967.

7. J. Katzenelson, "AEDNET: A simulator for nonlinear networks," *Proc. IEEE*, vol. 54, pp. 1536-1552, November 1966.

8. J. J. Ebers and J. L. Moll, "Large-signal behavior of junction transistors," *Proc. IRE*, vol. 42, pp. 1761-1778, December 1954.

9. R. Beaufoy and J. J. Sparkes, "The junction transistor as a charge-controlled device," *Automatic Telephone and Electric Journal*, vol. 13, pp. 310-327, April, 1957.

10. J. G. Linvill, "Lumped models of transistors and diodes," *Proc. IRE*, vol. 46: pp. 949-957, June, 1958.

11. F. F. Kuo, *Network Analysis and Synthesis*, 2nd Ed., New York: Wiley, pp. 111-126, 1966.

12. S. Seshu and M. B. Reed, *Linear Graphs and Electrical Networks*, Reading, Mass.: Addison Wesley, pp. 315-330, 1961.

13. T. R. Bashkow, "The *A* matrix, a new network description," *IRE Trans. on Circuit Theory*, vol. CT-4: pp. 117-120, September, 1957.

14. P. R. Bryant, "The explicit form of Bashkow's *A* matrix," *IRE Trans. on Circuit Theory,* vol. CT-9, pp. 303-306, September 1962.

15. R. L. Wilson and W. A. Massena, "An extension of Bryant-Bashkow *A* matrix," *IRE Trans. on Circuit Theory*, vol. CT-12, pp. 120-122, March 1965.

16. F. B. Hildebrand, *Introduction to Numerical Analysis,* New York: McGraw-Hill Book Co., 1956.

17. F. H. Branin, Jr., "D-C and transient analysis of networks using a digital computer," *IRE Inter. Conv. Rec.*, vol. 2: pp. 236-256, 1962.

18. R. W. Hamming, *Numerical Methods for Scientists and Engineers,* New York: McGraw-Hill Book Co., 1962.

19. R. M. Warten, "Automatic step-size control for Runge-Kutta integration," *IBM J. Res. Dev.*, vol. 7: pp. 340-341, 1963.

20. D. A. Pope, "An exponential method of numerical integration of ordinary differential equations," *Comm. of the ACM*, vol. 6, no. 8, pp. 491-493, August 1963.

21. M. E. Fowler and R. M. Warten, "A numerical technique for ordinary differential equations with widely separated eigenvalues," *IBM J. Res. Dev., pp.* 537-543, September 1967.

22. J. Shekel, "LINGO - A programming technique for linear network analysis at a remote terminal," *Conf. Rec. Tenth Midwest Symp. on Circuit Theory*, pp. I-5-1 to I-5-8, 1967.

23. G. A. Michael and R. A. Cralle, "A survey of graphic data processing," Chapter 9, this book.

24. W. H. Ninke, "Man-computer graphical communication," in *System Analysis by Digital Computer*, F. F. Kuo & J. F. Kaiser, eds, New York: Wiley, 1966, Chapter 12.

25. H. C. So, "Analysis and iterative design of networks using on-line simulation," in *System Analysis by Digital Computer*, F. F. Kuo & J. F. Kaiser, eds., New York: Wiley, 1966, Chapter 2.

26. C. J. Fisk, D. L. Caskey, and L. E. West, "ACCEL: Automated Circuit Card Etching Layout," *Proc. IEEE*, vol. 55 No. 11, pp. 1971-1981, November 1967.

27. A. Spitalny and M. J. Goldberg, "On-line graphics applied to layout design of integrated circuits," *Proc. IEEE*, vol. 55 No. 11, pp. 1982-1988, November 1967.

28. C. Y. Lee, "An algorithm for path connections and its applications," *IRE Trans. Electronic Computers*, vol. EC-10, pp. 346-365, September 1961.

29. R. C. Prim, "Shortest connection network and some generalizations," *Bell Sys. Tech. J.*, vol. 3, pp. 1389-1401, November 1957.

30. J. W. Lathrop, et al., "A discretionary wiring system as the interface between design automation and semiconductor array manufacture," *Proc. IEEE*, vol. 55 No. 11, pp. 1988-1997, November 1967.

31. "Automated digital computer program for determining responses of electronic systems to transient nuclear radiation (PREDICT)," vol. II, IBM Space Guidance Center, Owego, N. Y., IBM File 64-521-5, July 1964.

32. R. A. Rohrer, "Fully automated network design by digital computer: preliminary considerations," *Proc. IEEE*, vol. 55 No. 11, pp. 1929-1939, November 1967.

CHAPTER 2

THE TOPOLOGICAL APPROACH TO COMPUTER-AIDED ANALYSIS

D. F. DAWSON, Lawrence Radiation Laboratory, Livermore

INTRODUCTION

Network analysis is one of the fundamental tools of the electrical engineer. It is also considered one of the basic parts of engineering education, and the typical undergraduate program in electrical engineering consists of three or four courses in network analysis. In these courses the students are given an introduction to classical mesh and node analysis, and a great number of two and three mesh network problems are solved. Network analysis is taught as part of an idealized design process. The analysis of the model of the physical network being designed provides information as to the suitability or need for modification of the design. However, in practice the electronic design engineer almost completely ignores the analysis approach. The reason for this is a practical one, for even a simple two-stage electronic amplifier circuit, with its associated bias and load circuitry, can have over 10 nodes and 10 meshes. This means that an analysis of such a circuit involves the evaluation of at least a tenth order determinant and the factorization of up to a tenth order polynomial. As this would have to be done for each change of parameter, it is obvious that exact analysis by hand calculation is not practical. In general the necessity of prohibitive amounts of hand calculations has prevented the electronic design engineer from applying the tools and techniques of network analysis to practical electronic design problems. Thus the experienced design engineer uses alternate techniques to aid in the design of networks. He makes an approximate analysis neglecting bias circuitry, or he uses previous knowledge of other networks to improve an existing design. As a result, design procedure has become somewhat of an art, and the better networks are developed somewhat intuitively.

The use of large digital computers has produced a change in emphasis in the above approach to network analysis. The tremendous calculating capability of such computers removes the restriction that analysis be limited to simple networks. As an example of one use of computers, a FORTRAN program can be written to set up and solve the equations corresponding to a lumped linear model or a multi-stage electronic amplifier. This technique, while providing a method of extending the analysis procedure, creates additional complications. A new burden is placed on the designer, as he now must understand computer programming and have at least a rudimentary idea about numerical analysis in order to solve this program. In addition it may take several weeks or months of programming time to set up and debug the program and solve a particular problem on the digital computer.

In order to provide the designer with the capability of analysis of a multi-stage network without the burden of time involved in computer programming, an intermediate step has been developed. It has been implemented during the past few years in a number of elaborate network analysis programs. While these programs are sophisticated and have many special features, they are limited by inherent numerical problems and are difficult to adapt to a given installation. The general nonlinear capabilities of these programs are needed for applications where, for example, the active devices are being used in a switching mode. They can, in effect, simulate the actual operation of a physical device and can be considered as an advanced form of bread-board checkout for reliability analysis and other applications. Over small operating ranges it may be possible to obtain a linear model for the active device. For the linear active and passive circuit, different analysis programs can be used that bypass some of the numerical problems of the general purpose programs. In addition, the linear programs can be used to provide frequency-domain information, and can be extended to provide root-locus plots and sensitivity information.

The linear network analysis program serves a useful role in small-signal analysis of active networks and in the complete solution of passive filters.

2.1 LINEAR NETWORK ANALYSIS PROGRAMS

Different approaches can be used to obtain both frequency and time domain information about linear networks. Two techniques described here treat the circuit as a two-port and obtain a system function as the ratio of two polynomials in s.

Typical of the first approach is the computer analysis program written by Pottle [1]. The state-variable method is still used to formulate the set of first-order differential equations; then either the Sourian-Frame algorithm may be used to provide a transfer function in terms of the s-domain, or a Taylor-series approximation to the state transition matrix may be used to obtain a time response.

The second approach is denoted as the *topological approach*. This technique formulates the equations in terms of the usual mesh or node methods, but bypasses the matrix inversion problems by a topological technique in the solution of these equations. Typical of this topological approach is a program written by Calahan [2]. This program has evolved out of a general interest in topological analysis at the University of Illinois by Mayeda [3], Van Valkenburg [4], and others. This topological method circumvents many of the previously listed objections to network analysis programs, but several new problems are introduced by the topological approach. These new problems will be discussed in the brief description of the topological approach which follows.

2.1.1 Brief Description of the Topological Approach

In this section the topological approach to network analysis is defined as the method based on the Binet-Cauchy theorem. This method effectively evaluates determinants and cofactors of the node admittance matrix by finding the related n-tree products of the corresponding linear graph.

The basic concepts of topological analysis are not new; they originated with Kirchhoff in 1847 and Maxwell in 1892. However, topological analysis was originally restricted to passive networks. For small networks and hand calculations, topological techniques offered no real advantage over the usual determinant solution of either loop or node equations. The topological technique does eliminate the terms in the evaluation of a determinant that ultimately cancel out, but this computational advantage is counterbalanced by the disadvantage of having to learn a different approach. However, the topological approach is conveniently programmed on a digital computer, and increasing use of digital computers provided a new interest in topological techniques. Additional emphasis was given to the interest in topological analysis when Percival [5] introduced artificial two-terminal elements, so-called "current and voltage elements," extending topological analysis to active and mutually coupled networks.

This topological technique bypasses the usual numerical matrix inversion problems encountered in the solution of node voltage equations. The basic analysis problem of solving the node voltage equations by either determinant or matrix methods is transformed by the Binet-Cauchy theorem to the topological and combinatorial problem of finding the trees and associated n-tree products of a linear graph. The tree-finding problem becomes more difficult in the case of the active network. Since in this case the tree-finding routine must select the trees common to both the voltage and current graph of the network. The correct sign for the n-tree products must also be determined by the tree-finding routine. The tree-finding problem is the basic limitation in the topological approach to network analysis. The limitation is primarily in the amount of computer time required to solve a particular problem.

2.2 BASIC THEORY OF THE TWO-GRAPH APPROACH

The basic concepts of the topological two-graph approach to the analysis of linear active networks are given in this section. The presentation is given first with reference to a passive network and then extended to the active case by the two-graph technique. Further details of this approach can be found in Mayeda [3].

The physical elements of a network are represented by a model consisting of passive and active elements. The passive elements are considered to be lumped, linear, and finite. The active elements are ideal, controlled or independent sources. It is assumed that this representation leads to a useful equivalence, and the term *electrical network* is used to refer to such a representation. The equations defining this model are derived from physical properties of the elements, and a knowledge of these equations, along with a knowledge of the Kirchoff voltage and current laws, is assumed in this chapter.

Particular attention can be directed to the topological properties of the linear model shown in Fig. 2.1 by replacing the schematic of the electrical network by a linear graph as shown in Fig. 2.2. The basic theory relating to such a linear graph is given in Seshu and Reed [6] and Kim and Chien [7]. To provide continuity a few basic definitions of linear graph theory are given in the following section.

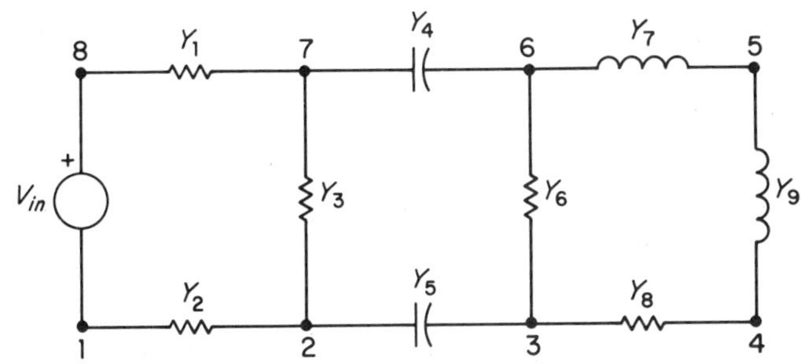

FIG. 2.1 A schematic representation of a linear network

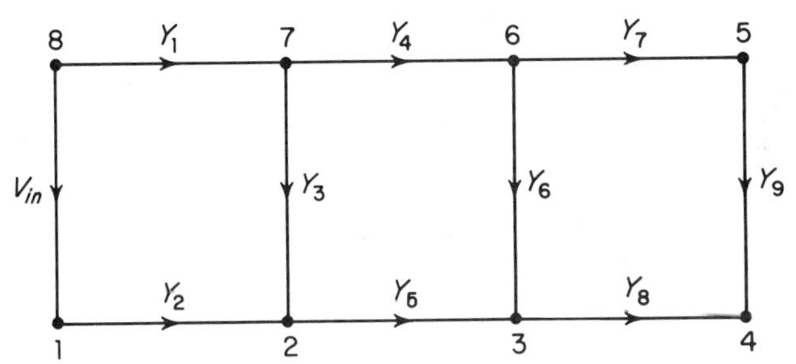

FIG. 2.2 Linear graph representation of a linear network

2.2.1 Basic Definitions of Linear Graph Theory

DEFINITION 1: An *edge* or *element* is a line segment with two end points. The end points are called nodes. A node and an edge are "incident" with each other if the node is an end point of the edge. The edge is called "self-looped" if its nodes are identical.

DEFINITION 2: A *linear graph* is a collection of edges with no self-looped edges. An oriented graph is a graph in which the direction of each edge is specified. Otherwise, the graph is nonoriented.

DEFINITION 3: The *degree* of a node is the number of edges incident at that node.

DEFINITION 4: A *path* between nodes i and j in a graph is an ordered sequence of edges in the graph in which every node is of degree two, except nodes i and j, which are of degree one.

DEFINITION 5: A *connected* graph has a path between every pair of nodes in the graph. It therefore cannot contain any isolated edges or nodes.

DEFINITION 6: A *circuit* is a connected graph of nodes all of which are of degree two.

DEFINITION 7: A *tree* is a connected graph or subgraph containing all nodes of the graph but no circuits. A tree *branch* is an edge of a tree, and a *chord* is an edge of the complement of a tree.

DEFINITION 8: The *incidence matrix* denoted by $\mathbf{A_a} = [a_{ij}]$, of an oriented graph with n nodes and b edges, is a matrix with n rows and b columns. Each row corresponds to a node, and each column corresponds to an edge, such that

$a_{ij} = 1$, if edge j is incident at node i and directed away from node i;
$a_{ij} = -1$, if edge j is incident at node i and directed toward node i;
$a_{ij} = 0$, if edge i is not incident at node i.

The matrix obtained by deleting any row from $\mathbf{A_a}$ is denoted as the *reduced incidence matrix* $\mathbf{A}$. Some important properties of $\mathbf{A_a}$ and $\mathbf{A}$ follow [8]:

PROPERTY 1: The rank of the incidence matrix $\mathbf{A_a}$ of n rows and b columns is $(n - 1)$. The rank of $\mathbf{A}$ is also $n - 1$.

PROPERTY 2: A square submatrix of $\mathbf{A}$ of order $n - 1$ is nonsingular if, and only if, the columns of this submatrix correspond to the branches of a tree. In the original formulation of the $\mathbf{A}$ matrix each edge of the graph determines a column of the $\mathbf{A}$ matrix.

PROPERTY 3: The determinant of a nonsingular sub-matrix of $\mathbf{A}$ is equal to $+1$ or -1.

2.2.2 Topological Analysis of Passive Networks

This section provides an introduction to topological analysis. The concepts presented are based on the node-voltage equations and the reduced incidence matrix $\mathbf{A}$. A dual approach could be taken in terms of mesh-current equations and the circuit matrix $\mathbf{B}$. The latter technique is quite similar but introduces one additional complication, as the resulting topological process requires a determination of the chords of a tree. This is an extra step beyond the determination of the tree branches which is required for the node-voltage approach. For this reason the dual approach has not received much use.

A passive RLC network such as previously shown in Fig. 2.1 can be represented as a linear graph as shown in Fig. 2.2. Each edge of the linear graph is weighted by an admittance of the network. In the network each admittance satisfies the following equation:

$$Y_{ab} V_{ab} = i_{ab} \tag{2.1}$$

where v_{ab} is the edge voltage with the positive reference at a, and i_{ab} is the edge current which flows from a to b. Expressed in matrix form for the complete network and graph equation (2.1) becomes

$$\mathbf{Y_b V = I} \tag{2.2}$$

where $\mathbf{Y_b}$ is a $b \times b$ diagonal matrix, $\mathbf{V}$ is the $b \times 1$ branch voltage column matrix, and $\mathbf{I}$ is the $b \times 1$ branch current matrix.

Kirchoff's current law is expressed as

$$\mathbf{AI = J} \tag{2.3}$$

where $\mathbf{J}$ is the column matrix of current sources, $\mathbf{I}$ is the column matrix of edge currents, and $\mathbf{A}$ is the reduced incidence matrix. Substitution of (2.2) into (2.3) yields

$$\mathbf{AY_b V = J}. \tag{2.4}$$

Also

$$\mathbf{V = A^t V_n} \tag{2.5}$$

where $\mathbf{V_n}$ is the column matrix of node voltages with respect to the reference node. Next (2.5) is substituted into (2.4) and the basic node voltage equation is obtained:

$$\mathbf{AY_b A^t V_n = Y_m V_n = J}. \tag{2.6}$$

A complete solution for equation (2.6) involves the inversion of the matrix $\mathbf{Y_m}$. This is difficult to implement except for some special cases, for in general the individual members of the matrix $\mathbf{Y_m}$ consist of polynomials in s, and must be carried in literal form until after the matrix inversion. The determination of a system function requires only a partial solution of the node voltage equation. The driving point impedance of a network requires only that a voltage across a pair of terminals be expressed in terms of a current source applied to the terminals. Other system functions, either driving point or transfer functions, require only a partial solution of the node voltage equation. In an attempt to avoid the problems of matrix inversion, Cramer's rule is used to find the required variables in (2.6). The driving point function Z_{ij} is defined as the ratio of V_{ij} to J_{ij}, where j is the reference node and J_{ij} is a current source connected between nodes i and j. In solving for V_{ij} by applying Cramer's rule to (2.6) and letting all other current sources except J_{ij} equal zero, the following equation is obtained:

$$V_{ij} = \frac{J_{ij}\Delta_{11}}{\Delta} \tag{2.7}$$

where $\Delta = |\mathbf{Y}_\mathrm{m}| = |\mathbf{A}\mathbf{Y}_\mathrm{b}\mathbf{A}^\mathrm{t}|$ and Δ_{11} is the 1, 1, cofactor of $\mathbf{Y}_\mathrm{m}$. A direct determination of Δ and Δ_{11} by conventional determinant techniques leads to the same problem as matrix inversion. An alternate approach is taken to the problem of determinant evaluation by using the Binet-Cauchy Theorem [8]. This theorem provides a method for evaluating the determinant of the product of two matrices. It can be stated as follows:

BINET-CAUCHY THEOREM: Let $\mathbf{P}$ of order (m, n) and $\mathbf{Q}$ of order (n, m) be matrices of elements. Let $m < n$. Then the determinant of the product $\mathbf{PQ}$ is given by:

$$\det \mathbf{PQ} = \Sigma \text{ (Products of corresponding majors of } \mathbf{P} \text{ and } \mathbf{Q}) \tag{2.8}$$

where the summation is over all such majors. A *major* or a *major determinant* means the determinant of a largest square submatrix (in this case of order m). The phrase *corresponding major* implies that if columns $i_1, i_2, ..., i_m$ of $\mathbf{P}$ are chosen for a major, rows $i_1, i_2, ..., i_m$ of $\mathbf{Q}$ should be chosen to form the corresponding major.

The Binet-Cauchy theorem is used to calculate the determinant of $\mathbf{A}\mathbf{Y}_\mathrm{b}\mathbf{A}^\mathrm{t}$. $\mathbf{A}\mathbf{Y}_\mathrm{b}$ is related with the matrix $\mathbf{P}$ and $\mathbf{A}^\mathrm{t}$ with the matrix $\mathbf{Q}$. As $\mathbf{Y}_\mathrm{b}$ is a diagonal matrix, the product $\mathbf{A}\mathbf{Y}_\mathrm{b}$ has the same structure as the matrix $\mathbf{A}$ except that column j is multiplied by Y_j, $j = 1, 2, ..., \mathrm{b}$. Therefore the nonsingular submatrices of $(\mathbf{A}\mathbf{Y}_\mathrm{b})$ still correspond to the trees of the network, but the value of the determinant of each nonsingular submatrix instead of being ±1 is now

$$(\pm 1)\,(Y_{i1})\,(Y_{i2})\,...\,(Y_{ik}) \qquad (2.9)$$

where $k = n$ and $i_1, i_2, ..., i_k$ are the branches of a tree. The corresponding submatrix of $\mathbf{A}^t$ is merely the transpose of the submatrix of **A**, and so it is also nonsingular, and has the same determinant (±1) as the submatrix of **A**. The product of the determinant of the two submatrices is $(Y_{i1})(Y_{i2})\,...\,(Y_{ik})$, which is defined as the tree admittance product. For passive networks the tree admittance product is always prefixed by a positive sign. All the other majors, which do not correspond to trees, are zero and do not contribute to the solution of $\mathbf{A}\mathbf{Y}_b\mathbf{A}^t$ · Therefore, by using the Binet-Cauchy theorem, the following expression for Δ is obtained:

$$W(Y) = \Delta = |\mathbf{A}\mathbf{Y}_b\mathbf{A}^t| = \sum_{\substack{\text{all}\\\text{trees}}} \text{tree admittance products of the networks.} \qquad (2.10)$$

This result is known as Maxwell's rule or formula. To calculate the node admittance determinant, each tree of the network must be found, then the tree admittance product must be found for each tree, and finally all the tree admittance products must be summed.

The use of Maxwell's rule in hand calculations produces a computational advantage, since the partial calculation of terms that occurs in determinant evaluation does not occur in the topological approach. For this reason (2.10) is sometimes called a minimum effort formula.

In computer calculations there is a more significant advantage than in hand calculation. Literal matrix inversion can be achieved by simple computer orientated operations. A network function can be conveniently determined in several different forms. It can be expressed in terms of its poles and zeros in the s domain, an operation that would be very difficult to do if the attempt were made to invert the matrix $\mathbf{Y}_m$ directly. A network function may also be obtained in terms of the individual admittances of the network; or, another formulation can be made expressing the network function as a linear function of a single element.

By an extension of the approach shown in evaluating Δ, a cofactor Δ_{ij} can be evaluated, leading to expressions for network functions and parameters in terms of tree products and n-tree products. This extension is developed in the next section.

2.2.3 Evaluation of Cofactors by Use of the N-Tree Concept

Symmetrical cofactors of the form Δ_{jj} are considered first. To obtain a cofactor Δ_{jj}, both row j and column j of the node admittance matrix $\mathbf{Y_m}$ must be deleted. Since $\mathbf{AY_bA^t} = \mathbf{Y_m}$, the same result can be obtained by deleting row j of the first matrix and column j of the last matrix. But column j of $\mathbf{A^t}$ is row j of $\mathbf{A}$. If $\mathbf{A_j}$ is defined as the matrix $\mathbf{A}$ with row j deleted the result is

$$\Delta_{jj} = \det(\mathbf{A_jY_bA_j^t}). \tag{2.11}$$

The relationship of $\mathbf{Aj}$ to the original network N is now examined. First node j is shorted to the reference node. Then the incidence matrix of the new network N_1 is rewritten. None of the other nodes are affected, so the reduced incidence matrix of N_1 will be $\mathbf{A_j}$. Therefore

$$\Delta_{jj} = \sum \text{ tree admittance products of } N_1. \tag{2.12}$$

Δ_{jj} is also related to the original network N. Since N_1 has n-1 nodes, trees of N_1 cannot be trees of N. They are in two pieces (one "piece" may simply be an isolated node). The node j and the reference node are in the two different pieces (otherwise shorting them will produce a loop). Such a structure is called a 2-tree and in this case a 2-tree (j, r), where r stands for the reference node.

Given a network N shown in Fig. 2.3(a), the network N_1 corresponding to a cofactor Δ_{11} is shown in Fig. 2.3(b). Node 5 is the refer-

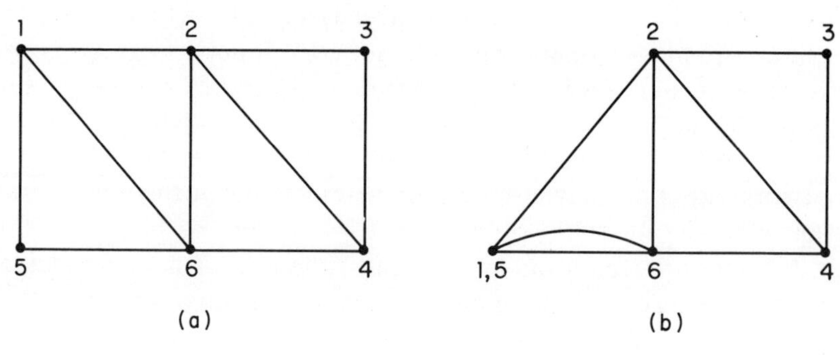

FIG. 2.3. Networks N and N_1

ence node. In Fig. 2.4 each set shown is a 2-tree of N with node 1 and node 5 in different parts. Each of these terms contributes to the evaluation of Δ_{jj}. Maxwell's formula for a symmetrical cofactor becomes

$$\Delta_{jj} = \sum_{\substack{\text{all} \\ \text{2-trees}}} \text{2-tree } (j, r) \text{ admittance products} \tag{2.13}$$

where r is the reference node.
Defined at this point is

$$W_{i,\,j} = \sum_{\substack{\text{all} \\ \text{2-trees}}} \text{2-tree } (i, j) \text{ admittance products} \tag{2.14}$$

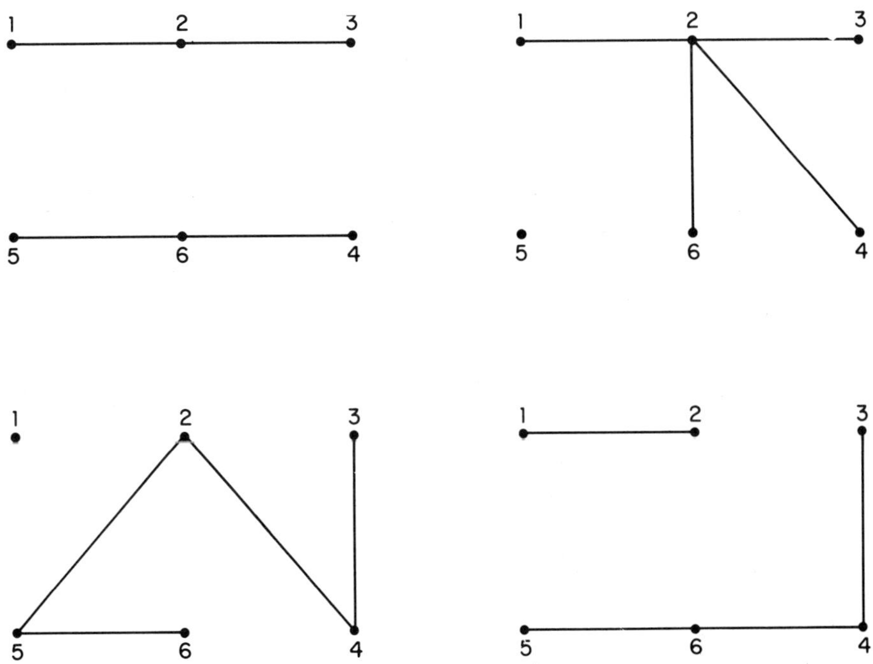

FIG. 2.4. Some 2-Trees of N

The subscripts in $W_{i,j}$ are separated by a comma. This indicates that the nodes corresponding to the subscripts are in different parts of the network.

Next, the asymmetrical cofactors of the node admittance matrix are considered. If row i and column j are deleted, the following equation is obtained.

$$\Delta_{ij} = \det\,(\mathbf{A_i Y_b A_j^t}) \tag{2.15}$$

Again the Binet-Cauchy theorem is applied to get

$$\Delta_{ij} = \Sigma \text{ products of the corresponding majors of } \mathbf{A_i Y_b} \text{ and } \mathbf{A_j^t}. \tag{2.16}$$

As before, the nonzero majors of $\mathbf{A_i Y_b}$ correspond to 2-trees (i, r) where r is the reference node. The nonzero majors of $\mathbf{A_j^t}$ correspond to 2-trees (j, r). The only terms that contribute to Δ_{ij} must be both 2-trees of (j, r) and (i, r), and therefore the intersection of $W_{i,r}$ and $W_{j,r}$. A 2-tree has only two parts; since r must be in one part, both i and j must be in the other. Therefore

$$\Delta_{ij} = W_{ij,r} = W_{i,r} \,\Omega\, W_{j,r}. \tag{2.17}$$

There are now formulas for Δ, Δ_{jj}, and Δ_{ij} of the node admittance matrix. Any transfer function of network parameter originally expressed in terms of the node admittance matrix can now be expressed in terms of tree or n-tree products of the network N. For example, an open-circuit voltage transfer function is evaluated as

$$H(S) = \frac{V_2\,(S)}{V_1(S)} = \frac{\Delta_{12}\,(S)}{\Delta_{11}\,(S)} = \frac{W_{12,3}(S)}{W_{1,3}\,(S)} \tag{2.18}$$

for the three-terminal network N shown in Fig. 2.5. Some network functions require more than values of the determinant and the first cofactors of $\mathbf{Y_m}$. Second and higher cofactors lead to 3-tree and n-tree products which are an extension of the method used here [3].

2.2.4 Extension of the Topological Method of Active Networks

Most of the linear networks that the design engineer is required to evaluate have some active device involved; therefore, interest in the topological approach had been dormant until a strategy was developed to include the active elements. This was done by Percival [5]. He intro-

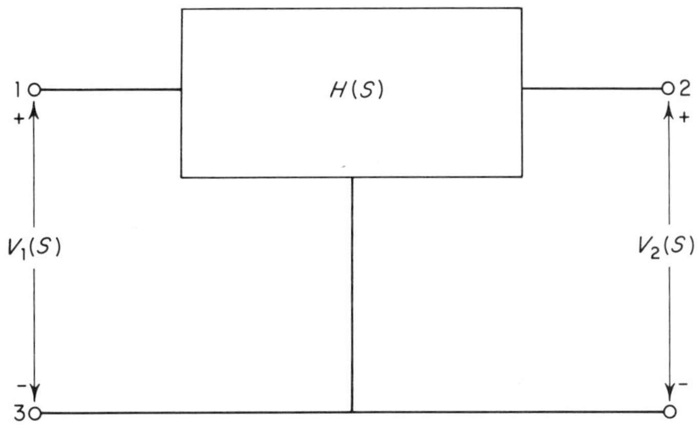

FIG. 2.5 Three-terminal network

duced artificial two-terminal elements, so-called "current and voltage elements," and made it possible for the theory of linear graphs to be extended to the analysis of active networks. Percival's papers [5], [9] initiated most of the concepts and ideas that have generated the active topological papers of the last decade.

Percival's method can best be illustrated by the following example. Notation used here follows that of Kim and Chien [7]. Given a linear network as shown in Fig. 2.6 with node pairs (pq) and (mn)

$$Y_{pq} V_{pq} + Y_{pq,mn} V_{mn} = I_{pq} \tag{2.19}$$

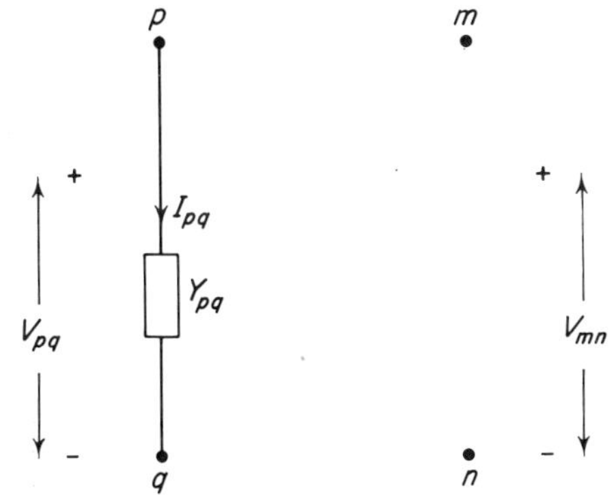

FIG. 2.6 Linear Network

where the Y's are proportional factors and $I = \mathcal{L}(i(t))$, $V = \mathcal{L}(v(t))$. In (2.19) there is a self admittance term Y_{pq} and a mutual admittance term $Y_{pq,mn}$. I_{pq}, which depends on both V_{pq} and V_{mn} can be decomposed into two components: one due to V_{pq} denoted by I'_{pq}, and the other, due to V_{mn}, denoted by I''_{pq}. Then

$$Y_{pq} V_{pq} = I'_{pq} \tag{2.20}$$

$$Y_{pq,mn} V_{mn} = I''_{pq} \tag{2.21}$$

$$I'_{pq} + I''_{pq} = I_{pq} . \tag{2.22}$$

Linear graph theory is extended to active devices by adding a "voltage element" between nodes m and n to sense a voltage difference in the node pair. The element introduced in node pair (pq), called a "current element," produces a current of magnitude I''_{pq} flowing from node p to node q due to the voltage difference in node-pair mn. The voltage and current elements are related by the mutual admittance $Y_{pq,mn}$. The voltage and current elements will always occur in pairs, but will be between different node pairs. If they occur in the same node-pair, that is, $p = m$ and $q = n$, the mutual admittance $Y_{pq,mn}$ is reduced to the self-admittance Y_{pq}. The schematic representation of voltage and current elements for dependent node-pairs (pq) and (mn) is shown in Fig. 2.7. Since the current and voltage elements are two-terminal elements, they can be represented by the edges of a linear graph as "current and voltage" edges respectively. The weight of the edges is the mutual admittance. The orientation of a current edge is determined by the direction of the current-flow in the corresponding current element, the orientation

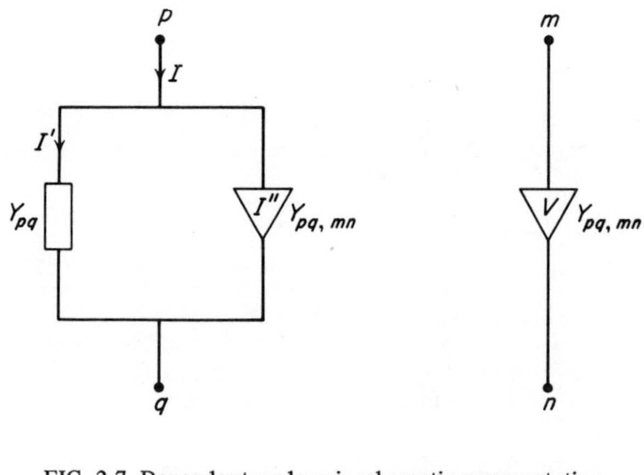

FIG. 2.7 Dependent node pair schematic representation

of a voltage edge by the polarity of the corresponding voltage element. The graphical representation for the network N of Fig. 2.7 is shown in Fig. 2.8. The current edge is indicated by double arrows, and the voltage

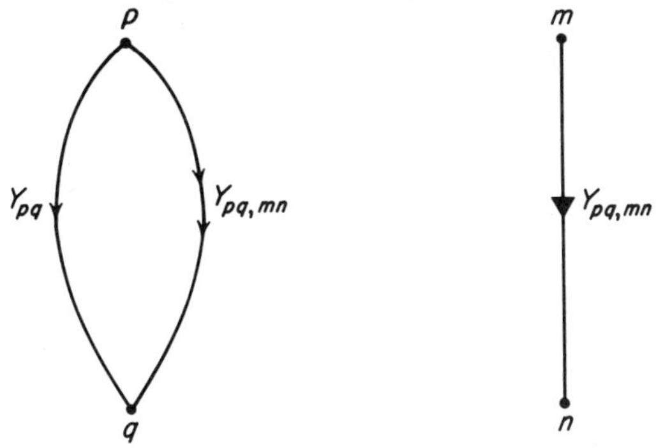

FIG. 2.8 Graphical representation of dependent node pairs

edge by a triangular-shaped arrow. The single arrow represents a passive element.

Next, the node-voltage equations are developed in terms of the current and voltage graph of a network. $\mathbf{A_v}$ and $\mathbf{A_i}$ denote respectively the incidence matrices for the voltage and current graph of a network. The same reference node is used in each graph. Kirchhoff's current law gives

$$\mathbf{A_i I} = \mathbf{J} \tag{2.23}$$

where $\mathbf{I}$ is the column matrix of edge currents and $\mathbf{J}$ is the column matrix of current sources. The node transformation for the voltage graph is

$$\mathbf{V} = \mathbf{A_v^t V_n} \tag{2.24}$$

where $\mathbf{V}$ is the column matrix of edge voltages and $\mathbf{V_n}$ is the column matrix of node voltage with respect to the reference vertex. Equation (2.20) can be expressed in matrix form as

$$\mathbf{Y_b V} = \mathbf{I} \tag{2.25}$$

where $\mathbf{Y_b}$ is a diagonal matrix in which each diagonal entry gives an admittance in the network. A combination of (2.23), (2.24), and (2.25) yields

$$\mathbf{Y_n V_n} = \mathbf{J} \tag{2.26}$$

where

$$\mathbf{Y_n} = \mathbf{A_i Y_b A_v^t}. \tag{2.27}$$

As in the previous case of passive networks the Binet-Cauchy theorem is used to solve for the determinant and cofactors of $\mathbf{Y_n}$.

$$\Delta = |\mathbf{A_i Y_b A_v^t}| = \Sigma \text{ corresponding majors of } \mathbf{A_i Y} \text{ and } \mathbf{A_v^t}. \tag{2.28}$$

A major of $\mathbf{A_i Y_b}$ is a non-singular if, and only if, the columns of $\mathbf{A_i}$ correspond to a tree of the current graph. A nonsingular major of $\mathbf{A_i Y_b}$ has the value + $(Y_{i1})(Y_{12}), \ldots, (Y_{ik})$ where $k = n - 1$ and $i_1, i_2, \ldots, i_k$ are the branches of the tree.

The corresponding major $\mathbf{A_v^t}$ has the value of ± 1 if the columns of $\mathbf{A_v}$ correspond to a tree of the voltage graph. The term *common tree* is used to denote a tree that is common to both the current and the voltage graph. Δ is now given in terms of the tree products of the common trees of a network:

$$\Delta = W(Y) = \Sigma \pm(\text{tree products of the common trees of a network}) \tag{2.29}$$

The addition of active elements to the graph extends the usefulness of the procedure, but also complicates the tree-finding procedure. Determination of the common trees and the associated sign are difficult topological problems.

The effective solution of these problems must lead to algorithms that can be efficiently implemented on a digital computer. The algorithms must determine common n-tree products as well as common tree products in order to obtain system functions for active networks. Details of the complete program are found in section 2.3, and additional aspects of the tree-finding problem are discussed in section 2.4.

Several examples are given to illustrate the concepts and use of current and voltage elements in characterizing a network containing linear active devices as well as mutually coupled multiterminal passive elements.

2.2.5 Examples of the Topological Method Applied to Active Networks

EXAMPLE 1: A small-signal linear model of a transistor is shown in Fig. 2.9. The graphical representation of the same model is shown in Fig. 2.10. There the voltage controlled current source has been replaced by a current edge from node e to c and a voltage edge from node b to node e.

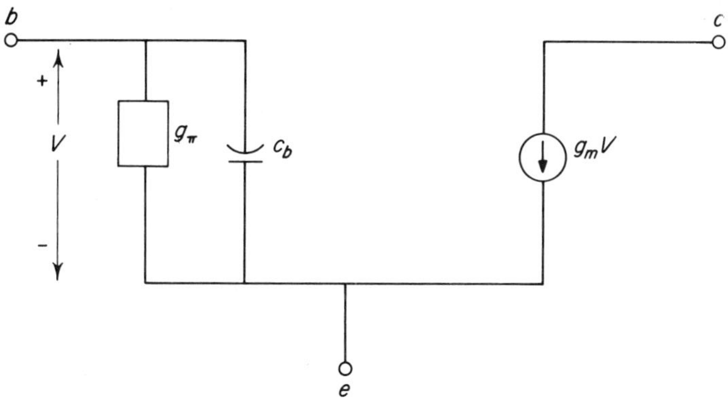

FIG. 2.9. Small-signal model of a transistor

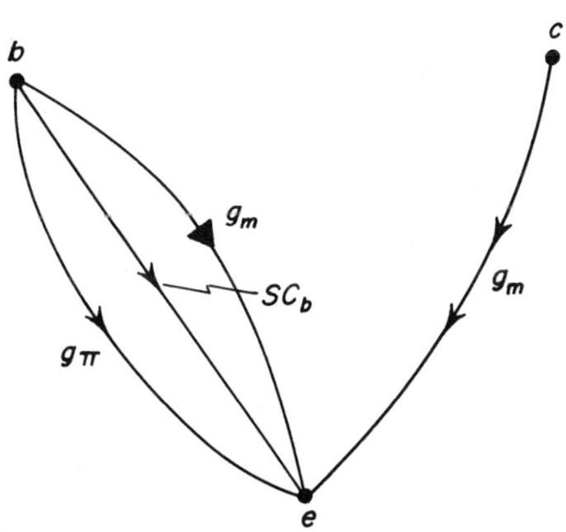

FIG. 2.10. Graphical representation of a transistor

EXAMPLE 2: The characteristics of the gyrator shown in Fig. 2.11 are given by the following equations

$$V_1 = a(-I_2) \tag{2.30}$$

$$I_1 = bV_2. \tag{2.31}$$

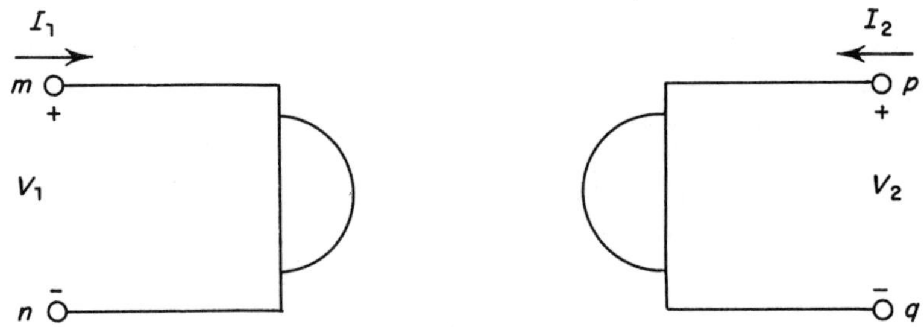

FIG. 2.11 Gyrator

The gyrator fits very neatly into this current and voltage graphical representation.

Equation (2.30) is rewritten

$$I_2 = (-\frac{1}{a})V_1. \tag{2.32}$$

Now both (2.30) and (2.31) fit into the form of $I''_{pq} = Y_{pq,mn} V_{mn}$, and the graph representation of Fig. 2.11 is shown in Fig. 2.12.

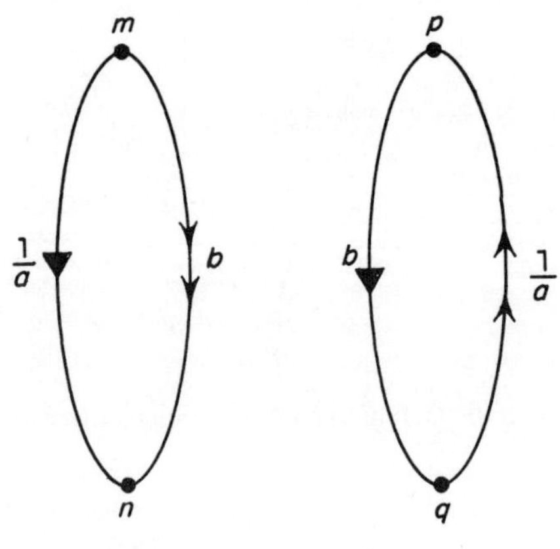

FIG. 2.12 Graphic representation of a gyrator.

EXAMPLE 3: A simple equivalent network for a two transistor cascade amplifier is shown in Fig. 2.13. The graph for the complete net-

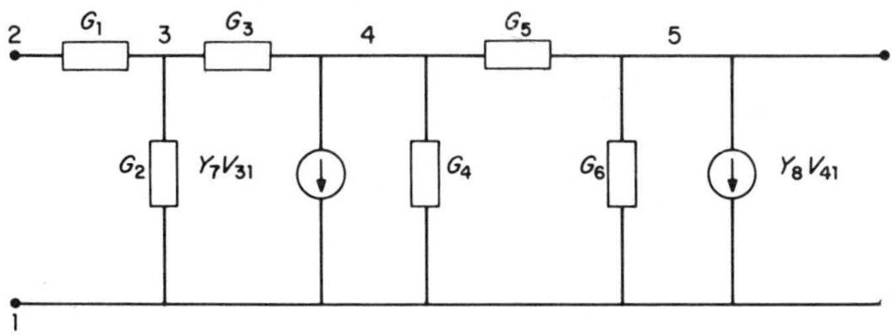

FIG. 2.13 Equivalent network of a two transistor network

work, including both ordinary edges and the voltage and current edges, is shown in Fig. 2.14. Arbitrary directions were assigned to the passive elements, and a relative sign convention was assigned to relate the voltage and current edges of the graph. Each edge of the graph is associated with the corresponding element of the network.

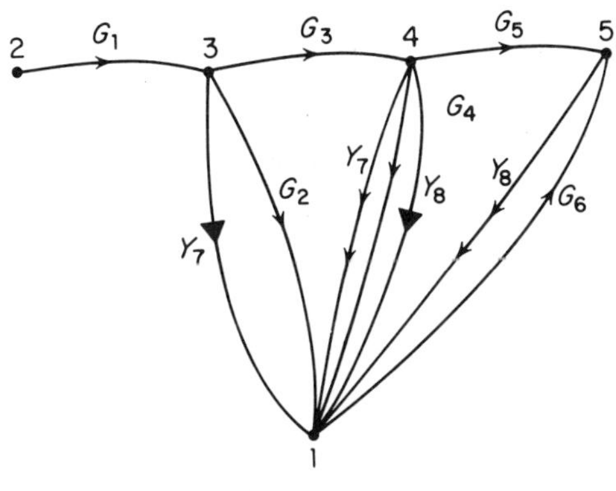

FIG. 2.14. Complete graph of the two transistor network

All the passive elements and the current edges are used to construct the current graph. The passive elements and the voltage edges form the voltage graph. The voltage and current graphs are shown in Fig. 2.15. $\mathbf{A}_{i1}$ and $\mathbf{A}_{v1}$, the reduced incident matrices with row 1 deleted, follow the graphs of Fig. 2.15. $\mathbf{Y}_b$, the branch admittance, is also given below.

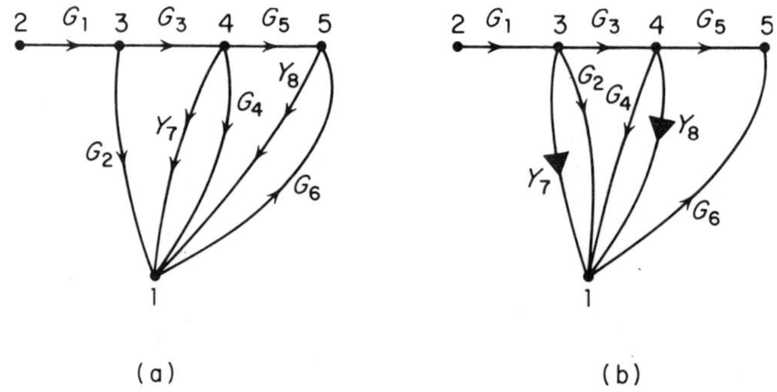

FIG. 2.15 Current and voltage graphs of the two transistor network

$$\mathbf{A}_{i1} = \begin{array}{c} \\ 2 \\ 3 \\ 4 \\ 5 \end{array} \begin{array}{c} \begin{matrix} 1 & 2 & 3 & 4 & 5 & 6 & 7 & 8 \end{matrix} \\ \begin{bmatrix} 1 & 0 & 0 & 0 & 0 & 0 & 0 & 0 \\ -1 & 1 & 1 & 0 & 0 & 0 & 0 & 0 \\ 0 & 0 & -1 & 1 & 1 & 0 & 1 & 0 \\ 0 & 0 & 0 & 0 & -1 & -1 & 0 & 1 \end{bmatrix} \end{array} \tag{2.33}$$

$$\mathbf{A}_{v1} = \begin{array}{c} \\ 2 \\ 3 \\ 4 \\ 5 \end{array} \begin{array}{c} \begin{matrix} 1 & 2 & 3 & 4 & 5 & 6 & 7 & 8 \end{matrix} \\ \begin{bmatrix} 1 & 0 & 0 & 0 & 0 & 0 & 0 & 0 \\ -1 & 1 & 1 & 0 & 0 & 0 & 1 & 0 \\ 0 & 0 & -1 & 1 & 1 & 0 & 0 & 1 \\ 0 & 0 & 0 & 0 & -1 & -1 & 0 & 0 \end{bmatrix} \end{array} \tag{2.34}$$

$$\mathbf{Y}_b = \begin{array}{c} \begin{matrix} 1 & 2 & 3 & 4 & 5 & 6 & 7 & 8 \end{matrix} \\ \begin{bmatrix} G_1 & & & & & & & \\ & G_2 & & & & & & \\ & & G_3 & & & & & \\ & & & G_4 & & & & \\ & & & & G_5 & & & \\ & & & & & G_6 & & \\ & & & & & & Y_7 & \\ & & & & & & & Y_8 \end{bmatrix} \end{array} \tag{2.35}$$

The node voltage equation for this network and its corresponding graph is

$$\mathbf{J} = \mathbf{Y}_m \mathbf{V} = (\mathbf{A}_{i1} \mathbf{Y}_b \mathbf{A}_{v1}^t)\mathbf{V} \tag{2.36}$$

or

$$\begin{bmatrix} J_{21} \\ J_{31} \\ J_{41} \\ J_{51} \end{bmatrix} = (\mathbf{A}_{i1} \mathbf{Y}_b \mathbf{A}_{v1}^t) \begin{bmatrix} V_{21} \\ V_{31} \\ V_{41} \\ V_{51} \end{bmatrix} \tag{2.37}$$

At this point one of the network parameters will be selected to demonstrate the topological technique. The transfer impedance Z_{52} for this network is defined as the ratio of the response voltage V_{51} with respect to the excitation current generator J_{21}. In terms of the determinant of $\mathbf{Y}_m$, if V_{51} is found in terms of J_{21}, with $J_{31} = J_{41} = J_{51} = 0$, the result, by Cramer's rule, would be

$$V_{51} = \frac{\Delta_{14}}{\Delta} J_{21} \tag{2.38}$$

where $\Delta = |\mathbf{Y}_m|$ and Δ_{14} is the cofactor of $\mathbf{Y}_m$. If $\mathbf{A}_{i1}$ is first multiplied by $\mathbf{Y}_b$, the outcome is:

$$\mathbf{A}_{i1}\mathbf{Y}_b = \begin{bmatrix} G_1 & 0 & 0 & 0 & 0 & 0 & 0 & 0 \\ -G_1 & G_2 & G_3 & 0 & 0 & 0 & 0 & 0 \\ 0 & 0 & -G_3 & G_4 & G_5 & 0 & Y_7 & 0 \\ 0 & 0 & 0 & 0 & -G_5 & -G_6 & 0 & Y_8 \end{bmatrix} \tag{2.39}$$

The product of $\mathbf{A}_{i1}\mathbf{Y}_b$ and $\mathbf{A}_{v1}^t$ is $\mathbf{Y}_m$.

$$\mathbf{Y}_m = \begin{bmatrix} G_1 & -G_1 & 0 & 0 \\ -G_1 & G_1+G_2+G_5 & -G_3 & 0 \\ 0 & -G_3+Y_7 & G_3+G_4+G_5 & -G_5 \\ 0 & 0 & -G_5+Y_8 & G_5+G_6 \end{bmatrix} \tag{2.40}$$

The direct non-numerical solution of (2.36) as stated previously requires the inversion of $\mathbf{Y_m}$. In the topological approach (2.40) is never obtained. Instead, all trees common to both the current graph of Fig. 2.15(a) and the voltage graph of Fig. 2.15(b) are obtained. Edges 1, 2, 3, 5 form one of the common trees. The tree admittance product is given by the sum of the tree admittance products associated with the intersection of trees of $\mathbf{A_{i1}}$ and $\mathbf{A_{vl}}$.

$$\begin{aligned} \Delta = \; & G_1 G_2 G_3 G_5 + G_1 G_2 G_3 G_6 + G_1 G_2 G_4 G_5 \\ & + G_1 G_2 G_4 G_6 + G_1 G_2 G_5 G_6 + G_1 G_2 G_5 Y_8 \\ & + G_1 G_3 G_4 G_5 + G_1 G_3 G_4 G_6 + G_1 G_3 G_5 G_6 \\ & + G_1 G_3 G_5 Y_7 + G_1 G_3 G_5 Y_8 + G_1 G_3 G_6 Y_7 \end{aligned} \tag{2.41}$$

The 12 tree products listed above correspond to the 12 trees common to both $\mathbf{A_{i1}}$ and $\mathbf{A_{vl}}$. This intersection of trees left out the trees 1238, 1248, 1348 of the current graph, and 1268, 1368 of the voltage graph.

The evaluation of Δ_{14} involves an evaluation of 2-tree-products for this network giving

$$\Delta_{14} = G_1 G_3 G_5 - G_1 G_3 Y_8 - G_1 G_5 Y_7 + G_1 Y_8 Y_7. \tag{2.42}$$

The number of trees of a network grows rapidly with the addition of extra branches and nodes. The evaluation of networks larger than the above example obviously requires the use of a digital computer. The complete digital computer program for topological analysis is given in the next section.

2.3 THE BASIC COMPUTER PROGRAM

The program described in this section is basically the program as developed by Calahan [2]. A new version of this program is described in the CADAR Bulletin [10]. The program has been in use for several years at LRL and has proved to be very dependable. It has been adapted to time-sharing [11] and serves as the analysis portion of the GINA program [12].

2.3.1 Basic Topological Program

A block diagram of the program is shown in Fig. 2.16. A description of the active and passive elements and program options selected is required as input data. This data must be in a fixed field FORTRAN format or, as indicated in section 2.6, additional programming can be used to provide a field-free input format.

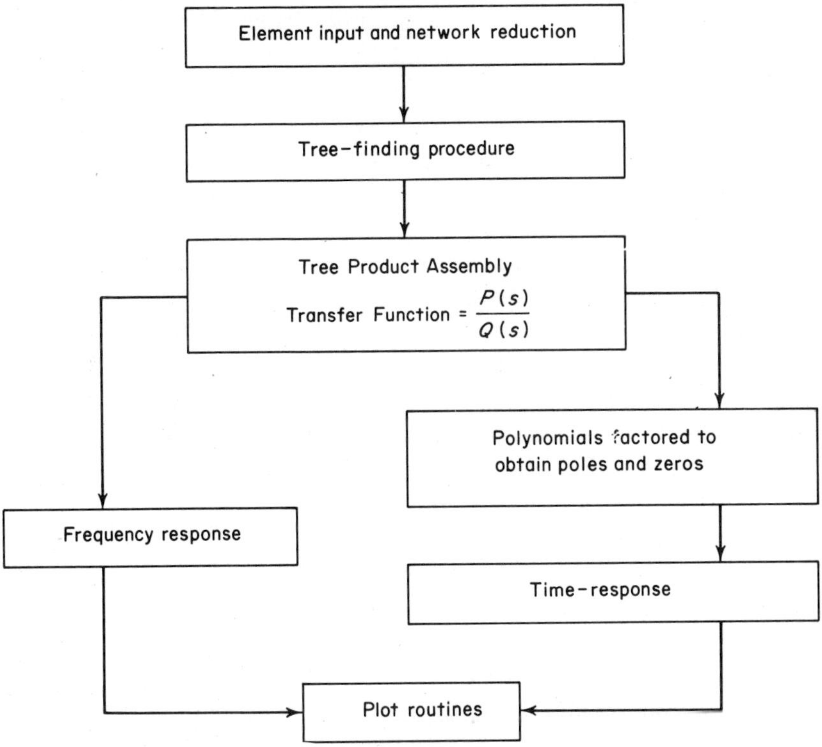

FIG. 2.16 Basic topological program

The network is set up and solutions obtained in terms of two-port system functions. Transfer functions or driving point functions can be obtained.

Each of the operations shown is independent, and thus the program can be interrupted to obtain output or to change running options. This provides the flexibility for adaptation to time-sharing systems, a point which is discussed in further detail in Section 2.6.

All passive elements in parallel are lumped together and considered as a single admittance. The tree-finding procedure is used to find and assemble the n-tree product terms used to form the numerator and denominator polynomials. A combinatorial technique is used that checks each possible combination of branches for the tree property. If the network contains active devices, then an active tree of the current graph is also checked to see if it is a tree of the voltage graph. A sign algorithm is then used to check the sign of a common tree. The tree product associated with this tree is then added to polynomials being assembled. The tree test algorithms are modified when 2-trees or n-trees are being tested. Test branches are inserted and checked to allow the general algorithm to find n-trees. The above procedure is used to express a system function as a ratio of two polynomials in s. The system function can also be assembled as the ratio of two polynomials, each of which is a linear combination of a single element. This variation leads to applications in root-locus or sensitivity work. Further details of this option are givèn in Section 2.5. Also a literal solution for the system function can be obtained by printing out the symbolic n-tree product.

If the selection is made to assemble the transfer function as a ratio of polynomials in s, then the location of the roots of the numerator and denominator is determined by a root-finding routine.

There are a number of standard root-finding routines available in the SHARE library. Evaluation of these routines by Lawrence [13] has shown that for most applications the MULLER root-finding routine is superior in speed and accuracy.

Frequency Response—The frequency response option calculates and plots the gain, phase, and group delay of a system function between the minimum and maximum values of frequency selected by the user. Since the system function is expressed as a ratio of polynomials in s, the even and odd parts can be found simply. For $s = jw$, the even and odd parts of a system function relate respectively to the real and imaginary parts of the system function. Thus gain, phase, and delay can be calculated as a function of the even and odd parts of the system function. In this part of the program there is no trouble with multiple roots since it is not necessary to obtain the roots of a polynomial to get the even and odd parts of the polynomial.

Time Response – The system function is the ratio of the Laplace transform of the response to the Laplace transform of excitation. If the input to a network is a simple function, then the output can be found by conventional inverse Laplace transform techniques. For example, if the input is a step function, then the step function response can be obtained in an analytical form. Numerical answers and plots are obtained by substituting values of time into the analytical expression. One difficulty in this procedure arises when poles of the system function are very close to each other. The residues obtained in the partial fraction expansion are obtained by a Heavisides procedure. When the roots of the function are very close to each other, the numerical value of a residue depends on the subtraction of several nearly equal values, and large errors can be introduced despite the accuracy of the digital computer. This problem could be partially solved by pre-examination of the roots. Any root that was within a certain percentage of another root would be considered a multiple root and a special case of the Heaviside technique used.

In the event that the input is not a simple function, a numerical integration or convolution procedure is used to obtain the numerical output.

2.3.2 The Solution of a Typical Circuit

The artifice of the current and voltage graph allows the inclusion of a voltage controlled source as the basic active element in the modeling process. The dual topological approach based on the mesh-currents equation uses a current-controlled voltage source as the basic active element. Therefore, any device that can be modeled in terms of either of the above controlled sources and passive lumped *RLC* elements can be easily modeled for analysis by the topological approach. Most physical devices can be directly modeled either by such an approach or by a rearrangement of elements by Thevenin's or Norton's theorem.

The simplified Hybrid π model shown in Fig. 2.9 is a useful and accurate model for transistors such as the 2N3251, operating in linear mode. Hatch [14] used the topological approach for the computer solution of the four transistor loop shown in Fig. 2.17. The results of this

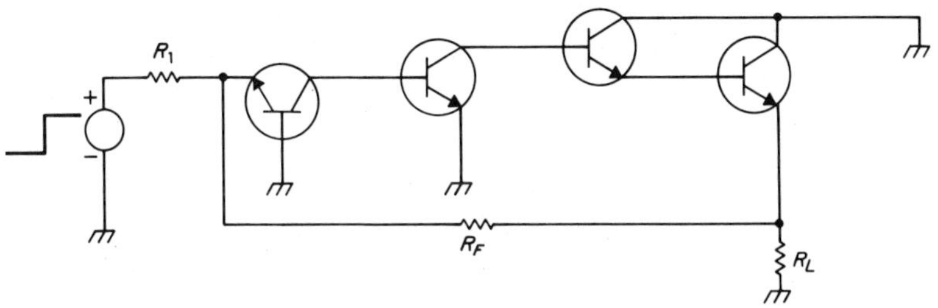

FIG. 2.17 Four transistor loop

problem were compared with the actual physical network and the answers obtained were accurate to a few percent, which agrees with the accuracy of the modeling procedure. Small signal models were selected for the transistors, and the modeled circuit is shown in Fig. 2.18. A normalized set of units is selected before the input data is prepared. The normalized values are shown in Fig. 2.18. The input data is shown

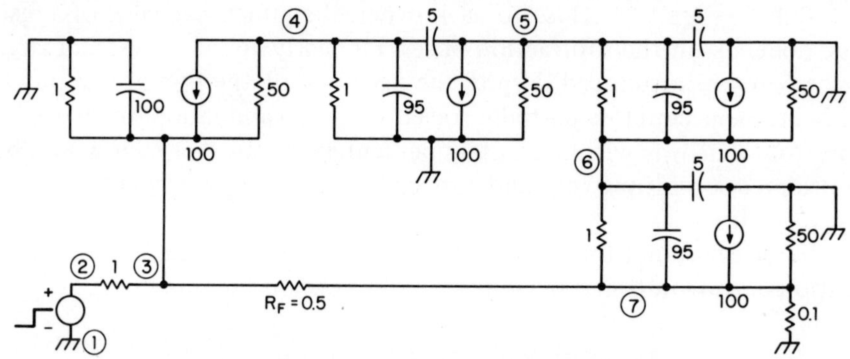

FIG. 2.18. Equivalent circuit and scaled elements

in Fig. 2.19. The response to a unit step input is shown in Fig. 2.20. The magnitude response is shown in Fig. 2.21, where the results are as would be expected with this unity-gain feedback loop.

1	2	3	4	5	6	7	8	9	10	11	12	13	14	15	16	17	18	19	20	21	22	23	24	25	26	27	28	29	30	31	32	33	34	35	36	37	38	39	40	41	42	43	44	45
0	1																																										
1	8		0	4		0	7		0	2		0	1		0	1		0	7		0	1		0	4		1	8		0	4												
0	2		0	3		R		1	.	0																																	
0	3		0	1		R		1	.	0																																	
0	3		0	1		C		1	0	0	.	0																															
0	3		0	4		R		5	0	.	0																																
0	4		0	1		R		1	.	0																																	
0	4		0	1		C		9	5	.	0																																
0	5		0	4		C		5	.	0																																	
0	5		0	1		R		5	0	.	0																																
0	5		0	6		R		1	.	0																																	
0	5		0	6		C		9	5	.	0																																
0	1		0	5		C		5	.	0																																	
0	1		0	6		R		5	0	.	0																																
0	6		0	7		R		1	.	0																																	
0	6		0	7		C		9	5	.	0																																
0	1		0	6		C		5	.	0																																	
0	1		0	7		R		5	0	.	0																																
0	7		0	1		R		0	.	1																																	
0	7		0	3		R		0	.	5																																	
0	4		0	3		0	1		0	3		G		1	0	0	.	0																										
0	5		0	1		0	4		0	1		G		1	0	0	.	0																										
0	1		0	6		0	5		0	6		G		1	0	0	.	0																										
0	1		0	7		0	6		0	7		G		1	0	0	.	0																										
0	.	5									1	.	0										2	.	0										5	.	0												
0	1																																										
1	.	0									0	.	0										1	.	0										7	4	.	0								0	.	5		

FIG. 2.19. Input Data

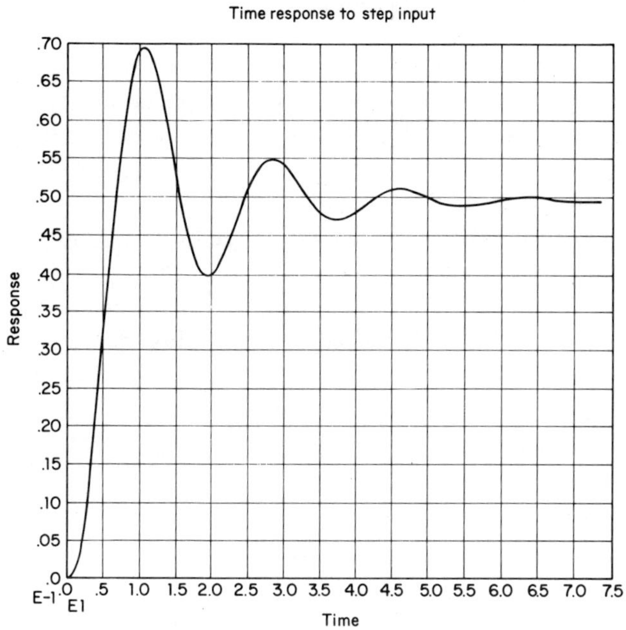

FIG. 2.20. Time response to a step input.

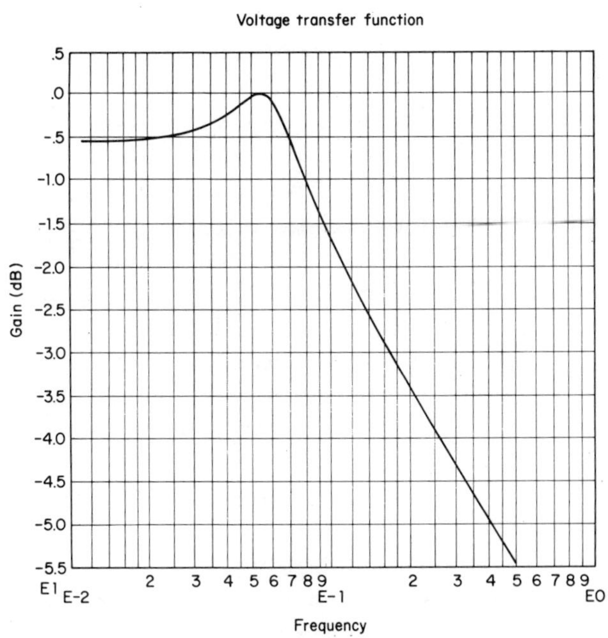

FIG. 2.21. Magnitude response

2.4 THE TREE GENERATION PROBLEM

As shown in Section 2.2.4 the interjection of active elements into a network through the current and voltage edge concept produces two new problems in topological analysis, i.e., the common trees and their associated sign must now be found. A survey is given of some of the attempts at the solution of these problems.

This problem has been solved by Mayeda and others in the sense that algorithms exist that accurately generate the common trees for an active graph. In order for a new approach to the problem to be more useful it should retain the accuracy of the Mayeda approach and increase the speed and the convenience of the selection process. Unfortunately most of the new techniques reported have not been programmed for computer use and therefore they cannot be judged by this criterion.

2.4.1 Previous Investigations of the Problem

In addition to Mayeda a number of authors have investigated the problems of the common tree and the correct sign, and many papers have appeared as a result of these studies.

Watanabe [15] introduced the idea of a spread transformation to generate the trees for a passive graph. However this required a large amount of computer memory space for even a small number of trees. In a later paper [16] he attempted to decompose the original network into subgraphs and connecting cut sets. Later Roe [17] showed that Watanabe's decomposition method was incorrect as it would produce terms that were not actually trees of the original network. Mayeda [18] also investigated the idea of network decomposition. This method was later programmed by Kim and others [19], and was restricted to passive networks. Hakimi [20], and later Deo [21], investigated the problem along a different line. Using the concept of distance introduced by Watanabe, Hakimi picked an arbitrary tree t_0, and found all trees at a given distance from t_0. Then he investigated the methods of generating all trees in terms of t_0. His method proved useful for passive circuits where the corresponding matrix has "large" ranks and "small" nullities. In this method the choice of t_0 is an important one, and if t_0 is a "central tree," the iterations required in this technique can be optimized. However, Deo and Hakimi have not found a way of finding the central tree except by inspection for very simple networks.

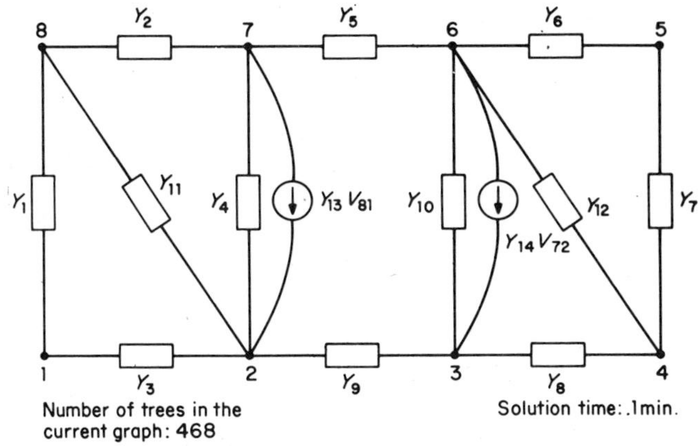

FIG. 2.22 Test network 1

Mayeda [22] investigated a method of elementary complete tree transformation in an effort to obtain the common trees with their sign.

Talbot [23] took another line of attack to the problem. In an attempt to reduce the size of the graph, he did not use the current and voltage tree concept, but carried a more general $\mathbf{Y}_b$ matrix into the problem. The elements of $\mathbf{Y}_b$ themselves could be matrices with both diagonal and off diagonal elements. This requires a great number of rules in order to correlate the trees and the $\mathbf{Y}_b$ matrix. Talbot's method requires the generation and storage of all trees of the graph and then all possible pairs of trees. Such a procedure makes a prohibitively large computer memory necessary.

None of the above techniques seems to offer a significant advantage over the basic approach taken by Mayeda, where a combinatorial technique is used along with test branches inserted to select the n-tree products that are also required.

The number and type of trees provide guidelines for the selection of an efficient algorithm. Several examples are given next to illustrate the relative number of trees involved in the various graphs of a network.

2.4.2 A Correlation Between the Size of a Network, the Number of Trees and Computer Solution Time.

The first network is shown in Fig. 2.22. It consists of 12 *RLC* elements, 8 nodes, and 2 voltage-controlled current sources. The term *passive trees* is used to indicate the trees of a given network with all active

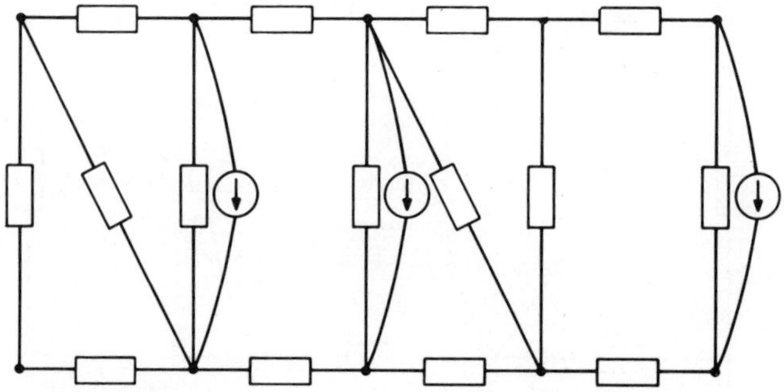

FIG. 2.23 Test network 2

edges deleted. The network of Fig. 2.22 has 208 passive trees. There are 468 trees of the current graph associated with the network and 505 trees of the voltage graph. There are 252 common trees.

The second network is shown in Fig. 2.23, and consists of 3 active elements, 10 nodes, and 15 *RLC* elements. It has 777 passive trees, 1058 common trees, 2953 trees of the current graph, and 2851 trees of the voltage graph.

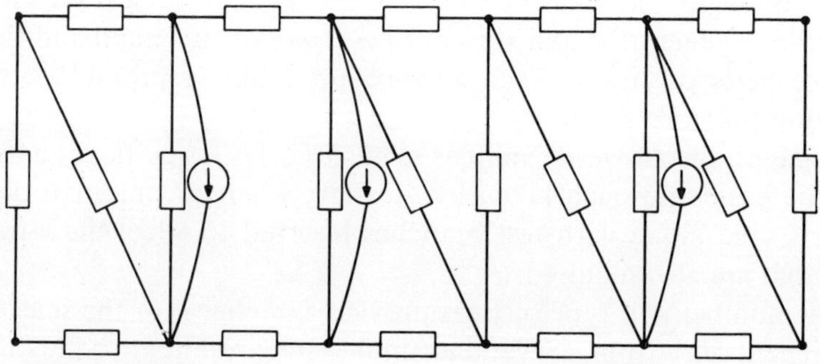

FIG. 2.24 Test network 3

The general conclusion that can be drawn from these examples is that the addition of active elements produces a large number of additional trees in the current and voltage graph. The actual number of common trees is very close to the number of passive trees. A sidelight is given by further reference to the CALAHAN program. The combinatorial approach requires the consideration of 3432 groups of 7 edges for the first example and 48620 groups of 9 edges in the second example. The addition of active trees has greatly increased the number of combinations of n-1 edges that must be tested.

Two other networks are shown in Fig. 2.24 and 2.25 to illustrate the tremendous increase in the number of trees as additional nodes and

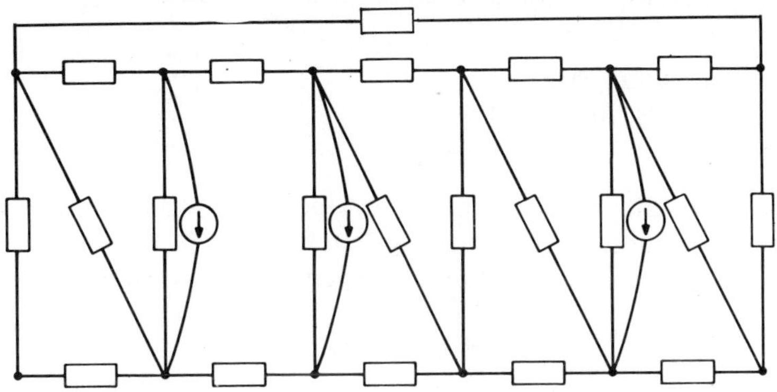

FIG. 2.25 Test network 4

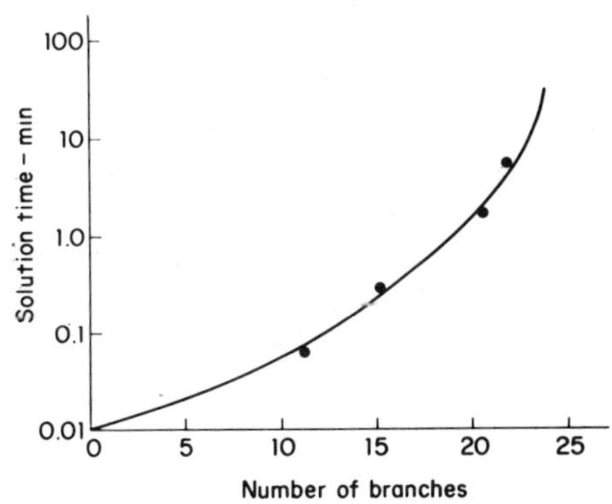

FIG. 2.26 Solution times as a function of the number of branches

branches are added. Each of these test networks was run on the CALAHAN program and solution times obtained. The results are shown in Fig. 2.26.

The conclusion is that, regardless of the efficiency of an algorithm used, there is an upper bound on the complexity of a circuit that the topological approach can solve in a practical amount of computer time.

A certain minimum amount of time is required for the selection and assembly of each tree product regardless of the efficiency of an algorithm. This limit is approximately 12-15 nodes and 20-25 branches.

However, if the network is below this upper bound, then the topological program of the previous section can be used to accurately find the system function for a large class of linear networks.

2.5 EXTENSIONS OF THE TOPOLOGICAL APPROACH

Many interesting applications of the topological technique are due to the fact that the polynomials of the system function of the 2-port can be assembled as a linear combination of a given element as shown in (2.43).

$$\frac{V_2(S)}{V_1(S)} = \frac{Y_i A_1(S) + A_2(S)}{Y_i A_1(S) + B_2(S)} \tag{2.43}$$

Thus the lengthy tree-finding procedure need be used only once to assemble a system function. The element Y_i can then be stepped through a number of values in an iterative procedure and the results used to provide a generalized root-locus plot where any element in a 2-port network can be used as a parameter. In addition, changes in both frequency domain and time-domain responses can be obtained with respect to changes in element values – variation response sensitivity.

Further details of these procedures are given along with an additional technique that allows the inclusion of nonlinear or time-varying elements. These extensions, as well as the basic topological program, can be implemented with a relatively small program, thus fitting in with a time-sharing or graphical input package as shown in section 2.6.

2.5.1 A Generalized Root-Locus Approach to System and Network Problems

A large portion of the effort in system theory during the past decade has been in the analysis and design of linear feedback systems. One standard approach to the problem has been to idealize and represent the system in block-diagram form shown in Fig. 2.27. The root locus approach

[24] to the solution is to attempt to relate the closed-loop performance of the system to the transfer functions $G(s)$ and $H(s)$. This is done by finding the locus of poles and zeros in the s-plane as a function of the gain K of the system or a time constant of $G(s)$ or $H(s)$. From this information in many cases it is possible to predict the overall system stability and performance. The first limitation to this approach is that it is often difficult to match the block diagram representation of Fig. 2.27 to the physical network under test. The second limitation is that locus has been obtained in terms of the gain K of the system, while, in fact, it may be desired to predict system performance in terms of other system element values.

The topological approach can be used to remove the second limitation and to relax the first limitation to the extent that if a model can be obtained in terms of an electrical network, then the generalized root-locus can be obtained.

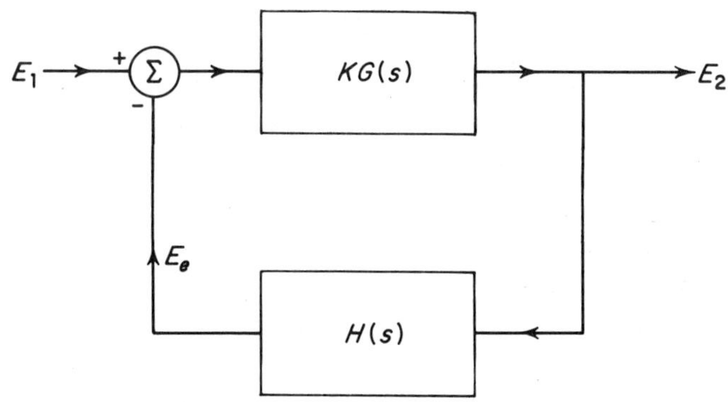

FIG. 2.27. System function in block diagram form

After $H(s)$ is obtained in the form shown in (2.43), a particular value of $Y_i(s)$ is selected; the root-finding routine is used to find the zeros and poles of $H(s)$. Then a plot routine is used to give a graphical presentation which is the generalized root-locus. The lengthy tree-finding algorithm is used only once, the root-finding and plot routines are relatively fast, and the required plot is obtained as the various values of Y_i are selected.

The outline of the generalized root-locus procedure is shown in Fig. 2.28. Both the numerator and the denominator polynomial are linear in a given element which is denoted as Y_i. Therefore, the assembly procedure of the tree generation algorithm is used to separate both the num-

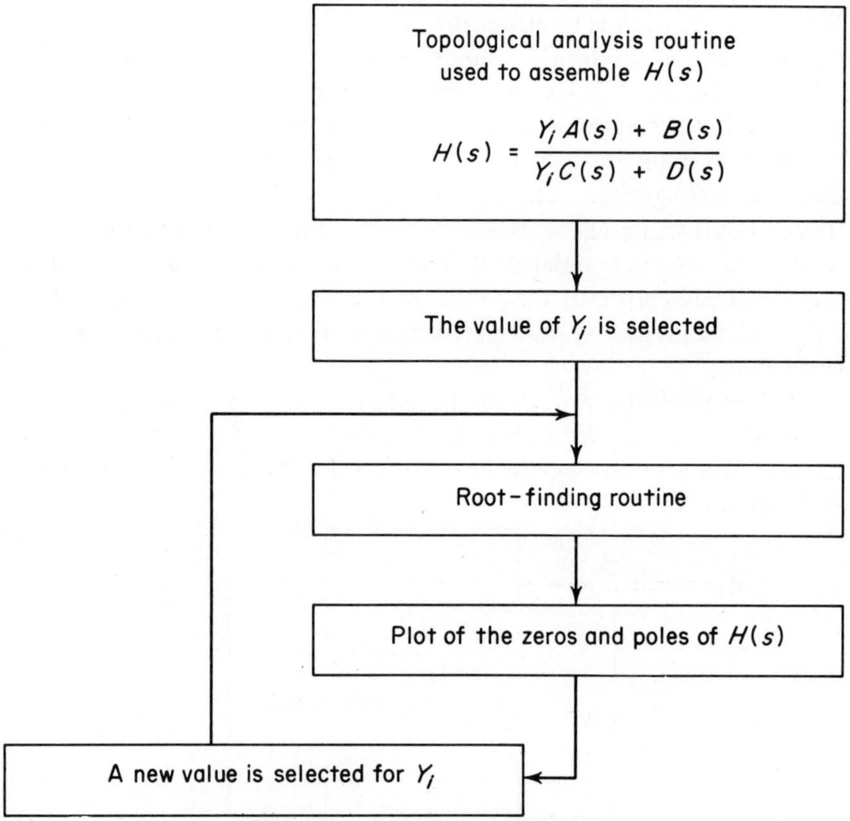

FIG. 2.28. Generalized root locus procedure

erator and the denominator into terms that contain Y_i or do not contain Y_i. This is shown in (2.43). The variables $V_1(s)$ and $V_2(s)$ are selected for examples; any two current or voltage variables could be used.

The sensitivity of the responses of the network to element variation can also be obtained by a similar procedure.

2.5.2 Response Sensitivity Procedure

The extension here requires only that either the frequency domain or time-domain response be obtained as an additional step in the repetitive process. This procedure is shown in Fig. 2.29. Again the lengthy tree-assembly routine is used only once, and the remaining steps of the program are relatively fast.

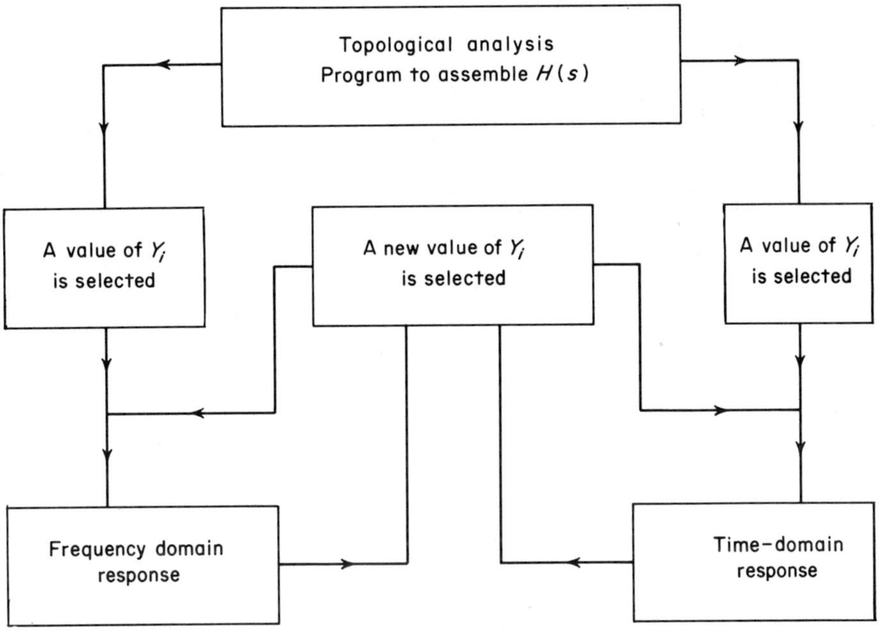

FIG. 2.29 Response sensitivity procedure

2.5.3 Extension to Time-Varying and Non-Linear Elements

The topological approach can be simply expanded to include one time-varying or non-linear element. This is done by using the same concepts that were used in the generalized root-locus approach. An elaboration of this approach could be used to extend the idea to several elements, but two or three elements are probably the practical limit in this type of approach. Then in the case of a time-varying element an update procedure is used to find the time response. This cyclic procedure is shown in Fig. 2.30. $Y_t(t)$ can be specified either by mathematical expression or by a table of values.

The extension to a non-linear element involves a dual analysis procedure. Two system functions are now assembled. The first system function is as before the ratio of response to excitation.

$$\frac{E_2(s)}{E_1(s)} = H_1(s) = \frac{A_1 Y_L + A_2}{B_1 Y_L + B_2}. \tag{2.44}$$

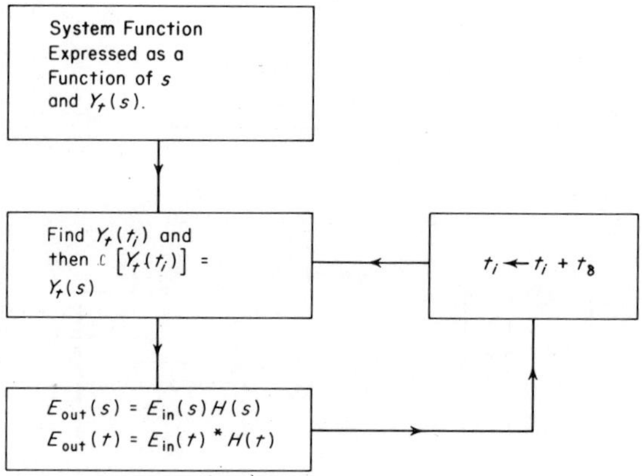

FIG. 2.30 Cyclic analysis procedure

The second system function expresses

$$\frac{E_L(s)}{E_1(s)} = H_2(s) = \frac{A_3 Y_L + A_4}{B_3 Y_L + B_4} \tag{2.45}$$

where Y_L is a function of E_L. Now a piece-wise linear approximation is made for Y_L. A current value is selected for Y_L, and the dual procedure solves for both $E_2(t)$ and $E_L(t)$. The value of E_L is used to find the new value of Y_L and the procedure is repeated. This assumes that the piece-wise approximation for the non-linear element leads to a convergent solution [25]. This dual procedure is shown in Fig. 2.31. The

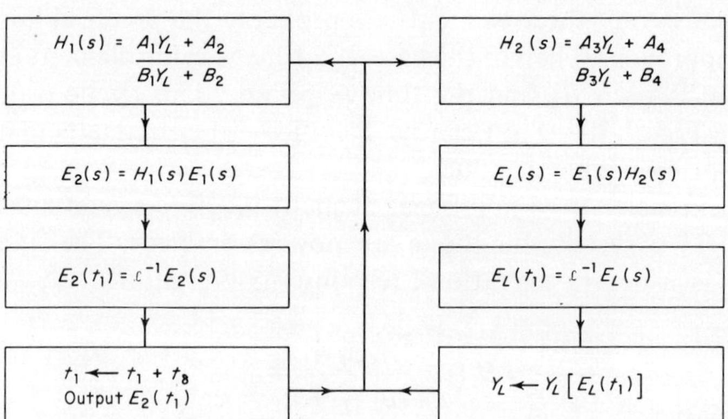

FIG. 2.31 Nonlinear analysis procedure

approaches used in this section to find the sensitivity of a network to element variation and to extend the topological approach to non-linear and time-varying networks meet the dual criterion of segmentation and small program size of a time-sharing computer system. The topological approach is useful in the area of man and machine interaction, as shown in the next section.

2.6 TIME-SHARING APPLICATIONS

The simplicity and relatively small size of a program based on topological analysis are particularly advantageous in the new generation of time-sharing computer systems such as OCTOPUS [26]. Most present computer installations use a batch monitor type procedure which in effect provides input and output queues of computer jobs. This means that even if the actual computer running time is only a few minutes for a problem, there is a built-in waiting period of several hours before the results of the problem are available to the user. This procedure, coupled with the fact that a minor programming error will probably prevent the problem from running, can add days or even weeks to the time it may take for a successful computer run on a given problem. The OCTOPUS system provides fast input-output capability through a time-sharing system where the user is allowed direct access to his problem through a teletype. This allows a user to correct quickly any minor errors and also to make part of the decisions as the problem is being run. In order to efficiently run on a time-sharing system a program must be relatively small and segmented so that only a portion of the program is needed at any one time.

The basic topological program and the extensions presented are easily segmented and work well in a time-sharing system.

The CALAHAN program has been adapted and has been running on the OCTOPUS system for two years [11] [27]. Utility routines are used to create the input data file on teletype. The compiled program exists in a public file, and a simple command starts the program running for the user. The output of the program can be put out directly on the teletype or stored in an output file. The output file can either be examined for the teletype or printed out on one of several printers. Options can be selected to provide teletype plots or CRT plots. The conversational operating mode of the OCTOPUS system allows the user to change parameters or correct mistakes quickly. Additional programming can be used to provide additional flexibility. A format free language removes some of the restrictions of a FORTRAN input format [28]. The new input languages also check data lines for correctness before a computer run is made.

REFERENCES

1. Pottle, C., "Comprehensive Active Network Analysis by Digital Computer: A State-Space Approach," Research Report EERL 59, Cornell University, Ithaca, New York, Electrical Engineering Research Laboratory.
2. Calahan, D. A., "Linear Network Analysis and Realization Digital Computer Programs: An Instruction Manual," University of Illinois Bulletin, Vol. 62, No. 58.
3. Mayeda, W., "Applications of Linear Graphs to Electrical Networks, Switching Networks, and Communication Networks," Report R-203, Coordinated Science Laboratory, University of Illinois, Urbana, Illinois, April 1964.
4. Mayeda, W., and Van Valkenburg, M. E., "Network Analysis and Synthesis by Digital Computers," *WESCON Convention Record*, pp. 137-149, 1957.
5. Percival, W. S., "The Graphs of Active Networks," *Proceedings of the IEE*, (London) Monograph No. 129R, Vol. 102, pt 3, pp. 270-278, April 1955.
6. Seshu, S., and Reed, M. B., *Linear Graphs and Electrical Networks.* Reading, Massachusetts: Addison Wesley, 1961.
7. Kim, Wan Hee, and Chien, Robert Tien-Wen, *Topological Analysis and Synthesis of Communication Networks.* New York: Columbia University Press, 1962.
8. Seshu, S., and Balabanian, N., *Linear Network Analysis.* New York: John Wiley and Sons, 1959.
9. Percival, W. S., "The Solution of Passive Electrical Networks by Means of Mathematical Trees," *Proceedings of the IEE*, (London) Paper No. 1429 R (100, part II) page 143, May 1953.
10. *CADAR Bulletin*, Issue No. 5, January/February 1968. IEEE.
11. Dawson, D. F., and Magnuson, Jr., W. G., "Computer-Aided Design Program Adapted to Time-Sharing," *Electro-Technology*, May 1967.
12. Magnuson, Jr., W. G., Kuo, F. F., and Walsh, W. J., "On-Line Graphical Circuit Design," Lawrence Radiation Laboratory, Livermore, Report UCRL-70796.

13. Lawrence, J. D., "Polynomial Root Finder," Computer Information Center, Report C2.2-001, LRL, University of California, Livermore, California, Dec. 9, 1966.

14. Hatch, Kenneth F., "Analysis and Description of a High Stability Nuclear Pulse Amplifier," *IEEE Transactions on Nuclear Science*, Vol. NS-le, No. 1, pp. 351-370, February 1966.

15. Watanabe, Hitoshi, "A Computational Method for Network Topology," *IRE Transactions on Circuit Theory*, Vol. CT-7, pp. 296-303, Sept. 1960.

16. Watanabe, H., "Method of Tree Expansion in Network Topology," *IRE Transaction on Circuit Theory*, Vol. CT-8, pp. 4-10, March 1961.

17. Roe, P. H., "On a Tree Expansion Theorem," *IRE Transactions on Circuit Theory*, CT-10 (1961) December 1961, pp. 496-500.

18. Mayeda, W., "Reducing Computation Time in the Analysis of Networks by Digital Computer," *Transactions of IRE*, CT-6, 136-7, March 1959.

19. Kim, W. H., Young, D. H., Freiman, C. V., and Mayeda, W., "On Iterative Factorization in Network Analysis by Digital Computer," *Proceedings Eastern Joint Computer Conference*, December 1960.

20. Deo, Narsingh, *Topological Analysis of Active Network and Generalization of Hamiltonian Tree*, Ph.D. Dissertation, Evanston, Illinois: Northwestern University, June 1965.

21. Hakimi, S. L., "On Trees of a Graph and Their Generation," J. Franklin Inst., 270 pp. 347-359, 1961.

22. Mayeda, W., "Generation of Trees and Complete Trees," Report R-284, Coordinated Science Laboratory, University of Illinois, Urbana, Illinois.

23. Talbot, A., "Topological Analysis of General Linear Networks," *IEEE Transactions on Circuit Theory*, Vol. CT-12, p. 170, June 1965.

24. Truxal, *Control System Synthesis.* New York: McGraw-Hill, 1955.

25. Katzenelson, J., "AED NET: A Simulator for Nonlinear Networks," *Proceedings of the IEEE*, Vol. 54, No. 11, November 1966.

26. Wyman, R. H., "Octopus System History to Date," UCRL 700-84, LRL, University of California, August 15, 1966.

27. Dawson, D. F., "Computational Aspects of the Topological Approach to Active Linear Network Analysis," Lawrence Radiation Laboratory, Livermore, Report UCRL-70710, given at the Hawaii International Conference on System Sciences, January 1968.

28. Dawson, D. F., and Magnuson, Jr., W. G., "Field-Free Input Languages for Network Analysis Programs," Report UCRL-70931, given at the 11th Midwest Symposium on Circuit Theory, May 1968.

CHAPTER 3

COMPUTER METHODS OF NETWORK ANALYSIS†

F. H. Branin, Jr., IBM Corporation, Kingston, N. Y.

It is obvious from recent publications [1], [2] and from the various topics treated in this book that the computer is exerting a profound influence on the *modus operandi* of circuit design. With the aid of existing computer programs, the circuit designer may carry out "numerical experiments" which would be difficult or even impossible to duplicate with actual hardware on the bench. Often, he can do such experiments faster and cheaper on the computer. He may analyze circuits for which certain components have not yet been manufactured – or even invented – and he may vary parameters, either singly or in combinations, seeking for an optimum design. Moreover, he may do all this without the risk of burning up expensive or hard-to-get components even if he inadvertently overdrives the circuit with a megavolt input signal.

† This Chapter is based on papers published by the IEEE. (See references [2], [7], and [36].)

Thus, the digital computer provides circuit designers with an analytical tool of great power and versatility. Generally speaking, the computer can be trusted to produce reliable solutions to an arbitrary network problem, except when the problem involves pathological numerical difficulties or when the method of analysis has been poorly chosen. It is the user's responsibility, however, to pose a problem that accurately represents the circuit which the designer has in mind, since the computer does only what it is explicitly instructed to do and has no way of second-guessing the designer's real intent. For this reason, high-fidelity network models for diodes, transistors, and other devices are essential to the practical application of almost every network analysis program. Accordingly, device modeling is a fast growing art that has evolved in response to the use of computers in circuit design [1], [3].

In addition to these influences, the computer is also having a more subtle influence on network theory itself. For example, because matrix notation is so convenient in programming a digital computer, many existing network analysis programs are based on a matrix formulation of the network problem. This, together with the inherent elegance of the matrix approach, is helping to promote its acceptance among network theorists.

In another direction, computers are helping to hasten the recognition of the inherently broad scope of network theory by drawing attention to the fact that a network analysis program is theoretically capable of converting a digital computer into a veritable analog computer of great versatility and power. While it may not be practical to write a single program to do this, at least two programs have been written for solving nonelectrical problems by network analytical methods: DYANA, [4] which handles either electrical or mechanical problems, and STRESS, [5] which is based on a network approach to structural analysis [6].

Computers are also influencing network theory by demanding methods of analysis adapted to the solution of computer-sized problems. As the engineer will appreciate, maximum "power transfer" from the computer to his network problem requires a method which properly matches the computer to the problem. Traditional methods for "hand solution" of networks are not necessarily best for use on a computer with networks of much greater size. The Laplace transform techniques fit this category and should at least be supplemented, if not supplanted, by numerical methods better adapted to the computer. One such method, [7] described below, gives essentially all of the information obtainable by Laplace transforms.

The objective of this paper is to highlight the importance of keeping the computer and its peculiarities in mind when developing the theory on which to base network analysis programs. Since matrix notation is the tool *par excellence* for describing the network problem in a form suitable for programming, we shall use a matrix approach throughout the following discussion of various aspects of dc, ac, and transient analysis of networks on a digital computer.

3.1 MATRIX FORMULATION OF THE NETWORK PROBLEM

It is an interesting historical coincidence that just prior to the advent of the high-speed electronic computer, the pioneering work of Gabriel Kron [8] demonstrated convincingly the superior organizational powers of the matrix-tensor notation in network theory. Kron emphasized not only the conceptual elegance of this notation but also its virtually automatic way of handling the tedious bookkeeping chores of network analysis. He thereby laid some of the logical and procedural foundations necessary for programming a digital computer to analyze networks – that is, both to compile and to solve the network equations automatically.

The following matrix formulation of the network problem [9], [10] is a direct outgrowth of Kron's work [8], [11] and that of Roth [12]-[14] who provided an algebraic-topological explanation of the network problem as an abstraction. This formulation has been used at least in part by such programs as TAP, [15], [16] NET-1, [17] and ECAP [18]. Kron's work has also been used as the basis for much of the computer analysis of networks encountered in the electrical power industry [19]-[21].

From a physical point of view, the network problem is concerned with predicting the behavior of a system of interconnected elements in terms of the element characteristics and the manner in which these elements are interconnected. Viewed as a mathematical entity, the network problem involves the properties of an underlying *topological* structure, called a linear graph, and a superimposed *algebraic* structure consisting of the interrelations between certain quantities associated with the nodes, branches, and meshes of the graph.

Obviously, the very act of interconnecting the network elements introduces certain constraints on the physical variables of the network problem and this gives rise to part of the bookkeeping chore. But the bookkeeping can be systematized in terms of matrices which describe the connectivity properties of a linear graph. Here is where topology is introduced, albeit in a very rudimentary form. Finally, the physical variables of the network problem correspond to the quantities which enter into the algebraic structure referred to above. The matrix approach nicely correlates these two points of view and has the additional merit of identifying precisely the separate roles of the element characteristics and of the interconnections in determining the overall behavior of the network.

The ABCD's of Network Topology

There are four topological matrices relating to a directed (or oriented) linear graph which are useful in network theory. We shall describe them briefly. (More detailed descriptions are given elsewhere [6]-[10].)

Consider the graph of Fig. 3.1(a) for which the branch-node connections may be tabulated as follows:

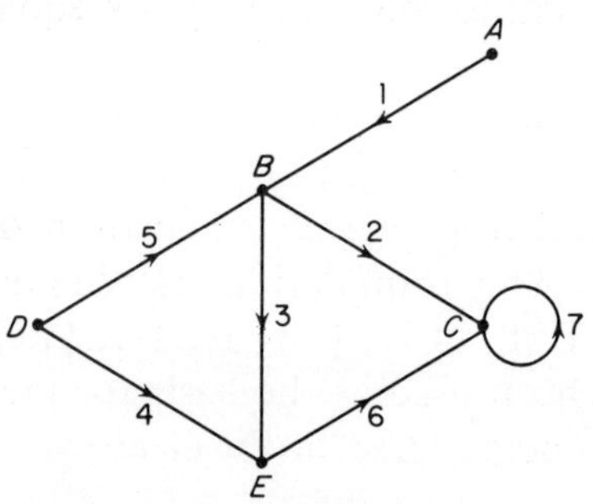

(a)

$\bar{\mathbf{A}}$ Matrix

Branch \ Node	A	B	C	D	E
1	1	-1	0	0	0
2	0	1	-1	0	0
3	0	1	0	0	-1
4	0	0	0	1	-1
5	0	-1	0	1	0
6	0	0	-1	0	1
7	0	0	0	0	0

(b)

FIG. 3.1. Linear graph

Branch	Initial Node	Final Node
1	*A*	*B*
2	*B*	*C*
3	*B*	*E*
4	*D*	*E*
5	*D*	*B*
6	*E*	*C*
7	*C*	*C*

This connection table, which is easily obtained from input data describing the network configuration, is a convenient starting point for compiling the network equations.

An alternative way of storing this connectivity information in a computer is in matrix form as displayed in Fig. 3.1(b), where the columns show which branches are connected to each node and the polarity of these connections. The way in which this $\overline{\mathbf{A}}$ matrix is obtained from the connection table is obvious by inspection. (Branch 7, being short-circuited, requires both a + 1 and a − 1 entry in column C so that both entries cancel.)

Since each row of the $\overline{\mathbf{A}}$ matrix contains both a + 1 and a − 1 as its nonzero entries, its columns are linearly dependent – that is, their sum is zero so that any column is equal to the negative sum of all the other columns. Hence, any one column may be deleted since it contains redundant information. We choose to delete the column corresponding to node E in Fig. 3.1 and regard this node as the *datum* or ground node. The resulting $\mathbf{A}$ matrix, called the *branch-node* matrix, is shown in Fig. 3.2(b).

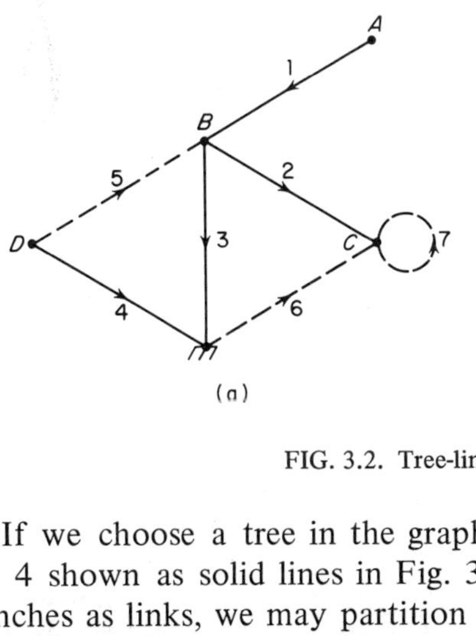

(a)

A Matrix

Branch \ Node	A	B	C	D	
1	1	-1	0	0	$\mathbf{A}_T$
2	0	1	-1	0	
3	0	1	0	0	
4	0	0	0	1	
5	0	-1	0	1	$\mathbf{A}_L$
6	0	0	-1	0	
7	0	0	0	0	

(b)

FIG. 3.2. Tree-link structure of graph.

If we choose a tree in the graph – for example, branches 1, 2, 3, and 4 shown as solid lines in Fig. 3.2(a) – and classify the remaining branches as links, we may partition the $\mathbf{A}$ matrix into the submatrices $\mathbf{A}_T$ and $\mathbf{A}_L$ referring to tree-branches and links, respectively. (An algorithm for performing a tree-link sort on the computer is given in Appendix II of Branin [16]. The sequence of branches may need to be rearranged before partitioning the $\mathbf{A}$ matrix, but this is a simple task.)

It is easily shown that the submatrix $\mathbf{A}_T$ is square and has an inverse which we may determine directly from the graph of the tree in terms of the *node-to-datum path* matrix $\mathbf{B}_T$ for the tree [10].

For the tree shown in Fig. 3.3(a) the columns of $\mathbf{B}_T$, shown in Fig. 3.3(b), indicate which branches are included in the paths from each node and datum node. The desired relation between $\mathbf{B}_T$ and the inverse of $\mathbf{A}_T$ is

$$\mathbf{A}_T^{-1} = \mathbf{B}_T^t \tag{3.1}$$

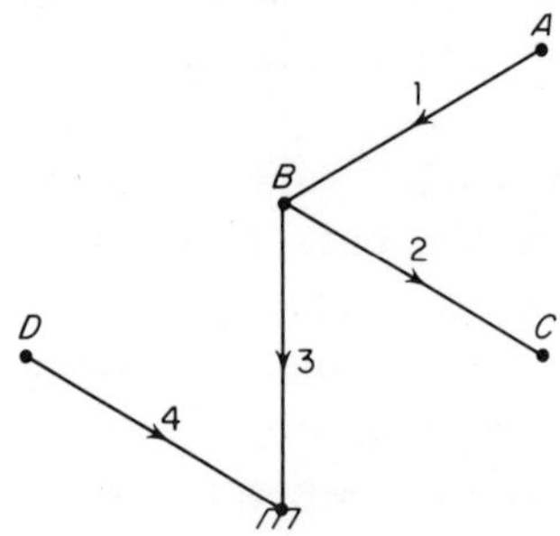

(a)

$\mathbf{B}_T$ Matrix

Branch \ Node	A	B	C	D
1	1	0	0	0
2	0	0	-1	0
3	1	1	1	0
4	0	0	0	1

(b)

FIG. 3.3. Tree of graph.

where $\mathbf{B}_T^t$ is the transpose of the $\mathbf{B}_T$ matrix. (An algorithm for determining the $\mathbf{B}_T$ matrix from the $\bar{\mathbf{A}}$ matrix is given in Appendix III of Branin [16].)

The next topological matrix of interest is the *branch-mesh* or *circuit* matrix, **C**, shown in Fig. 3.4(b). The columns of this matrix show which branches are included in each mesh.

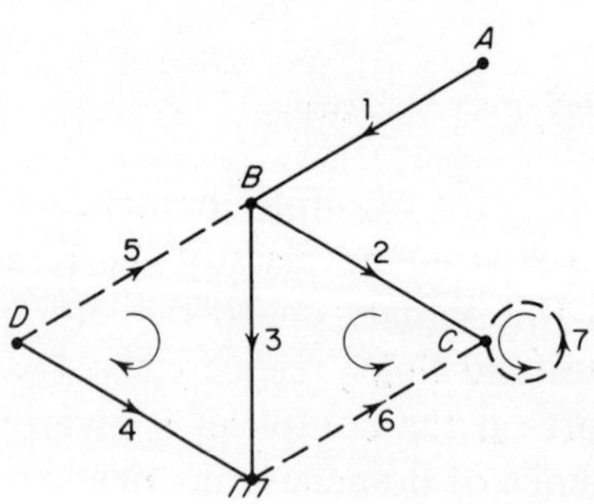

(a)

C Matrix

Branch \ Mesh	5	6	7	
1	0	0	0	$\mathbf{C}_T$
2	0	-1	0	
3	1	1	0	
4	-1	0	0	
5	1	0	0	$\mathbf{C}_L$
6	0	1	0	
7	0	0	1	

(b)

FIG. 3.4. Basic meshes of graph.

If we adopt the convention that each link, together with its contiguous tree branches, defines a basic mesh whose orientation coincides with that of its defining link, then each basic mesh will contain only one link, oriented positively, as shown in Fig. 3.4(a). Consequently, the submatrix $\mathbf{C}_L$, referring to the links, is a *unit* matrix as shown in Fig. 3.4(b) and it need not be stored explicitly in the computer. Only the submatrix $\mathbf{C}_T$, which gives the path-in-tree for each basic mesh, is needed and this may be obtained as follows.

It is a standard theorem that, for any linear graph, the branch-node and branch-mesh matrices obey the fundamental relations,

$$\mathbf{A}^t\mathbf{C} = \mathbf{O} \text{ and/or } \mathbf{C}^t\mathbf{A} = \mathbf{O} \tag{3.2}$$

where the product is a null matrix. As a consequence, we may write the first of these expressions in partitioned form.

$$[\mathbf{A}_T^t \mathbf{A}_L^t]\begin{bmatrix}\mathbf{C}_T\\ \mathbf{U}_L\end{bmatrix} = \mathbf{A}_T^t\mathbf{C}_T + \mathbf{A}_L^t = \mathbf{O} \tag{3.3}$$

where the unit matrix $\mathbf{U}_L$ replaces $\mathbf{C}_L$. We may solve this equation for $\mathbf{C}_T$ by inverting $\mathbf{A}_T^t$ and using (3.1) to obtain the relation [10].

$$\mathbf{C}_T = -\mathbf{B}_T\mathbf{A}_L^t \tag{3.4}$$

The final topological matrix we need is the *cutset* matrix **D**. Just as each link, together with a certain set of tree branches, defines a basic mesh, so each tree branch together with a certain set of links defines a basic cutset. This is shown in Fig. 3.5(a).

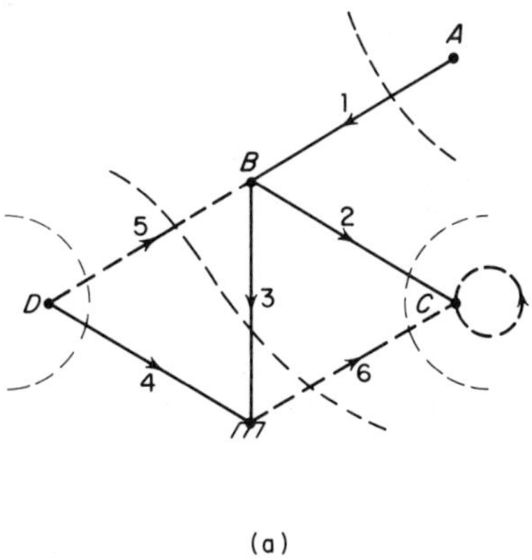

(a)

D Matrix

Branch \ Cutset	1	2	3	4	
1	1	0	0	0	$\mathbf{D}_T$
2	0	1	0	0	
3	0	0	1	0	
4	0	0	0	1	
5	0	0	-1	1	$\mathbf{D}_L$
6	0	1	-1	0	
7	0	0	0	0	

(b)

FIG. 3.5. Basic cutsets of graph.

The columns of the **D** matrix show which branches are included in each cutset. With each tree branch positively oriented in its corresponding basic cutset, the submatrix $\mathbf{D}_T$, referring to the tree branches, is a unit matrix while the submatrix $\mathbf{D}_L$, referring to the links, turns out to be the negative transpose of $\mathbf{C}_T$. (See Branin [10], for example.) This is shown in Fig. 3.5(b). As a result, it follows that

$$\mathbf{D} = \begin{bmatrix} \mathbf{D}_T \\ \mathbf{D}_L \end{bmatrix} = \begin{bmatrix} \mathbf{U}_T \\ -\mathbf{C}_T^t \end{bmatrix} = \begin{bmatrix} \mathbf{A}_T \\ \mathbf{A}_L \end{bmatrix} \mathbf{B}_T^t = \mathbf{A}\mathbf{B}_T^t \tag{3.5}$$

so that the cutset matrix **D** is readily computed in terms of **A**, $\mathbf{B}_T$, and/or $\mathbf{C}_T$. Except for pointing out that

$$\mathbf{D}^t\mathbf{C} = \mathbf{O} \text{ and/or } \mathbf{C}^t\mathbf{D} = \mathbf{O} \tag{3.6}$$

This completes our description of topological matrices. Their use in performing the bookkeeping tasks in network analysis will be explained in the next section.

The Electrical Network Laws

Consider as a typical branch in an electrical network the configuraion shown in Fig. 3.6. (This is Kron's branch convention [8].)

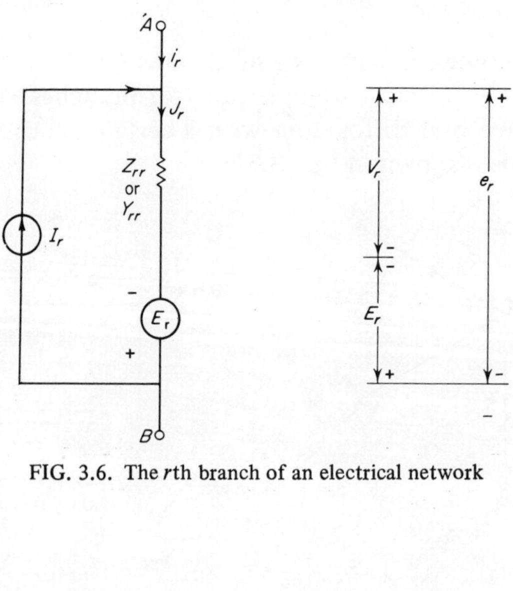

FIG. 3.6. The rth branch of an electrical network

Here, the rth branch is shown consisting of an impedance (or admittance) element Z_{rr} (or Y_{rr}) in series with a voltage source E_r and in parallel with a current source I_r, either of which sources may have any value. Thus, there are three distinct voltage and current variables to identify in this branch. As shown in the drawing, branch current i_r enters the branch at its initial node but the element current J_r actually passing through Z_{rr} is, according to the sign conventions depicted,

$$J_r = I_r + i_r \tag{3.7}$$

Similarly, although the branch voltage, measured from initial to final node, is e_r, the element voltage V_r actually appearing across Z_{rr} is

$$V_r = E_r + e_r \tag{3.8}$$

with the sign conventions taken as shown.

Clearly, since E_r and I_r may be nonzero, Ohm's law for this branch must be written in terms of the element voltage and the element current variables V_r and J_r. Accordingly, we may express Ohm's law for the entire network as the matrix equations

$$\mathbf{V} = \mathbf{ZJ} \tag{3.9}$$

and/or

$$\mathbf{J} = \mathbf{YV} \tag{3.10}$$

where the vectors **V** and **J** consist of the entire set of element voltage and element current variables for the network and where the admittance matrix **Y** is the inverse of the impedance matrix **Z**. The diagonal terms of **Z** (or **Y**) are the self-impedances (or self-admittances) of each branch, where as the off-diagonal terms are the transimpedances (or transadmittances) between pairs of branches. If corresponding off-diagonal terms are equal, they correspond to mutual impedances or admittances. Active devices, or so-called "dependent sources," may be represented by off-diagonal terms in the **Z** or **Y** matrix.

In order to account for the constraints on the branch voltages **e** and branch currents **i** imposed by the interconnections, we invoke Kirchhoff's voltage and current laws. In their original form, these laws correspond to the matrix relations

$$\mathbf{C}^t\mathbf{e} = \mathbf{O} \tag{3.11}$$

and

$$\mathbf{A}^t\mathbf{i} = \mathbf{O} \tag{3.12}$$

where the topological matrix $\mathbf{C}^t$ acts as an operator which sums all the branch voltages around each basic mesh while the matrix $\mathbf{A}^t$ sums all the branch currents leaving each node.

Although we may not always be aware of the fact, we are actually using alternative and mathematically equivalent forms of Kirchhoff's voltage and current laws when we say "the branch voltages are a linear combination of the node voltages" and "the branch currents are a linear combination of the mesh currents." These two statements correspond to the matrix equations,

$$\mathbf{e} = \mathbf{A}\mathbf{e}' \tag{3.13}$$

and

$$\mathbf{i} = \mathbf{C}\,\mathbf{i}' \tag{3.14}$$

where $\mathbf{e}'$ is the vector of node-to-datum voltages and $\mathbf{i}'$ is the vector of mesh currents. Here, as in (3.11) and (3.12), the topological matrices automatically do the bookkeeping for us. Incidentally, it is easy to derive (3.11) and (3.12) from (3.13) and (3.14) by premultiplying the latter by $\mathbf{C}^t$ and $\mathbf{A}^t$, respectively, and then invoking (3.2). The reverse derivations can also be made, proving that both forms of Kirchhoff's laws are equivalent.

The basic interrelations between the network variables and the topological matrices may be summarized by a transformation diagram, given by Roth, [12] as shown in Fig. 3.7. This diagram, whose relation to (3.9)-(3.13) is evident, characterizes the algebraic structure associated with a linear graph. The only additional relations are $\mathbf{C}^t\mathbf{E} = \mathbf{E}'$, which defines the equivalent mesh emfs $\mathbf{E}'$, and $\mathbf{A}^t\mathbf{I} = \mathbf{I}'$, which defines the equivalent nodal current sources $\mathbf{I}'$.

By a proper interpretation of this algebraic structure, it can be shown that a wide variety of nonelectrical systems – e.g., mechanical and structural systems – fit into this algebraic pattern and so can be treated as

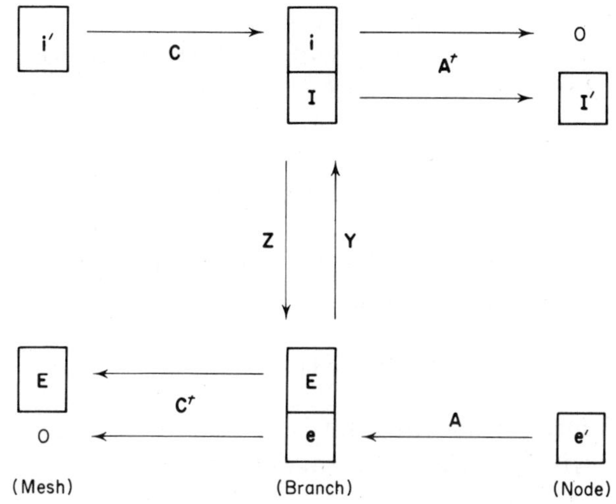

FIG. 3.7. Algebraic structure of the network problem.

network problems. Moreover, by extending this algebraic structure to characterize topological structures containing surface and volume elements as well as points and lines, a direct correspondence with the operational structure of the vector calculus emerges [22]. As a result, the use of network analogies for a wide class of partial differential equations – including Maxwell's equations – can be justified. The exploration and exploitation of this interesting topic is one of the broader aspects of network theory whose development the computer is helping to stimulate.

The Electrical Network Problem

A formal statement of the electrical network problem may be expressed as follows: "given a network whose configuration determines the topological matrices **A** and **C**, given its impedance matrix **Z** (or admittance matrix **Y**), and given the voltage source vector **E** and current source vector **I**, find the branch voltages **e** and branch currents **i** such that (3.7)-(3.12) hold true."

We shall discuss four distinct ways of attacking this problem. In each case, it will be necessary to introduce auxiliary variables such as the node voltages **e**′ or mesh currents **i**′ appearing in (3.13) and (3.14). In each case, it will also be necessary to invoke both of Kirchhoff's laws and, since each of these laws has two equivalent forms, the use of one form is tantamount to invoking the other.

The Mesh Method: Starting with the impedance form of Ohm's law given in (3.9), we introduce (3.7) and (3.8) and rearrange terms to obtain the expression

$$(\mathbf{E} - \mathbf{ZI}) + \mathbf{e} = \mathbf{Zi} \tag{3.15}$$

If we premultiply by $\mathbf{C}^t$ and take into account Kirchhoff's voltage law in the form of (3.11), we find that

$$\mathbf{C}^t(\mathbf{E} - \mathbf{ZI}) = \mathbf{C}^t\mathbf{Zi} \tag{3.16}$$

We may then incorporate Kirchhoff's current law by using (3.14) and writing the matrix equivalent of the mesh equations

$$\mathbf{C}^t(\mathbf{E} - \mathbf{ZI}) = \mathbf{C}^t\mathbf{ZCi}' \tag{3.17}$$

where $\mathbf{C}^t\mathbf{ZC}$ is the *mesh impedance* matrix.

If we solve this system of equations for the mesh currents $\mathbf{i}'$ – or write the formal solution for $\mathbf{i}'$ as

$$\mathbf{i} = (\mathbf{C}^t\mathbf{ZC})^{-1}\,\mathbf{C}^t(\mathbf{E} - \mathbf{ZI}) \tag{3.18}$$

where $(\mathbf{C}^t\mathbf{ZC})^{-1}$ is the *mesh solution* matrix – we may compute the branch currents $\mathbf{i}$ by using (3.14) and finally obtain the branch voltages $\mathbf{e}$ by reverting to (3.15).

The Node Method: If we start with the admittance form of Ohm's law given in (3.10), we may proceed to treat the network problem via the node method, following steps quite similar to those just described for the mesh method. Without describing the steps in detail, we simply list the corresponding equations:

$$(\mathbf{I} - \mathbf{YE}) + \mathbf{i} = \mathbf{Ye} \tag{3.19}$$

$$\mathbf{A}^t(\mathbf{I} - \mathbf{YE}) = \mathbf{A}^t\mathbf{Ye} \tag{3.20}$$

$$\mathbf{A}^t(\mathbf{I} - \mathbf{YE}) = \mathbf{A}^t\mathbf{YAe}' \tag{3.21}$$

$$\mathbf{e}' = (\mathbf{A}^t\mathbf{YA})^{-1}\,\mathbf{A}^t\,(\mathbf{I} - \mathbf{YE}) \tag{3.22}$$

In (3.21), $\mathbf{A}^t\mathbf{YA}$ is the *nodal admittance* matrix while in (3.22) $(\mathbf{A}^t\mathbf{YA})^{-1}$ is the *nodal solution* matrix. The desired variables **e** and **i** may be computed from (3.13) and (3.19).

The Cutset Method: In both of the foregoing methods of analysis, the auxiliary variables $\mathbf{i}'$ and $\mathbf{e}'$ constitute a "basis" for all the branch currents **i** or branch voltages **e**, respectively. A basis such as $\mathbf{e}'$ is a linearly independent set of voltages in terms of which all the branch voltages may be computed, using (3.13). Any other basis besides the node-to-datum voltages will serve as well. In particular, we may use the set of tree-branch voltages $\mathbf{e}_T$ which are linearly independent and in terms of which the branch voltages may be computed using the expression

$$\mathbf{e} = \mathbf{D}\mathbf{e}_T \tag{3.23}$$

Here **D** is the cutset matrix. This is a statement of Kirchhoff's voltage law and it is the counterpart of (3.13). The counterpart of (3.12) for Kirchhoff's current law is

$$\mathbf{D}^t\mathbf{i} = \mathbf{O} \tag{3.24}$$

which follows directly from (3.6) and (3.14).

Using these relations, the steps involved in the cutset method of analysis are readily seen to be

$$(\mathbf{I} - \mathbf{YE}) + \mathbf{i} = \mathbf{Ye} \tag{3.25}$$

$$\mathbf{D}^t(\mathbf{I} - \mathbf{YE}) = \mathbf{D}^t\mathbf{Ye} \tag{3.26}$$

$$\mathbf{D}^t(\mathbf{I} - \mathbf{YE}) = \mathbf{D}^t\mathbf{Y}\,\mathbf{D}\mathbf{e}_T \tag{3.27}$$

$$\mathbf{e}_T = (\mathbf{D}^t\mathbf{YD})^{-1}\,\mathbf{D}^t(\mathbf{I} - \mathbf{YE}) \tag{3.28}$$

Here $\mathbf{D}^t\mathbf{YD}$ is the *cutset admittance* matrix and $(\mathbf{D}^t\mathbf{YD})^{-1}$ the *cutset solution* matrix. The variables **e** and **i** may be evaluated using (3.23) and (3.25)

The Hybrid Method: In addition to these three standard methods of analysis, there is a fourth which is beginning to be more widely used, namely the *hybrid* (or "mixed") method. This method, which is the basis for the writer's extension [16] of Bashkow's *A-matrix* formulation of the transient problem [23] and for a new method of ac analysis, [7] can also be used for dc analysis. Moreover, it underlies both Bryant's treatment [24] of Bashkow's formulation and the so-called state-variable approach [25] to transient analysis.

The idea behind the hybrid method is that Ohm's law may be expressed in a hybrid form involving the admittance matrix for certain branches of the network and the impedance matrix for the rest of the branches. Thus, we may write Ohm's law in the form

$$\begin{bmatrix} \mathbf{J}_y \\ \mathbf{V}_z \end{bmatrix} = \begin{bmatrix} \mathbf{Y}_y & \mathbf{O} \\ \mathbf{O} & \mathbf{Z}_z \end{bmatrix} \begin{bmatrix} \mathbf{V}_y \\ \mathbf{J}_z \end{bmatrix} \tag{3.29}$$

where the subscripts y and z refer to the branches which are treated as admittances or as impedances, respectively. With this equation, or its equivalent

$$\begin{bmatrix} \mathbf{I}_y + \mathbf{i}_y \\ \mathbf{E}_z + \mathbf{e}_z \end{bmatrix} = \begin{bmatrix} \mathbf{Y}_y & \mathbf{O} \\ \mathbf{O} & \mathbf{Z}_z \end{bmatrix} \begin{bmatrix} \mathbf{E}_y + \mathbf{e}_y \\ \mathbf{I}_z + \mathbf{i}_z \end{bmatrix} \tag{3.30}$$

as a starting point, we may analyze the admittance branches by the cut-set method and the impedance branches by the mesh method – virtually considering the two sets of branches as comprising separate subnetworks. We then couple the resulting two sets of equations together by means of an appropriate topological matrix. The procedure is as follows:

Taking all the y-branches (admittances) first, a tree-link sort is performed. Any y-links thus identified will define basic meshes which contain *only* y-tree branches since no z-branches (impedances) have been considered up to this point. We will designate the paths-in-tree for these y- meshes by the submatrix $\mathbf{C}_{Tyy}$.

The tree-link sort is then resumed, taking the z-branches into account. Now, however, the z-links define basic meshes whose paths-in-tree may involve *both* y-tree branches and z-tree branches, designated by the submatrices $\mathbf{C}_{Tyz}$ and $\mathbf{C}_{Tzz}$, respectively. Thus, the $\mathbf{C}_T$ matrix for the entire network has the form

$$\mathbf{C}_T = \begin{bmatrix} \mathbf{C}_{Tyy} & \mathbf{C}_{Tyz} \\ \mathbf{O} & \mathbf{C}_{Tzz} \end{bmatrix} \tag{3.31}$$

If we express the y-branch voltages $\mathbf{e}_y$ in terms of the y-tree branch voltages, $\mathbf{e}_{Ty}$, and recall (3.5) and (3.23) we have

$$\mathbf{e}_y = \begin{bmatrix} \mathbf{e}_{Ty} \\ \mathbf{e}_{Ly} \end{bmatrix} = \begin{bmatrix} \mathbf{U}_{Ty} \\ -\mathbf{C}^t_{Tyy} \end{bmatrix} \mathbf{e}_{Ty} = \mathbf{D}_y \mathbf{e}_{Ty} \tag{3.32}$$

where $\mathbf{D}_y$ is the cutset matrix for the y-branches alone with the z-branches regarded as being open-circuited. For the entire network, the corresponding relation is

$$\begin{bmatrix} \mathbf{e}_{Ty} \\ \mathbf{e}_{Tz} \\ \mathbf{e}_{Ly} \\ \mathbf{e}_{Lz} \end{bmatrix} = \begin{bmatrix} \mathbf{U}_{Ty} & \mathbf{O} \\ \mathbf{O} & \mathbf{U}_{Tz} \\ -\mathbf{C}_{Tyy}^t & \mathbf{O} \\ -\mathbf{C}_{Tyz}^t & -\mathbf{C}_{Tzz}^t \end{bmatrix} \begin{bmatrix} \mathbf{e}_{Ty} \\ \mathbf{e}_{Tz} \end{bmatrix} \tag{3.33}$$

where the $\mathbf{D}$ matrix for *all* of the y- and z-branches is required.

Similarly, we may express the z-branch currents $\mathbf{i}_z$ as a linear combination of the z-link currents $\mathbf{i}_{Lz}$ which are identical with the mesh currents, obtaining the relation

$$\mathbf{i}_z = \begin{bmatrix} \mathbf{i}_{Tz} \\ \mathbf{i}_{Lz} \end{bmatrix} = \begin{bmatrix} \mathbf{C}_{Tzz} \\ \mathbf{U}_{Lz} \end{bmatrix} \mathbf{i}_{Lz} = \mathbf{C}_z \mathbf{i}_{Lz} \tag{3.34}$$

Here, $\mathbf{C}_z$ is the circuit matrix for the z-branches alone with the y-branches regarded as being short-circuited. Again, for the entire network, we have

$$\begin{bmatrix} \mathbf{i}_{Ty} \\ \mathbf{i}_{Tz} \\ \mathbf{i}_{Ly} \\ \mathbf{i}_{Lz} \end{bmatrix} = \begin{bmatrix} \mathbf{C}_{Tyy} & \mathbf{C}_{Tyz} \\ \mathbf{O} & \mathbf{C}_{Tzz} \\ \mathbf{U}_{Ly} & \mathbf{O} \\ \mathbf{O} & \mathbf{U}_{Lz} \end{bmatrix} \begin{bmatrix} \mathbf{i}_{Ly} \\ \mathbf{i}_{Lz} \end{bmatrix} \tag{3.35}$$

where the $\mathbf{C}$ matrix for *all* of the y- and z-branches is required.

Now it is true for the *entire* network that $\mathbf{D}^t\mathbf{i} = \mathbf{0}$ and the $\mathbf{C}^t\mathbf{e} = \mathbf{0}$. As a consequence of these relations and (3.32) – (3.35), it may be shown that

$$\mathbf{D}_y^t\, \mathbf{i}_y = \begin{bmatrix} \mathbf{U}_{Ty} - \mathbf{C}_{Tyy} \end{bmatrix} \begin{bmatrix} \mathbf{i}_{Ty} \\ \mathbf{i}_{Ly} \end{bmatrix} = \mathbf{C}_{Tyz}\, \mathbf{i}_{Lz} \tag{3.36}$$

and

$$\mathbf{C}_z^t\, \mathbf{e}_z = \begin{bmatrix} \mathbf{C}_{Tzz}^t & \mathbf{U}_{Lz} \end{bmatrix} \begin{bmatrix} \mathbf{e}_{Tz} \\ \mathbf{e}_{Lz} \end{bmatrix} = -\mathbf{C}_{Tyz}^t\, \mathbf{e}_{Ty} \tag{3.37}$$

Returning to (3.30), we may rearrange terms and write

$$\begin{bmatrix} (\mathbf{I}_y - \mathbf{Y}_y \mathbf{E}_y) \\ (\mathbf{E}_z - \mathbf{Z}_z \mathbf{I}_z) \end{bmatrix} = \begin{bmatrix} \mathbf{Y}_y & \mathbf{O} \\ \mathbf{O} & \mathbf{Z}_z \end{bmatrix} \begin{bmatrix} \mathbf{e}_y \\ \mathbf{i}_z \end{bmatrix} - \begin{bmatrix} \mathbf{i}_y \\ \mathbf{e}_z \end{bmatrix} \tag{3.38}$$

The next step is to premultiply the first row of this equation by $\mathbf{D}_y^t$ and the second row by $\mathbf{C}_z^t$ and to replace $\mathbf{e}_y$ and $\mathbf{i}_z$ by the expressions given in (3.32) and (3.34). The result thus obtained,

$$\begin{bmatrix} \mathbf{D}_y^t(\mathbf{I}_y - \mathbf{Y}_y \mathbf{E}_y) \\ \mathbf{C}_z^t(\mathbf{E}_z - \mathbf{Z}_z \mathbf{I}_z) \end{bmatrix} = \begin{bmatrix} \mathbf{D}_y^t \mathbf{Y}_y \mathbf{D}_y & \mathbf{O} \\ \mathbf{O} & \mathbf{C}_z^t \mathbf{Z}_z \mathbf{C}_z \end{bmatrix} \begin{bmatrix} \mathbf{e}_{Ty} \\ \mathbf{i}_{Lz} \end{bmatrix} - \begin{bmatrix} \mathbf{D}_y^t \mathbf{i}_y \\ \mathbf{C}_z^t \mathbf{e}_z \end{bmatrix} \tag{3.39}$$

is – *except* for the second term on the right – exactly what would follow from using the cutset method of analysis of the y-branches, with the z-branches open-circuited, and the mesh method on the z-branches, with the y-branches short-circuited.

The final step is to replace the terms involving $\mathbf{i}_y$ and $\mathbf{e}_z$ by taking advantage of the relations developed in (3.36) and (3.37). When this is done, we find that

$$\begin{bmatrix} \mathbf{D}_y^t(\mathbf{I}_y - \mathbf{Y}_y \mathbf{E}_y) \\ \mathbf{C}_z^t(\mathbf{E}_z - \mathbf{Z}_z \mathbf{I}_z) \end{bmatrix} = \begin{bmatrix} \mathbf{D}_y^t \mathbf{Y}_y \mathbf{D}_y & -\mathbf{C}_{Tyz} \\ \mathbf{C}_{Tyz}^t & \mathbf{C}_z^t \mathbf{Z}_z \mathbf{C}_z \end{bmatrix} \begin{bmatrix} \mathbf{e}_{Ty} \\ \mathbf{i}_{Lz} \end{bmatrix} \tag{3.40}$$

where the topological matrix $\mathbf{C}_{Tyz}$ effectively couples the two systems of cutset and mesh equations.

3.2 METHODS OF SOLVING THE NETWORK PROBLEM

The foregoing discussion relates to four general methods of network analysis – that is, "separation of (the network problem) into constituent parts or elements."† The present section refers to methods primarily numerical, for solving network problems of three main types: dc, ac, and transient.

† Webster's Collegiate Dictionary.

Dc Network Problems

The most straightforward way of solving linear dc network problems is by Gaussian elimination [26], [27] of the variables in the mesh or node equations. For a general system of n equations in n unknowns having a full matrix of coefficients, the amount of computation required to carry out this elimination process is about $n^3/3$ operations (multiplications and additions). Indeed, it has recently been proved [28] that "no general system of linear algebraic equations can be solved in fewer operations than are required by Gaussian elimination." [27] However, for systems of equations having sparse coefficient matrices, which is the case in most network problems, a marked improvement in computational efficiency can be obtained by taking advantage of this sparsity [29], [30]. More about this later.

When several solutions are desired for a given network problem (or system of equations) with different voltage and/or current sources (right-hand sides), the usual method used is to compute the mesh or nodal solution matrix (inverse matrix) and then to multiply it by each different source vector (right-hand side). For a full matrix, the cost of inversion is about n^3 operations while the cost of each matrix-vector multiplication is n^2 operations.

A better and cheaper way to accomplish the same result, however, is to use the so-called **LU** *decomposition* [27]. This involves finding a lower triangular matrix **L** and an upper triangular matrix **U** whose product **LU** is equal to the coefficient matrix **A** in the general system of equations $\mathbf{Ax} = \mathbf{b}$. But the objective of the Gaussian elimination procedure is to produce explicitly the upper triangular matrix **U** with all its diagonal elements equal to 1. At the same time, the elements of the lower triangular matrix **L** are implicitly generated and need only be saved – instead of being discarded, as is usually the case.

Fortunately, the essential elements of **L** can be stored *on* the diagonal, in place of the diagonal 1's of the upper triangular matrix **U**, and *below* the diagonal in place of the zeros generated during the elimination process. Moreover, this can be done with negligible extra cost so that the **LU** decomposition requires only $n^3/3$ operations. This is 1/3 the cost of matrix inversion. Then, since $\mathbf{LU} = \mathbf{A}$, the original system of equations $\mathbf{Ax} = \mathbf{b}$ can be replaced by the two triangular systems $\mathbf{Ly} = \mathbf{b}$ and $\mathbf{Ux} = \mathbf{y}$ which can both be solved in n^2 operations. This is equal to the cost of matrix-vector multiplication. Thus, the **LU** decomposition technique is considerably cheaper than the conventional matrix inversion procedure for obtaining solutions for a succession of several right-hand side vectors.

There are several refinements, such as scaling and pivoting (interchanging rows and columns), which can improve the accuracy of the **LU** decomposition [26], [27]. Iterative improvement of the first solution $\mathbf{x}_1$ can also be achieved by a double-precision computation of the corresponding residual $\mathbf{r}_1 = \mathbf{b} - \mathbf{A}\mathbf{x}_1$ followed by a single-precision solution of the system $\mathbf{A}\mathbf{x}_2 = \mathbf{r}_1$. The improved solution then is $\mathbf{x} = \mathbf{x}_1 + \mathbf{x}_2$. A program containing all these features has recently been published by Forsythe and Moler [27] (in ALGOL 60, FORTRAN, and PL/1) for solving the general system of equations $\mathbf{A}\mathbf{x} = \mathbf{b}$ using **LU** decomposition.

In most network problems the coefficient matrix is so sparse, however, that it is worthwhile to take full advantage of this sparsity in programming the solution technique. Unfortunately, we electronic practitioners of the network analysis programming art have not been as alert to this possibility as our confreres in the electrical power industry. This may be due to the fact that the power engineer is being forced to consider much larger problems – in the order of 1000 to 2000 variables – than is the electronics engineer, for whom 150 to 200 variables seems to be an upper limit at present. With the advent of large-scale integrated circuits, though, there will soon be problems of ten times this size to be solved by the electronics engineer!

In two significant papers, Sato and Tinney [29] and more recently Tinney and Walker [30] have described techniques for modifying the **LU** decomposition procedure to take advantage of the sparsity of the nodal admittance matrix. These techniques, which store and process only the *nonzero* elements of the matrices involved, greatly reduce both the computation time and the memory space required to solve network problems. In particular, these authors show that the *order* in which the variables are processed during Gaussian elimination strongly affects the amount of computation needed to triangularize a sparse matrix. This question has also been considered by Parter. [31]

One rule which considerably reduces the computation time is to eliminate, at each stage, that variable for which the corresponding row in the still-to-be-triangularized matrix has fewest nonzero elements. Although this simple rule may not produce an optimum strategy, it does yield improvements in the order of 4 to 1 in computing speed [29].

This is because the rule helps to preserve sparsity in the upper triangular matrix which results from the elimination procedure. A better rule, proposed by Bree [32], is based on a graph theoretical representation of the problem similar to that described by Parter.

As Sato and Tinney point out, the **LU** decomposition procedure can be represented as a produce of elementary matrices. These are matrices that differ from a unit matrix by either one diagonal or one off-diagonal entry [26]. And the (nonzero) elements of the **L** and **U** triangular matrices correspond to the nontrivial entries of these elementary matrices. Thus, the **LU** decomposition can be used to provide a "product form of the inverse," [33] in effect, in which only the essential terms are retained For a very sparse matrix and with a proper ordering of the variables, the number of essential terms can be kept much smaller than that required for a full matrix so that a significant economy in both storage space and computation time can be effected. The price to be paid for this economy, however, is a very intricate indexing scheme to keep track of the row and column indices of the nonzero elements of the **L** and **U** matrices.

The order of elimination of variables affects not only the amount of computation in the **LU** decomposition of sparse matrices, but also the accuracy of the solution – for both full and sparse matrices. To minimize round-off errors, the rule generally followed is to interchange rows and columns at each step of the elimination procedure so as to bring the largest possible term into the "pivotal" position [26], [27], [33]. This is the position on the diagonal corresponding to the next variable to be eliminated. In many network problems, the matrices are diagonally dominant so that pivoting may not be necessary. Even so, for best accuracy, this technique – or at least partial pivoting, in which only the rows are interchanged – is worthwhile and not too expensive in machine time. The program by Forsythe and Moler [27] previously mentioned incorporates the effect of row interchanges by indexing rather than by moving the data in memory. Whether or not this technique can be combined with that of Sato and Tinney remains to be explored.

Sometimes, inaccuracies arise which are not related to the round-off errors introduced in the process of solving equations or inverting a matrix. For example, suppose the node method is being used and it is desired to compute the current in a branch whose node voltages are almost equal. Clearly, a severe loss of accuracy will occur when these two node voltages are subtracted during the evaluation of (3.13), even if $\mathbf{e}'$ is known to full machine precision. This kind of problem, then, is a function of the source vectors **E** and **I**.

One way of avoiding this difficulty, which was actually used in TAP [16], is to compute currents using the mesh method and voltages using the node method. An alternative which has not been tried, but merits investigation, is the hybrid method.

Kron's Method: Kron's method of interconnecting solutions [9] – [11] provides some tools which are very useful for certain applications. In particular, if it is desired to study the effect of varying a single resistance, letting it take on a succession of values, Kron's method provides a much more efficient way of handling this problem than solving each case from scratch. This method, which is the basis of the MODIFY feature of ECAP, permits the nodal solution matrix for the original network to be updated in accordance with changes in one or more branch resistances. The amount of computation involved is trivial compared to that of solving the nodal equations or inverting the nodal admittance matrix all over again unless a sparse matrix version of the **LU** decomposition is used. But Kron's method can also be used in conjunction with sparse matrix techniques, if desired.

A simple way of understanding the use of Kron's method is the following. Consider the network configuration shown in Fig. 3.8(a) which consists of a single link connected from node p to node q of a tree all of whose branches converge on a datum node. The path-in-tree for the single mesh in this network is given by the $\mathbf{C}_T$ matrix depicted in Fig. 3.8(b).

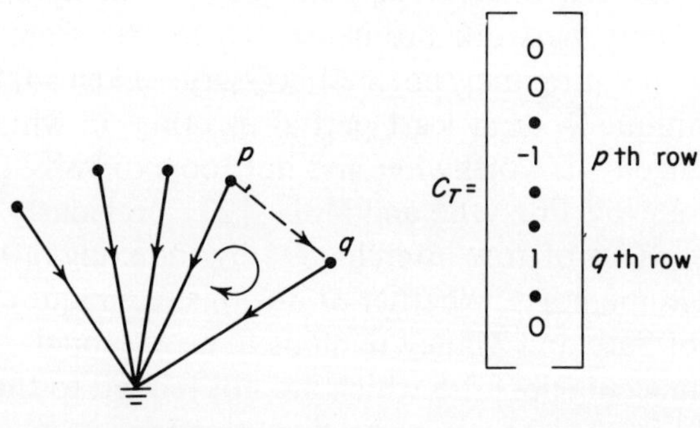

FIG. 3.8. Connection of single link to tree.

The mesh impedance matrix $\mathbf{C}^t\mathbf{Z}\mathbf{C}$ for this network is easily shown to be

$$\mathbf{C}^t\mathbf{Z}\mathbf{C} = \mathbf{C}_T^t\mathbf{Z}_T\mathbf{C}_T + \mathbf{Z}_L \tag{3.41}$$

where $\mathbf{Z}_T$ is the impedance matrix for the tree while $\mathbf{Z}_L$ is the impedance "matrix" of the single link being added to the tree.

Now by substituting (3.14) and (3.18) into (3.15), it can be shown that

$$\mathbf{e} = [\mathbf{Z} - \mathbf{ZC}(\mathbf{C}^t\mathbf{ZC})^{-1}\,\mathbf{C}^t\mathbf{Z}]\,(\mathbf{I} - \mathbf{YE}) \tag{3.42}$$

Then, by partitioning this equation into submatrices and subvectors relating to trees and links, it can further be shown that [10]

$$\mathbf{e}_T = [\mathbf{Z}_T - \mathbf{Z}_T\mathbf{C}_T(\mathbf{C}^t\mathbf{ZC})^{-1}\,\mathbf{C}_T^t\mathbf{Z}_T]\,\mathbf{D}^t(\mathbf{I} - \mathbf{YE}) \tag{3.43}$$

Finally, by comparing this result with (3.28) and using (3.41), it follows that

$$(\mathbf{D}^t\mathbf{YD})^{-1} = \mathbf{Z}_T - \mathbf{Z}_T\mathbf{C}_T(\mathbf{C}_T^t\mathbf{Z}_T\mathbf{C}_T + \mathbf{Z}_L)^{-1}\,\mathbf{C}_T^t\mathbf{Z}_T \tag{3.44}$$

Thus, we have derived from the *mesh* method an expression for updating the cutset solution matrix for any tree to which a single link has been added. (Remember that for a *tree*, the impedance matrix $\mathbf{Z}_T$ is identical with the cutset solution matrix $(\mathbf{D}^t\mathbf{YD})^{-1}$ since $\mathbf{D} = \mathbf{D}_T$ is a unit matrix.) But in view of (3.4) and (3.5) we can extend this to an expression for updating the nodal solution matrix as well, [10] namely

$$(\mathbf{A}^t\mathbf{YA})^{-1} = \mathbf{Z}_1 - \mathbf{Z}_1\mathbf{A}_L^t\,(\mathbf{A}_L^t\,\mathbf{Z}_1\,\mathbf{A}_L + \mathbf{Z}_L)^{-1}\,\mathbf{A}_L\,\mathbf{Z}_1 \tag{3.45}$$

where $\mathbf{Z}_1 = \mathbf{B}_T^t\mathbf{Z}_T\mathbf{B}_T$.

Returning now to our original problem, that of updating the nodal solution matrix when a single branch resistance (between nodes p and q) is changed, we assume that this change is effected by adding another resistance in parallel, either positive or negative in value. We assume also that the nodal solution matrix for the original network has been computed. This matrix may then be interpreted as the impedance matrix $\mathbf{Z}_T$ of a tree having the "star" configuration shown in Fig. 3.8(a). Since $\mathbf{B}_T$ for this tree is a unit matrix, it follows that we may substitute $\mathbf{Z}_T$ for $\mathbf{Z}_1$ in (3.45) which thus enables us to update the nodal solution matrix one link at a time.

Equation (3.45) may be used recursively for adding several links, one after the other [10]. If this technique is used, however, to compute a sequence of solutions as a given resistance is stepped through several different values, roundoff errors will accumulate from one step to the next.

If the range of variation of resistance is large enough and many steps are taken, the errors may eventually become serious enough to vitiate the results. A better technique, but one requiring more memory space, would be to update only the original nodal solution matrix each time. This would preclude the accumulation of errors from one step to the next.

Sensitivities: By suitably extending this idea of updating the nodal solution matrix to take into account changes of a single resistance value, differential changes in resistance can be dealt with. In other words, the partial derivative or sensitivity of the nodal solution matrix with respect to any resistance change can be computed explicitly [10].

Rather than describing how this may be done starting from (3.45), we may derive a similar and broader relation by differentiating the nodal equations themselves. Thus, from (3.21) it follows that

$$\mathbf{A}^t(d\mathbf{I} - \mathbf{Y}d\mathbf{E} - d\mathbf{Y}\mathbf{E}) = \mathbf{A}^t d\mathbf{Y}\mathbf{A}\mathbf{e}' + \mathbf{A}^t\mathbf{Y}\mathbf{A}d\mathbf{e}' \qquad (3.46)$$

or

$$d\mathbf{e}' = (\mathbf{A}^t\mathbf{Y}\mathbf{A})^{-1}\,\mathbf{A}^t[d\mathbf{I} - \mathbf{Y}d\mathbf{E} - d\mathbf{Y}(\mathbf{E} + \mathbf{e})] \qquad (3.47)$$

The latter expression gives the variation in node voltages as a function not only of changes in resistance (specified as $d\mathbf{Y}$) but also of changes in the **I** and **E** vectors. This equation is used in ECAP for computing sensitivities.

Nonlinear Networks: One standard method for solving nonlinear problems is the Newton-Raphson iteration method [34]. This method may be described in terms of the nodal equations, but is applicable to all of the other methods of analysis as well.

We assume that all nonlinearities are voltage dependent and so write the admittance matrix as $\mathbf{Y}(\mathbf{V})$. The nonlinear network problem then amounts to solving the system of equations.

$$\mathbf{f}(\mathbf{e}') \equiv \mathbf{A}^t[\mathbf{Y}(\mathbf{V})(\mathbf{A}\mathbf{e}' + \mathbf{E}) - \mathbf{I}] = 0 \qquad (3.48)$$

where $\mathbf{f}(\mathbf{e}')$ is a (nonlinear) vector function of the vector variable $\mathbf{e}'$. For any given vector $\mathbf{e}'$, the function $\mathbf{f}(\mathbf{e}')$ may, of course, differ from zero and so we call **f** the *error vector*. Any change $d\mathbf{e}'$ in the vector $\mathbf{e}'$ will give rise to a corresponding change $d\mathbf{f}$ in the error vector according to

$$d\mathbf{f} = \left(\frac{\partial \mathbf{f}}{\partial \mathbf{e}'}\right) d\mathbf{e}' \qquad (3.49)$$

Here $(\partial \mathbf{f}/\partial \mathbf{e}')$ is the Jacobian matrix whose jkth element is defined by

$$\left(\frac{\partial \mathbf{f}}{\partial \mathbf{e}'}\right)_{jk} = \frac{\partial f_j}{\partial e'_k} \tag{3.50}$$

where f_j is the jth element of the vector $\mathbf{f}(\mathbf{e}')$ and e'_k is the kth element of the vector $\mathbf{e}'$.

Suppose that some iterative scheme is to be used for solving (3.48) and that, corresponding to the ith approximation $\mathbf{e}'_{(i)}$ to the solution vector, the function vector $\mathbf{f}_{(i)} = \mathbf{f}(\mathbf{e}'_{(i)})$ is not zero. According to (3.50), the approximate change in $\mathbf{e}'_{(i)}$ required to reduce $\mathbf{f}_{(i)}$ to zero may be determined by solving

$$d\mathbf{f}_{(i)} = -\mathbf{f}_{(i)} = \left(\frac{\partial \mathbf{f}}{\partial \mathbf{e}'}\right)_{(i)} d\mathbf{e}'_{(i)} \tag{3.51}$$

Hence, the $(i + 1)$th approximation to the solution vector $\mathbf{e}'$ is given by

$$\mathbf{e}'_{(i+1)} = \mathbf{e}'_{(i)} - \left(\frac{\partial \mathbf{f}}{\partial \mathbf{e}'}\right)^{-1}_{(i)} \mathbf{f}_{(i)} \tag{3.52}$$

This is the Newton-Raphson recursion formula.

For well-behaved problems, this method converges quadratically. However, it may, under certain circumstances, diverge or oscillate about a solution. It also requires the inverse of the Jacobian matrix of the system of equations at each iteration and this may give rise to a sizeable computational burden.

Broyden [35] has recently developed a modification of the Newton-Raphson technique which avoids both of these major drawbacks. Broyden's method *cannot* diverge because it never allows the length of vector $\mathbf{f}_i$ to increase. It also provides a method of updating the inverse Jacobian matrix at each iteration, starting with an arbitrary approximation to this inverse. Branin and Wang [36] have applied this method to the solution of nonlinear dc network problems and find that it is fast and reliable. They have also extended it to statistical analysis of both linear and nonlinear networks.

Before describing Broyden's method and its adaptation to nonlinear dc analysis, we will derive an expression for the Jacobian matrix by a method that is simpler and more general than that given in [36]. Noting that $\mathbf{J} = \mathbf{Y}(\mathbf{V})\mathbf{V}$, we first define a *differential* (or small signal) admittance matrix as follows:

$$\widetilde{\mathbf{Y}} = \frac{\partial \mathbf{J}}{\partial \mathbf{V}} \tag{3.53}$$

wherein $\widetilde{Y}_{ij} = \partial J_i/\partial V_j$. Then using $\mathbf{V} = \mathbf{A}\mathbf{e}' + \mathbf{E}$, we rewrite (3.48) in the form

$$\mathbf{f}(\mathbf{e}') = \mathbf{A}^t(\mathbf{J} - \mathbf{I}) \tag{3.54}$$

Finally, by differentiating (3.54) we get

$$\frac{\partial \mathbf{f}}{\partial \mathbf{e}'} = \mathbf{A}^t \frac{\partial \mathbf{J}}{\partial \mathbf{e}'} = \mathbf{A}^t \frac{\partial \mathbf{J}}{\partial \mathbf{V}} \frac{\partial \mathbf{V}}{\partial \mathbf{e}'} = \mathbf{A}^t\widetilde{\mathbf{Y}}\mathbf{A} \tag{3.55}$$

so that (3.52) becomes

$$\mathbf{e}'_{(i+1)} = \mathbf{e}'_{(i)} - (\mathbf{A}^t\widetilde{\mathbf{Y}}\mathbf{A})^{-1}\mathbf{f}_{(i)} \tag{3.56}$$

As indicated in (3.52) or (3.56), the Newton-Raphson formula prescribes that a certain *correction vector* be applied to the ith approximation to the desired solution vector. Dropping the parentheses enclosing the iteration indices, we may define this correction vector as

$$\mathbf{p}_i = -\left(\frac{\partial \mathbf{f}}{\partial \mathbf{e}'}\right)_i^{-1} \mathbf{f}_i \tag{3.57}$$

so that (3.52) becomes $\mathbf{e}'_{i+1} = \mathbf{e}'_i + \mathbf{p}_i$.

But Broyden does not take this full correction. Instead, he takes only a *fraction* of it according to

$$\mathbf{e}'_{i+1} = \mathbf{e}'_i + t_i\mathbf{p}_i \tag{3.58}$$

where the scalar t_i is chosen so as to *minimize* (or at least to reduce) the length of the error vector $\mathbf{f}_{i+1}$ relative to that of $\mathbf{f}_i$. In this way, divergence of the iteration scheme is absolutely prevented – but *convergence* is not necessarily guaranteed.

Broyden next derives an algorithm for computing an approximation to the inverse Jacobian matrix, an approximation which gets better and better as the iteration scheme converges to a solution. First, he considers the vector $\mathbf{e}'$ on the left of (3.58) as a continuous function of the scalar variable t in the expression

$$\mathbf{e}'(t) = \mathbf{e}'_i + t\mathbf{p}_i \tag{3.59}$$

This, along with (3.48), implies that the error vector $\mathbf{f}(\mathbf{e}')$ is also a function of t. And since the Jacobian matrix is assumed to exist, it follows that

$$\frac{d\mathbf{f}}{dt} = \left(\frac{\partial \mathbf{f}}{\partial \mathbf{e}'}\right)\frac{d\mathbf{e}'}{dt} = \left(\frac{\partial \mathbf{f}}{\partial \mathbf{e}'}\right)\mathbf{p}_i \tag{3.60}$$

where the relation $d\mathbf{e}'/dt = \mathbf{p}_i$ derived from (3.59) has been used.

Letting $\mathbf{H}_i$ represent an approximation to the negative of the inverse Jacobian matrix, (3.60) can be rewritten in the form

$$\mathbf{p}_i = -\mathbf{H}_i \frac{d\mathbf{f}}{dt} \tag{3.61}$$

This is a relation which, if $d\mathbf{f}/dt$ as well as $\mathbf{p}_i$ were known, would have to be satisfied by any approximating matrix $\mathbf{H}_i$. But of course (3.61) is insufficient to *determine* $\mathbf{H}_i$.

Pursuing this idea, Broyden next computes $d\mathbf{f}/dt$ by differencing. He defines $\mathbf{f}_{i+1} = \mathbf{f}(t_i)$ and then expands $\mathbf{f}(t_i - s)$ as a Taylor series about t_i, keeping only the first term in the expansion. In this way, he obtains the approximation

$$\frac{d\mathbf{f}}{dt} \approx \frac{\mathbf{f}_{i+1} - \mathbf{f}(t_i - s_i)}{s_i} = \frac{\mathbf{y}_i}{s_i} \tag{3.62}$$

where s_i is chosen to be a small increment in the scalar variable t and y_i is the corresponding increment in the error vector $\mathbf{f}$. Combining (3.61) and (3.62) then yields

$$\mathbf{H}_i\mathbf{y}_i = -s_i\mathbf{p}_i \tag{3.63}$$

This is another way of describing the condition that $\mathbf{H}_i$ must satisfy – but again, it does not define $\mathbf{H}_i$ uniquely.

Finally, to determine $\mathbf{H}_i$ uniquely, Broyden assumes the following auxiliary condition: Since no information is obtainable regarding the change in the error vector $\mathbf{f}$ due to changes in $\mathbf{e}'$ in any direction *other than* $\mathbf{p}_i$, it is assumed there is no change in $\mathbf{f}$ due to changes of $\mathbf{e}'$ in any direction *orthogonal* to $\mathbf{p}_i$. The net result of the ensuing calculation (which we omit) is the recursion formula

$$\mathbf{H}_{i+1} = \mathbf{H}_i - \frac{(s_i \mathbf{p}_i + \mathbf{H}_i \mathbf{y}_i)\,\mathbf{p}_i^t \mathbf{H}_i}{\mathbf{p}_i^t \mathbf{H}_i \mathbf{y}_i} \tag{3.64}$$

This is the expression by means of which Broyden directly updates the inverse Jacobian matrix without ever having to compute the Jacobian itself or invert it numerically. (Broyden also derives two other recursion formulas, but these are of no interest in the present discussion.)

An algorithm for choosing t_i (to minimize $|\mathbf{f}_i^2 + 1|$)most economically and for choosing s_i appropriately is discussed later. But a simple-minded procedure is to let $t = 0.1, 0.2, 0.3, \ldots$ etc. and evaluate the magnitude of each corresponding error vector, stopping when a minimum is reached. The value of s_i is then chosen to be 0.1 and $\mathbf{y}_i$ is taken as the difference betweeen the minimum $\mathbf{f}$ and the $\mathbf{f}$ corresponding to the next smaller value of t. This algorithm is useful in studying the $|\mathbf{f}^2|$ *versus* t curve. But the procedure is inefficient and is not recommended for general use.

An alternative choice of s_i is to let $s_i = t_i$, regardless of how t_i is determined. Broyden calls this the "full step" method and the other one the "incremental" method. For solving a single (nominal) case the incremental method is better; but, for statistical analysis, we describe below an adaptation of the full step method which is particularly effective.

Algorithm 1 [36] for solving nonlinear dc network problems, is essentially as follows.

Starting Procedure:

1) Choose an initial estimate for the solution vector $\mathbf{e}'_0$.

2) Compute $\mathbf{V}_0 = \mathbf{A}\mathbf{e}'_0 + \mathbf{E}$, evaluate $\widetilde{\mathbf{Y}}(\mathbf{V}_0)$ and $\mathbf{Y}(\mathbf{V}_0)$, compute $\mathbf{H}_0 = -(\mathbf{A}^t\widetilde{\mathbf{Y}}\mathbf{A})^{-1}$, $\mathbf{f}_0 = \mathbf{A}^t[\mathbf{Y}(\mathbf{V}_0)\mathbf{V}_0 - \mathbf{I}]$ and $|\mathbf{f}_0^2|$. Then set $i = 0$.

Iteration Procedure (ϵ is the convergence criterion):

1) Compute $\mathbf{p}_i = \mathbf{H}_i\,\mathbf{f}_i$, $|\mathbf{p}_i^2|$, and $|\mathbf{e}'^2_i|$.

2) If $|\mathbf{p}_i^2| < \epsilon|\mathbf{e}'^2_i|$, exit.

3) Otherwise, find an optimum value of t_i. This step yields $\mathbf{e}'_{i+1}$, $\mathbf{V}_{i+1}$, $\mathbf{Y}(\mathbf{V}_{i+1})$, $\mathbf{f}_i+1$, and $|\mathbf{f}_{i+1}^2|$.

4) Compute s_i using the incremental method and compute $\mathbf{y}_i = \mathbf{f}_{i+1} - \mathbf{f}(t_i - s_i)$.

5) Compute $\mathbf{H}_{i+1}$ according to (3.64)

6) Advance the iteration index to $i + 1$ and go to Step 1.

Because the inverse Jacobian matrix is full, no special programming techniques are used in Step 1) of the iteration procedure. This step, involving a matrix-vector multiplication, takes N^2 operations (multiply-add) where N is the number of nodal equations. The test for convergence in Step 2) succeeds when the length of the correction vector $\mathbf{p}_i$ relative to that of the solution vector $\mathbf{e}'_i$ is negligible.

In Step 3), we may use a Fibonacci search technique [37] for determining an optimum value of t_i. This method of search, which we shall not describe in detail, is based on subdividing the interval $t = (0, 1)$ into successively smaller subintervals and directing the next search to the left or right of the current value of t according to the outcome of the evaluation of (3.48).

As the first step in this process, we set $t = 1$ and if the corresponding value of $|\mathbf{f}_{i+1}^2|$ is less than $|\mathbf{f}_i^2|$, then we settle for the value $t_i = 1$ as optimum, bypassing any further search. This process allows the immediate detection of the "Newtonian mode" which characterizes the iteration close to the point of convergence. And even if a smaller value of $|\mathbf{f}_{i+1}^2|$ could be found by further searching, it pays to go to the next iteration immediately.

When the choice $t = 1$ results in $|\mathbf{f}_{i+1}^2|$ being larger than $|\mathbf{f}_i^2|$, a multistep Fibonacci search is undertaken. A three-step or four-step search is adequate even though the "interval of uncertainty" is $t = 1/3$ or $t = 1/5$, respectively. For it is preferable to at least reduce the length of the error vector $\mathbf{f}_{i+1}$ and pass on to the next iteration rather than spending more time in minimizing it. Moreover, a more refined Fibonacci search does not noticeably decrease the total number of iterations required.

If the Fibonacci search does not result in a reduction in length of the error vector, we take this to indicate that the minimum error vector corresponds to $t = 0$, in which case the iteration cannot procede. This situation may be due to the evolution of a poor approximation to the inverse Jacobian during successive applications of (3.64). One remedy is simply to restart the procedure by computing $(\mathbf{A}^t\widetilde{\mathbf{Y}}\mathbf{A})^{-1}$. If this remedy should fail, the only other recourse would be to choose an entirely new starting vector $\mathbf{e}'_0$.

In Step 4), s_i is set equal to the last increment in t when the Fibonacci search is used; otherwise, the value chosen is $s_i = t_i = 1$. In either case, the (approximate) inverse Jacobian is updated using (3.64), in Step 5). This equation requires the computation of two matrix-vector products, namely $\mathbf{H}_i\mathbf{y}_i$ and $\mathbf{p}_i^t\mathbf{H}_i$, which requires $2N^2$ operations (multiply-add) where again N is the number of nodal equations. Neglecting the other operations of order N, the computation of the second term on the right of (3.64) requires an additional N^2 operation (multiply-add) due to the postmultiplication of the column vector $(s_i\mathbf{p}_i + \mathbf{H}_i\mathbf{y}_i)$ by the row vector $\mathbf{p}_i^t\mathbf{H}_i$. Thus, the updating of $\mathbf{H}_{i+1}$ requires roughly $3N^2$ operations.

Finally, in Step 6), the iteration index is advanced by one and Step 1) is resumed until convergence is obtained.

Although the substance of Broyden's method is the avoidance of computing and inverting the Jacobian matrix, the contrary strategy makes really good sense within the context of the network problem. The reason for this statement is the fact that the Jacobian matrix – like the nodal admittance matrix – for most networks is very sparse. It would be considerably faster to invert the matrix $\mathbf{A}^t\tilde{\mathbf{Y}}\mathbf{A}$ by the Sato-Tinney method each iteration than to update the inverse by (3.64). It would be even better to solve (3.51) directly.

There would be no necessity to compute both $\mathbf{Y(V)}$ and $\tilde{\mathbf{Y}}\mathbf{(V)}$, however. Only $\tilde{\mathbf{Y}}\mathbf{(V)}$ would be needed. The way around the requirement of using $\mathbf{Y(V)}$ in (3.48) is the following. Referring to Fig. 3.9, we see that

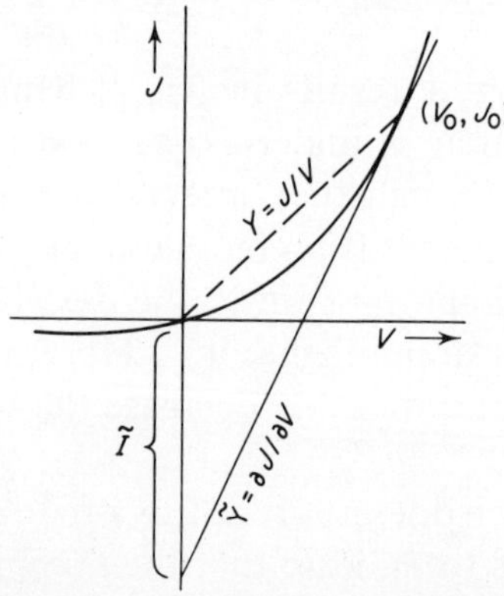

FIG. 3.9. Differential admittance of a nonlinear network element.

the tangent line at the point (V_0, J_0) is actually a better approximation to the curve over a small range than the secant line through the origin. The tangent line, of course, defines the differential or dynamic admittance $\tilde{\mathbf{Y}}\mathbf{(V)}$ while the secant line defines the static admittance $\mathbf{Y(V)}$.

But if the tangent line is used in place of the V-J curve, then clearly the intercept $\tilde{I}$ must be used along with the slope $\tilde{Y}$ to define this line. Accordingly, if the V-J curve is represented by a differentiable function, the formulas for both $\tilde{Y}(V)$ and $\tilde{I}(V)$ can be derived and programmed. Alternatively, if a table is used to represent the V-J characteristic, then this table can be supplanted by a corresponding table of values of $\tilde{Y}(V)$

and $\widetilde{I}(V)$. In either case, the $\widetilde{I}$ vector giving the intercepts of all the nonlinear functions would have to be added to the I vector of current sources specified in the actual problem.

Algorithm 2, using the sparse-matrix technique, would be as follows.

Starting Procedure:

1) Choose an initial estimate for $\mathbf{e}'_0$.

2) Compute $\mathbf{V}_0 = \mathbf{A}\mathbf{e}'_0 + \mathbf{E}$, evaluate $\widetilde{\mathbf{Y}}(\mathbf{V}_0)$ and $\widetilde{\mathbf{I}}(\mathbf{V}_0)$, compute $\mathbf{A}^t\widetilde{\mathbf{Y}}\mathbf{A}$, $\mathbf{f}_0 = \mathbf{A}^t[\widetilde{\mathbf{Y}}(\mathbf{V}_0)\mathbf{V}_0 - \mathbf{I} - \widetilde{\mathbf{I}}]$ and $|\mathbf{f}_0^2|$.

3) Set $i = 0$.

Iteration Procedure:

1) Solve $(\mathbf{A}^t\widetilde{\mathbf{Y}}\mathbf{A})\mathbf{p}_i = -\mathbf{f}_i$ (using sparse-matrix method); compute $|\mathbf{p}_i^2|$ and $|\mathbf{e}'^2_i|$.

2) If $|\mathbf{p}_i^2| < \epsilon |\mathbf{e}'^2_i|$, exit.

3) Otherwise, find an optimum value of t_i. This step yields $\mathbf{e}'_{i+1}$, $\mathbf{V}_{i+1}$, $\widetilde{\mathbf{Y}}(\mathbf{V}_{i+1})$, $\widetilde{\mathbf{I}}(\mathbf{V}_{i+1})$, $\mathbf{f}_{i+1}$, and $|\mathbf{f}_{i+1}^2|$.

4) Compute $\mathbf{A}^t\widetilde{\mathbf{Y}}\mathbf{A}$.

5) Advance iteration index to $i + 1$ and go to Step 1).

Statistical (Monte Carlo) Analysis: Statistical analysis by the Monte Carlo method involves solving several thousand cases of a given network problem with its parameters (such as resistance values and voltage and current sources) chosen at random over certain specified tolerance ranges. Obviously, for this technique to be practical an extremely rapid method of solving each such case must be used. Starting with the nominal solution as the initial estimate $\mathbf{e}'_0$ for each case, Algorithm 1 converges in a very few iterations and provides an economical method for statistical analysis.

Evidently, each solution lies within a relatively small neighborhood of the nominal solution, judging from the rapid convergence of this algorithm. By the same token, the inverse Jacobian matrix $\mathbf{H}_0$ corresponding to the *nominal* solution is not likely to differ markedly from the successive iterates $\mathbf{H}_1$, $\mathbf{H}_2$, . . ., etc., obtained by using (3.64). On the strength of this argument, we can modify Algorithm 1 so as to reduce the operation count to N^2 in place of the $4N^2$ operations required by Steps 1) and 5) together in the original algorithm.

The basic idea behind this improved technique is to *apply the correction* to $\mathbf{H}_0$, as prescribed by the second term on the right side of (3.64), but not to update $\mathbf{H}_0$. In other words, since the ultimate use of the updated $\mathbf{H}_{i+1}$ matrix is to produce a new $\mathbf{p}_{i+1}$ vector according to the relation $\mathbf{p}_{i+1} = \mathbf{H}_{i+1}\mathbf{f}_{i+1}$, the right side of (3.64) is multiplied by $\mathbf{f}_{i+1}$ *directly* without going through the updating process at all. Thus, the benefit of the correction is obtained without the full expense of applying it to the actual updating of $\mathbf{H}_i$.

Here, the full step technique is used in establishing s_i, for when $s_i = t_i$ it follows from (3.62) that

$$\mathbf{y}_i = \mathbf{f}_{i+1} - \mathbf{f}(0) = \mathbf{f}_{i+1} - \mathbf{f}_i \tag{3.65}$$

Using this relation in (3.64) and postmultiplying by $\mathbf{f}_{i+1}$ yields

$$\mathbf{p}_{i+1} = \mathbf{H}_{i+1}\mathbf{f}_{i+1} = \left[\mathbf{H}_i - \frac{(t_i\mathbf{p}_i + \mathbf{H}_i\mathbf{f}_{i+1} - \mathbf{H}_i\mathbf{f}_i)}{\mathbf{p}_i^t\mathbf{H}_i\mathbf{f}_{i+1} - \mathbf{p}_i^t\mathbf{H}_i\mathbf{f}_i}\mathbf{p}_i^t\mathbf{H}_i\right]\mathbf{f}_{i+1} \tag{3.66}$$

Then, since $\mathbf{p}_i = \mathbf{H}_i\mathbf{f}_i$, it follows that

$$\mathbf{p}_{i+1} = \mathbf{H}_i\mathbf{f}_{i+1} - \frac{[(t_i - 1)\mathbf{p}_i + \mathbf{H}_i\mathbf{f}_{i+1}]\,\mathbf{p}_i^t\mathbf{H}_i\mathbf{f}_{i+1}}{\mathbf{p}_i^t\mathbf{H}_i\mathbf{f}_{i+1} - \mathbf{p}_i^t\mathbf{p}_i} \tag{3.67}$$

If we make the assumption that $\mathbf{H}_i = \mathbf{H}_0$ and define $\mathbf{q}_{i+1} = \mathbf{H}_0\mathbf{f}_{i+1}$, then (3.67) reduces to

$$\mathbf{p}_{i+1} = \frac{(\mathbf{p}_i^t\mathbf{p}_i)\mathbf{q}_{i+1} + (\mathbf{p}_i^t\mathbf{q}_{i+1})(t_i - 1)\mathbf{p}_i}{\mathbf{p}_i^t\mathbf{p}_i - \mathbf{p}_i^t\mathbf{q}_{i+1}} \tag{3.68}$$

Thus, the bulk of the computation involved is the N^2 operations required to calculate $\mathbf{q}_{i+1}$.

On the basis of (3.68), Algorithm 3 for statistical analysis is as follows.

Starting Procedure:

Solve the nominal case using the Algorithm 1 or 2. This step yields $\mathbf{e}'_0$ and $\mathbf{H}_0$.

Iteration Procedure (*for each case*):

1) Select the network parameters at random (within tolerance limits specified for each parameter).

2) Compute $\mathbf{V}_0 = \mathbf{A}\mathbf{e}'_0 + \mathbf{E}$, evaluate $\mathbf{Y}(\mathbf{V}_0)$, and compute $\mathbf{f}_0 = \mathbf{A}^t[\mathbf{Y}(\mathbf{V}_0)\mathbf{V}_0 - \mathbf{I}]$ and $|\mathbf{f}_0^2|$.

3) Compute $\mathbf{p}_0 = \mathbf{H}_0\mathbf{f}_0$, $|\mathbf{p}_0^2|$, and $|\mathbf{e}'^2_0|$.

4) Set $i = 0$.

5) If $|\mathbf{p}_i^2| < \epsilon\,|\mathbf{e}'^2_i|$ go to Step 1) and take the next case.

6) Otherwise, find an optimum value of t_i. This step yields $\mathbf{e}'_{i+1}$, $\mathbf{V}_{i+1}$, $\mathbf{Y}(\mathbf{V}_{i+1})\mathbf{f}_{i+1}$, and $|\mathbf{f}_{i+1}^2|$.

7) Compute $\mathbf{q}_{i+1} = \mathbf{H}_0\mathbf{f}_{i+1}$ and evaluate $\mathbf{p}_{i+1}$ according to (3.68).

8) Advance the iteration index to $i + 1$; go to Step 5).

Now on the very first pass through Steps 5) and 6), the first and *largest* correction ($t_0\mathbf{p}_0$) is made to the initial approximate solution vector $\mathbf{e}'_0$. Moreover, at Step 7), there is no approximation involved in setting $\mathbf{H}_i = \mathbf{H}_0$, as is assumed in deriving (3.68), since $i = 0$. Thus $\mathbf{p}_i$ is evaluated to the full accuracy provided by Broyden's method, just as if $\mathbf{H}_0$ had actually been updated once, according to (3.64). Therefore, the second and the *next to largest* correction ($t_1\mathbf{p}_1$) to $\mathbf{e}'$ also is made with full accuracy.

It is only beyond this point that the assumption $\mathbf{H}_i = \mathbf{H}_0$ begins to degrade the potential accuracy of the algorithm. In cases where only three or four iterations are required to obtain convergence, a solution may be reached before this degradation can take effect. But when five or six iterations are required the degradation may be significant enough to favor using (3.64) for updating the $\mathbf{H}$ matrix as in Algorithm 1.

Using a sparse matrix technique, Algorithm 2 would probably be superior to Algorithm 3. For performance evaluation of Algorithms 1 and 3, see Branin and Wang [36].

Ac Network Problems

Since all of the theoretical developments up through (3.47) apply to complex as well as to real numbers, steady-state ac network problems may be handled using any of the four basic methods of analysis described above or using Kron's method. For example, by programming the Gaussian elimination procedure for complex arithmetic, the mesh or nodal equations for any ac problem may be solved directly, at a computational cost of approximately four times that for real arithmetic. However, since the coefficient matrix is frequency-dependent, the equations must be solved at each new frequency. This approach is usually quite reliable, although it may involve reduced numerical accuracy in the vicinity of a resonant frequency.

A new approach to the ac problem [7], which is based on the hybrid method of analysis, tackles the job of matrix inversion by first solving the much more difficult problem of computing the eigenvalues and eigenvectors [26] of the matrix. At first glance, solving the eigenvalue problem appears to be "doing it the hard way," since the amount of computational effort required is an order of magnitude greater than that for inverting a matrix. This would indeed be the case if only a single inverse matrix were needed. But in calculating frequency response, a new inverse, or more precisely, a new solution vector, is needed at each frequency. In this case, using the eigenvalues and eigenvectors turns out to be just the right thing to do because it reduces the problem to one of inverting a diagonal matrix at each new frequency.

Starting with (3.40) of the hybrid method, we can derive the differential equations for an RLC network in first-order form as follows. All capacitors are treated as admittance branches and hence may have a nonzero conductance in parallel. Thus, the admittance matrix becomes

$$\mathbf{Y}_y = p\mathbf{K} + \mathbf{G} \tag{3.69}$$

where $p = d/dt$ and the symbol **K** is used for capacitance so as not to conflict with the symbol **C** for the circuit matrix. On the other hand, all inductors are regarded as impedance branches with (possibly nonzero) resistance in series so that the impedance matrix is

$$\mathbf{Z}_z = p\mathbf{L} + \mathbf{R} \tag{3.70}$$

(Branches consisting of resistance only can also be handled. But, since these give rise to algebraic equations [16] which unduly complicate the discussion, we shall not consider them here.)

Using (3.69) and (3.70) together with (3.40), we may write the differential equations (for constant **K** and **L**) in the form

$$\begin{bmatrix} \mathbf{D}_y^t\mathbf{K}\mathbf{D}_y & \mathbf{O} \\ \mathbf{O} & \mathbf{C}_z^t\mathbf{L}\mathbf{C}_z \end{bmatrix} \begin{bmatrix} p\mathbf{e}_{Ty} \\ p\mathbf{i}_{Lz} \end{bmatrix} + \begin{bmatrix} \mathbf{D}_y^t\mathbf{G}\mathbf{D}_y & -\mathbf{C}_{Tyz} \\ \mathbf{C}_{Tyz}^t & \mathbf{C}_z^t\mathbf{R}\mathbf{C}_z \end{bmatrix} \begin{bmatrix} \mathbf{e}_{Ty} \\ \mathbf{i}_{Lz} \end{bmatrix} = \begin{bmatrix} \mathbf{I}_T \\ \mathbf{E}_L \end{bmatrix} \tag{3.71}$$

where the symbols $\mathbf{I}_T$ and $\mathbf{E}_L$ designate the driving currents and voltages.

In more compact form, (3.71) may be written as

$$\mathbf{P}\dot{\mathbf{x}} + \mathbf{Q}\mathbf{x} = \mathbf{f}(t) \tag{3.72}$$

where $\dot{\mathbf{x}} = d\mathbf{x}/dt$ and **x** is the *state-variable vector* of capacitive tree-branch voltages and inductive link currents. By inverting **P** we may rewrite (3.72)

$$\dot{\mathbf{x}} = \mathbf{A}\mathbf{x} + \mathbf{g}(t) \tag{3.73}$$

where $\mathbf{g} = \mathbf{P}^{-1}\mathbf{f}$ and $\mathbf{A} = -\mathbf{P}^{-1}\mathbf{Q}$, which corresponds to Bashkow's **A**-matrix [23]. (This matrix is not to be confused with the branch-node matrix A, which will not be referred to any further.)

For steady-state ac problems, the driving function may be written as

$$\mathbf{g}(t) = \mathbf{g}_0 e^{st} \tag{3.74}$$

with s a complex number. The steady-state response then is

$$\mathbf{x}(t) = \mathbf{x}_0 e^{st} \tag{3.75}$$

Substituting these relations in (3.73) we have the linear system of equations

$$(s\mathbf{U} - \mathbf{A})\mathbf{x}_0 = \mathbf{g}_0 \tag{3.76}$$

where $\mathbf{g}_0$ is a known vector and $\mathbf{x}_0$ is to be determined.

The formal solution for (3.76) is

$$\mathbf{x}_0 = (s\mathbf{U} - \mathbf{A})^{-1}\mathbf{g}_0 \tag{3.77}$$

and so we need to find an effective way of computing $(s\mathbf{U} - \mathbf{A})^{-1}$. This may be done using the eigenvalues λ_i and the eigenvectors χ_i of the **A**-matrix as follows. (We assume that the λ_i are distinct.)

Let Λ be the diagonal matrix of eigenvalues and **X** the matrix whose columns are the corresponding eigenvectors. The eigenproblem, then, is characterized by

$$\mathbf{A}\mathbf{X} = \mathbf{X}\Lambda \tag{3.78}$$

which may be recast into the form

$$\mathbf{X}^{-1}\mathbf{A}\mathbf{X} = \Lambda \tag{3.79}$$

This defines a similarity transformation [26] which diagonalizes the matrix **A**.

The same transformation also diagonalizes $(s\mathbf{U} - \mathbf{A})$ giving

$$\mathbf{X}^{-1}(s\mathbf{U}-\mathbf{A})\mathbf{X}=s\mathbf{U}-\mathbf{A} \tag{3.80}$$

Solving this expression for (sU − A) and then inverting, we find that

$$(s\mathbf{U}-\mathbf{A})^{-1}=\mathbf{X}(s\mathbf{U}-\Lambda)^{-1}\mathbf{X}^{-1}\mathbf{g}_0 \tag{3.81}$$

so that (3.77) becomes

$$\mathbf{x}_0=\mathbf{X}(s\mathbf{U}-\Lambda)^{-1}\mathbf{X}^{-1} \tag{3.82}$$

This is the basic result of the new method, requiring the inversion of the diagonal matrix ($s\mathbf{U}-\Lambda$) at each frequency.

If we let $\mathbf{d}=\mathbf{X}^{-1}\mathbf{g}_0$, which we may calculate once for all, then the ith element of the $\mathbf{x}_0$ vector may be computed using the relation

$$x_i=\sum_{j=1}^{n}\frac{X_{ij}d_j}{s-\lambda_j} \tag{3.83}$$

This *partial fraction expansion* of the state variable x_i may be evaluated at a computational cost proportional to n operations per frequency as opposed to a maximum of $n^3/3$ operations if (3.76) were solved by Gaussian elimination.

Now the cost of solving the eigenproblem is not negligible since it is equivalent to solving (3.76) at roughly 10 frequencies. Accordingly, the present method would not pay off if solutions were required at only a few frequencies. But for a frequency response calculation, this method can be very effective.

One of the more interesting features of the method is the fact that it is far better suited for computer-sized problems than the traditional Laplace transform techniques involving ratios of polynomials and the poles and zeros thereof. In particular, the task of computing the coefficients of the polynomials in a network function $P(s)/Q(s)$ is not only time-consuming but also prone to serious numerical inaccuracies, especially when the polynomials are of high degree. The so-called "topological formula" approach to computing these network functions involves finding all the trees of a network and then computing the sum of the corresponding tree-admittance products. But the number of trees may run into millions for a network with only 20 nodes and 40 branches. And even if this were not enough of an impediment, the computation of the roots of the polynomials $P(s)$ and $Q(s)$ is hazardous because these roots may be extremely sensitive to errors in the coefficients [26].

In the writer's judgment, therefore, the polynomial approach is just not matched to the network analysis tasks which the computer is called upon to handle. The eigenvalue approach is much better suited and gives all of the theoretical information that the Laplace transform methods are designed to provide. For example, the eigenvalues are identical with the poles of the network functions, as (3.83) indicates. Moreover, as shown below, any network function desired may be computed straightforwardly and its sensitivity obtained, either with respect to frequency or with respect to any network parameter. Finally, even the pole sensitivities can be calculated.

This is not to say the eigenvalue method is free of all difficulties. Even with the *Q-R* method of Francis [38] which, according to Wilkinson [26], is "the most effective of known methods for the solution of the general algebraic eigenvalue problem" it is possible that numerical errors of a crippling nature may arise when networks having a very large ratio of greatest to smallest eigenvalue are treated. The difficulty with this kind of "pathological" network is that although the large eigenvalues can be computed accurately, the small ones cannot. But it is the small ones which are principally responsible for determining the behavior of the network, as (3.83) implies.

One possible remedy for this situation is to solve the eigenvalue problem for the inverse matrix $\mathbf{A}^{-1}$. Then the situation is reversed, for the large eigenvalues of $\mathbf{A}^{-1}$, which will be obtained accurately, are the reciprocals of the small eigenvalues of $\mathbf{A}$. Computing $\mathbf{A}^{-1}$ may be error-prone in its own right because of the large ratio of eigenvalues. But if the matrix $\mathbf{Q}$ in (3.72) is nonsingular, then we may use the relation

$$\mathbf{A}^{-1} = -\mathbf{Q}^{-1}\mathbf{P} \tag{3.84}$$

to compute a more accurate inverse for $\mathbf{A}$. This question remains to be investigated.

Fortunately, the *Q-R* method is able to compute multiple eigenvalues accurately, but this is no guarantee that the corresponding eigenvectors can be accurately determined even if they exist – which they may not. In some cases, it may be acceptable to perturb the matrix $\mathbf{A}$ slightly to remove the multiplicities and yield distinct eigenvalues and eigenvectors. However, this is not a genuine solution to the basic problem.

Network Response Functions: If in place of (3.74) we use the corresponding expression for $\mathbf{f}(t)$ in (3.72), (3.82) becomes

$$\mathbf{x}_0 = [\mathbf{X}(s\mathbf{U} - \Lambda)^{-1} \mathbf{X}^{-1} \mathbf{P}^{-1}] \mathbf{f}_0 = \mathbf{T}\mathbf{f}_0 \tag{3.85}$$

which defines a new matrix $\mathbf{T}$. The elements of $\mathbf{T}$ will be called the *basic response functions* of the network because they relate the elements of $\mathbf{x}_0$, the (basic) response vector, to the elements of $\mathbf{f}_0$, the (basic) excitation vector. Accordingly, the diagonal elements T_{ii} are *driving-point impedances* when they relate an element of $\mathbf{e}_{Ty}$ to the corresponding element of $\mathbf{I}_T$ in (3.71); they are *driving-point admittances* when they relate an element of $\mathbf{i}_{Lz}$ to the corresponding element of $\mathbf{E}_L$. Similarly, the off-diagonal elements $\mathbf{T}_{ij}$ will be *transfer impedances, admittances* or *functions*, depending on which elements of the subvectors $\mathbf{e}_{Ty}$, $\mathbf{i}_{Lz}$, $\mathbf{I}_T$, and $\mathbf{E}_L$ are involved.

Since $\mathbf{I}_T$ and $\mathbf{E}_L$ account for the voltage and/or current sources in every branch, and since every branch voltage and/or branch current can be computed in terms of $\mathbf{e}_{Ty}$ and $\mathbf{i}_{Lz}$, it follows that the elements of the matrix $\mathbf{T}$ in (3.85) suffice to determine the usual driving point and transfer functions of the network. This may be shown explicitly as follows.

Let the admittance (capacitive) branch voltages be designated as $\mathbf{e}_y$ and the corresponding current sources as $\mathbf{I}_y$; also, let the impedance (inductive) branch currents be $\mathbf{i}_z$ and the corresponding voltage sources $\mathbf{E}_z$. Then it is easily shown that

$$\begin{bmatrix} \mathbf{e}_y \\ \mathbf{i}_z \end{bmatrix} = \begin{bmatrix} \mathbf{D}_y & \mathbf{O} \\ \mathbf{O} & \mathbf{C}_z \end{bmatrix} \begin{bmatrix} \mathbf{e}_{Ty} \\ \mathbf{i}_{Lz} \end{bmatrix} \tag{3.86}$$

and

$$\begin{bmatrix} \mathbf{I}_T \\ \mathbf{E}_L \end{bmatrix} = \begin{bmatrix} \mathbf{D}_y^t & \mathbf{O} \\ \mathbf{O} & \mathbf{C}_z^t \end{bmatrix} \begin{bmatrix} \mathbf{I}_y \\ \mathbf{E}_z \end{bmatrix} \tag{3.87}$$

where the vectors $\mathbf{I}_T$ and $\mathbf{E}_L$ have the same meaning as in (3.71), except that now $\mathbf{E}_y = 0$ and $\mathbf{I}_z = 0$.

Combining (3.71), (3.85), (3.86), and (3.87), it follows that

$$\begin{bmatrix} \mathbf{e}_y \\ \mathbf{i}_z \end{bmatrix} = \begin{bmatrix} \mathbf{D}_y & \mathbf{O} \\ \mathbf{O} & \mathbf{C}_z \end{bmatrix} \begin{bmatrix} \mathbf{T}_{yy} & \mathbf{T}_{yz} \\ \mathbf{T}_{zy} & \mathbf{T}_{zz} \end{bmatrix} \begin{bmatrix} \mathbf{D}_y^t & \mathbf{O} \\ \mathbf{O} & \mathbf{C}_z^t \end{bmatrix} \begin{bmatrix} \mathbf{I}_y \\ \mathbf{E}_z \end{bmatrix} \tag{3.88}$$

or

$$\begin{bmatrix} \mathbf{e}_y \\ \mathbf{i}_z \end{bmatrix} = \begin{bmatrix} \mathbf{D}_y \mathbf{T}_{yy} \mathbf{D}_y^t & \mathbf{D}_y \mathbf{T}_{yz} \mathbf{C}_z^t \\ \mathbf{C}_z \mathbf{T}_{zy} \mathbf{D}_y^t & \mathbf{C}_z \mathbf{T}_{zz} \mathbf{C}_z^t \end{bmatrix} \begin{bmatrix} \mathbf{I}_y \\ \mathbf{E}_z \end{bmatrix} \tag{3.89}$$

where the basic response function matrix **T** has been appropriately partitioned. Equation (3.89) now defines many, but not all, of the usual driving-point and transfer functions explicitly. In particular, the elements of $\mathbf{D}_y \mathbf{T}_{yy} \mathbf{D}_y^t$ correspond to the driving-point and transfer impedances of all the capacitive branches. The elements $\mathbf{D}_y \mathbf{T}_{yz} \mathbf{C}_z^t$ correspond to transfer functions relating unit voltage excitations in each inductive branch to the response voltages in each capacitive branch. All the other response functions may be determined by a straightforward extension of (3.89)

Using the relation $\mathbf{W} = \mathbf{X}^{-1}\mathbf{P}^{-1}$, which may be computed once for all, any desired basic response function may be evaluated in terms of the partial fraction expansion:

$$T_{ij} = \sum_{k=1}^{n} \frac{X_{ik} W_{kj}}{s - \lambda_k} \tag{3.90}$$

the computational cost again being proportional to n operations per frequency.

Since the eigenvalues have already been computed, the poles of $\mathbf{T}_{ij}$ are known. Numerical computation of the zeroes, if desired, could be carried out by any of a number of root-finding procedures.

Sensitivity Computations: Eq. (3.82) provides a convenient springboard for determining sensitivity of the network response to variations in frequency and/or component values. It is readily shown that for any nonsingular matrix, the differential of its inverse is

$$d\mathbf{A}^{-1} = \mathbf{A}^{-1} d\mathbf{A}\mathbf{A}^{-1} \tag{3.91}$$

Accordingly, the frequency sensitivity of $\mathbf{x}_0$ turns out to be

$$\frac{d\mathbf{x}_0}{ds} = -\mathbf{X}(s\mathbf{U} - \Lambda)^{-2} \mathbf{X}^{-1} \mathbf{g}_0 \tag{3.92}$$

where $(s\mathbf{U} - \Lambda)^{-2}$ is simply the square of $(s\mathbf{U} - \Lambda)^{-1}$ and hence trivial to compute. Similarly, it follows from (3.90) that the frequency sensitivity of T_{ij} is

$$\frac{dT_{ij}}{ds} = \sum_{k=1}^{n} \frac{-X_{ik} W_{kj}}{(s - \lambda_k)^2} \tag{3.93}$$

Sensitivity of the network response to component variations may be computed as follows. In place of (3.85), write

$$\mathbf{x}_0 = (s\mathbf{U} - \mathbf{A})^{-1} \mathbf{P}^{-1} \mathbf{f}_0 \tag{3.94}$$

and then it can be shown that

$$\frac{d\mathbf{x}_0}{dr} = (s\mathbf{U} - \mathbf{A})^{-1} \left[\frac{d\mathbf{A}}{dr} (s\mathbf{U} - \mathbf{A})^{-1} - \mathbf{P}^{-1} \frac{d\mathbf{P}}{dr} \right] \mathbf{P}^{-1} \mathbf{f}_0 \tag{3.95}$$

where dr represents an arbitrary parameter variation. Since $\mathbf{A} = -\mathbf{P}^{-1}\mathbf{Q}$ it follows that

$$\frac{d\mathbf{A}}{dr} = \mathbf{P}^{-1} \frac{d\mathbf{P}}{dr} \mathbf{P}^{-1} \mathbf{Q} - \mathbf{P}^{-1} \frac{d\mathbf{Q}}{dr} \tag{3.96}$$

and, since all the topological matrices are constant, that,

$$\frac{d\mathbf{P}}{dr} = \begin{bmatrix} \mathbf{D}_y^t \dfrac{d\mathbf{K}}{dr} \mathbf{D}_y & \mathbf{O} \\ \mathbf{O} & \mathbf{C}_z^t \dfrac{d\mathbf{L}}{dr} \mathbf{C}_z \end{bmatrix} \tag{3.97}$$

and

$$\frac{d\mathbf{Q}}{dr} = \begin{bmatrix} \mathbf{D}_y^t \dfrac{d\mathbf{G}}{dr} \mathbf{D}_y & \mathbf{O} \\ \mathbf{O} & \mathbf{C}_z^t \dfrac{d\mathbf{R}}{dr} \mathbf{C}_z \end{bmatrix} \tag{3.98}$$

Thus, in terms of (3.95) through (3.98) the sensitivity of any response variable to changes in component or parameter values can be computed explicitly.

Finally, the pole-sensitivity may be computed using a standard formula from matrix theory

$$\frac{d\lambda_i}{dr} = \frac{\zeta_i^t \frac{d\mathbf{A}}{dr} \chi_i}{\zeta_i^t \chi_i} \tag{3.99}$$

where λ_i is the ith eigenvalue of $\mathbf{A}$ and where χ_i and ζ_i are the corresponding eigenvectors of $\mathbf{A}$ and $\mathbf{A}^t$, respectively.

Transient Network Problems

Of the three types of network problems under discussion, transient problems present the greatest computational difficulty. This is because the numerical integration of the differential equations is burdensome for nonlinear networks and extremely slow when the network has a wide spread of eigenvalues. The nature of these problems will be discussed presently.

Analytic Solution of Linear Problems: To fix ideas, let us first consider the analytic solution of the state-variable equations for a linear network with constant parameters. Referring to (3.73), we assume a solution of the form

$$\mathbf{x}(t) = e^{\mathbf{A}t}\,\mathbf{v}(t) \tag{3.100}$$

where $\mathbf{v}(t)$ is a vector function of time and where $e^{\mathbf{A}t}$ is defined as the matrix function

$$e^{\mathbf{A}t} = \mathbf{U} + \mathbf{A}t + \frac{(\mathbf{A}t)^2}{2!} + \frac{(\mathbf{A}t)^3}{3!} + \cdots \tag{3.101}$$

By differentiating (3.100) and using the definition of $e^{\mathbf{A}t}$, it follows that

$$\dot{\mathbf{x}} = \mathbf{A}e^{\mathbf{A}t}\mathbf{v} + e^{\mathbf{A}t}\dot{\mathbf{v}} = \mathbf{A}\mathbf{x} + e^{\mathbf{A}t}\dot{\mathbf{v}} \tag{3.102}$$

Combining this with (3.73), we obtain a differential equation for $\mathbf{v}(t)$ in the form

$$\dot{\mathbf{v}} = e^{-\mathbf{A}t}\,\mathbf{g}(t) \tag{3.103}$$

whose solution may be written as

$$\mathbf{v}(t+h) = \mathbf{v}(t) + \int_0^h e^{-\mathbf{A}(t+\tau)}\ \mathbf{g}(t+\tau)\,d\tau \tag{3.104}$$

Multiplying this equation by $e^{\mathbf{A}(t+h)}$ and using (3.100) we find that the solution to (3.73) is

$$\mathbf{x}(t+h) = e^{\mathbf{A}h}\,\mathbf{x}(t) + \int_0^h e^{\mathbf{A}(h-\tau)}\,\mathbf{g}(t+\tau)\,d\tau \tag{3.105}$$

This solution is exact, but its utility depends on the computation of $e^{\mathbf{A}h}$.

Liou [1], [39] has proposed the use of a finite number of terms in the series expansion of $e^{\mathbf{A}h}$ according to (3.101), asserting that this method "completely avoids eigenvalue problems". In one sense, this assertion is true in that Liou's method does not require the calculation of eigenvalues. On the other hand, the method runs headlong into another kind of eigenvalue problem which seriously limits its usefulness; namely, that when the matrix **A** has a large eigenvalue, the integration step size h must be kept small, roughly less than the reciprocal of $|\lambda_{max}(\mathbf{A})|$, in order to permit rapid convergence of (3.101). Precisely the same limitation arises in numerical integration using, say, the Runge-Kutta or Adam's method, where h must be kept small to avoid numerical instability [34]. More about this later.

If, instead of avoiding the computational eigenvalue problem, we *solve* it, then (3.105) becomes as practical to use as (3.82) and (3.83). From (3.101) it can be shown that the transformation which diagonalizes **A** also diagonalizes $e^{\mathbf{A}h}$, yielding the relation

$$\mathbf{X}^{-1}\,e^{\mathbf{A}h}\,\mathbf{X} = e^{\Lambda h} \tag{3.106}$$

where $e^{\Lambda h}$ is a diagonal matrix whose terms $e^{\lambda_i h}$ are trivial to compute. Solving this expression for $e^{\mathbf{A}h}$ we may recast (3.105) into the form

$$\mathbf{x}(t+h) = \mathbf{X}e^{\Lambda h}\mathbf{X}^{-1}\mathbf{x}(t) + \int_0^h \mathbf{X}e^{\Lambda(h-\tau)}\mathbf{X}^{-1}\mathbf{g}(t+\tau)\,d\tau \qquad (3.107)$$

and our computational problem reduces to one of evaluating the integral expression. If the driving function $g(t + \tau)$ is approximated as a polynomial in τ, the integration can be carried out exactly, term by term. This provides a very satisfactory way of evaluating the integral on a digital computer.

Equation (3.107) is instructive from a theoretical point of view. For example, if $\mathbf{g}(t) = 0$ and we apply initial conditions to the network which correspond to, say, the ith eigenvector χ_i, then we may write the solution as

$$\mathbf{x}(t) = \mathbf{X}\,e^{\Lambda t}\,\mathbf{X}^{-1}\chi_i \qquad (3.108)$$

But since $\mathbf{X}^{-1}\mathbf{X}$ is a unit matrix, $\mathbf{X}^{-1}\chi_i$ is equal to the ith column of a unit matrix. (Remember that the columns of $\mathbf{X}$ are the eigenvectors χ_i.) Thus, the eigenfunction $e^{\lambda_i t}$ is *singled out* and (3.108) reduces to

$$\mathbf{x}(t) = \chi_i e^{\lambda_i t} \qquad (3.109)$$

In other words, the ith eigenvector prescribes the *particular* initial conditions required to excite the ith characteristic mode (eigenmode) of response and suppress all the others. For arbitrary initial conditions, given by a vector $\mathbf{x}_0$, all of the eigenmodes will be excited, in general. But the matrix-vector product $\mathbf{X}^{-1}\mathbf{x}_0$ automatically produces the appropriate *linear combination* of eigenfunctions in terms of which to express the solution.

For the case where $\mathbf{g}(t) \neq 0$ the "initial conditions" are continually changing, so to speak, and so an additional term involving a convolution integral of the eigenfunctions and driving functions is required, as in (3.107).

Numerical Integration: The numerical methods of integration currently in use are, for the most part, based on the use of finite difference approximations to the actual differential equations to be solved. For a single differential equation of the form

$$dx = f(x,t) \qquad (3.110)$$

a solution may be effected by the general finite difference expression [34]

$$x_n = \sum_{i=1}^{M} a_i x_{n-i} + h \sum_{j=0}^{N} b_j f_{n-j} \tag{3.111}$$

where h is the integration interval, $x_n = x(nh)$, and $f_{n-j} = (x_{n-j}, (n-j)h)$. The number and values of the coefficients a_i and b_i are determined by the number of terms of the Taylor expansion of $x(nh)$ which are to be matched by the integration formula so defined.

If $b_0 = 0$, the integration formula is called a "predictor," since the new value of x_n is predicted in terms of previous values (x_{n-i}) and derivatives (f_{n-j}) only. If $b_0 \neq 0$, the formula is called a "corrector," since it is generally used in conjunction with a predictor formula to provide a corrected value of x_n. This is accomplished by using the predicted value of x_n to compute an approximate value of the "new" derivative f_n, which is then used in the corrector formula.

Now, in the simple case of one homogeneous linear differential equation with constant coefficient, where $f(x, t) = gx(t)$, (3.111) may be written in the form

$$(1 - hgb_0)x_n - (a_1 + hgb_1)x_{n-1} - \dots - (a_r + hgb_r)x_{n-r} = 0 \tag{3.112}$$

Here, at least one of the coefficients a$_r$ or b_r is nonzero and all coefficients of higher index are zero. That is, $r = \max(M, N)$. If we assume a particular solution of the form $x_n = p^n$, then (3.112) may be reduced to the polynomial equation [34]

$$(1 - hgb_0)p^r - (a_1 + hgb_1)p^{r-1} - \dots - (a_r + hgb_r) = 0 \tag{3.113}$$

Obviously, any value of p which is a root of this *characteristic equation* is a candidate for a particular solution $x_n = p^n$, but the general solution takes the form of a linear combination of particular solutions, namely

$$x_n = \sum_{k=1}^{r} c_r p_k^n \tag{3.114}$$

where the p_k are all the roots of (3.113)

One of these roots, lying near the point $1 + j0$ in the complex plane, will be the *principal* root since it primarily determines the desired approximation to the solution of the differential equation in question.

All the other roots are *parasitic roots* and it is clear from (3.114) that if any parasitic root lies outside the unit circle, x_n will eventually diverge. This situation is called "numerical instability" because it vitiates the solution.

In order to prevent numerical instability, the integration step h must be kept small enough to insure that all of the parasitic roots of (3.113) remain inside the unit circle. In the case of a *system* of linear differential equations with constant coefficients, as typified by (3.73), the requisite condition on h for most integration formulas is

$$h|\lambda_{\max}(\mathbf{A})| \leqslant c \tag{3.115}$$

Here, $\lambda_{\max}(\mathbf{A})$ is the largest eigenvalue of the coefficient matrix $\mathbf{A}$ while the constant c is in the order of 1 or 2, depending on the integration method used. Even in the case of non-linear differential equations, an equivalent restriction, involving the (instantaneously) largest eigenvalue of the Jacobian matrix, must be imposed to prevent numerical instability.

It is just this constraint on the integration interval which makes most integration techniques so painfully slow in the case of network problems having a wide spread of eigenvalues. For the largest eigenvalue (or, equivalently, its reciprocal, the smallest time constant) controls the permissible size of the integration step h. But the smallest eigenvalues (largest time constants) control the network response and so determine the total length of time over which the integration must be carried out to characterize the response. In the case of a network with a 1000 to 1 ratio of largest to smallest eigenvalue, for instance, it might be necessary to take the order of 1000 times as many integrations steps with, say, the Runge-Kutta method as with some method which was free of the restriction given by (3.115). This so-called "minimum time constant" barrier is without doubt the biggest obstacle to be overcome in the transient analysis of networks.

There are, however, some integration formulas in common use for which the constraint of h given in (3.115). does not apply; for example,

$$x_n = x_{n-1} + hf_n \tag{3.116}$$

and

$$x_n = x_{n-1} + \frac{h}{2}(f_n + f_{n-1}) \tag{3.117}$$

Both of these have characteristic equations of the first degree and so have no parasitic roots. But the main disadvantage in using these formulas is that they are of such low-order accuracy that the integration interval must often be shortened anyway to insure sufficient accuracy. Another disadvantage is that both are implicit formulas because they involve f_n, which cannot be computed until x_n, is known. When applied to (3.73), therefore, these formulas require the solution of a linear system of equations (or matrix inversion) which complicates their use.

Schemes for retaining the numerical stability inherent in (3.116) and (3.117) without the need to solve simultaneous equations have been reported by Stineman [40] and by Richards, Lanning, and Torrey [41] but the resulting methods are again of low-order accuracy. Treanor [42] on the other hand, describes a method of high (fourth-order) accuracy, which is identical with the Runge-Kutta method for short integration intervals but retains numerical stability even for very large integration intervals. His method, however, involves computation of the matrix function $e^{\mathbf{A}h}$ and so presents no real advantage over the eigenvalue approach for linear problems. In the case of nonlinear problems, the difficulty in using Treanor's method is compounded by the fact that $e^{\mathbf{A}h}$ must be recomputed at each time step.

Quasi-Analytic Solutions: A quite different attack on this problem has been reported by Cohen and Flatt [43], [44], and by Certaine [45]. (Incidentally, the title of Certaine's paper is misleading since what he calls "time constant" is the reciprocal of the time constant usually referred to by electrical engineers.) The essence of this attack is the following: in (3.73) the *diagonal* terms of the coefficient matrix $\mathbf{A}$ are handled analytically, as was $\mathbf{A}$ itself in (3.105), while the off-diagonal terms of $\mathbf{A}$ are lumped together with $\mathbf{g}(t)$ and considered as part of the driving function.

In other works, letting $\mathbf{A} = \mathbf{D} + \mathbf{B}$, where $\mathbf{D}$ contains all the diagonal terms of $\mathbf{A}$ while $\mathbf{B}$ contains all the off-diagonal terms, we may write (3.73) as

$$\dot{\mathbf{x}} = \mathbf{D}\mathbf{x} + [\mathbf{B}\mathbf{x} + \mathbf{g}(t)] \tag{3.118}$$

with $\mathbf{B}\mathbf{x} + \mathbf{g}(t)$ serving as a driving function. The analytic solution to this equation can be obtained by modifying (3.105) to read

$$\mathbf{x}(t+h) = e^{\mathbf{D}h}\mathbf{x}(t) + \int_0^h e^{\mathbf{D}(h-\tau)}[\mathbf{Bx}(t+\tau) + \mathbf{g}(t+\tau)]\,d\tau \qquad (3.119)$$

Here, $e^{\mathbf{D}h}$ is easy to compute, since **D** is a diagonal matrix. The integration can be carried out by using a polynomial approximation of any desired degree for $\mathbf{Bx} + \mathbf{g}$. (The method of integration used in NET-1 is based on a modification of this approach.)

When the matrix **A** is dominantly diagonal, the use of (3.119) may permit a much larger integration step size than that allowed in the usual methods. However, for network problems, this property of diagonal dominance of the **A** matrix does not generally obtain and so (3.119) is not of itself sufficient to break through the "time-constant" barrier.

The writer has found that (3.119) has its own peculiar numerical instability which seems to be avoidable by choosing the integration interval subject to a restriction of the form

$$h|\lambda_{max}(\mathbf{B})| \leqslant c \qquad (3.120)$$

where c is a small constant. In other words, the largest eigenvalue of the *off-diagonal* matrix **B** controls the stability of the numerical integration process used in applying (3.119).

From a theoretical point of view, it is possible to choose a similarity transformation on **A** of the form

$$\mathbf{T}^{-1}\mathbf{AT} = \overline{\mathbf{A}} = \overline{\mathbf{D}} + \overline{\mathbf{B}} \qquad (3.121)$$

where $\overline{\mathbf{D}}$ is the diagonal component and $\overline{\mathbf{B}}$ the off-diagonal component of $\overline{\mathbf{A}}$, such that $\lambda_{max}(\overline{\mathbf{B}})$ is arbitrarily small. Indeed, if **T** is so chosen that $\lambda_{max}(\mathbf{B}) = 0$, then **D** contains all the eigenvalues of **A**. (In effect, the *Q-R* algorithm [38] involves an interative process for finding a succession of similarity transformations each of which reduces $\lambda_{max}(\mathbf{B})$ until it becomes vanishingly small.)

If we choose a transformation of variables of the form

$$\mathbf{y} = \mathbf{T}^{-1}\mathbf{x} \qquad (3.122)$$

to be used in conjunction with (3.121) – for the purpose of making $\lambda_{max}(\mathbf{B})$ small, but not necessarily zero – then (3.73) becomes

$$\mathbf{T}\dot{\mathbf{y}} = \mathbf{ATy} + \mathbf{g}(t) \tag{3.123}$$

or

$$\dot{\mathbf{y}} = \mathbf{T}^{-1}\mathbf{ATy} + \mathbf{T}^{-1}\mathbf{g}(t) \tag{3.124}$$

Accordingly, in view of (3.118), (3.119), and (3.121), we may write the solution as

$$\mathbf{y}(t+h) = e^{\overline{\mathbf{D}}h}\mathbf{y}(t) + \int_0^h e^{\overline{\mathbf{D}}(h-\tau)}[\overline{\mathbf{B}}\mathbf{y}(t+\tau) + \mathbf{w}(t+\tau)]\,d\tau \tag{3.125}$$

where $\mathbf{w}(t) = \mathbf{T}^{-1}\mathbf{g}(t)$. In effect, then, we solve the original problem by transforming to the domain of the new variables y before integrating.

From (3.125) and the relation $\mathbf{x} = \mathbf{Ty}$, a solution to the problem $\dot{\mathbf{x}} = \mathbf{Ax} + \mathbf{g}(t)$ may be computed. Moreover, if $\mathbf{T}$ has been chosen so as to reduce $\lambda_{max}(\overline{\mathbf{B}})$ sufficiently, a large integration step h may be used without risking numerical instability. The validity of this procedure has been verified in handling the following simple problem.

Consider the RC network shown in Fig. 3.10. With all resistances equal and with all capacitances equal, the maximum integration step which could be taken safely using (3.119) was found to be about equal

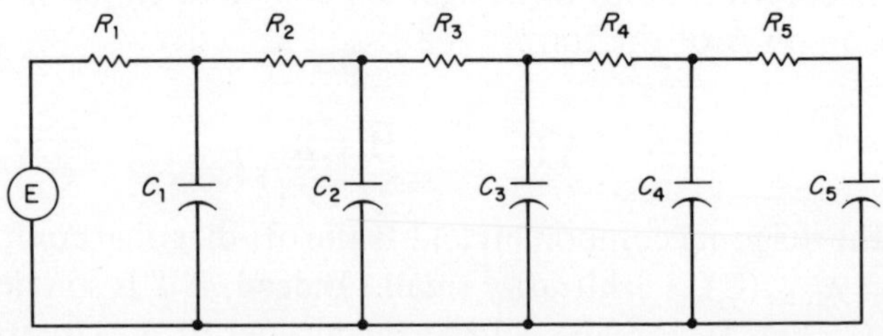

FIG. 3.10. *RC* ladder network.

to the time constant, RC. In the case of the modified network of Fig. 3.11, with C_3 split in half and bridging resistor r having a very small value relative to R, the integration step was found to be limited by the time constant rC.

Originally, the state-variable vector used for both networks consisted of all the capacitor voltages. However, when a different combination of voltages was used in treating the second network – namely, the voltages

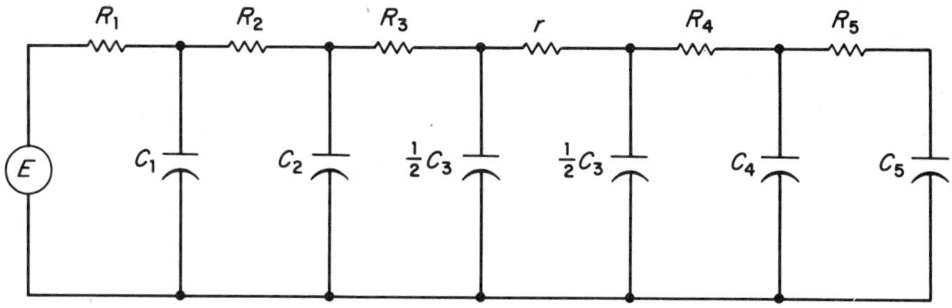

FIG. 3.11. Modified RC ladder network.

across capacitors C_1, C_2, C_4, and C_5 but the *sum* and *difference* of the voltages across the two half-size capacitors – it became possible to increase the integration step size until it was equal to that used in the first network. This simple change of variables defined a transformation which was sufficient to reduce significantly the maximum eigenvalue of $\bar{\mathbf{B}}$ in (3.121), though not sufficient to expose all the eigenvalues explicitly.

The lesson to be learned from the foregoing considerations is evidently this: one key to the minimum time constant problem seems to lie in operating on the coefficient matrix **A** in some meaningful way. But if the **A** matrix is left intact, then we must accept the limitations implicit in (3.115). The simplest kind of operation on the **A** matrix is to extract its diagonal terms to be handled analytically as in (3.119). The next more difficult operation is to transform the matrix as in (3.121) – (3.125). The ultimate, of course, is to solve the eigenproblem completely.

Thus, there are clear guidelines to follow in tackling the integration problem for nonlinear systems where the Jacobian matrix now plays the role of the **A** matrix. Obviously, solving the eigenproblem repeatedly is inappropriate, since the Jacobian matrix changes during the course of the solution. However, finding a "quick and dirty" transformation **T** to be used in conjunction with (3.121) – (3.125) does seem plausible. Moreover, it should not be necessary to update **T** at each integration step since it is likely that a given transformation could be used for many integrations before becoming degraded to the point of uselessness.

The feasibility of this approach would depend not only on minimizing the frequency of updating **T** but also on developing an efficient algorithm for computing this transformation. If the average computational cost per integration step can be kept proportional to the square of the number of equations, then the minimum time-constant barrier can be breached for nonlinear problems as well as for linear.

In the case of linear systems, the superiority of the eigenvalue approach for both ac and transient analysis is unmistakable, since it goes directly, and with reasonable computational efficiency, to the very heart of the problem. This is not too surprising in view of the well known facts that the eigenproblem is the most fundamental problem in matrix theory and that its solution provides essentially all there is of real value to be learned about a matrix. Fortunately, the advent of the digital computer, and the *Q-R* algorithm, now brings this powerful theoretical tool down to a practical working level.

3.3 FUTURE DEVELOPMENTS

It is evident from the topics discussed in this chapter that much work needs to be done in the immediate future to upgrade the network analysis programs currently available to circuit designers. In addition to the incorporation of new and improved analytical techniques, these upgraded programs will need to provide facilities for statistical analysis of circuits, for storage and retrieval of device models, for handling transmission lines, [46] and for treating much larger problems. Graphical presentation of output variables in a variety of forms will be essential and a much closer degree of man-machine interaction will become possible through the use of CRT displays and remote access to the computer. A good example of this kind of development is Katzenelson's AEDNET program [1], [47].

In another direction, there is a growing need to handle integrated and monolithic circuits. This new technology is already placing strong demands on our analytical, numerical, and programming techniques and it will inevitably force the development of more powerful methods for solving partial differential equations. This, in turn, is likely to stimulate the exploration and exploitation of such techniques as Kron's method of interconnecting solutions, [9] – [11] the use of higher dimensional network models, [22] and a novel numerical method for treating partial differential equations recently described by Polozhii [48].

A step beyond the production of more powerful and effective network analysis programs will be the development of *optimization* techniques--in which the analysis tasks will likely be relegated to subroutine status. Significant steps are already being taken in this direction [49] – [51]. Moreover, automatic network synthesis is also being carried out to some extent [49]. It is probable, however, that for a long time to come a majority of computer-aided design will be based primarily on analysis and optimization techniques, with the engineer himself not only playing an essential role in the "design loop" but also controlling the entire process of design. Thus, while the computer will certainly take over the routine computational tasks of the engineer, this will enhance rather than supplant his creative abilities – which are really his *raison d'etre.*

REFERENCES

1. F. F. Kuo, "Network analysis by digital computer," *Proc. IEEE*, vol. 54, pp. 820-829, June 1966.

2. Special Issue on Computer-Aided Design, *Proc. IEEE*, vol. 55, pp. 1775-2018, Nov. 1967.

3. H. N. Ghosh, F. H. de la Moneda, and N. R. Dono, "Computer-aided transistor design, characterization and optimization," *Proc. IEEE*, vol. 55, pp. 1897-1912, Nov. 1967.

4. J. J. Olsztyn and T. L. Theodoroff, "GMR DYANA Programming Manuals I and II," General Motors Research Labs., Warren, Mich., 1959.

5. S. J. Fenves, "STRESS (structural engineering system solver) a computer programming system for structural engineering problems," M.I.T., Cambridge, Mass., Tech. Rept. T63-2, June 1963.

6. S. J. Fenves and F. H. Branin, Jr., "A network-topological formulation of structural analysis," *ASCE J. Struct. Div.*, Vol. 89, pp. 483-514, August 1963.

7. F. H. Branin, Jr., "A new method for steady state A-C analysis of RLC networks," *IEEE Internat'l Conv. Rec.*, pt. 7, pp. 218-223, March 1966.

8. G. Kron, *Tensor Analysis of Networks.* New York: Wiley, 1939.

9. F. H. Branin, Jr., "Kron's method of tearing and its applications," *Proc. 2nd Midwest Symp. on Circuit Theory* (Michigan State University, East Lansing, Mich.), December 1956.

10. F. H. Branin, Jr., "The relation between Kron's method and the classical methods of network analysis," *IRE WESCON Conv. Rec.*, pt. 2, pp. 3-28, August 1959.

11. G. Kron, "A set of principles to interconnect the solutions of physical systems," *J. Appl. Phys.*, vol. 24, pp. 965-980, 1953.

12. J. P. Roth, "An application of algebraic topology to numerical analysis: on the existence of a solution to the network problem," *Proc. Nat'l Acad. Sciences*, vol. 41, pp. 518-521, 1955.

13. J. P. Roth, "The validity of Kron's method of tearing," *Proc. Nat'l Acad. Sciences*, vol. 41, pp. 599-600, 1955.

14. J. P. Roth, "An application of algebraid topology: Kron's method of tearing," *Quart. Appl. Math.*, vol. 17, pp. 1-24, 1959.

15. N. G. Brooks and H. S. Long, "A program for computing the transient response of transistor switching circuits – PE TAP," IBM Development Lab., Poughkeepsie, N. Y., Tech. Rept. 00.700, 1959.

16. F. H. Branin, Jr., "D-c and transient analysis of networks using a digital computer," *IRE Internat'l Conv. Rec.*, pt. 2, pp. 236-256, March 1962.

17. A. F. Malmberg, F. L. Cornwell, and F. N. Hofer, "NET-1 network analysis program," Los Alamos Scientific Lab., Los Alamos, N. Mex., Rept. LA-3119, 1964.

18. "1620 electronic circuit analysis program (ECAP)," IBM Corporation, Data Processing Div., White Plains, N. Y., Application Program 1620-EE-02X.

19. J. B. Ward and H. W. Hale, "Digital computer solution of power-flow problems" *Trans, AIEE (Power Apparatus and Systems)*, Vol. 75, pt. III, pp. 398-404, June 1956.

20. A. H. El-Abiad, M. Watson, and G. W. Stagg, "The load-flow problem: its formulation and solution, pt. I," AIEE Paper GICP1054, 1961.

21. H. E. Brown, G. K. Carter, H. H. Happ, and C. E. Person, "Power flow solution by impedance matrix iterative method," *IEEE Trans. Power and Apparatus*, vol. 82, pp. 561-564, August 1963.

22. F. H. Branin, Jr., "The algebraic-topological basis for network analogies and the vector calculus," *Proc. Symp. on Generalized Networks*, vol. 16, Microwave Research Inst. Symposia Ser., Polytechnic Inst. of Brooklyn, pp. 453-491, 1966.

23. T. R. Bashkow, "The *A* matrix, new network concept," *IRE Trans. Circuit Theory*, vol. CT-4, pp. 117-119, September 1957.

24. P. R. Bryant, "The explicit form of Bashkow's *A* matrix," *IRE Trans. Circuit Theory (Correspondence)*, vol. CT-9, pp. 303-306, September 1962.

25. E. S. Kuh and R. A. Rohrer, "The state-variable approach to network analysis, *Proc. IEEE*, vol. 53, pp. 672-686, July 1965.

26. J. H. Wilkinson, *The Algebraic Eigenvalue Problem.* London: Oxford University Press, 1965.

27. G. E. Forsythe and C. B. Moler, *Computer Solution of Linear Algebraic Systems.* Englewood Cliffs, N. J.: Prentice-Hall, 1967.

28. V. V. Klyuyev and N. I. Kokovkin-Shcherbak, "On the minimization of the number of arithmetic operations for the solution of linear algebraic systems"

(translated by G. J. Tee), Computer Science Dept., Stanford University, Stanford, Calif., Tech. Rept. CS24, 1965.

29. N. Sato and W. F. Tinney, "Techniques for exploiting the sparsity of the network admittance matrix," *IEEE Trans. Power and Apparatus*, vol. 82, pp. 944-950, December 1963.

30. W. F. Tinney and J. W. Walker, "Direct solutions of sparse network equations by optimally ordered triangular factorization," *Proc. IEEE*, vol. 55, pp. 1801-1809, November 1967.

31. S. Parter, "The use of linear graphs in gauss elimination," *SIAM Rev.*, vol. 3, pp. 119-130, April 1961.

32. D. Bree, Jr., "Some remarks on the application of graph theory to the solution of sparse systems of linear equations," thesis, Dept. of Math., Princeton University, Princeton, N. J., May 1965.

33. A. Orden, "Matrix inversion and related topics by direct methods," in *Mathematical Methods for Digital Computers*, A. Ralston and H. S. Wilf, Eds. New York: Wiley, 1960, ch. 2.

34. F. B. Hildebrand, *Introduction to Numerical Analysis*, New York: McGraw-Hill, 1956.

35. C. G. Broyden, "A class of methods for solving nonlinear simultaneous equations." *Math. Comp.* vol. 19, pp. 577-593, October 1965.

36. F. H. Branin, Jr., and H. H. Wang, "A fast reliable iteration method for dc analysis of nonlinear networks," *Proc. IEEE*, vol. 55, pp. 1819-1826, November, 1967.

37. D. J. Wilde, *Optimum Seeking Methods*, Englewood Cliffs, N. J.: Prentice-Hall 1964.

38. J. G. F. Francis, "The Q-R transformation, pt. I," *Computer J.*, vol. 4, pp. 265-271, October, 1961; "The Q-R transformation, pt. II," *Computer J.*, vol. 4, pp. 332-345, January 1962.

39. M. L. Liou, "A numerical solution of linear time-invariant systems," *Proc. 3rd Allerton Conf. on Circuit and System Theory*, pp. 669-676, 1965.

40. R. W. Stineman, "Digital time-domain analysis of systems with widely separated poles," *J. Assoc. Comp. Mach.*, vol. 12, pp. 286-293, April 1965.

41. P. I. Richards, W. D. Lanning, and M. D. Torrey, "Numerical integration of large, highly-damped, nonlinear systems," *SIAM Rev.*, vol. 7, pp. 376-380, July 1965.

42. C. E. Treanor, "A method for the numerical integration of coupled first-order differential equations with greatly different time constants," *Math. Comp.*, vol. 20, pp. 39-45, January 1966.

43. E. R. Cohen, "Some topics in reactor kinetics," *Proc. 2nd UN Internat'l Conf.*

on Peaceful Uses of Atomic Energy (Geneva, Switzerland), vol. 11, pp. 302-309, 1958.

44. E. R. Cohen and H. P. Flatt, "Numerical solution of quasi-linear equations," *Proc. Seminar on Codes for Reactor Computations* (Vienna, Austria), pp. 461-484, 1960.
45. J. Certaine, "The solution of ordinary differential equations with large time constants," in *Mathematical Methods for Digital Computers*, A. Ralston and H. S. Wilf, Eds. New York: Wiley, 1960, ch. 11.
46. F. H. Branin, Jr., "Transient analysis of lossless transmission lines," *Proc. IEEE*, vol. 55, pp. 2012-2013, November 1967.
47. J. Katzenelson, "AEDNET: a simulator for nonlinear networks," *Proc. IEEE*, vol. 54, pp. 1536-1552, November 1966.
48. G. N. Polozhii, *The Method of Summary Representation for Numerical Solution of Problems of Mathematical Physics.* New York: Pergamon, 1965.
49. *System Analysis by Digital Computer*, F. F. Kuo and J. F. Kaiser, Eds. New York: Wiley, 1966.
50. G. C. Temes and D. A. Calahan, "Computer-aided network optimization – The state-of-the-art," *Proc. IEEE*, vol. 55, pp. 1832-1864, November 1967.
51. G. D. Hachtel and R. A. Rohrer, "Techniques for the optimal design and synthesis of switching circuits," *Proc. IEEE*, vol. 55, pp. 1864-1877, Nov. 1967.

CHAPTER 4

COMPUTER-AIDED ANALYSIS OF NONLINEAR NETWORKS

L. O. CHUA, Purdue University

INTRODUCTION

Most practical electronic circuits are nonlinear to a greater or lesser extent. In the case of signal transmission networks, the nonlinearity is usually undesirable, but is nevertheless present due to the inherent nonlinearity of the devices. In most other cases (such as signal processing, digital, and switching circuits), the nonlinearity is intentionally introduced. The subject of this chapter is concerned mainly with the analysis of the latter class of nonlinear networks.

Unlike the case of linear networks where the solution can always be found in closed form, it is extremely rare that the solution of a nonlinear network can be found analytically. Hence, to analyze a nonlinear network of reasonable complexity, the use of a computer is virtually a necessity. Consequently, a great deal of effort has gone into developing user-oriented computer programs, over the past few years, for analyzing nonlinear networks automatically.† Most of these programs are written

† Some of these programs are described in Chapter 8. In particular, see the discussion on the CIRCUS and SCEPTRE programs.

specifically for analyzing common large-signal electronic circuits where the nonlinear elements are usually diodes, transistors, and other solid state devices. For this type of common electronic circuits, these programs work very well, and are widely used today. They certainly meet their intended objectives. However, there are occasions where none of these programs would work.† Since the user's manuals for these programs seldom state precisely the limitations of the program, the user is usually frustrated whenever such an occasion arises. The reason behind such possible failures is far from being trivial. It is due neither to any programming error nor numerical inaccuracies – but something much more fundamental in nature. In fact, the entire foundation of nonlinear network analysis must be re-examined in order to uncover the source of the difficulty.

The main objective of this chapter is to provide a careful introduction to the foundation of nonlinear network analysis. The second objective is to present a survey of various numerical techniques useful in nonlinear analysis. The final objective is to show that in analyzing new problems, a preliminary *qualitative* analysis should precede the quantitative analysis. This leads to the inevitable observation that one must be able to solve the multivalued resistive subnetwork.

4.1 REPRESENTATION OF NONLINEAR NETWORK ELEMENTS

A network element may be defined as a black box with 2, 3, or n terminals accessible to the external world (Fig. 4.1). To each terminal j, we assign a current variable i_j, a voltage variable v_j, a charge variable q_j, and a flux-linkage variable φ_j. Without loss of generality, we may assume that each terminal voltage v_j is measured with respect to the common terminal n. The standard references for voltage polarities and current directions are chosen as in Fig. 4.1 so that the power $p_j \equiv v_j i_j$ *enters*

† This is especially true when one is analyzing a *new* nonlinear network containing hypothetical elements. As will be shown in Sec. 4.2, the source of trouble usually comes from the presence of non-monotonic elements. This observation leads one to conjecture that provided only monotonic elements, (such as diodes, transistors, etc.), are allowed, no trouble should be expected. Unfortunately, this conjecture is false because it is easy to obtain an equivalent non-monotonic – or even multivalued – element by combining monotonic active elements such as transistors, with monotonic elements [1].

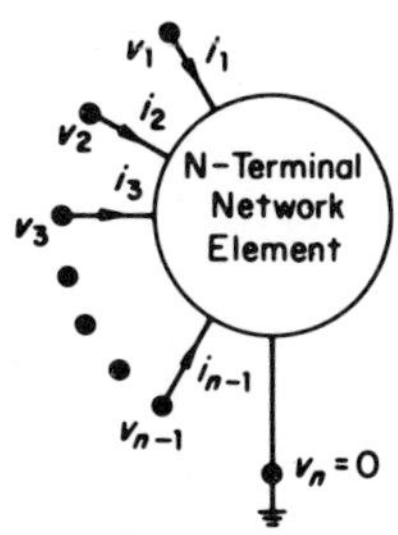

FIG. 4.1. Symbolic representation of an n-terminal network element.

the element through terminal j whenever $p_j > 0$. The variables q_j and φ_j are related to the terminal currents and voltages by†

$$q_j(t) = \int_{-\infty}^{t} i_j(\tau)d\tau = q_j(t_0) + \int_{t_0}^{t} i_j(\tau)d\tau \tag{4.1}$$

$$\varphi_j(t) = \int_{-\infty}^{t} v_j(\tau)d\tau = \varphi_j(t_0) + \int_{t_0}^{t} v_j(\tau)d\tau \tag{4.2}$$

4.1.1 Two-Terminal Network Elements

There are three basic types of 2-terminal network elements – a resistor, an inductor, or a capacitor. Any 2-terminal black box which is characterized by a v-i curve, an i-φ curve, or a v-q curve is called a *resistor*, an *inductor*, or a *capacitor* respectively [1, 2, 3]. This curve may either be measured experimentally – as in the case of all practical devices – or postulated as in the case of a hypothetical device. If the curve representing a resistor, inductor, or capacitor consists of a straight line *through* the origin, the corresponding element is said to be a *linear element*. Otherwise, it is called a nonlinear resistor, a nonlinear inductor, or a nonlinear capacitor.

If the curve characterizing a nonlinear element is *not* symmetrical with respect to the origin it is said to be a *non-bilateral* element. This term is motivated by the fact that if we inadvertently interchange the two terminals without redefining the voltage and current variables, a different curve (obtained by revolving the curve 180° with respect to the

† Eqs. (4.1) and (4.2) also define the means for measuring the charge q_j and the flux-linkage φ_j in practice; namely, by means of a current or a voltage integrator.

origin) will be measured. To avoid this common source of error, the *unsymmetrical* symbols shown in Figs. 4.2(a), (b), and (c) are proposed for a nonlinear resistor, inductor, or capacitor, respectively.

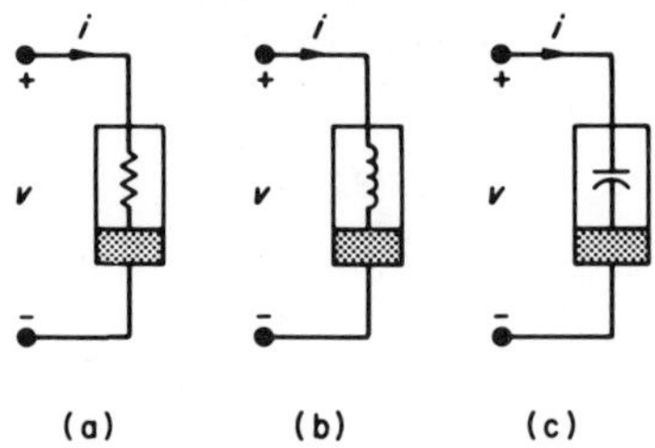

FIG. 4.2. Symbolic representation of a 2-terminal nonlinear resistor, inductor, or capacitor.

A 2-terminal element may be characterized by a curve which varies as a function of one or more physical variables such as light intensity (e.g., a photo diode), temperature (e.g., thermistors), etc. These elements are called controlled 2-terminal elements. If the value of the controlling variable is prescribed a-priori, the above element reduces to an ordinary 2-terminal element. Sometimes, the controlling variable is the *time t,* in which case, the element is said to be a *time-varying* element. Due to the limitation of space, we will not consider time-varying elements in this chapter.

4.1.2 *N*-Terminal Network Elements

By analogy, we define three basic types of n-terminal ($n > 2$) network elements – an *n-terminal resistor*, an *n-terminal inductor*, or an *n-terminal capacitor*. An n-terminal resistor is characterized by a set of relationships involving *only* the voltage variables $v_1, v_2, \ldots, v_{n-1}$ and the current variables $i_1, i_2, \ldots, i_{n-1}$. An n-terminal inductor is characterized by a set of relationships involving *only* the current variables i_1, $i_2, \ldots, i_{n-1}$ and the flux-linkage variables $\varphi_1, \varphi_2, \ldots, \varphi_{n-1}$. An n-terminal capacitor is characterized by a set of relationships inolving *only* the voltage variables $v_1, v_2, \ldots, v_{n-1}$ and the charge variables q_1, q_2, $\ldots, q_{n-1}$. An n-terminal element is said to be linear if the relationships characterizing it consist of a linear combination of the defining variables. Otherwise it is nonlinear.

When $n = 3$, the relationships characterizing a nonlinear element are specified in general by two sets of curves. Each set involves two variables and a parameter representing one of the two variables of the second set.

For example, a transistor is a 3-terminal resistor whereby one set of curves involves the emitter current and the emitter-to-base voltage with the collector-to-base voltage as the parameter. The second set involves the collector current and the collector-to-base voltage with the emitter current as a parameter.

An n-terminal element with $n > 3$ is generally used as a controlled 3-terminal element, by constraining the voltages or currents of the remaining terminals to some prescribed values (e.g., by connecting the terminals to fixed voltage or current sources). An n-terminal element can therefore be characterized by two sets of curves; however, these curves will change as we change the values of the controlling parameters. For example, a pentode is a 5-terminal resistor; but it is usually used in practice as a 3-terminal resistor with the screen grid constrained to positive potential and the suppressor grid to ground potential. The reason for operating an n-terminal element as a controlled 3-terminal element is mainly practical expediency – for to completely describe an n-terminal element would in theory require an infinite set of curves.

In the analysis of nonlinear networks containing n-terminal elements, *the standard procedure is to replace each n-terminal element by a network model containing only 2-terminal elements, independent sources, and controlled sources.* There are two fundamental reasons for doing this. First, most of the general systematic procedures available for formulating nonlinear network equations are restricted primarily to networks containing *only* 2-terminal elements, independent sources, and controlled sources [4-8]. Second, except in very special cases, the relationships characterizing an n-terminal element are generally given in terms of families of curves. To input these curves directly into a computer, a sophisticated interpolation subroutine is necessary. This is often a very time-consuming and inaccurate process for functions of more than two variables.

The procedure for modeling an n-terminal nonlinear element is a problem all by itself and is beyond the scope of this chapter. Although a general solution to the modeling problem is not yet in sight, a number of special techniques are currently available [1, 9, 10]. In particular, the class of linear n-terminal or n-port elements such as ideal transformers, gyrators, negative-impedance converters, mutual inductances, etc. can be easily modeled by means of controlled sources and linear elements. Hence, without any great loss of generality, we will assume in the following that the nonlinear networks to be analyzed have already been transformed to one containing only 2-terminal resistors, inductors, capacitors, independent sources, and controlled sources. In view of this assumption, we need only consider how to represent the "curve" characterizing a 2-terminal element.

4.1.3 Basic Forms of Representation

The first step in the analysis of a nonlinear network is to find an appropriate representation of an arbitrary curve Γ. It is important to classify curves of various complexities into several different types. The curve Γ shown in Fig. 4.3(a) has the property that if (x_1, y_1) and (x_2, y_2) are any two points on Γ such that $x_2 > x_1$, then $y_2 > y_1$. It is called a *strictly monotonically-increasing curve.* If the strict inequality signs are replaced by the "greater-than-or-equal" sign ($\geqslant$), the curve is said to be a *monotonically-increasing curve.* An example of a monotonically-increasing (not strictly) curve is shown in Fig. 4.3(b). The curve

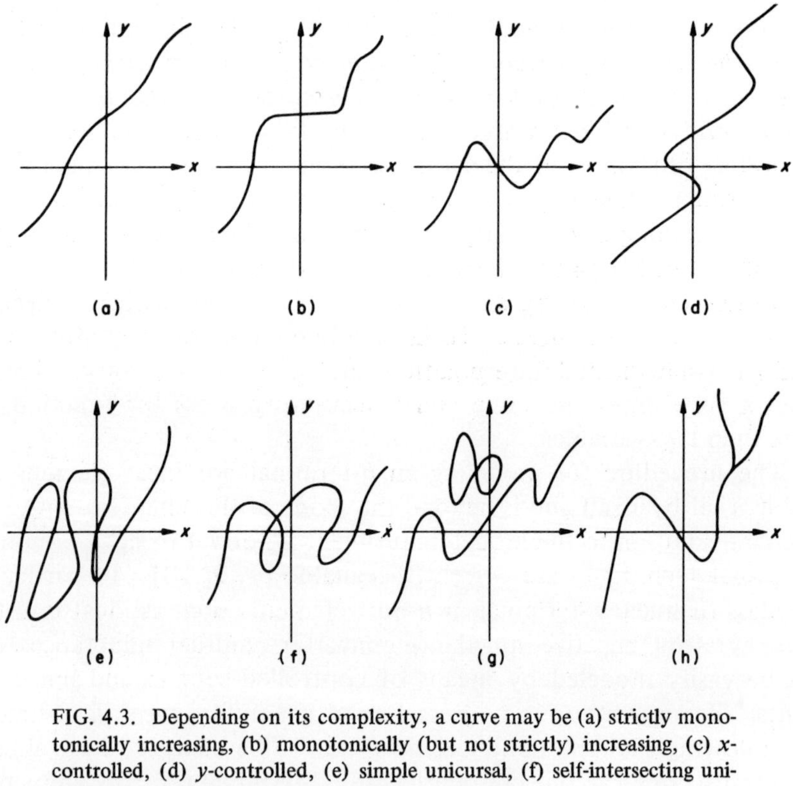

FIG. 4.3. Depending on its complexity, a curve may be (a) strictly monotonically increasing, (b) monotonically (but not strictly) increasing, (c) x-controlled, (d) y-controlled, (e) simple unicursal, (f) self-intersecting unicursal, (g) separately unicursal, or (h) non-unicursal.

shown in Fig. 4.3(c) is a function of x *only* and is called an *x-controlled curve*.† The curve shown in Fig. 4.3(d) is a function of y *only* and is called a *y-controlled curve*. The curves shown in Figs. 4.3(e) to (h) are neither x-controlled nor y-controlled; they are *multivalued curves*. The multivalued curves shown in Figs. 4.3(e) and (f) are said to be *unicursal* in the sense that each curve can be traced from one end to the other end in *a single stroke* [11]. The curve shown in Fig. 4.3(g) consists of two separate parts, each qualifies as a unicursal curve. An example of a multivalued curve which is not unicursal is shown in Fig. 4.3(h). Multivalued curves are of practical interest because they occur frequently in networks containing two or more resistors characterized by non-monotonic curves.

The significance of a strictly monotonically-increasing curve lies in the property that it is a one-to-one function and therefore possesses an *inverse* [12]. As such, it can be represented either by $y = f(x)$ or $x = g(y)$, where $g = f^{-1}$ or $f = g^{-1}$ (the superscript -1 denotes the inverse function). The curves shown in Figs. 4.3(b) and (c) are both x-controlled and can be represented by $y = f(x)$, but not $x = g(y)$, because the latter is not a function. For the same reason, the curve shown in Fig. 4.3(d) is y-controlled and can be represented by $x = g(y)$, but not $y = f(x)$.

The curves shown in Figs. 4.3(e) to (h) are multivalued and therefore cannot be represented by a single function. However, they *may be* represented by *two functions* of a third parameter ρ; namely, $x = x(\rho)$ and $y = y(\rho)$. In particular, a unicursal curve *can* always be represented in this manner by choosing the parameter ρ to be the arc-length [1, 12]. To see this, one may imagine passing a measuring tape along the contour of the curve and marking off the values at various points corresponding to the length from one end of the curve to the points in question.* An element is said to be unicursal if it can be characterized by a unicursal curve.

There are various reasons for choosing the parametric representation. The most immediate reason is due to the fact that the v-i curve of a

† Recall that a function f with domain A and range B (denoted by $f\colon A \to B$) is a set of ordered pairs (a, b) such that for each point $a \in A$, there corresponds one, and only one, point $b \in B$. We will often denote the *value* b of the function f at a by the usual symbol $b = f(a)$.

* In practice, it is sometimes convenient to pick an arbitrary point on Γ as the origin. The arc-length x on one side of the origin is then assumed to be positive while the other considered to be negative.

new resistor formed by connecting two or more non-monotonic resistors is in general a multivalued curve. Another reason is due to its more general nature – a pair of parametric equations $x = x(\rho)$ and $y = y(\rho)$ can be reduced to the conventional representation (in the case of a function) by simply setting the parameter ρ equal to one of the two independent variables. For example, if the curve is x-controlled, then upon setting $\rho = x$, the parametric equations reduce to $y = y(x)$. A less obvious but perhaps a more compelling reason for choosing the parametric representation – even in the case of functions – is its *analytical flexibilities*. This is because the parameter x can be chosen in an infinitely many distinct ways. In particular, any one-to-one transformation $\rho' = f(\rho)$ of a given parameter ρ leads to another parameter ρ'. In this sense, a given v-i curve admits of an infinitely many different but *equivalent* representation. This flexibility allows one to alter the representation if the computer iteration of a given problem fails to converge.

In terms of the parametric representation, a nonlinear resistor is completely specified by

$$\begin{aligned} i &= i(\rho) \\ v &= v(\rho) \end{aligned} \tag{4.3}$$

In the special cases, a *voltage-controlled resistor* may be represented by $i = i(v)$ (where $\rho = v$) and a *current-controlled resistor* may be represented by $v = v(i)$ (where $\rho = i$). A nonlinear inductor is completely specified by

$$\begin{aligned} \varphi &= \varphi(\rho) \\ i &= i(\rho) \\ v &= \frac{d\varphi}{dt} = \frac{d\varphi(\rho)}{d\rho}\frac{d\rho}{dt} \end{aligned} \tag{4.4}$$

In the special cases, a *current-controlled inductor* may be represented by $\varphi = \varphi(i)$ (where $\rho = i$) and a *flux-linkage-controlled inductor* may be represented by $i = i(\varphi)$ (where $\rho = \varphi$). A nonlinear capacitor is completely specified by

$$q = q(\rho)$$

$$v = v(\rho) \tag{4.5}$$

$$i = \frac{dq}{dt} = \frac{dq(\rho)}{d\rho}\,\frac{d\rho}{dt}$$

In the special cases, a *voltage-controlled capacitor* may be represented by $q = q(v)$ (where $\rho = v$) and a *charge-controlled capacitor* may be represented by $v = v(q)$ (where $\rho = q$).

Regardless of whether a curve is represented by the conventional or the parametric representation, it must be specified in an appropriate form before a computer analysis can be initiated. In the case where the equation is given *a*-priori, we only need to supply a subroutine specifying this equation. In most practical cases, however, only data points are available. It is therefore necessary to provide an *interpolation subroutine* for calculating the intermediate points whenever they are needed by the main program. If the interpolation routine consists of only first order terms, the result is equivalent to a *piecewise-linear representation* with the data points coinciding with the breakpoints. If the curve is highly nonlinear, a higher order interpolation subroutine may be used for improved accuracy. An alternate method is to provide a subroutine for fitting a polynomial through the data points. In this case, only the polynomial coefficients need be stored, thus resulting in a more efficient use of computer storage space.

In order to implement the parametric approach, several computer programs are presently available for obtaining both the parametric curves, and a polynomial approximation for each curve. To illustrate the ease for which a unicursal curve may be transformed into a pair of parametric curves, consider the multivalued v-i curve shown in Fig. 4.4(a). To show the flexibility of the parametric representation, a point on the v-i curve is arbitrarily chosen to be the origin. Hence, the arc-length x at this point is assigned the value zero. Since there are two possible directions, it is necessary to assume (arbitrarily) one of them to be the positive direction. Accordingly, the arc-length parameter at any point on the curve measured from the origin in the positive direction is taken to be positive. Conversely, the parameter measured in the opposite direction is negative. The program is written to accept the curve in various forms. A smooth curve may be entered by a listing of points on the curve. A piecewise-linear curve is entered by a listing of its breakpoints, plus a point on each of the end segments. In any case, the computer

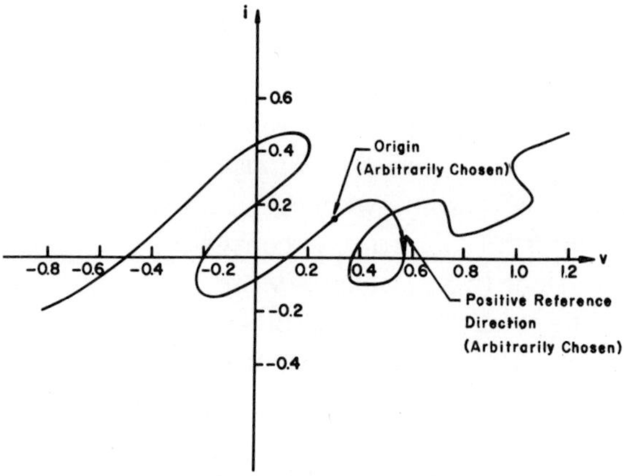

(a)

X	I	V
-2.6013	-0.1900	-0.8000
-2.5454	-0.1650	-0.7500
-2.4916	-0.1450	-0.7000
-2.4305	-0.1100	-0.6500
-2.3712	-0.0780	-0.6000
-2.3263	-0.0500	-0.5650
-2.2623	-0.0100	-0.5150
-2.2021	0.0300	-0.4700
-2.1489	0.0650	-0.4300
-2.0919	0.1000	-0.3850
-2.0209	0.1450	-0.3300
-1.9502	0.1950	-0.2800
-1.8795	0.2450	-0.2300
-1.8263	0.2800	-0.1900
-1.7732	0.3150	-0.1500
-1.7166	0.3550	-0.1100
-1.6611	0.3900	-0.0670
-1.6037	0.4230	-0.0200
-1.5648	0.4400	0.0150
-1.5063	0.4600	0.0700
-1.4553	0.4700	0.1200
-1.4253	0.4700	0.1500
-1.3889	0.4600	0.1850
-1.3639	0.4400	0.2000
-1.3423	0.4200	0.2080
-1.3113	0.3900	0.2000
-1.2778	0.3600	0.1850
-1.2495	0.3400	0.1650
-1.2245	0.3200	0.1500
-1.1820	0.2900	0.1200
-1.1320	0.2600	0.0800
-1.0960	0.2400	0.0500
-1.0599	0.2200	0.0200
-1.0349	0.2050	0.
-0.9779	0.1700	-0.0450
-0.9279	0.1400	-0.0850
-0.8818	0.1100	-0.1200
-0.8428	0.0800	-0.1450
-0.8037	0.0500	-0.1700
-0.7717	0.0250	-0.1900
-0.7314	-0.0100	-0.2100
-0.6902	-0.0500	-0.2200
-0.6696	-0.0700	-0.2250
-0.6391	-0.1000	-0.2200
-0.6185	-0.1200	-0.2150
-0.5973	-0.1350	-0.2000
-0.5750	-0.1450	-0.1800
-0.5543	-0.1500	-0.1600
-0.5227	-0.1400	-0.1300
-0.4911	-0.1300	-0.1000
-0.4530	-0.1150	-0.0650
-0.4195	-0.1000	-0.0350
-0.3748	-0.0800	0.0050
-0.3207	-0.0500	0.0500
-0.2707	-0.0200	0.0900
-0.2166	0.0100	0.1350
-0.1666	0.0400	0.1750
-0.1205	0.0700	0.2100
-0.0673	0.1050	0.2500
-0.0283	0.1300	0.2800
0.	0.1500	0.3000
0.0610	0.1850	0.3500
0.0902	0.2000	0.3750
0.1487	0.2200	0.4300
0.1787	0.2200	0.4600
0.2011	0.2100	0.4800
0.2371	0.1900	0.5100
0.2762	0.1600	0.5350
0.3209	0.1200	0.5550
0.3612	0.0800	0.5600
0.3916	0.0500	0.5650
0.4220	0.0200	0.5700
0.4537	-0.0100	0.5600
0.4841	-0.0400	0.5550
0.5091	-0.0600	0.5400
0.5374	-0.0800	0.5200
0.5597	-0.0900	0.5000
0.5803	-0.0950	0.4800
0.6107	-0.1000	0.4500
0.6607	-0.1000	0.4000
0.6831	-0.0900	0.3800
0.7011	-0.0800	0.3650
0.7328	-0.0500	0.3550
0.7632	-0.0200	0.3600
0.7948	0.0100	0.3700
0.8309	0.0400	0.3900
0.8532	0.0600	0.4000
0.8848	0.0900	0.4100
0.9209	0.1100	0.4400
0.9492	0.1300	0.4600
0.9852	0.1500	0.4900
1.0076	0.1600	0.5100
1.0411	0.1750	0.5400
1.0792	0.1900	0.5750
1.1061	0.2000	0.6000
1.1474	0.2100	0.6400
1.1778	0.2150	0.6700
1.2078	0.2150	0.7000
1.2328	0.2000	0.7200
1.2663	0.1700	0.7350
1.2967	0.1400	0.7400
1.3272	0.1100	0.7450
1.3522	0.0900	0.7600
1.3922	0.0900	0.8000
1.4226	0.0950	0.8300
1.4653	0.1100	0.8700
1.5191	0.1300	0.9200
1.5619	0.1450	0.9600
1.6066	0.1650	1.0000
1.6426	0.1850	1.0300
1.6746	0.2050	1.0550
1.6905	0.2200	1.0600
1.7111	0.2400	1.0550
1.7361	0.2600	1.0400
1.7681	0.2800	1.0150
1.7931	0.3000	1.0000
1.8200	0.3250	0.9900
1.8380	0.3400	0.9800
1.8635	0.3650	0.9850
1.8989	0.3900	1.0100
1.9436	0.4100	1.0500
1.9995	0.4350	1.1000
2.0534	0.4550	1.1500
2.1072	0.4750	1.2000

(b)

FIG. 4.4. The computer printed output of the unicursal curve in (a) consists of (b) a table listing the parameter value x associated with each point (v,i) on the curve, (c) a graph showing the i vs. x curve, and (d), a graph showing the v vs. x curve.

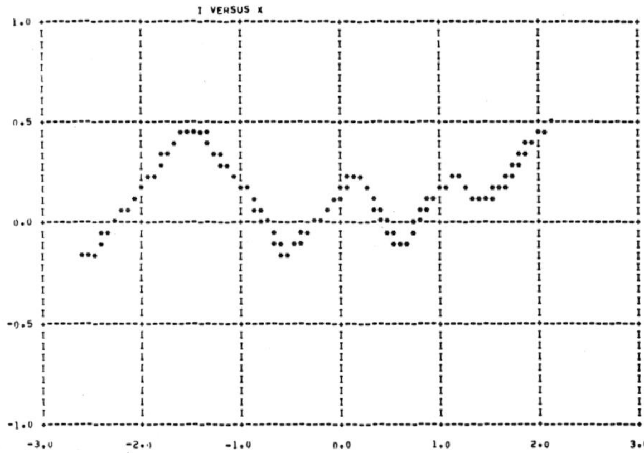


```
.                                                                      SMOTH004
A COMPUTER PROGRAM TO OBTAIN A PARAMETRIC REPRESENTATION OF A          SMOTH005
MULTIVALUED UNICURSAL SMOOTH CURVE.                                    SMOTH006
                                                                       SMOTH007
                                                                       SMOTH008
                                                                       SMOTH009
SYMBOLS                                                                SMOTH010
-----------------------------------------------------------------------
N = NUMBER OF POINTS CHOSEN FROM THE CURVE TO REPRESENT THE CURVE      SMOTH012
AS N GOES UP, THE REPRESENTATION WOULD BE MOHE ACCUHATE.               SMOTH013
CHANGE OF N HAS TO BE FOLLOWED BY THE CORRESPONDING CHANGE OF          SMOTH014
DIMENSION .   FOR N POINTS, THE DIMENSION STATEMENT IS AS FOLLOWS      SMOTH015
DIMENSION I(N), V(N), X(N), DS(N-1)                                    SMOTH016
                                                                       SMOTH017
THE STARTING POINT IN THE PARAMETERIZATION PROCESS HAS TO BE           SMOTH018
ONE OF THE POINTS  IN THE INPUT INFORMATION.                           SMOTH019
                                                                       SMOTH020
                                                                       SMOTH021
INFORMATION REQUIRED TO OBTAIN A PARAMETRIZED REPRESENTATION           SMOTH022
SI, SV THE IV COORDINATES OF THE STARTING POINT AND                    SMOTH023
NS THE INDEX NUMBEH OF THE STARTING POINT.                             SMOTH024
-----------------------------------------------------------------------
```

```
HE FOLLOWING PAGES ARE THE POLYNOMIAL REGRESSION OF THE CURVE FOR I(X)


POLYNOMIAL REGRESSION OF DEGREE 10

INTERCEPT (A VALUE)...     0.07034

REGRESSTON COEFFICIENTS
    0.10575X  +   -0.46243X**2  +   -0.08746X**3  +    0.80001X**4  +   -0.08007X**5  +
  -0.34373X**6  +    0.04760X**7  +    0.07558X**8  +   -0.00563X**9  +   -0.00540X**10

STD. ERROR OF REG. COEF.
    0.04528X  +    0.07894X**2  +    0.11184X**3  +    0.09588X**4  +    0.08697X**5  +
   0.04759X**6  +    0.02605X**7  +    0.01101X**8  +    0.00264X**9  +    0.00100X**10
```

(c)

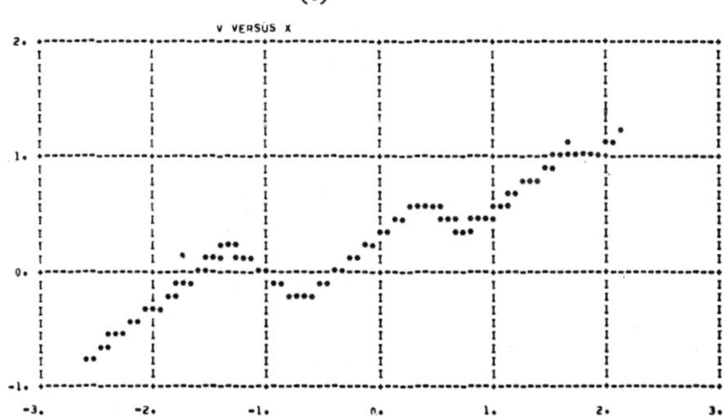


```
HE FOLLOWING PAGES ARE THE POLYNOMIAL REGRESSION OF THE CURVE FOR V(X)


POLYNOMIAL REGRESSION OF DEGREE 10

INTERCEPT (A VALUE)...     0.31961

REGRESSION COEFFICIENTS
    0.88666X  +   -0.68394X**2  +   -1.23732X**3  +    0.81738X**4  +    0.79484X**5  +
  -0.25792X**6  +   -0.20501X**7  +    0.01869X**8  +    0.01879X**9  +    0.00135X**10

STD. ERROR OF REG. COEF.
    0.03238X  +    0.05645X**2  +    0.07997X**3  +    0.06856X**4  +    0.06219X**5  +
   0.03403X**6  +    0.01863X**7  +    0.00787X**8  +    0.00189X**9  +    0.00071X**10
```

(d)

FIG. 4.4

would automatically interpolate between the points and calculate the parameter x associated with each point. The printed output takes the form of a listing of points with its parameter x (Fig. 4.4(b)), and a plot of the curves v vs. x, and i vs. x, as shown in Figs. 4.4(c) and (d), respectively. An option is also provided for the computer to approximate each of these curves by a polynomial of some prescribed degree. For this example, a tenth degree polynomial is specified in each case, and the printout of the regression coefficients is shown below each figure.

A few remarks are perhaps appropriate here concerning the practicality of such a seemingly complex curve as the one chosen for the example above. Certainly, no physical device would ever exhibit such a v-i curve. The only occasion where we may encounter such a multivalued curve is when two, or more, non-monotonic resistors are combined to form a composite element. But why do we want to combine nonlinear resistors? Why can't we analyze the original network *per se*? There are many practical reasons whereby it is desirable to combine two or more resistors into a composite element. The most important reason is that it reduces the number of network variables that must be solved. As will be obvious in Sec. 4.3.1, the weakest link in the computer analysis of a nonlinear problem is *not* the solution of nonlinear differential equations, but the solution of *nonlinear functional equations*. In general, the degree of difficulty in solving this type of equations increases rapidly as the number of variables increases. The difficulty is even more serious when the equations – such as those describing the class of resistive switching networks in Sec. 4.5.3 – possess more than one solution.

The simplest technique for overcoming the above difficulty is to break up the network into an interconnection of as many 2-terminal black boxes as possible. Each black box is represented by a composite v-i curve – which is generally multivalued. This partitioning procedure greatly reduces the number of resistors – and hence the number of network variables to be solved – thereby making an efficient computer analysis possible. Once the solutions are found, the corresponding solutions inside each of the equivalent black boxes can be easily obtained [1]. In short, *while multivalued v-i curves do not occur in nature, they are far from being academic because they do arise in the process of analyzing practical nonlinear networks.*

4.2 FORMULATION OF STATE EQUATIONS

The first step for analyzing a nonlinear network by a digital computer is to formulate its state equations in the normal form

$$\dot{\mathbf{x}} = \mathbf{f}(t, \mathbf{x}) \tag{4.6}$$

This particular form of equations of motion is chosen because it is the only form whereby a systematic theory is available for its solution. In fact, virtually all numerical techniques for solving nonlinear differential equations are based on the normal form [13-15].

One naive approach to computer network analysis consists of deriving (4.6) for each network manually and then let the computer obtain the numerical solution. This approach is fine for simple networks. However, if a network contains more than four or five elements, the formulation of the normal form equations is at best a tedious task. The main difficulty is due to the fact that the expression $\mathbf{f}(t, \mathbf{x})$ – even if it exists – is generally not obtainable in analytic form. The only general approach is to provide the computer with an algorithm for calculating the numerical value of $\mathbf{f}(t, \mathbf{x})$ – whenever it is needed in the subsequent numerical iterations. For example, corresponding to an initial state $\mathbf{x}_0 = \mathbf{x}(t_0)$, the algorithm calculates the value of $\mathbf{f}(t_0, \mathbf{x}_0)$. This value can then be substituted into an appropriate integration formula (to be discussed in Sec. 4.3) to calculate the value of $\mathbf{x}(t)$ at $t_1 = t_0 + \Delta t$. This procedure can obviously be repeated to find $\mathbf{x}(t)$ at $t_2, t_3, \ldots$ etc. The basic idea is to calculate the solution $\mathbf{x}(t)$ from some recursive formula

$$\mathbf{x}(t_{j+1}) = T[\mathbf{f}(t_j, \mathbf{x}(t_j))] \tag{4.7}$$

where $\mathbf{T}$ is an operator. Observe that at each iteration j, the value of $\mathbf{f}(t, \mathbf{x})$ corresponding to $t = t_j$ and $\mathbf{x} = \mathbf{x}(t_j)$ is required in order to calculate $\mathbf{x}(t_{j+1})$. Hence, the crucial problem that we must solve is to formulate a *general algorithm* for calculating $\mathbf{f}(t, \mathbf{x})$.

4.2.1 Choice of State Variables

In order to point out the difficulties that one may encounter in trying to formulate an algorithm for calculating $\mathbf{f}(t, \mathbf{x})$, let us consider first the simple network shown in Fig. 4.5(a). The equations of motion are given by

$$dq_2/dt = i_2 = -i_1 \tag{4.8}$$

$$v_2 = v_1 \tag{4.9}$$

To recast (4.8) and (4.9) into the normal form, we must make use of the v-i curve of the resistor and the v-q curve of the capacitor. It is instructive to investigate the effect of the complexity of these curves upon the normal form equations.

Case I: Consider the case where both resistor and capacitor are characterized by strictly-monotonically-increasing curves; e.g., let $i_1 = v_1^3$ and $q_2 = e^{v_2} - 1$. In this case, (4.8) becomes

$$dq_2/dt = -v_1^3 = -v_2^3 \tag{4.10}$$

Since the $q_2 - v_2$ curve is one-to-one, its inverse exists; namely, $v_2 = \ln(q_2 + 1)$. Hence (4.10) becomes

$$dq_2/dt = -[\ln(q_2 + 1)]^3 \tag{4.11}$$

Equation (4.11) is now in the normal form with q_2 as the state variable. Instead of q_2, we may choose v_2 as the state variable by rewriting (4.10) in the form

$$(dq_2/dv_2)(dv_2/dt) = -v_2^3$$

or

$$dv_2/dt = -v_2^3/e^{v_2} \tag{4.12}$$

Therefore, in this case, either the capacitor charge or voltage may be chosen to be the state variable. The same conclusion is easily seen to be true if the resistor is voltage-controlled.

Case 2: Consider next the case where the resistor is voltage-controlled; e.g., let $i_1 = 1 - v_1^2 + v_1^3$ and $v_2 = q_2^3 - 3q_2$. If we choose the charge again as the state variable, we obtain the desired normal form

$$dq_2/dt = -1 + (q_2^3 - 3q_2)^2 - (q_2^3 - 3q_2)^3 \tag{4.13}$$

On the other hand, if we were to choose the voltage v_2 as the state variable, we would obtain

$$dv_2/dt = (-1 + v_2^2 - v_2^3)/[dq_2(v_2)/dv_2] \tag{4.14}$$

To calculate $dq_2(v_2)/dv_2$, we must first solve for q_2 in terms of v_2, i.e., we must find the inverse function of the v-q curve. But this is impossible because the v-q curve is not a one-to-one function. Hence, the normal form in this case does not exist if the voltage is chosen as the state variable. This simple example demonstrates that *it may be necessary to choose the charge as the state variable,* even though the voltage is a more convenient variable to use in practice.

Case 3: Suppose now both the resistor and the capacitor are voltage-controlled; e.g., let $i_1 = 1 - v_1^2 + v_1^3$ and $q_2 = v_2^3 - 3v_2$. If we choose q_2 as the state variable, we obtain

$$dq_2/dt = -1 + v_2^2 - v_2^3 \tag{4.15}$$

In order to recast (4.15) into the normal form, it is necessary to express v_2 as a function of q_2. But this is impossible for the same reason as above. Therefore, the normal form does not exist with q_2 as the state variable. However, if we choose the voltage instead of charge, we obtain the normal form

$$dv_2/dt = (-1 + v_2^2 - v_2^3)/(3v_2^2 - 3) \tag{4.16}$$

Therefore, in this case, it is necessary to choose the voltage as the state variable.

If we compare the normal form equation (4.13) of case 2 with that of (4.16) of case 3, we observe an important difference. In (4.13), the normal form exists for all finite values of the state variable q_2. However, in (4.16), the normal form does not exist at $v_2 = 1$ and $v_2 = -1$ because the right hand side of (4.16) is discontinuous at $v_2 = \pm 1$. In order to differentiate between these two cases, the normal form given by (4.13) is said to exist in the *entire state space* whereas the normal form given by (4.16) is said to exist at all points in the state space for which the denominator is not zero.

Case 4: Suppose the resistor is current-controlled and the capacitor is strictly-monotonically-increasing; e.g., let $v_1 = i_1^3 - 3i_1$ and $q_2 = Cv_2$. If we choose q_2 to be the state variable, we obtain

$$dq_2/dt = -i_1 \tag{4.17}$$

In order to recast (4.17) into the normal form, we must express i_1 as a function of $v_1 = v_2 = q_2/C$. But this is impossible because i_1 is not a single-valued function of v_1. Our experience suggests that we choose the voltage v_2 instead of q_2. In this case,

$$dv_2/dt = -i_1/[dq_2(v_2)/dv_2] = (-i_1)/C \tag{4.18}$$

Again, we must express i_1 as a function of v_2. Hence in this case, the normal form does not exist with either state variable!

What significance can we attach to the case where the normal form fails to exist? Usually, it signifies trouble in the sense that unless we postulate the existence of discontinuous solutions, no solution can be found. To clarify this statement, let us rewrite (4.18) into the form

$$dv_1/dt = (-i_1)/C \tag{4.19}$$

and examine a typical current-controlled v_1-i_1 curve (Fig. 4.5(b)). Since $dv_1/dt > 0$ whenever $i_1 < 0$ and $dv_1/dt < 0$ whenever $i_1 > 0$, the *dynamic route* (the locus of all permissible solutions) [1] must be as indicated in Fig. 4.5(b). The vertical routes indicate the presence of instantaneous jumps in the current waveform $i_1(t)$ as shown in Fig. 4.5(e). Since this jump phenomenon cannot be justified mathematically, its existence can only be postulated. It must be emphasized however that this postulate is only a gimmick introduced to obtain an approximate solution for a non-physical network.

If we were to solve the above network by a computer, we would immediately get into trouble because starting at any initial point between Q_1' and Q_2 or Q_2' and Q_1, there correspond three values of i_1 for each value of v_1. The computer cannot decide which of the three solutions actually apply at a given time and hence will either stop or yield a pos-

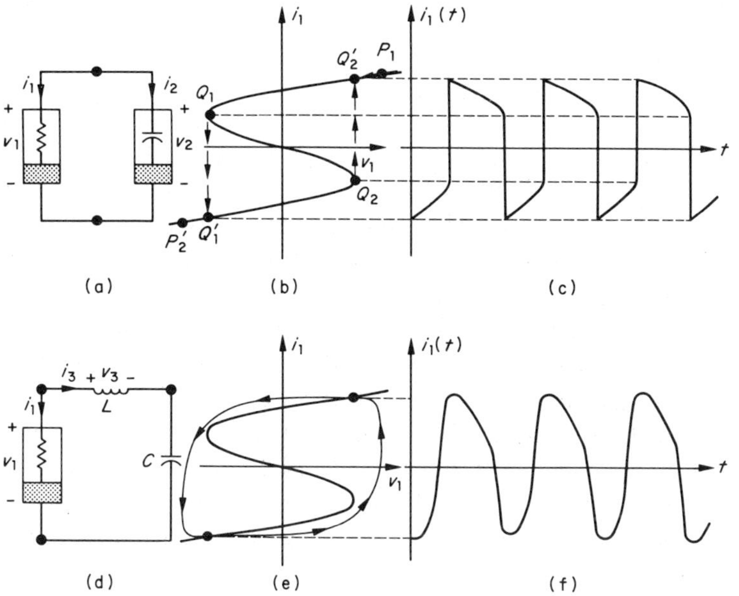

FIG. 4.5. The non-existence of a normal form equation for the network in (a) leads to a jump discontinuity in its dynamic route (b), and its current wave form (c). By augmenting the network with an inductor (d), the normal form equations exist with a continuous dynamic route (e) and a continuous current wave form (f).

sibly erroneous solution. It should now be clear that *one fundamental reason for choosing the normal form is to preclude the existence of any jump phenomenon.*

From a strictly mathematical standpoint, the multivalued character of (4.19) implies that the network does not possess a solution. The introduction of a jump phenomenon is one way to overcome the difficulty. A better way is to realize that since every physical network must possess a continuous solution, the non-existence of the normal form must be due to an over-idealization of the network models. Hence, by "refining" these models, we should be able to obtain the normal form equation of any network. One obvious procedure consists of inserting appropriate parasitic inductors and capacitors at strategic positions [6]. For example, if we add a linear inductor with a small linear inductance (simulating the actual line inductance) to the network in Fig. 4.5(a), we obtain the "modified" network shown in Fig. 4.5(d). The normal form equations can now be easily obtained in terms of v_2 and i_3 as the state variables; namely,

$$dv_2/dt = i_3/C$$
$$di_3/dt = [(-i_3^3 + 3i_3) - v_2]/L \tag{4.20}$$

In this case, the dynamic route and current waveform for $i_1(t)$ can be easily obtained as shown in Figs. 4.5(e) and (f). Observe that the jump phenomenon is no longer present.

For most practical networks, the existence of normal form equations is usually sufficient for a successful computer analysis. For more general networks, however, the question of existence and uniqueness of solutions must be considered. To illustrate the possible difficulties that may arise, suppose we let the resistor of Fig. 4.5(a) be a 1 Ω linear resistance and let the capacitor be characterized by a square-law capacitance curve $C(v_2) = 2v_2^2$. This corresponds to case 3 above and hence has a normal form

$$dv_2/dt = -1/2v_2^2 \tag{4.21}$$

Equation (4.21) can be easily integrated to obtain the solution

$$v_2(t) = \sqrt{v_0^2 - t} \tag{4.22}$$

where v_0 is the initial capacitor voltage at $t = 0$. Observe that if we choose the initial voltage to be $v_0 = 0$, then $v_2(t) = \sqrt{-t}$ is undefined for $t > 0$. This is equivalent to saying that the solution of (4.21) does not exist with this initial condition! An examination of (4.21) might suggest the source of trouble namely; the normal form does not exist at $v_2(0) = 0$. However, even if we pick a nonzero initial voltage v_0, the solution would exist only for a short time interval $0 \leqslant t \leqslant v_0^2$. In either case, we have a highly non-physical situation.

Another non-physical situation occurs whenever the system possesses more than one solution. For example, suppose the capacitor of the network in Fig. 4.5(a) is linear with a capacitance equal to $C = 1$ farad and let the nonlinear resistor be characterized by $v_1 = i_1^3$. In this case, the normal form exists for all values of v_2; namely,

$$dv_2/dt = -v_2^{1/3} \tag{4.23}$$

However, corresponding to the initial voltage $v_2(0) = 0$, (4.23) admits of an infinite number of distinct solutions:

$$\begin{aligned} v_2(t) &= 0 \qquad 0 \leqslant t \leqslant k \\ &= -\left[\frac{2}{3}(t-k)\right]^{3/2} \qquad \text{for } t > k \end{aligned} \tag{4.24}$$

where k is any real number.

An examination of (4.23) shows that the slope of $v_2^{1/3}$ at $v_2 = 0$ is infinite. This leads us to suspect that if we exclude this type of infinite slope elements, we might obtain a unique solution. In a sense, this conjecture is true (see Sec. 4.3.3). However, even if we know the solution is unique, we may still face another non-physical situation. Consider the same example as above, but this time let the nonlinear resistor be characterized by a voltage-controlled curve $i_1 = -v_1^2$. The normal form equation becomes

$$dv_2/dt = v_2^2 \tag{4.25}$$

Corresponding to the initial condition $v_2(0) = 1$, the solution to (4.25) is given by

$$v_2(t) = \frac{1}{(1-t)}$$

Observe that this solution approaches infinity as t approaches 1. A system with this property is said to possess a *finite-escape-time* and is clearly non-physical.

The above examples are contrived to illustrate the various non-physical situations that may occur. It is comfortable to know that not many networks fall under this category. In fact, we will present several theorems in Sec. 4.3.3 for excluding this type of non-physical networks.

4.2.2 Two General Normal Form Algorithms

Let N be a connected network containing 2-terminal resistors, inductors, capacitors, independent sources, and controlled sources.† For

† We will assume that N contains neither current-source-only-cut-sets nor voltage-source-only-loops. These pathological cases are of no practical interest.

complete generality, we will assume that each 2-terminal element is associated with a voltage source and a current source (either independent or controlled sources), thereby forming a composite network branch as shown in Fig. 4.6. If a network is not already in this form, we can always apply the *v-shift theorem* or the *i-shift theorem* [1] to transform the network into the proper equivalent form. The normal form algorithm consists of two parts.

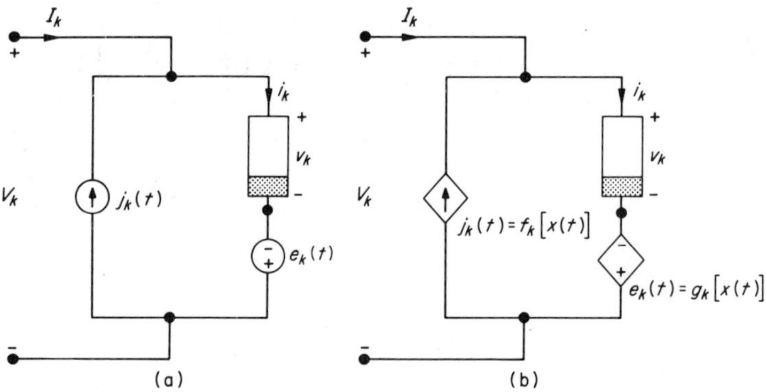

FIG. 4.6. A composite nonlinear 2-terminal element; each element is accompanied by a series voltage source and a parallel current source. Independent sources are denoted by circles (a), whereas controlled sources are denoted by diamonds.

I. Topological Algorithm. This algorithm is concerned with the systematic formulation of Kirchhoff Voltage Law (KVL) and Kirchhoff Current Law (KCL) and is therefore independent of the network elements. Therefore, it suffices to consider only the linear graph G associated with N. To be specific, let n be the total number of nodes and b the total number of branches of G. Among the various methods for writing topological network equations, Bryant's algorithm is the most appropriate for our purpose [16]. This algorithm consists of picking a special tree – called the *normal tree* – consisting of as many capacitor branches as possible and as few inductor branches as possible.† Let the branches of the normal tree be denoted by the set $\mathcal{T}$ and let the remaining branches (the links) of the associated co-tree be denoted the set $\mathcal{L}$.

† Clearly, the choice of a normal tree is not unique. However, it can be shown that for the purpose of writing normal form equations, one choice is as good as another.

We will further classify the links in $\mathcal{L}$ into three subgroups α, β, and γ, where α consists of all link capacitors, β consists of all link resistors, and γ consists of all link inductors. Similarly, the tree-branches in $\mathcal{T}$ are classified into three subgroups, δ, ϵ, and ξ, where δ consists of all tree-branch capacitors, ϵ consists of all tree-branch resistors, and ξ consists of all tree-branch inductors. Let us next label the branches in the order α, β, γ, δ, ϵ, and ξ. Hence, if G has $n_{\mathcal{L}}$ links and $n_{\mathcal{T}}$ tree-branches, then $n_{\mathcal{L}} = n_\alpha + n_\beta + n_\gamma$ and $n_{\mathcal{T}} = n_\delta + n_\epsilon + n_\xi$, where n_x denotes the number of branches in the set "x".

To illustrate the above definitions, consider the network shown in Fig. 4.7(a) and its graph G as shown in Fig. 4.7(b). The solid branches

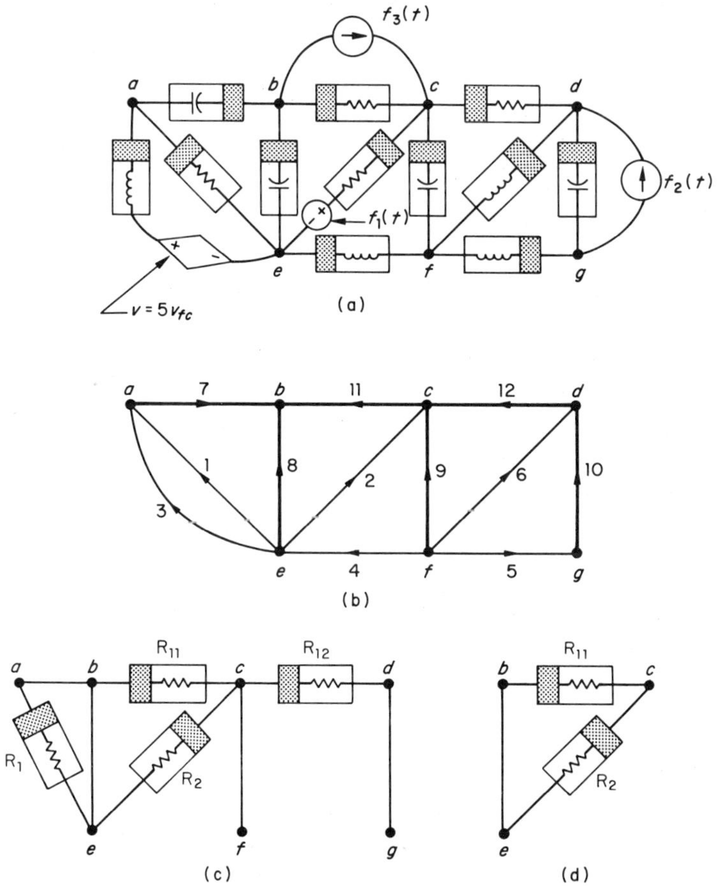

FIG. 4.7. A typical nonlinear network N is represented by its graph G. The heavy branches represent a normal tree. The resistive subnetworks N_R and $N_{\beta\epsilon}$ are shown respectively in (c) and (d).

qualify as a normal tree. Following our labeling procedure, we first label the link capacitors belonging to α. But there are no link capacitors, hence α in this case is the empty set. We proceed to label the link resistors in β; namely, R_1 and R_2. We continue to label the link inductors L_3, L_4, L_5, and L_6 in γ. Next, we label the capacitor-tree branches C_7, C_8, C_9, and C_{10} in the set δ and the resistor branches R_{11} and R_{12} in the set ϵ. The last set to be labeled is ξ. But since there are no tree-branch inductors, ξ is the empty set.

The next step in the topological algorithm is to obtain the *fundamental circuit matrix* **B** [17]. In view of the labeling convention (with the links first), this matrix is easily seen to be of the form

$$\mathbf{B} = [\mathbf{U}_{\mathcal{L}} \vdots \mathbf{B}_{\mathcal{T}}] \tag{4.26}$$

where $\mathbf{U}_{\mathcal{L}}$ is an $n_{\mathcal{L}} \times n_{\mathcal{L}}$ unit matrix, $\mathbf{B}_{\mathcal{T}}$ is an $n_{\mathcal{L}} \times n_{\mathcal{T}}$ matrix, and **B** is an $n_{\mathcal{L}} \times b$ matrix whose elements are given by

$b_{ij} = 1$ whenever branch j belongs to the fundamental circuit formed by link i and has the same orientation as link i.

$= 0$ whenever branch j does not belong to the fundamental circuit i.

$= -1$ whenever branch j belongs to the fundamental circuit formed by link i and has opposite orientation as link i.

From the computer programming point of view, the observation that $\mathbf{U}_{\mathcal{L}}$ is a unit matrix makes it unnecessary to store the entire **B** matrix. Instead, only $\mathbf{B}_{\mathcal{T}}$ needs to be found and a simple subroutine can be written for this purpose [18]. As a direct consequence of Bryant's labeling convention, the submatrix $\mathbf{B}_{\mathcal{T}}$ – when partitioned in accordance with the sets α, β, and γ – is easily seen to be always of the following form:

$$\mathbf{B}_{\mathcal{T}} = \begin{bmatrix} \mathbf{F}_{\alpha\delta} & \mathbf{0} & \mathbf{0} \\ \mathbf{F}_{\beta\delta} & \mathbf{F}_{\beta\epsilon} & \mathbf{0} \\ \mathbf{F}_{\gamma\delta} & \mathbf{F}_{\gamma\epsilon} & \mathbf{F}_{\gamma\xi} \end{bmatrix} \tag{4.27}$$

Each submatrix $\mathbf{F}_{xy}$ denotes an $n_x \times n_y$ matrix representing the elements in "y" which also belong to the fundamental circuit formed by the links in "x". For example, the submatrix $F_{\beta\delta}$ denotes an $n_\beta \times n_\delta$ matrix representing the tree-branch capacitors (δ) which belong to the fundamental circuit formed by a link-resistor (β). With this interpretation, it is easy to see that the submatrices $\mathbf{F}_{\alpha\epsilon}$, $\mathbf{F}_{\alpha\xi}$, and $\mathbf{F}_{\beta\xi}$ are zero matrices. This accounts for the three zero sub-matrices in $\mathbf{B}_{\mathcal{T}}$.† Clearly, we can take advantage of this property to reduce storage requirements.

The next topological matrix that we need is the *fundamental cut-set matrix* $\mathbf{C}$, which is always of the form (assuming the same branch labeling as before):

$$\mathbf{C} = [\mathbf{C}_{\mathcal{L}} \vdots \mathbf{U}_{\mathcal{T}}] \tag{4.28}$$

where $\mathbf{C}_{\mathcal{L}}$ is an $(n-1) \times n_{\mathcal{L}}$ matrix, $\mathbf{U}_{\mathcal{T}}$ is an $n_{\mathcal{T}} \times n_{\mathcal{T}}$ unit matrix, and $\mathbf{C}$ is an $(n-1) \times b$ matrix whose elements are given by

$C_{ij} = 1$ whenever branch j belongs to the fundamental cut-set formed by tree-branch i and has the same orientation as tree-branch i.

$= 0$ whenever branch j does not belong to the fundamental cut-set i.

$= -1$ whenever branch j belongs to the fundamental cut-set formed by tree-branch i and has the opposite orientation as tree-branch i.

An interesting relationship exists between the submatrices $\mathbf{C}_{\mathcal{L}}$ and $\mathbf{B}_{\mathcal{T}}$ [17]; namely,

$$\mathbf{C}_{\mathcal{L}} = -\mathbf{B}'_{\mathcal{T}} \tag{4.29}$$

where the prime denotes "transpose" of a matrix. In view of (4.29), the fundamental cut-set matrix $\mathbf{C}$ can be trivially obtained from (4.27) and (4.29):

$$\mathbf{C}_{\mathcal{L}} = -\begin{bmatrix} \mathbf{F}'_{\alpha\delta} & \mathbf{F}'_{\beta\delta} & \mathbf{F}'_{\gamma\delta} \\ \mathbf{0} & \mathbf{F}'_{\beta\epsilon} & \mathbf{F}'_{\gamma\xi} \\ \mathbf{0} & \mathbf{0} & \mathbf{F}'_{\gamma\xi} \end{bmatrix} \tag{4.30}$$

As an example, the fundamental circuit matrix $\mathbf{B}$ of the network graph shown in Fig. 4.7(b) is given by

† The elements of the submatrix $\mathbf{F}_{\alpha\epsilon}$ are zero because they correspond to tree-branch resistors (ϵ) which belong to a fundamental circuit formed by a link capacitor α. But the way a normal tree is constructed precludes such a possibility. The other submatrices $\mathbf{F}_{\alpha\xi}$ and $\mathbf{F}_{\beta\xi}$ can be interpreted in a similar manner.

$$
\mathbf{B} = \left[\begin{array}{cccccc:cccc:cc}
1 & 0 & 0 & 0 & 0 & 0 & 1 & -1 & 0 & 0 & 0 & 0 \\
0 & 1 & 0 & 0 & 0 & 0 & 0 & -1 & 0 & 0 & 1 & 0 \\
\hdashline
0 & 0 & 1 & 0 & 0 & 0 & 1 & -1 & 0 & 0 & 0 & 0 \\
0 & 0 & 0 & 1 & 0 & 0 & 0 & 1 & -1 & 0 & -1 & 0 \\
0 & 0 & 0 & 0 & 1 & 0 & 0 & 0 & -1 & 1 & 0 & 1 \\
0 & 0 & 0 & 0 & 0 & 1 & 0 & 0 & -1 & 0 & 0 & 1
\end{array}\right]
$$

The six partitioned submatrices of $\mathbf{B}_{\mathcal{T}}$ are easily identified to be:

$$
\mathbf{F}_{\alpha\delta} = \mathbf{0}, \qquad \mathbf{F}_{\gamma\xi} = \mathbf{0}
$$

$$
\mathbf{F}_{\beta\delta} = \begin{bmatrix} 1 & -1 & 0 & 0 \\ 0 & -1 & 0 & 0 \end{bmatrix}, \qquad \mathbf{F}_{\beta\epsilon} = \begin{bmatrix} 0 & 0 \\ 1 & 0 \end{bmatrix} \tag{4.31}
$$

$$
\mathbf{F}_{\gamma\delta} = \begin{bmatrix} 1 & -1 & 0 & 0 \\ 0 & 1 & -1 & 0 \\ 0 & 0 & -1 & 1 \\ 0 & 0 & -1 & 0 \end{bmatrix}, \qquad \mathbf{F}_{\gamma\epsilon} = \begin{bmatrix} 0 & 0 \\ -1 & 0 \\ 0 & 1 \\ 0 & 1 \end{bmatrix}
$$

If we define the vectors $\mathbf{V} = (V_1, V_2, \ldots, V_b)'$ and $\mathbf{I} = (I_1, I_2, \ldots, I_b)'$, then the system of KVL and KCL equations can be written compactly as:

$$
\begin{aligned} \mathbf{BV} &= \mathbf{0} \\ \mathbf{CI} &= \mathbf{0} \end{aligned} \tag{4.32}
$$

Observe that the variables $\mathbf{V}$ and $\mathbf{I}$ pertain to the composite branch. If we define $\mathbf{v} = (v_1, v_2, \ldots, v_b)'$, $\mathbf{i} = (i_1, i_2, \ldots, i_b)'$, $\mathbf{e}(t) = (e_1(t), e_2(t), \ldots, e_b(t))'$, and $j(t) = (j_1(t), j_2(t), \ldots, j_b(t))'$, then in view of Fig. 4.6, (4.32) can be rewritten more explicitly as:

$$
\begin{aligned} \mathbf{Bv} &= \mathbf{Be}(t) \\ \mathbf{Ci} &= \mathbf{Cj}(t) \end{aligned} \tag{4.33}
$$

For those readers not familiar with the above development, it is instructive to carry out the matrix multiplication in (4.33) and verify that they correspond to the system of independent equations obtained from KVL and KCL respectively.

Equation (4.33) constitutes b linear equations in $2b$ unknowns. This means that we need an additional b independent relationships to complete the equations of motion. These relationships are given precisely by the b characteristic curves of the nonlinear elements. For the sake of generality, we will represent each nonlinear element by its parametric equations. This establishes a one-to-one correspondence between each nonlinear element b_j and the associated parameter x_j.† Hence, if we define the vectors

$$\mathbf{x}_\mu \equiv (x_{\mu_1}, x_{\mu_2}, \ldots, x_{\mu_\mu})'$$

$$\mathbf{I}_\mu(\mathbf{x}_\mu) \equiv (I_{\mu_1}(x_{\mu_1}), I_{\mu_2}(x_{\mu_2}), \ldots, I_{\mu_\mu}(x_{\mu_\mu}))'$$

$$\mathbf{V}_\mu(\mathbf{x}_\mu) \equiv (V_{\mu_1}(x_{\mu_1}), V_{\mu_2}(x_{\mu_2}), \ldots, V_{\mu_\mu}(x_{\mu_\mu}))'$$

where μ is a running index which may take on any of the letters α, β, γ, δ, ϵ, and ξ whenever appropriate. For example, when $\mu = \alpha$, $\mathbf{x}_\alpha$ denotes the vector with elements $(x_{\alpha_1}, x_{\alpha_2}, \ldots, x_{\alpha_\alpha})$ of the link resistors, and $\mathbf{I}_\alpha(\mathbf{x}_\alpha)$ denotes the vector with elements $I_{\alpha_1}(x_{\alpha_1}), I_{\alpha_2}(x_{\alpha_2}), \ldots, I_{\alpha_\alpha}(x_{\alpha_\alpha})$ representing the parametric equations for each resistor branch current i_{α_j} as a function of the parameter x_{α_j}. Similarly, $\mathbf{V}_\alpha(\mathbf{x}_\alpha)$ denotes a vector with elements $V_{\alpha_1}(x_{\alpha_1}), V_{\alpha_2}(x_{\alpha_2}), \ldots, V_{\alpha_\alpha}(x_{\alpha_\alpha})$ representing the parametric equation for each resistor branch voltage v_{α_j} as a function of the parameter x_{α_j}. Substituting (4.27) and (4.30) for $\mathbf{B}_\mathcal{L}$ and $\mathbf{C}_\mathcal{L}$ in (4.33) and rearranging the terms, we obtain the complete system of independent equations of motion:

$$\begin{bmatrix} \mathbf{V}_\alpha(\mathbf{x}_\alpha) \\ \mathbf{V}_\beta(\mathbf{x}_\beta) \\ \dot{\varphi}_\gamma(\mathbf{x}_\gamma) \end{bmatrix} = -\begin{bmatrix} \mathbf{F}_{\alpha\delta} & 0 & 0 \\ \mathbf{F}_{\beta\delta} & \mathbf{F}_{\beta\epsilon} & 0 \\ \mathbf{F}_{\gamma\delta} & \mathbf{F}_{\gamma\epsilon} & \mathbf{F}_{\gamma\xi} \end{bmatrix} \begin{bmatrix} \mathbf{V}_\delta(\mathbf{x}_\delta) \\ \mathbf{V}_\epsilon(\mathbf{x}_\epsilon) \\ \dot{\varphi}_\xi(\mathbf{x}_\xi) \end{bmatrix} + \begin{bmatrix} \mathbf{E}_C(t) \\ \mathbf{E}_R(t) \\ \mathbf{E}_L(t) \end{bmatrix} \tag{4.34}$$

and

$$\begin{bmatrix} \dot{\mathbf{q}}_\delta(\mathbf{x}_\delta) \\ \mathbf{I}_\epsilon(\mathbf{x}_\epsilon) \\ \mathbf{I}_\xi(\mathbf{x}_\xi) \end{bmatrix} = \begin{bmatrix} \mathbf{F}'_{\alpha\delta} & \mathbf{F}'_{\beta\delta} & \mathbf{F}'_{\gamma\delta} \\ 0 & \mathbf{F}'_{\beta\epsilon} & \mathbf{F}'_{\gamma\epsilon} \\ 0 & 0 & \mathbf{F}'_{\gamma\xi} \end{bmatrix} \begin{bmatrix} \dot{\mathbf{q}}_\alpha(\mathbf{x}_\alpha) \\ \mathbf{I}_\beta(\mathbf{x}_\beta) \\ \mathbf{I}_\gamma(\mathbf{x}_\gamma) \end{bmatrix} + \begin{bmatrix} \mathbf{J}_C(t) \\ \mathbf{J}_R(t) \\ \mathbf{J}_L(t) \end{bmatrix} \tag{4.35}$$

† For convenience, we may assume the parameter x_j to be the arc-length.

where

$$\begin{aligned}
\mathbf{E}_C(t) &= \mathbf{e}_\alpha(t) + \mathbf{F}_{\alpha\delta}\,\mathbf{e}_\delta(t) \\
\mathbf{E}_R(t) &= \mathbf{e}_\beta(t) + \mathbf{F}_{\beta\delta}\,\mathbf{e}_\delta(t) + \mathbf{F}_{\beta\epsilon}\mathbf{e}_\epsilon(t) \\
\mathbf{E}_L(t) &= \mathbf{e}_\gamma(t) + \mathbf{F}_{\gamma\delta}\,\mathbf{e}_\delta(t) + \mathbf{F}_{\gamma\epsilon}\mathbf{e}_\epsilon(t) + \mathbf{F}_{\gamma\xi}\mathbf{e}_\xi(t) \\
\mathbf{J}_C(t) &= \mathbf{j}_\delta(t) - \mathbf{F}'_{\alpha\delta}\,\mathbf{j}_\alpha(t) - \mathbf{F}'_{\beta\delta}\,\mathbf{j}_\beta(t) - \mathbf{F}'_{\gamma\delta}\,\mathbf{j}_\gamma(t) \\
\mathbf{J}_R(t) &= \mathbf{j}_\epsilon(t) - \mathbf{F}'_{\beta\epsilon}\mathbf{j}_\beta(t) - \mathbf{F}'_{\gamma\epsilon}\mathbf{j}_\gamma(t) \\
\mathbf{J}_L(t) &= \mathbf{j}_\xi(t) - \mathbf{F}'_{\gamma\xi}\mathbf{j}_\gamma(t)
\end{aligned} \tag{4.36}$$

and where the "dot" denotes differentiation with respect to time; namely,

$$\dot{\varphi}_\gamma(\mathbf{x}_\gamma) = \frac{d}{dt}[\varphi_\gamma(\mathbf{x}_\gamma)] = \mathbf{V}_\gamma \tag{4.37}$$

$$\dot{\varphi}_\xi(\mathbf{x}_\xi) = \frac{d}{dt}[\varphi_\xi(\mathbf{x}_\xi)] = \mathbf{V}_\xi \tag{4.38}$$

$$\dot{\mathbf{q}}_\delta(\mathbf{x}_\delta) = \frac{d}{dt}[\mathbf{q}_\delta(\mathbf{x}_\delta)] = \mathbf{I}_\delta \tag{4.39}$$

$$\dot{\mathbf{q}}_\alpha(\mathbf{x}_\alpha) = \frac{d}{dt}[\mathbf{q}_\alpha(\mathbf{x}_\alpha)] = \mathbf{I}_\alpha \tag{4.40}$$

Observe that the left hand side of (4.34) expresses the link voltages in terms of the tree-branch voltages and the algebraic sum of the voltage sources around each fundamental circuit. Similarly, the left-hand side of (4.35) expresses the tree-branch current in terms of the link currents and the algebraic sum of the current sources in each fundamental cut-set.

Much insight is gained by expanding (4.34) and (4.35) and rearranging them in the following order:

$$\dot{\varphi}_\gamma(\mathbf{x}_\gamma) + \mathbf{F}_{\gamma\xi}\dot{\varphi}_\xi(\mathbf{x}_\xi) = -\mathbf{F}_{\gamma\delta}\,\mathbf{V}_\delta(\mathbf{x}_\delta) - \mathbf{F}_{\gamma\epsilon}\,\mathbf{V}_\epsilon(\mathbf{x}_\epsilon) + \mathbf{E}_L(t) \tag{4.41}$$

$$\dot{\mathbf{q}}_\delta(\mathbf{x}_\delta) - \mathbf{F}'_{\alpha\delta}\,\dot{\mathbf{q}}_\alpha(\mathbf{x}_\alpha) = \mathbf{F}'_{\beta\delta}\,\mathbf{I}_\beta(\mathbf{x}_\beta) + \mathbf{F}'_{\gamma\delta}\,\mathbf{I}_\gamma(\mathbf{x}_\gamma) + \mathbf{J}_C(t) \tag{4.42}$$

$$\mathbf{V}_\alpha(\mathbf{x}_\alpha) = -\mathbf{F}_{\alpha\delta}\,\mathbf{V}_\delta(\mathbf{x}_\delta) + \mathbf{E}_C(t) \tag{4.43}$$

$$\mathbf{I}_\xi(\mathbf{x}_\xi) = \mathbf{F}'_{\gamma\xi}\mathbf{I}_\gamma(\mathbf{x}_\gamma) + \mathbf{J}_L(t) \tag{4.44}$$

$$\mathbf{V}_\beta(\mathbf{x}_\beta) + \mathbf{F}_{\beta\epsilon}\mathbf{V}_\epsilon(\mathbf{x}_\epsilon) = -\mathbf{F}_{\beta\delta}\mathbf{V}_\delta(\mathbf{x}_\delta) + \mathbf{E}_R(t) \tag{4.45}$$

$$-\mathbf{F}'_{\beta\epsilon}\mathbf{I}_\beta(\mathbf{x}_\beta) + \mathbf{I}_\epsilon(\mathbf{x}_\epsilon) = \mathbf{F}'_{\gamma\epsilon}\mathbf{I}_\gamma(\mathbf{x}_\gamma) + \mathbf{J}_R(t) \tag{4.46}$$

Observe that (4.41) and (4.42) represent a system of nonlinear differential equations whereas (4.43)-(4.46) represent a system of nonlinear functional equations. Together, these equations completely characterize any nonlinear network containing *unicursal* elements. The solutions to these equations therefore represent the complete solutions of the network. Unfortunately, since these equations are not in the normal form, no *general* computational procedure is available for solving them. Hence, our next step is to find an appropriate algorithm to recast these equations into the normal form.

Before we do this, it is important to be able to interpret the relationship between each of these equations and the pertinent set of network elements which they represent.† A review of the preceding discussion on the topological algorithm will show that (4.41) represents the KVL equations written around each fundamental circuit formed by a link inductor and at least one tree-branch capacitor or resistor. Similarly, (4.42) represents the KCL equations written across each fundamental cut-set formed by a tree-branch capacitor and at least one link resistor or inductor. In contrast to this, (4.43) represents the KVL equations written around each fundamental circuit formed by a link capacitor and *tree-branch capacitors only*. This means that (4.43) will appear only if the network contains at least one capacitor-only-loop. Equation (4.44) represents the KCL equations written across each fundamental cut-set formed by a link inductor and *tree-branch inductors only*. This means that (4.44) will appear only if the network contains at least one inductor-only-cut-set. Equation (4.45) represents the KVL equations written around each fundamental circuit formed by a link resistor and at least one tree-branch resistor – in addition to possibly one or more tree-branch capacitors (observe that no tree-branch inductor will be present because of the construction of the normal tree). Finally, (4.46) represents the KCL equations written across each fundamental cut-set formed by a tree-branch resistor and at least one link resistor – in addition to possibly one, or more link inductors (observe that no link capacitors will be present in this case).

† These interpretations are made under the assumptions that the submatrices $\mathbf{F}_{\alpha\delta}$, $\mathbf{F}_{\beta\delta}$, $\mathbf{F}_{\gamma\delta}$, $\mathbf{F}_{\beta\epsilon}$, $\mathbf{F}_{\gamma\epsilon}$, and $\mathbf{F}_{\gamma\xi}$ are not null matrices.

II. *Equation Reduction Algorithm.* Depending on the choice of the state variables, several algorithms are available for transforming (4.41)-(4.46) to the normal form. Most of these algorithms can be considered to be special cases of two basic algorithms which we will now describe:

Basic Algorithm I – Chua-Rohrer Algorithm. In this algorithm, we define the state variables to be the parametric variables $\mathbf{x}_\gamma$ and $\mathbf{x}_\delta$ associated with the link inductors and tree-branch capacitors, respectively. Our main objective is to solve for the non-state variables $\mathbf{x}_\alpha$, $\mathbf{x}_\beta$, $\mathbf{x}_\epsilon$, and $\mathbf{x}_\xi$ in terms of the state variables $\mathbf{x}_\gamma$ and $\mathbf{x}_\delta$. The step-by-step procedure to be described shortly shows that under appropriate assumptions, these variables can be expressed as follows:

$$
\begin{aligned}
\mathbf{x}_\xi(t) &= \mathbf{f}_\xi(\mathbf{x}_\gamma(t)) \\
\mathbf{x}_\alpha(t) &= \mathbf{f}_\alpha(\mathbf{x}_\delta(t)) \\
\mathbf{x}_\beta(t) &= \mathbf{f}_\beta(\mathbf{x}_\gamma(t), \mathbf{x}_\delta(t)) \\
\mathbf{x}_\epsilon(t) &= \mathbf{f}_\epsilon(\mathbf{x}_\gamma(t), \mathbf{x}_\delta(t))
\end{aligned}
\tag{4.47}
$$

If we substitute these relations in place of the respective variables in (4.41)-(4.42), the term $\dot{\varphi}_\xi(\mathbf{x}_\xi)$ becomes a function of $\mathbf{x}_\gamma(t)$ and will be denoted simply by $\dot{\varphi}_\xi(\mathbf{x}_\gamma(t))$. Similarly, the term $\dot{\mathbf{q}}_\alpha(\mathbf{x}_\alpha)$ becomes a function of $\mathbf{x}_\delta(t)$ and will be denoted by $\dot{\mathbf{q}}_\alpha(\mathbf{x}_\delta(t))$. If we now apply the chain-rule to (4.41) and (4.42), we obtain: †

$$
\begin{aligned}
&[\Phi_\gamma^*(\mathbf{x}_\gamma)]\,\dot{\mathbf{x}}_\gamma + \mathbf{F}_{\gamma\xi}\left[\Phi_\xi^*(\mathbf{x}_\gamma)\dot{\mathbf{x}}_\gamma + \frac{\partial \varphi_\xi(\mathbf{x}_\gamma(t))}{\partial t}\right] \\
&\quad = -\mathbf{F}_{\gamma\delta}\mathbf{V}_\delta(\mathbf{x}_\delta) - \mathbf{F}_{\gamma\epsilon}\mathbf{V}_\epsilon(\mathbf{x}_\epsilon) + \mathbf{E}_L(t)
\end{aligned}
\tag{4.48}
$$

$$
\begin{aligned}
&[\mathbf{Q}_\delta^*(\mathbf{x}_\delta)]\,\dot{\mathbf{x}}_\delta - \mathbf{F}'_{\alpha\delta}\left[\mathbf{Q}_\alpha^*(\mathbf{x}_\delta)\,\dot{\mathbf{x}}_\delta + \frac{\partial \mathbf{q}_\alpha(\mathbf{x}_\delta(t))}{\partial t}\right] \\
&\quad = \mathbf{F}'_{\beta\delta}\mathbf{I}_\beta(\mathbf{x}_\beta) + \mathbf{F}'_{\gamma\delta}\mathbf{I}_\gamma(\mathbf{x}_\gamma) + \mathbf{J}_C(t)
\end{aligned}
\tag{4.49}
$$

† Observe that the jth element of $\varphi_\gamma(x_\gamma)$ is a function only of x_{γ_j}, hence $\Phi_\gamma^*(\mathbf{x}_\gamma)$ is a diagonal matrix. On the other hand, the jth element of $\varphi_\xi(\mathbf{x}_\gamma(t))$ is a function of $x_{\gamma_1}, x_{\gamma_2}, \ldots, x_{\gamma_\gamma}$ and t. Hence, $\Phi_\xi^*(x_\gamma)$ is a Jacobian matrix.

where $\Phi_\gamma^*(\mathbf{x}_\gamma)$ is a diagonal matrix whose jjth element is given by $d\varphi_{\gamma j}(x_{\gamma j})/dx_{\gamma j}$, $\mathbf{Q}_\delta^*(\mathbf{x}_\delta)$ is a diagonal matrix whose jjth element is given by $dq_{\delta j}(x_{\delta j})/dx_{\delta j}$. The matrices $\Phi_\xi^*(\mathbf{x}_\gamma)$ and $\mathbf{Q}_\alpha^*(\mathbf{x}_\delta)$ represent the Jacobian of $\varphi_\xi(\mathbf{x}_\gamma)$ and $\mathbf{q}_\alpha(\mathbf{x}_\delta)$, respectively. Equations (4.48) and (4.49) can be combined into the following compact forms:

$$\dot{\mathbf{x}}_\gamma = \mathbf{M}_\gamma^{-1}\mathbf{K}_\gamma \tag{4.50}$$

$$\dot{\mathbf{x}}_\delta = \mathbf{M}_\delta^{-1}\mathbf{K}_\delta \tag{4.51}$$

where

$$\mathbf{M}_\gamma \equiv \Phi_\gamma^*(\mathbf{x}_\gamma) + \mathbf{F}_{\gamma\xi}\Phi_\xi^*(\mathbf{x}_\gamma) \tag{4.52}$$

$$\mathbf{M}_\delta \equiv \mathbf{Q}_\delta^*(\mathbf{x}_\delta) - \mathbf{F}'_{\alpha\delta}\mathbf{Q}_\alpha^*(\mathbf{x}_\delta) \tag{4.53}$$

$$\mathbf{K}_\gamma \equiv -\mathbf{F}_{\gamma\delta}\mathbf{V}_\delta(\mathbf{x}_\delta) - \mathbf{F}_{\gamma\epsilon}\mathbf{V}_\epsilon(\mathbf{x}_\epsilon) + \mathbf{E}_L(t) - \mathbf{F}_{\gamma\xi}\frac{\partial\varphi_\xi(\mathbf{x}_\gamma(t))}{\partial t} \tag{4.54}$$

$$\mathbf{K}_\delta \equiv \mathbf{F}'_{\beta\delta}\mathbf{I}_\beta(\mathbf{x}_\beta) + \mathbf{F}'_{\gamma\delta}\mathbf{I}_\gamma(\mathbf{x}_\gamma) + \mathbf{J}_C(t) + \mathbf{F}'_{\alpha\delta}\frac{\partial\mathbf{q}_\alpha(\mathbf{x}_\delta(t))}{\partial t} \tag{4.55}$$

The Chua-Rohrer Algorithm can now be outlined as follows:

Step 1: The initial conditions at $t = t_0$ are given; namely $\mathbf{x}_\gamma(t_0)$ and $\mathbf{x}_\delta(t_0)$.

Step 2: Substitute the value of $\mathbf{x}_\gamma(t_0)$ and $\mathbf{x}_\delta(t_0)$ into (4.43)-(4.46) and obtain:

$$\mathbf{V}_\alpha(\mathbf{x}_\alpha(t_0)) = \mathbf{k}_\alpha(t_0) \tag{4.56}$$

$$\mathbf{I}_\xi(\mathbf{x}_\xi(t_0)) = \mathbf{k}_\xi(t_0) \tag{4.57}$$

$$\mathbf{V}_\beta(\mathbf{x}_\beta(t_0)) + \mathbf{F}_{\beta\epsilon}\mathbf{V}_\epsilon(\mathbf{x}_\epsilon(t_0)) = \mathbf{k}_\beta(t_0) \tag{4.58}$$

$$-\mathbf{F}'_{\beta\epsilon}\mathbf{I}_\beta(\mathbf{x}_\beta(t_0)) + \mathbf{I}_\epsilon(\mathbf{x}_\epsilon(t_0)) = \mathbf{k}_\epsilon(t_0) \tag{4.59}$$

where $\mathbf{k}_\alpha(t_0)$, $\mathbf{k}_\xi(t_0)$, $\mathbf{k}_\beta(t_0)$, and $\mathbf{k}_\epsilon(t_0)$ are constant vectors representing the numerical values of the expressions in the right-hand side of (4.43)-(4.46).

Step 3: Solve for $\mathbf{x}_\alpha(t_0)$ and $\mathbf{x}_\xi(t_0)$ from (4.56) and (4.57) respectively. This is equivalent to taking the inverse function at t_0; namely,

$$\mathbf{x}_\alpha(t_0) = \mathbf{V}_\alpha^{-1}(\mathbf{k}_\alpha(t_0)) \tag{4.60}$$

$$\mathbf{x}_\xi(t_0) = \mathbf{I}_\xi^{-1}(\mathbf{k}_\xi(t_0)) \tag{4.61}$$

Since the inverse function exists if, and only if, the functions $V_{\alpha_j}(x_{\alpha_j})$ and $I_{\xi_j}(x_{\xi_j})$ are one-to-one, *we must require all link capacitors to be voltage-controlled and all tree-branch inductors to be current-controlled.*† In this case, we may choose $\mathbf{x}_\alpha = \mathbf{V}_\alpha$ and $\mathbf{x}_\xi = \mathbf{I}_\xi$ in (4.47) and (4.48), thereby eliminating Step 3 completely.

Step 4: Solve for $\mathbf{x}_\beta(t_0)$ and $\mathbf{x}_\epsilon(t_0)$ from (4.58) and (4.59). These equations represent in general a system of nonlinear functional equations and may be solved by appropriate iteration techniques. For the time being, we will assume that a unique solution exists.

Step 5: Calculate the values of the diagonal matrices $\Phi_\gamma^*(\mathbf{x}_\gamma)$ and $\mathbf{Q}_\delta^*(\mathbf{x}_\delta)$ and the Jacobian matrices $\Phi_\xi^*(\mathbf{x}_\gamma)$ and $\mathbf{Q}_\alpha^*(\mathbf{x}_\delta)$.

Step 6: Substitute the value of $\mathbf{x}_\gamma(t_0)$, $\mathbf{x}_\delta(t_0)$, $\mathbf{x}_\beta(t_0)$, and $\mathbf{x}_\epsilon(t_0)$ into (4.52)-(4.55) and obtain $\mathbf{M}_\gamma^{-1}(t_0)$, $\mathbf{M}_\delta^{-1}(t_0)$, $\mathbf{K}_\gamma(t_0)$, and $\mathbf{K}_\delta(t_0)$.

Step 7: Apply appropriate numerical integration techniques to solve for $\mathbf{x}_\gamma(t_1)$ and $\mathbf{x}_\delta(t_1)$ from:

$$\begin{aligned}\dot{\mathbf{x}}_\gamma(t_0) &= \mathbf{M}_\gamma^{-1}(t_0)\mathbf{K}_\gamma(t_0)\\ \dot{\mathbf{x}}_\delta(t_0) &= \mathbf{M}_\delta^{-1}(t_0)\mathbf{K}_\delta(t_0)\end{aligned} \tag{4.62}$$

Step 8: Repeat Steps 2 through 7 as many times as desired to obtain $\mathbf{x}_\gamma(t_j)$ and $\mathbf{x}_\delta(t_j)$, $j = 1, 2, ..., n$.

The complete block diagram for this algorithm is shown in Fig. 4.8. Observe that in arriving at (4.62), we have made two assumptions: 1) (4.58) and (4.59) have a unique solution. 2) The inverse matrices $\mathbf{M}_\gamma^{-1}(t_j)$ and $\mathbf{M}_\delta^{-1}(t_j)$ exist for all t_j.

Let us now investigate how controlled sources may be handled: A careful study of (4.41)-(4.46) shows that any controlled sources must appear in either the current source terms ($\mathbf{J}_R$, $\mathbf{J}_L$, $\mathbf{J}_C$), or the voltage source terms ($\mathbf{E}_R$, $\mathbf{E}_L$, $\mathbf{E}_C$). In the light of the above algorithm, it is clear that

† It is easily shown that if $y = y(\rho)$ and $x = x(\rho)$ are the parametric equations of a curve Γ, then the function $x = x(\rho)$ is a one-to-one function if, and only if, the curve Γ is x-controlled [11].

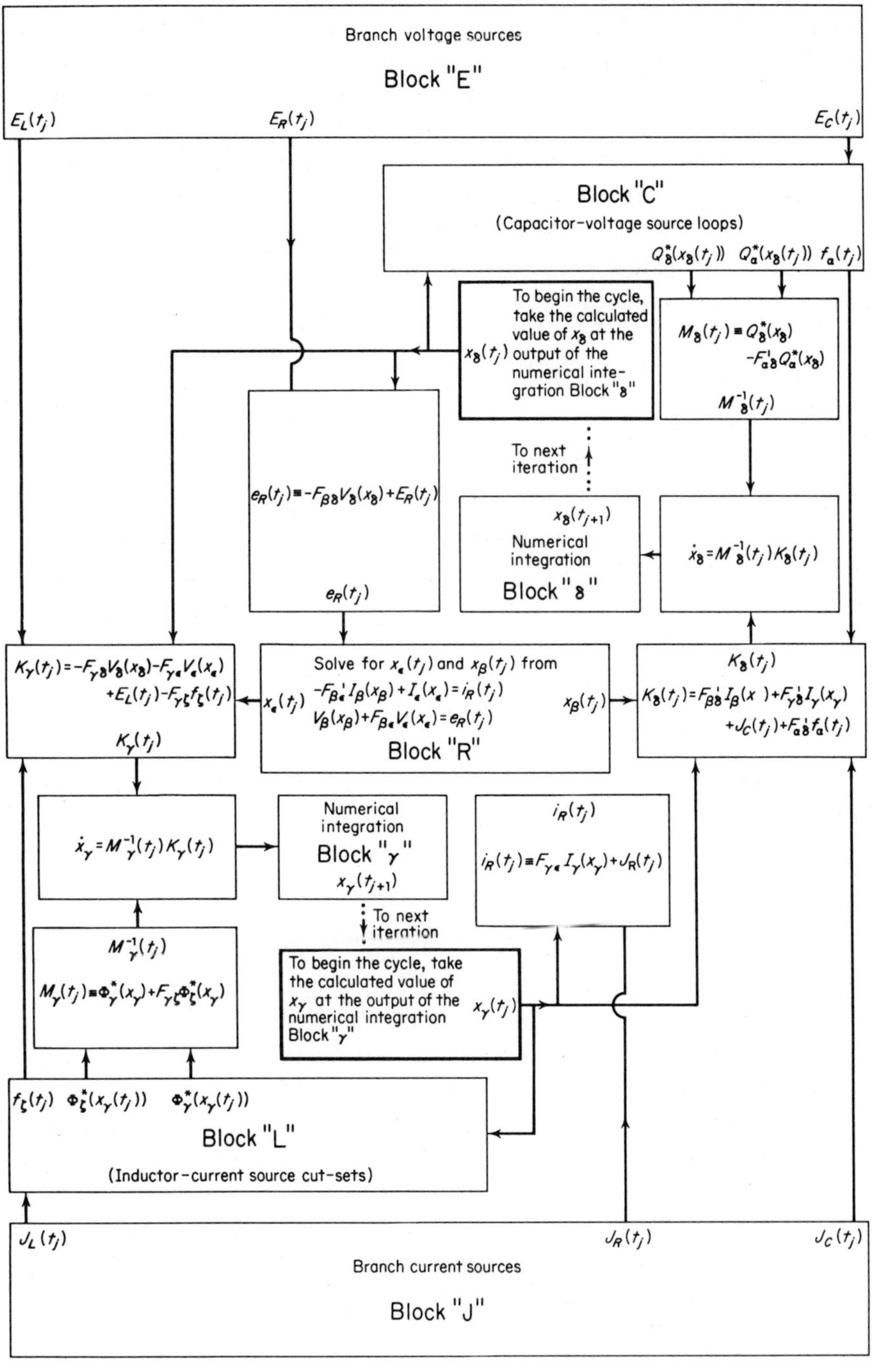

FIG. 4.8. The complete block diagram of the Chua-Rohrer Algorithm.

any controlled source is acceptable by this algorithm provided the value of the controlling variables can be calculated *before* the controlled source terminal voltage or current is required by the algorithm. For example, Fig. 4.8 shows that the controlled sources belonging to $\mathbf{E}_L$ or $\mathbf{J}_C$ are needed last and hence may depend on *any* controlling variables. The controlled sources belonging to $\mathbf{E}_R$ and $\mathbf{J}_R$ may depend on any variable other than a resistor voltage or a resistor current. If this condition is not satisifed; i.e., $\mathbf{E}_R = \mathbf{E}_R(\mathbf{x}_\beta, \mathbf{x}_\epsilon)$ and $\mathbf{J}_R = \mathbf{J}_R(\mathbf{x}_\beta, \mathbf{x}_\epsilon)$, then (4.45) and (4.46) must be transformed into the form

$$\begin{aligned} \mathbf{V}_\beta(\mathbf{x}_\beta) + \mathbf{F}_{\beta\epsilon}\mathbf{V}_\epsilon(\mathbf{x}_\epsilon) - \mathbf{E}_R(\mathbf{x}_\beta, \mathbf{x}_\epsilon) &= -\mathbf{F}_{\beta\delta}\mathbf{V}_\delta(\mathbf{x}_\delta) \\ -\mathbf{F}'_{\beta\epsilon}\mathbf{I}_\beta(\mathbf{x}_\beta) + \mathbf{I}_\epsilon(\mathbf{x}_\epsilon) - \mathbf{J}_R(\mathbf{x}_\beta, \mathbf{x}_\epsilon) &= \mathbf{F}'_{\gamma\epsilon}\mathbf{I}_\gamma(\mathbf{x}_\gamma) \end{aligned} \tag{4.63}$$

Clearly, if (4.63) has a unique solution, then the algorithm still applies. The same remarks apply to controlled sources belonging to $\mathbf{J}_L$ (or $\mathbf{E}_C$) which depend only on inductor (or capacitor) voltage or currents. However, if they depend on resistor voltages and currents, then (4.43)-(4.46) will become coupled to each other, and the entire set of equations must be solved together. Again, the uniqueness condition is a necessary condition here for the normal form to exist.

Basic Algorithm II – Desoer-Katzenelson-Stern Algorithm. In this algorithm, we define the state variables by the vectors $\boldsymbol{\varphi}$ and $\mathbf{q}$ according to the relationship

$$\boldsymbol{\varphi} \equiv \boldsymbol{\varphi}_\gamma(\mathbf{x}_\gamma) + \mathbf{F}_{\gamma\xi}\boldsymbol{\varphi}_\xi(\mathbf{x}_\xi) \tag{4.64}$$

$$\mathbf{q} \equiv \mathbf{q}_\delta(\mathbf{x}_\delta) - \mathbf{F}'_{\alpha\delta}\mathbf{q}_\alpha(\mathbf{x}_\alpha) \tag{4.65}$$

Observe that φ can be interpreted to be the algebraic sum of the inductor flux-linkages around each fundamental circuit formed by a link inductor. Similarly, $\mathbf{q}$ can be interpreted to be the algebraic sum of the capacitor charges across each cut-set formed by a tree-branch capacitor. Observe also that if there are no inductor-only-cut-sets, then $\mathbf{F}_{\gamma\xi} = \mathbf{0}$ and the state variables φ coincide with the flux-linkage of the link inductors. Similarly, if there are no capacitor-only loops, then $\mathbf{F}'_{\alpha\delta} = \mathbf{0}$ and the state variables $\mathbf{q}$ coincide with the charge of the tree-branch capacitors. This algorithm consists of rewriting (4.41)-(4.46) and (4.64)-(4.65) into the following four subgroups in accordance with the order of their occurrence:

Inductor Equations:

$$\begin{aligned} \boldsymbol{\varphi}_\gamma(\mathbf{x}_\gamma) + \mathbf{F}_{\gamma\xi}\boldsymbol{\varphi}_\xi(\mathbf{x}_\xi) &= \boldsymbol{\varphi} \\ -\mathbf{F}'_{\gamma\xi}\mathbf{I}_\gamma(\mathbf{x}_\gamma) + \mathbf{I}_\xi(\mathbf{x}_\xi) &= \mathbf{J}_L(t) \end{aligned} \tag{4.66}$$

Capacitor Equations:

$$\begin{aligned} \mathbf{V}_\alpha(\mathbf{x}_\alpha) + \mathbf{F}_{\alpha\delta}\mathbf{V}_\delta(\mathbf{x}_\delta) &= \mathbf{E}_C(t) \\ -\mathbf{F}'_{\alpha\delta}\mathbf{q}_\alpha(\mathbf{x}_\alpha) + \mathbf{q}_\delta(\mathbf{x}_\delta) &= \mathbf{q} \end{aligned} \tag{4.67}$$

Resistor Equations:

$$\begin{aligned} \mathbf{V}_\beta(\mathbf{x}_\beta) + \mathbf{F}_{\beta\epsilon}\mathbf{V}_\epsilon(\mathbf{x}_\epsilon) &= -\mathbf{F}_{\beta\delta}\mathbf{V}_\delta(\mathbf{x}_\delta) + \mathbf{E}_R(t) \\ -\mathbf{F}'_{\beta\epsilon}\mathbf{I}_\beta(\mathbf{x}_\beta) + \mathbf{I}_\epsilon(\mathbf{x}_\epsilon) &= \mathbf{F}'_{\gamma\epsilon}\mathbf{I}_\gamma(\mathbf{x}_\gamma) + \mathbf{J}_R(t) \end{aligned} \tag{4.68}$$

State Equations:

$$\begin{aligned} \dot{\boldsymbol{\varphi}} &= -\mathbf{F}_{\gamma\delta}\mathbf{V}_\delta(\mathbf{x}_\delta) - \mathbf{F}_{\gamma\epsilon}\mathbf{V}_\epsilon(\mathbf{x}_\epsilon) + \mathbf{E}_L(t) \\ \dot{\mathbf{q}} &= \mathbf{F}'_{\beta\delta}\mathbf{I}_\beta(\mathbf{x}_\beta) + \mathbf{F}'_{\gamma\delta}\mathbf{I}_\gamma(\mathbf{x}_\gamma) + \mathbf{J}_C(t) \end{aligned} \tag{4.69}$$

If we assume the right hand side of (4.66)-(4.68) as constants, then these equations are exactly identical in form -- they all represent a system of nonlinear functional equations of the form

$$\mathbf{f}(\mathbf{y}) = \mathbf{k} \tag{4.70}$$

where $\mathbf{y}$ represents respectively the vector $(\mathbf{x}_\gamma, \mathbf{x}_\xi)'$, $(\mathbf{x}_\alpha, \mathbf{x}_\delta)'$, or $(\mathbf{x}_\beta, \mathbf{x}_\epsilon)'$; and $\mathbf{k}$ represents the respective constants on the right hand side. The crucial part of this algorithm is to solve for $\mathbf{y}$ as a function of $\mathbf{k}$. (See Sec. 4.3.) For the time being, we will assume that a unique solution exists in each case. The Desoer-Katzenelson-Stern Algorithm can now be outlined as follows:

Step 1: The initial conditions at $t = t_0$ are given; namely, $\boldsymbol{\varphi}(t_0)$ and $\mathbf{q}(t_0)$.

Step 2: Solve for $\mathbf{x}_\gamma(t_0)$ and $\mathbf{x}_\xi(t_0)$ from (4.66).

Step 3: Solve for $\mathbf{x}_\alpha(t_0)$ and $\mathbf{x}_\delta(t_0)$ from (4.67).

Step 4: Substitute the values of $\mathbf{x}_\gamma(t_0)$ and $\mathbf{x}_\delta(t_0)$ for $\mathbf{x}_\gamma$ and $\mathbf{x}_\delta$ in the right hand side of (4.68), thereby reducing them to constant vectors.

Step 5: With the value of the right hand side of (4.68) known from step 4, solve for $\mathbf{x}_\beta(t_0)$ and $\mathbf{x}_\epsilon(t_0)$.

Step 6: Substitute the values of $\mathbf{x}_\delta(t_0)$, $\mathbf{x}_\epsilon(t_0)$, $\mathbf{x}_\beta(t_0)$, and $\mathbf{x}_\gamma(t_0)$ in place of $\mathbf{x}_\delta$, $\mathbf{x}_\epsilon$, $\mathbf{x}_\beta$, and $\mathbf{x}_\gamma$ in (4.69) to obtain

$$\begin{aligned} \dot{\boldsymbol{\varphi}}(t_0) &= \mathbf{f}_L(t_0) \\ \dot{\mathbf{q}}(t_0) &= \mathbf{f}_C(t_0) \end{aligned} \tag{4.71}$$

where $\mathbf{f}_L(t_0)$ and $\mathbf{f}_C(t_0)$ are now known constant vectors.

Step 7: Apply appropriate numerical integration techniques to (4.71) and calculate $\boldsymbol{\varphi}(t_1)$ and $\mathbf{q}(t_1)$.

Step 8: Repeat steps 1 through 7 as many times as are desired to calculate $\boldsymbol{\varphi}(t_j)$, $\mathbf{q}(t_j)$, $j = 2, 3, \ldots, n$.

The complete block diagram for implementing this algorithm is shown in Fig. 4.9. From this diagram, we can derive similar procedures for handling controlled sources. A computer program called AEDNET for implementing this algorithm has been developed [19]. It is by far the most general nonlinear network analysis program available.

A Comparison Between The Two Algorithms. The *Chua-Rohrer (C-R) Algorithm* and the *Desoer-Katzenelson-Stern (D-K-S) Algorithm* differ primarily on the choice of the state variables. The C-R Algorithm uses the parametric variables $\mathbf{x}_\gamma$ and $\mathbf{x}_\delta$ whereas the D-K-S Algorithm uses the loop flux-linkage $\boldsymbol{\varphi}$ and the cut-set charge $\mathbf{q}$. Since a wide variety of parameters can be chosen, the state variables in the C-R Algorithm are not unique. For example, if all elements in the network are strictly-monotonically-increasing, $\mathbf{x}_\gamma$ and $\mathbf{x}_\delta$ may be chosen to be the inductor currents or capacitor voltages, respectively. If the link inductors are flux-linkage-controlled and the tree-branch capacitors are charge-controlled, then $\mathbf{x}_\gamma$ and $\mathbf{x}_\delta$ may be chosen to be equal to $\boldsymbol{\varphi}_\gamma$ and $\mathbf{q}_\delta$ respectively. If the link inductors and tree-branch capacitors are multivalued, then $\mathbf{x}_\gamma$ and $\mathbf{x}_\delta$ may be chosen to be the arc-length, or any one-to-one transformation of arc-length. Since the convergence of many numerical iteration techniques depends upon the nature of the functional representations, it is clear that one choice of state variables may converge for a particular network, whereas another may not. Hence, the flexibility in the choice of state variables in the C-R Algorithm may help in the numerical iteration process. Although specific examples can be found to illustrate this point, a general criterion remains to be found.

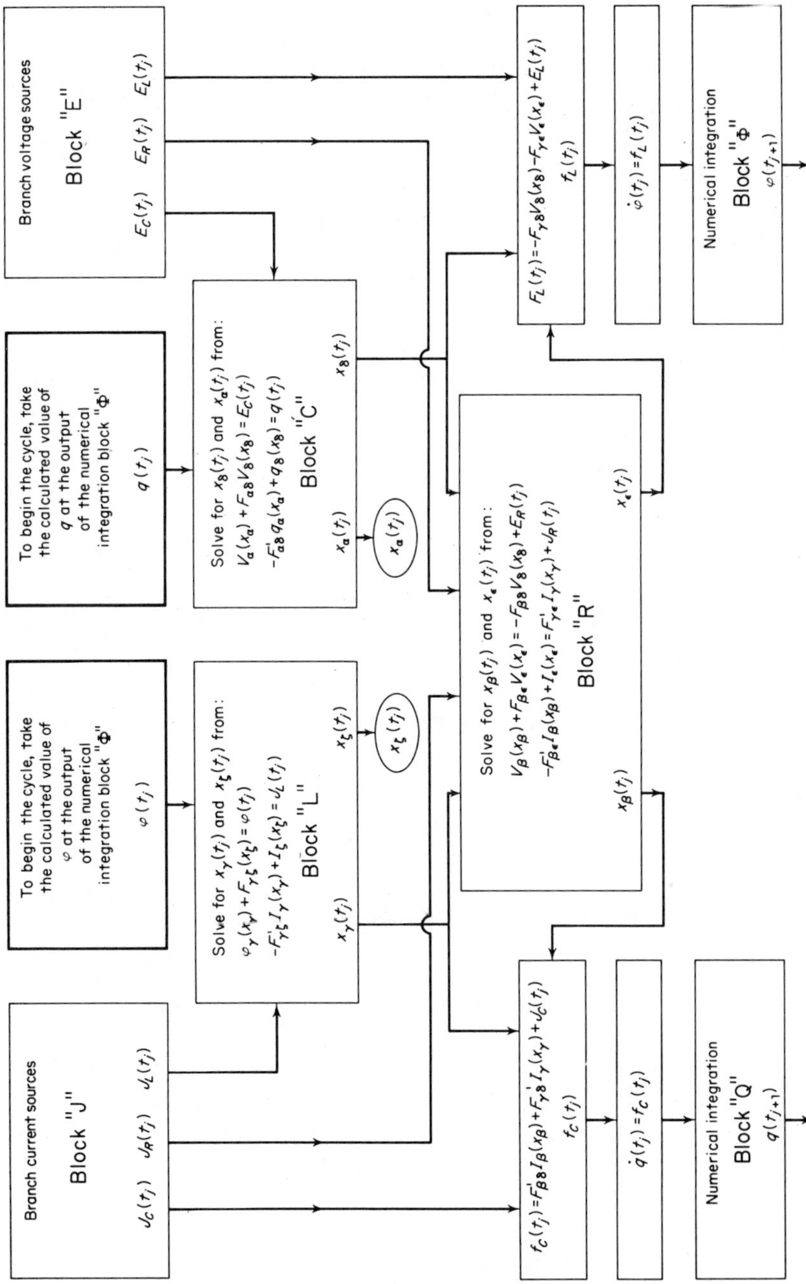

FIG. 4.9. The complete block diagram of the Desoer-Katzenelson-Stern Algorithm.

In contrast to this, the choice of the state variables in the D-K-S Algorithm is unique. In the special case where the network contains neither inductor-only cut-sets nor capacitor-only loops, the variables $\boldsymbol{\varphi}$ and $\mathbf{q}$ reduce to $\boldsymbol{\varphi}_\gamma$ and $\mathbf{q}_\delta$. Hence, for this class of networks, the D-K-S Algorithm becomes a special case of the C-R Algorithm.

Computationally, there are two major time-consuming tasks in the C-R Algorithm. The first task consists of solving *one set* of nonlinear functional equations (Step 4). The second task involves the iterative solution of a system of nonlinear differential equations (Step 7). In addition to these two time-consuming tasks, the calculation of the Jacobian matrix (Step 5) and the subsequent matrix-inversion (Step 6) also takes some time; however, unlike Steps 4 and 7, there is no convergence problem here. There are also two major time-consuming tasks in the D-K-S Algorithm. The first task consists of solving *three* separate sets of nonlinear functional equations (Steps 2, 3, and 5). The second task consists of the iterative solution of a system of nonlinear differential equations (Step 7). Observe that only one set of nonlinear functional equations needs to be solved in the C-R Algorithm in contrast to the three sets required by the D-K-S Algorithm. However, the C-R Algorithm requires the calculation of two Jacobian matrices (Step 5) and two inverse matrices (Step 6) which the D-K-S Algorithm does not have to do. In other words, any difference in computational times between the two algorithms will depend primarily on the difference in computation time between the calculation of the Jacobian and inverse matrices and the numerical solution of systems of nonlinear functional equations.

4.3 MATHEMATICAL FOUNDATIONS OF NONLINEAR NETWORK ANALYSIS

In the description of the two normal form reduction algorithms, we have made several important mathematical assumptions whose validity we will now investigate. Due to the limitation of space, only results of a foundational nature will be discussed and no mathematical proofs will be given. However, sufficient references will be provided for further reading.

4.3.1 Global Inversion of Functional Equations

One of the most crucial steps in both normal form algorithms consists of solving a system of nonlinear functional equations

$$\mathbf{f}(\mathbf{x}) = \mathbf{y} \tag{4.72}$$

where $\mathbf{y}$ is a constant vector. Since the value of $\mathbf{y}$ at any time t_j depends not only on the terminal voltage and current of the sources at time t_j, but also on the value of the state variable at the same time, we must be sure that (4.72) has a unique solution for all values that $\mathbf{y}$ may assume. Since the *range* of values of $\mathbf{y}$ is not known a-priori, it is safer to assume the range to be the entire n-dimensional space. Hence, what we are confronted with is not a local problem (which would be the case if only one specific value of $\mathbf{y}$ is assumed), but a global problem. Mathematically, this is equivalent to finding the global inverse of the nonlinear function $\mathbf{f}$; namely,

$$\mathbf{x} = \mathbf{f}^{-1}(\mathbf{y}) \tag{4.73}$$

Since the global inverse exists if, and only if, the transformation $\mathbf{f}(\mathbf{x})$ is a one-to-one function, it would be extremely useful to know the necessary and sufficient conditions for this to be true. The following *global inversion theorem* (which consists of two parts) can be used for this purpose.

Palais Theorem: Let $\mathbf{f}$: $R^n \to R^n$ be a function from R^n onto R^n.

I. If $\mathbf{f}(\mathbf{x})$ possesses *continuous* first order partial derivatives for every $\mathbf{x}$ in R^n, then the *necessary and sufficient conditions* for the existence of a global inverse function, $\mathbf{f}^{-1}(\mathbf{x})$ which also possesses continuous first order partial derivatives for every $\mathbf{x}$ in R^n are:

(1) The Jacobian of $\mathbf{f}$ is never zero

(2) $\lim\limits_{\|\mathbf{x}\| \to \infty} \|\mathbf{f}(\mathbf{x})\| = \infty$

where $\|\mathbf{x}\| \equiv \left[\sum_{i=1}^{n} \mathbf{x}_i^2\right]^{1/2}$

II. If $\mathbf{f}(\mathbf{x})$ is continuous (but not necessarily differentiable) then the *necessary and sufficient conditions* for the existence of a continuous (but not necessarily differentiable) global inverse function $\mathbf{f}^{-1}(\mathbf{x})$ are given by condition (2) above and the condition that $\mathbf{f}(\mathbf{x})$ be locally one-to-one for all $\mathbf{x}$ in R^n.

This theorem is a minor revision of a corollary originally stated and proved by Palais [20].† Observe that the conditions for part (I) are necessary and sufficient only *to the extent* that the inverse function $\mathbf{f}^{-1}$ not only exists, but is also differentiable. To show that the conditions are not necessary for the existence of a global inverse, consider the function:

$$
\begin{aligned}
y_1 &= x_1 \\
y_2 &= x_2 + (x_3^3 - x_3) \\
y_3 &= x_3 - x_2 \\
y_4 &= x_4
\end{aligned}
\tag{4.74}
$$

The determinant of the Jacobian of this function is easily found to be equal to $|\mathbf{J}(\mathbf{x})| = 3x_3^2$. Hence, condition (1) of part I is not satisfied for any vector $\mathbf{x}$ whose component $x_3 = 0$. However, the global inverse of (4.74) exists and is easily found to be:

$$
\begin{aligned}
x_1 &= y_1 \\
x_2 &= -y_3 + (y_2 + y_3)^{1/3} \\
x_3 &= (y_2 + y_3)^{1/3} \\
x_4 &= y_4
\end{aligned}
\tag{4.75}
$$

The reason that part I of the Palais Theorem is not applicable here can be traced to the fact that (4.75) is not differentiable everywhere in view of the exponent 1/3. However, the existence of the continuous global inverse in (4.75) is easily seen to satisfy part II of the Palais Theorem.

Some interesting results have recently been obtained by Hall [21] who translated the Palais Theorem in terms of the network topology and element characteristics. Many more sufficient conditions can be stated, some of these can be found in Stern [7]. We will single out one of these conditions for discussion because it is the simplest to apply.

Duffin's Theorem. If the v-i curve of all resistors of a resistive network containing no controlled sources is *strictly-monotonically-increasing*, and if $|i| \to \infty$ as $|v| \to \infty$, then there exists a unique solution for all values of the driving sources.

† This author is indebted to Professor Ruey-wen Liu for pointing out the existence of this theorem.

In spite of the fact that this theorem is overly restrictive, it is widely used because the conditions can be verified *a*-priori, by inspection. Recently, Desoer and Katzenelson [5] obtained an extension of Duffin's Theorem which would allow some resistors to be characterized by monotonically-increasing *v-i* curves which are not *strictly* monotonic [22].

4.3.2 Fixed-Point Methods

Once assured of a unique solution, the next step is to find an appropriate numerical iteration method to solve for the solution. The mathematical foundation of almost all iteration methods consists of finding the *fixed point* of an associated function $\mathbf{g}(\mathbf{x})$; namely, a point $\mathbf{x}^*$ which maps into itself, [23, 24, 25]. Hence, $\mathbf{x} = \mathbf{x}^*$ is a fixed-point of $\mathbf{g}(\mathbf{x})$ if, and only if,

$$\mathbf{x}^* = \mathbf{g}(\mathbf{x}^*) \tag{4.76}$$

If we define a sequence of iterates

$$\mathbf{x}_{n+1} = \mathbf{g}(\mathbf{x}_n) \tag{4.77}$$

then under appropriate conditions, this sequence may converge to a unique fixed point; namely,

$$\mathbf{x}^* = \lim_{n\to\infty} \mathbf{x}_n \tag{4.78}$$

A simple criterion for convergence is given by the following:

Principle of Contraction Mapping. If $g(x)$ is a contraction from the n-dimensional space R^n into R^n, i.e., if there exists a constant $L < 1$ such that

$$\| \mathbf{g}(x) - \mathbf{g}(y) \| \leqslant L \| \mathbf{x} - \mathbf{y} \|$$

for all $\mathbf{x} \in R^n$ and $\mathbf{y} \in R^n$, then $\mathbf{g}(\mathbf{x})$ has a *unique fixed print* $\mathbf{x} = \mathbf{x}^*$. Moreover, the sequence of iterates defined by (4.77) converges to this fixed point. An upper bound on the error for stopping at the pth iteration is given by

$$\| \mathbf{x}_p - \mathbf{x}^* \| \leqslant \frac{L^p}{1-L} \| \mathbf{x}_1 - \mathbf{x}_0 \| \tag{4.79}$$

Observe that when L is close to unity, convergence may be very slow. In fact, a major problem in numerical analysis is that of constructing an appropriate function $\mathbf{g}(\mathbf{x})$ which converges rapidly.

We will now show how to transform the problem of solving the system of nonlinear equations (4.72) into an equivalent fixed-point problem. First of all, at any time $t = t_j$, $\mathbf{y}$ is a constant, and hence we may rewrite (4.72) into the form

$$\mathbf{F}(\mathbf{x}) = \mathbf{0} \tag{4.80}$$

where $\mathbf{F}(\mathbf{x}) \equiv \mathbf{f}(\mathbf{x}) - \mathbf{y}$. The solution $\mathbf{x} = \mathbf{x}^*$ of (4.80) is therefore identical to the solution of (4.72). Let us now construct the associated function $\mathbf{g}(\mathbf{x})$ by defining

$$\mathbf{g}(\mathbf{x}) = \mathbf{x} - \mathbf{K}(\mathbf{x})\mathbf{F}(\mathbf{x}) \tag{4.81}$$

where $\mathbf{K}(\mathbf{x})$ is an $n \times n$ matrix function of $\mathbf{x}$ with the property that $\mathbf{K}(\mathbf{x})$ is non-singular at any fixed point of $\mathbf{g}(\mathbf{x})$. Clearly, any solution $\mathbf{x}^*$ of (4.80) is a fixed point of $\mathbf{g}(\mathbf{x})$ as can be seen by direct substitution of $\mathbf{x} = \mathbf{x}^*$ into (4.81). Conversely, if $\mathbf{x} = \mathbf{x}^*$ is a fixed point of (4.81), then

$$\mathbf{x}^* - \mathbf{K}(\mathbf{x}^*)\mathbf{F}(\mathbf{x}^*) = \mathbf{x}^*$$

and hence $\mathbf{F}(\mathbf{x}^*) = 0$ (since $\mathbf{K}^{-1}(\mathbf{x}^*)$ exists). We have therefore shown $\mathbf{x} = \mathbf{x}^*$ is a solution of (4.80) if, and only if, it is a fixed point of $\mathbf{g}(\mathbf{x})$. The problem of solving (4.80) reduces to that of finding an appropriate matrix $\mathbf{K}(\mathbf{x})$ so that the iterates defined by (4.77) converges to a unique fixed point. A common choice for $\mathbf{K}(\mathbf{x})$ is:

$$\mathbf{K}(\mathbf{x}) = \mathbf{J}^{-1}(\mathbf{x})$$

where $\mathbf{J}(\mathbf{x})$ is the Jacobian matrix of $\mathbf{F}(\mathbf{x})$. In this case, (4.77) reduces to

$$\mathbf{x}_{n+1} = \mathbf{x}_n - \mathbf{J}^{-1}(\mathbf{x}_n)\mathbf{F}(\mathbf{x}_n) \tag{4.82}$$

and is called the *Newton-Raphson* method. Depending on the choice of the initial guess $\mathbf{x}_0$, the Newton-Raphson method may, or may not converge. One sufficient condition to assure convergence is due to Ostrowski [23]:

Newton-Raphson Convergence Theorem. Let $\mathbf{x}_0 = (x_{0_1}, x_{0_2}, \ldots, x_{0_n})$ be the initial guess and $\mathbf{x}_1 = (x_{1_1}, x_{1_2}, \ldots, x_{1_n})$ be the first iteration obtained from (4.82). Let the inverse of the Jacobian at $\mathbf{x}_0$ be $\mathbf{J}^{-1}(\mathbf{x}_0) = A_{ij}(\mathbf{x}_0)/D(\mathbf{x}_0)$, where $A_{ij}(\mathbf{x}_0)$ is the cofactors of $\mathbf{J}(\mathbf{x}_0)$ and $D(\mathbf{x}_0)$ is the determinant of $\mathbf{J}(\mathbf{x}_0)$. Let the following conditions be satisfied:

1. $\sum_{i=1}^{n} (x_{1_i} - x_{0_i})^2 \leqslant A^2$

2. $\frac{1}{|D(\mathbf{x}_0)|}\left[\sum_{i,j=1}^{n} A_{ij}^2(\mathbf{x}_0)\right]^{1/2} \leqslant B$

3. $\sum_{i,j,k=1}^{n} \left[\frac{\partial^2 f_i(\mathbf{x})}{\partial x_j \partial x_k}\right]^2_{\mathbf{x}=\mathbf{x}_0} \leqslant K^2$

where A, B, and K are constants such that

4. $ABK \equiv C_0 \leqslant \frac{1}{2}$

Then the sequence (4.82) converges and the solution $\mathbf{x}^* = (\mathbf{x}_1^*, \mathbf{x}_2^*, \ldots, \mathbf{x}_n^*)$ satisfies

$$\|\mathbf{x}^* - \mathbf{x}_0\| \equiv \left[\sum_{i=1}^{n} (x_i^* - x_{0_i})^2\right]^{1/2} \leqslant \frac{1 - [1 - 2C_0]^{1/2}}{C_0} A$$

The most serious drawback of the Newton-Raphson method is that of finding an initial guess $\mathbf{x}_0$ which will satisfy the above theorem. Unfortunately, there is hardly any better method of comparable generality. There do exist several modifications of the Newton-Raphson method which lead to better computational efficiency. The latest method is due to Broyden [26]. This method avoids the need to compute the inverse of the Jacobian matrix at each point $\mathbf{x}_n$ as required by (4.82). It has been used quite extensively by Branin [27] and others (see Chapter 3).

4.3.3 Existence and Uniqueness of Normal Form Solutions

Consider the initial-value problem

$$\begin{aligned} \dot{\mathbf{x}} &= \mathbf{f}(t, \mathbf{x}) \\ \mathbf{x}(t_0) &= \mathbf{x}_0 \end{aligned} \tag{4.83}$$

We have already demonstrated in Sec. 4.2.1 the following possibilities:

(a) No solution exists. See (4.21).

(b) More than one solution exists. See (4.21) and (4.24).

(c) A unique solution $\mathbf{x}(t)$ exists but only for $t_0 < t < t_1$ where t_1 is a finite number. (See (4.21).

(d) A unique solution $\mathbf{x}(t)$ exists but has a finite escape time; i.e., $\mathbf{x}(t_1) = \infty$, where t_1 is a finite number. (See (4.25).

We will now state several mathematical theorems which preclude the above non-physical cases:

(a) *Cauchy-Peano Existence Theorem* [28, 29]. If each of the functions $f_i(t, x_1, x_2, ..., x_n)$ in $\mathbf{f}(t, \mathbf{x})$ is *continuous* in a domain D of the $(t, \mathbf{x})$ space, then (4.83) has at least one continuous solution which satisfies the given initial condition at t_0 and is valid during some finite time interval $t_0 \leqslant t \leqslant t_1$.

This theorem is the weakest – and hence the most general – existence theorem available. Only the continuity of $\mathbf{f}(t, \mathbf{x})$ is required. Observe also that this is a local theorem in the sense that it does not guarantee the solution to exist for all times $t_0 \leqslant t \leqslant \infty$. The local nature of this theorem is to be expected because one can hardly find a better behaving function than that of (4.25); and yet, we have seen that the solution of this equation cannot be extended beyond $t = 1$! The Cauchy-Peano theorem does not guarantee uniqueness either. For this, we need to impose an additional constraint. Before we state the constraint, we must define what is meant by a *Lipschitz* condition.

A function $f_i(x_1, x_2, ..., x_n)$ is said to satisfy a Lipschitz condition in a closed and bounded domain D of the n-dimensional space $(x_1, x_2, ..., x_n)$ if, and only if, there exists a number L (called the Lipschitz constant) such that

$$|f_i(x_1, x_2, ..., x_n) - f_i(x_1', x_2', ..., x_n')| \leqslant L \sum_{i=1}^{n} |x_i - x_i'| \tag{4.84}$$

for all $\mathbf{x}, \mathbf{x}' \in D$.

In the one-dimensional case, a function $f(x)$ is Lipschitz in a closed interval $a \leqslant x \leqslant b$ if, and only if $|f(x)-f(x')| \leqslant L|x-x'|$ for all $x, x' \in [a, b]$, where $y = f(x)$. Geometrically, this means that the curve $f(x)$ must lie between the two straight lines $y-f(x')=L(x-x')$ and $y-f(x') = -L(x-x')$ for all $x, x' \in [a, b]$. Hence, any function with an infinite slope at $x = x_0$ can not be Lipschitz at x_0. For example, the function $f(v_2) = -v_2^{1/3}$ in (4.23) is not Lipschitz. It is easy to prove that a Lipschitz function is continuous, but the converse is not true. It can also be shown that any function which possesses a bounded continuous derivative is Lipschitz. We are now ready to state the uniqueness theorem:

(b) *Picard's Existence and Uniqueness Theorem* [29]. If the function $\mathbf{f}(t, \mathbf{x})$ is Lipschitz on D (uniformly in t), then (4.83) has a unique solution on some time interval $t_0 \leqslant t \leqslant t_1$.

Again, we observe that this is a local theorem. For the solution to be valid for all time (i.e., no finite escape time), it is necessary to impose fairly strong conditions:

(c) *Wintner's Global Existence Theorem* [30]. Let $\mathbf{f}(t, \mathbf{x})$ be continuous in the $(n + 1)$-dimensional space $(t, \mathbf{x})$. If it is possible to find a continuous function $L(r)$ (where $r = x_1^2 + x_2^2 + ..., x_n^2$) such that

$$|f_i(t, x_1, x_2, ..., x_n)| < L(r), \qquad i = 1, 2, ..., n$$

for all values of $0 \leqslant r < \infty$, and

$$\int_0^\infty \frac{dr}{L(r)} = \infty$$

then all solutions of (4.83) are defined for all times $-\infty < t < \infty$.

As a corollary to this theorem, we have the following:

If there exists a constant $K < \infty$ such that

$$|f_i(t, x_1, x_2, ..., x_n)| < K \max(|x_1|, |x_2|, ..., |x_n|)$$

as $\max(|x_1|, |x_2|, |x_n|) \to \infty$, then all solutions are defined on the entire t-axis, $-\infty < t < \infty$.

Observe that the conditions required by this corollary are extremely strong. For example, it rules out all polynomials with degree greater than 1. But the conditions are certainly not superfluous for we have already seen an example of a second degree polynomial (4.25) which violates the above conditions. On the other hand the condition required by the corollary does allow a large class of non-elementary functions such as $f(x) = (1 + x^3)^{-1/2}$, $(1 + x^2)^{1/3}$, $x^3/(1 + x^2)$, $x \exp(\sin x)$, etc.

The three theorems above are introduced in order to provide some mathematical foundation for the numerical integration of the normal form equations. It must be emphasized that these theorems provide only sufficient conditions. Hence, even if (4.83) does not satisfy Picard's theorem, it may still possess a unique solution. However, in cases where a computer solution behaves strangely, then some kind of ill-conditioning due to the violation of one, or more conditions of the above theorems may be suspected. From a computational standpoint, it is probably more practical to assume that a unique solution exists and is defined for all times. This assumption is true in most cases where the normal form exists.

4.3.4 Numerical Integration of Normal Form Equations

Let us now discuss the various numerical techniques available for solving the normal form equation

$$\dot{\mathbf{x}} = \mathbf{f}(t, \mathbf{x}) \tag{4.85}$$

by a computer. Again, only a cursory treatment will be given and the reader is referred to the references for detailed proofs [13-15]. We will assume that (4.85) has a unique solution corresponding to each initial state $\mathbf{x}(t_0) = \mathbf{x}_0$. We will also assume that the solution $\mathbf{x}(t)$ is defined during the time interval $[t_0, T]$ of interest; i.e., no escape time within this interval.

Our first step is to divide the desired time interval $[t_0, T]$ into m points $t_i = t_0 + ih$, $i = 0, 1, 2, \ldots, m$ where h is called the *step size.* The step size may either be fixed or variable. However, for simplicity, we will assume it to be fixed. Corresponding to a fixed step size h, the integer m is chosen so that $t_m < T$ but $t_{m+1} > T$. It is a common practice to denote the exact solution $\mathbf{x}(t)$ at each point $t = t_i$ by $\mathbf{x}(t_i)$ and to denote an approximate solution (to be calculated by some numerical algorithm) by $\mathbf{x}_i$. In this case, each component j of $\mathbf{x}_i$ is denoted by $\mathbf{x}_{ij}$. The error between the exact solution and the approximate solution is usually called the *truncation error* and will be denoted by

$$\epsilon_i = \| \mathbf{x}(t_i) - \mathbf{x}_i \| \tag{4.86}$$

Since the formula used in calculating the approximate solution $\mathbf{x}_i$ differs from algorithm to algorithm, the truncation error ϵ_i is a good figure of merit for comparing the accuracy of each algorithm. In general, it is possible to derive a universal error bound for each algorithm of the form

$$\epsilon_i = \| \mathbf{x}(t_i) - \mathbf{x}_i \| \leqslant L h^n \tag{4.87}$$

for all $t_i \epsilon [t_0, T]$. The symbols L and n are constants *independent* of the step size h. Equation (4.87) is usually denoted by the 0-notation

$$\epsilon_i = 0(h^n) \tag{4.88}$$

An algorithm which satisfies (4.88) is said to be an *nth-order algorithm*. In general, higher order algorithms lead to more accurate solutions – but at the cost of increased computation time.

We will now describe a number of well-known algorithms for calculating the approximate solution $\mathbf{x}_i$, $i = 0, 1, 2, ..., m$. Most algorithms can be classified into a *single-step algorithm*, or a *multi-step algorithm*. An algorithm is said to be a single-step algorithm if the value of $\mathbf{x}_{i+1}$ can be calculated from its immediately preceding value $\mathbf{x}_i$; namely,

$$\mathbf{x}_{i+1} = \mathbf{g}(\mathbf{x}_i) \tag{4.89}$$

where the function $\mathbf{g}$ may depend on the function $\mathbf{f}$ in (4.85) and its time derivatives, $\mathbf{f}'$, $\mathbf{f}''$, ..., $\mathbf{f}^{(n)}$. Hence, once the initial state $\mathbf{x}_0$ is given, a single-step algorithm can be programmed to calculate $\mathbf{x}_1$, $\mathbf{x}_2$, ..., etc; in other words, it is self-starting. A *multi-step algorithm* differs from a *single-step algorithm* in that the formula for calculating $\mathbf{x}_{i+1}$ generally requires more than one previously calculated values; namely,

$$\mathbf{x}_{i+1} = \mathbf{g}(\mathbf{x}_{i-k}, \mathbf{x}_{i-k+1}, ..., \mathbf{x}_i) \tag{4.90}$$

Just as in (4.89), the function $\mathbf{g}$ may depend on $\mathbf{f}$ and its time derivatives. Observe however that an important difference exists; a *multi-step algorithm* is *not* self-starting. This is clear from (4.90) since more than one value ($\mathbf{x}_k$, $\mathbf{x}_{k-1}$, ..., $\mathbf{x}_0$) must be known initially to calculate $\mathbf{x}_{k+1}$. Since a multi-step algorithm is generally more efficient, a practical algorithm usually consists of a *single-step algorithm* to provide the initial states and then switch over to a *multi-step* algorithm.

The simplest example of a single-step algorithm is the *Euler Algorithm*:

$$\mathbf{x}_{i+1} = \mathbf{x}_i + h\mathbf{f}(t_i, \mathbf{x}_i) \tag{4.91}$$

In the one-dimensional case, the reader can easily verify geometrically that Euler's Algorithm consists of approximating the solution $\mathbf{x}(t_{i+1})$ by the value of its tangent line drawn from $\mathbf{x}(t_i)$. It can be shown that Euler's Algorithm is a *first order algorithm*; i.e.,

$$\epsilon_i = 0(h) \tag{4.92}$$

Hence, unless the function $\mathbf{f}(t, \mathbf{x})$ is sufficiently well-behaved, and unless a small step size is chosen, the Euler Algorithm is at best a gross approximation. An improved single-step method is given by the *Heun Algorithm*:

$$\mathbf{x}_{i+1} = \mathbf{x}_i + \frac{h}{2}\left\{\mathbf{f}(t_i, \mathbf{x}_i) + \mathbf{f}[t_{i+1}, \mathbf{x}_i + h\mathbf{f}(t_i, \mathbf{x}_i)]\right\} \tag{4.93}$$

Geometrically, (4.93) can be interpreted to be the trapezoidal rule with the ordinate of the second term approximated by (4.91). Hence, the function $\mathbf{f}(t_i, \mathbf{x}_i)$ is needed twice in the Heun Algorithm in contrast to only once in the Euler Algorithm. Of course, since the value $\mathbf{f}(t_i, \mathbf{x}_i)$ can be stored in the memory, only one evaluation is really necessary. It can be shown that the Heun Algorithm is a *second-order* algorithm; i.e.,

$$\epsilon_i = 0(h^2) \tag{4.94}$$

For improved accuracy, most practical algorithms are higher than second order. A general single-step nth order algorithm is given by *Taylor's Algorithm*:

$$\mathbf{x}_{i+1} = \mathbf{x}_i + h\mathbf{x}_i^{(1)} + \frac{h^2}{2!}\mathbf{x}_i^{(2)} + \dots + \frac{h^n}{n!}\mathbf{x}_i^{(n)} \tag{4.95}$$

where

$$\mathbf{x}_i^{(k)} \equiv \frac{d^{k-1}}{dt^{k-1}}\mathbf{f}(t, \mathbf{x}(t))\bigg|_{(t_i, \mathbf{x}_i)}$$

The reader can easily see that (4.95) is nothing but the Taylor Series Expansion. Clearly, this algorithm is very accurate if the higher-order time derivatives can be accurately calculated. This is true when the function $\mathbf{f}(t, \mathbf{x})$ is given *analytically*. Unfortunately, this is not the case in nonlinear network analysis; hence, higher order derivatives must be avoided, if at all possible. The most popular higher-order single-step algorithm which meets this requirement is the fourth-order *Runge-Kutta Algorithm*:

$$\mathbf{x}_{i+1} = \mathbf{x}_i + h\mathbf{R}(t_i, \mathbf{x}_i; h)$$

$$\mathbf{R}(t_i, \mathbf{x}_i; h) = \frac{1}{6}(\mathbf{k}_1 + 2\mathbf{k}_2 + 2\mathbf{k}_3 + \mathbf{k}_4)$$

where

$$\begin{aligned}
\mathbf{k}_1 &= \mathbf{f}(t_i, \mathbf{x}_i) \\
\mathbf{k}_2 &= \mathbf{f}(t_i + \frac{h}{2}, \mathbf{x}_i + \frac{h}{2}\mathbf{k}_1) \\
\mathbf{k}_3 &= \mathbf{f}(t_i + \frac{h}{2}, \mathbf{x}_i + \frac{h}{2}\mathbf{k}_2) \\
\mathbf{k}_4 &= \mathbf{f}(t_i + h, \mathbf{x}_i + h\mathbf{k}_3)
\end{aligned} \tag{4.96}$$

Geometrically, this algorithm computes first for the slope $\mathbf{k}_1$ at $\mathbf{x}_i$. Using this value, we move one-half step forward $(h/2)$ in time and evaluate the new slope there. Using this new slope $\mathbf{k}_2$, we move again *one-half step* forward from the same point $(t_i, \mathbf{x}_i)$ and calculate the slope $\mathbf{k}_3$. Finally, with the value of $\mathbf{k}_3$, we move one *full step* forward from $(t_i, \mathbf{x}_i)$ and calculate the slope $\mathbf{k}_4$. These four slopes are then averaged with respective weights $\frac{1}{6}$, $\frac{2}{6}$, $\frac{2}{6}$, and $\frac{1}{6}$. (Observe that if $\mathbf{f}(t, \mathbf{x})$ did not depend upon $\mathbf{x}$, then the above averaging process reduces to Simpson's rule). This average value is then used as the "slope" in place of Euler's Algorithm.

One disadvantage of the Runge-Kutta method is that it is not very efficient. This is due to the fact that since each step is independent of the calculations made in the previous steps, the Runge-Kutta Algorithm does not make use of *old information* – as is true with all single-step methods. Moreover, there is no mechanism for checking whether a given step size is too small or too large. To overcome these shortcomings, we must resort to a multi-step algorithm such as:

$$\mathbf{x}_{i+1} = \mathbf{x}_i + \frac{h}{2}[-\mathbf{f}(t_{i-1}, \mathbf{x}_{i-1}) + 3\mathbf{f}(t_i, \mathbf{x}_i)] \tag{4.97}$$

In this algorithm, two preceding values – $\mathbf{x}_{i-1}$ and $\mathbf{x}_i$ – are used to calculate $\mathbf{x}_{i+1}$. More generally, a multi-step algorithm may make use of not only previously calculated values, but also an appropriate guess at some present value; namely,

$$\begin{aligned}\mathbf{x}_{i+1} &= a_0\mathbf{x}_i + a_1\mathbf{x}_{i-1} + a_2\mathbf{x}_{i-2} + \ldots + \\ &a_k\mathbf{x}_{i-k} + \mathbf{g}(\mathbf{x}_{i+1}, \mathbf{x}_i, \mathbf{x}_{i-1}, \ldots, \mathbf{x}_{i-k})\end{aligned} \tag{4.98}$$

where $\mathbf{g}$ is some function of $\mathbf{f}(t, \mathbf{x})$. Notice that since $\mathbf{x}_{i+1}$ occurs also on the right-hand side, (4.98) is an implicit nonlinear functional equation. In fact, it is already in the desirable fixed-point form:

$$\mathbf{x}_{i+1} = \mathbf{F}(\mathbf{x}_{i+1}) \tag{4.99}$$

Therefore, $\mathbf{x}_{i+1}$ can be solved by some appropriate fixed-point iteration techniques. This means that we must initially make a guess at the actual value of $\mathbf{x}_{i+1}$. This is called the *predicted value*. The iteration formula is then used to obtain a *corrected value*. This is why the multi-step algorithm described above is often referred to as the *Predictor-Corrector Method*. The simplest example of a predictor-corrector algorithm is given by

$$\mathbf{x}_{i+1} = \mathbf{x}_i + \frac{h}{2}[\mathbf{f}(t_i, \mathbf{x}_i) + \mathbf{f}(t_{i+1}, \mathbf{x}_{i+1})] \tag{4.100}$$

Observe that given the initial state $\mathbf{x}(t_0) = \mathbf{x}_0$, we must predict first the value of $\mathbf{x}_1$ before we can calculate a corrected value. We may use the Euler Algorithm to make the following prediction:

$$\mathbf{x}_1 = \mathbf{x}_0 + h\mathbf{f}(t_0, \mathbf{x}_0) \tag{4.101}$$

Substituting this value for $\mathbf{x}_1$ in (4.100) with $i = 0$, we obtain a corrected value for $\mathbf{x}_1$; namely,

$$\mathbf{x}_1^{(1)} = \mathbf{x}_0 + \frac{h}{2}\{\mathbf{f}(t_0, \mathbf{x}_0) + \mathbf{f}[t_1, \mathbf{x}_0 + h\mathbf{f}(t_0, \mathbf{x}_0)]\} \tag{4.102}$$

where the superscript indicates the number of iterations made to obtain the corrected value. The reader should recognize that (4.102) is simply Heun's algorithm. Of course, we need not stop here; instead, we may continue to substitute $\mathbf{x}_1^{(1)}$ for $\mathbf{x}_1$ again in (4.100) to obtain $\mathbf{x}_1^{(2)}, \ldots,$ etc. until the fixed-point $\mathbf{x}_1$ is found. In practice, a finite number of iterations can be specified in the fixed-point determination routine. With the value of $\mathbf{x}_1$ found, we repeat the procedure all over again. First, we *predict* by applying Euler's Algorithm (or other simple algorithms), then we *correct* by applying the fixed-point determination routine.

The predictor-corrector method has several advantages over the Runge-Kutta method. First, in most cases, the function $\mathbf{f}(t_i, \mathbf{x}_i)$ needs to be calculated only twice in comparison to four for the Runge-Kutta algorithm. For high order networks, the saving in computing time could be significant. Second, the difference between the predicted and the corrected values is a good measure of the error committed in each step and hence this information can be used to control the subsequent step size. This feature is particularly important in networks having both very large and very small time constants [19, 27].

Another important subject that we have not yet discussed is the *stability* of an algorithm. The notion of numerical stability is related to a new kind of inevitable error – the round-off error – which we have not yet mentioned. The round-off error differs from the truncation error in that it is a computation error, whereas the truncation is an algorithmic error. In some algorithms, the round-off error, even though small to start with, gets amplified in each subsequent iteration. The end result can be so bad that it is useless. Such algorithms are said to be unstable. On the other hand, there are algorithms which are insensitive to round-off errors in the sense that the error either remains bounded or decays to zero. This type of algorithm is said to be stable. It can be shown that all algorithms described in this section are stable.

4.4 NETWORK CRITERIA FOR NORMAL FORM EQUATIONS

The two normal form algorithms in Sec. 4.2 were given under the *a*-priori assumption that the nonlinear functional equations have unique solutions. The theorems presented in Sec. 4.3 can be used to verify whether this assumption is satisfied. We will now combine the topological criteria in Sec. 4.2 with the mathematical criteria in Sec. 4.3 to obtain the following general theorem:

Main Existence Theorem. Let N be any time-invariant nonlinear network containing only 2-terminal elements, independent sources, and controlled sources. Let $\mathcal{T}$ be any normal tree of N and let the following conditions be satisfied:

1. Each link-inductor and tree-branch capacitor is characterized by a differentiable unicursal curve (Elements which are multi-valued functions of both terminal variables may be allowed).
2. Each link-capacitor is characterized by a differentiable voltage-controlled curve.
3. Each tree-branch inductor is characterized by a differentiable current-controlled curve.
4. Each controlled source is controlled by a variable whose value at any time can be calculated explicitly (by direct substitution) in terms of the state variables.
5. The nonlinear functional equations pertaining to the resistor portion of the network (Equations (4.45) and (4.46)) have a unique solution for all time.

Then the normal form equation $\dot{\mathbf{x}} = \mathbf{f}(t, \mathbf{x})$ can always be formulated by the Chua-Rohrer Algorithm. The function $\mathbf{f}(t, \mathbf{x})$ is continuous for all times except possibly for a finite set of discrete points. If the function $\mathbf{f}(t, \mathbf{x})$ also satisfies a Lipschitz condition, and if there are no finite-escape times, then the network has a unique solution for all time t.

The main existence theorem clearly includes most nonlinear networks of interest. Observe that the first three conditions can be verified *a*-priori by inspection. Condition 4 may be verified by the procedure described in the Chua-Rohrer Algorithm. Condition 5 may be verified by a number of sufficient conditions, (e.g., the Palais Theorem). It is also possible to find sufficient topological conditions for testing condition 5. First, we replace all capacitors and voltage sources by short circuits. Next, we replace all inductors and current sources by open circuits.

The resulting network N_R is purely resistive and may contain more than one separate part. Replace by open circuits all resistors in N_R which do not form *closed loops with at least one other resistor*. If there are any isolated nodes, delete them. Let us denote the resulting network by $N_{\beta\epsilon}$. If, (1) all link resistors which do not belong to $N_{\beta\epsilon}$ are voltage controlled, (2) all tree-branch resistors which do not belong to $N_{\beta\epsilon}$ are current controlled and (3) the functional equations describing $N_{\beta\epsilon}$ are a one-to-one transformation (such as the case when all resistors of $N_{\beta\epsilon}$ are characterized by strictly-monotonically-increasing curves which tend to $\pm\infty$ — Duffin's Theorem), then condition 5 is satisfied. The above sufficient conditions can be used to augment the parasitic elements so that condition 5 is always satisfied. The details for doing this are given in [6]. The Lipschitz condition requirements can also be translated in terms of the individual element characteristics [5, 31]. However, since this is only a sufficient condition, and since it does not preclude the existence of finite-escape times, it is not of much practical value. It is comfortable to know that if conditions 1 – 5 are satisfied, we can almost be certain that there exists a unique solution.

We will conclude this section by deriving the normal form equations of the nonlinear network shown earlier in Fig. 4.7. To be specific, let us pick the same tree and assign the following element characteristics: Let resistors R_1 and R_2 be voltage controlled with $i_k = I_k(v_k)$, $k = 1, 2$. Let the inductors be flux-linkage controlled with $i_k = I_k(\phi_k)$, $k = 3, 4$, 5, and 6. Let the capacitors be charge-controlled with $v_k = V_k(q_k)$, $k =$ 7, 8, 9, and 10. Let resistors R_{11} and R_{12} be current-controlled with $v_k = V_k(i_k)$, $k = 11$ and 12. In order to verify condition 5, we further assume R_2 to be a 1 ohm linear resistor and R_{11} to be characterized by $v_{11} = i_{11}^3 - i_{11}$.

It is easy to see that conditions 1 – 3 are satisfied. Since there are neither capacitor-only loops nor inductor-only cut-sets, we may choose the state variables $\mathbf{x}_\gamma$ and $\mathbf{x}_\delta$ to be the flux-linkage φ_γ and the charge $\mathbf{q}_\delta$. (In this case, the C-R Algorithm and the D-K-S Algorithm are identical.) There is only one controlled source whose terminal voltage depends on the voltage across capacitor C_9; namely $5v_{fc} = 5V_9(q_9)$. Therefore condition 4 is also satisfied. Next, we obtain the resistor subnetworks N_R and $N_{\beta\epsilon}$ as shown in Figs. 4.7 (c) and (d). Although R_{11} is not monotonic, it is easy to see that the equations describing $N_{\beta\epsilon}$ represent a one-to-one transformation. (In fact, the reader should verify that the equations for the resistive network are given by (4.74) and therefore have an inverse transformation, (4.75). Therefore condition 5 is also satisfied and the normal form exists. With $\mathbf{F}_{\alpha\delta} = \mathbf{F}_{\gamma\xi} = 0$ and with $\mathbf{x}_\gamma = \varphi_\gamma$,

$\mathbf{x}_\delta = \mathbf{q}_\delta$, the explicit normal form equations (4.48) and (4.49) can be simplified to:

$$\dot{\varphi}_\gamma = -\mathbf{F}_{\gamma\delta}\mathbf{V}_\delta(\mathbf{q}_\delta) - \mathbf{F}_{\gamma\epsilon}\mathbf{V}_\epsilon \circ [\mathbf{F}'_{\gamma\epsilon}\mathbf{I}_\gamma(\gamma) + \mathbf{J}_R(t)] + \mathbf{E}_L(t) \quad (4.103)$$

$$\dot{\mathbf{q}}_\delta = \mathbf{F}'_{\gamma\delta}\mathbf{I}_\gamma(\varphi_\gamma) + \mathbf{F}'_{\beta\delta}\mathbf{I}_\beta \circ [-\mathbf{F}_{\beta\delta}\mathbf{V}_\delta(\mathbf{q}_\delta) + \mathbf{E}_R(t)] + \mathbf{J}_C(t) \quad (4.104)$$

where the symbol $\circ$ denotes the "composition operation," i.e., taking the function of another function. We can now substitute the topological submatrices obtained earlier in (4.31) into (4.36), (4.103), and (4.104) to obtain the normal form equations. The reader should carry this through and verify that the normal form equations of this network are:†

$$\dot{\varphi}_3 = -V_7(q_7) + V_8(q_8) + 5V_9(q_9)$$

$$\dot{\varphi}_4 = V_9(q_9) + f_1(t) - I_4(\varphi_4) + f_3(t) - [V_8(q_8) + f_1(t) - I_4(\varphi_4) + f_3(t)]^{1/3}$$

$$\dot{\varphi}_5 = V_9(q_9) - V_{10}(q_{10}) - V_{12} \circ [I_5(\varphi_5) + I_6(\varphi_6)]$$

$$\dot{\varphi}_6 = V_9(q_9) - V_{12} \circ [I_5(\varphi_5) + I_6(\varphi_6)]$$

$$\dot{\mathbf{q}}_7 = I_3(\varphi_3) + I_1 \circ [V_8(q_8) - V_7(q_7)]$$

$$\dot{\mathbf{q}}_8 = -I_3(\varphi_3) - I_1 \circ [V_8(q_8) - V_7(q_7)] - f_3(t) - [V_8(q_8) + f_1(t) - I_4(\varphi_4) + f_3(t)]^{1/3}$$

$$\dot{\mathbf{q}}_9 = -I_4(\varphi_4) - I_5(\varphi_5) - I_6(\varphi_6)$$

$$\dot{\mathbf{q}}_{10} = I_5(\varphi_5) - f_2(t)$$

4.5 METHODS FOR ANALYZING MULTIVALUED NONLINEAR NETWORKS

So far, we have been concerned strictly with the computer solution of state equations for nonlinear networks. Corresponding to each initial state $\mathbf{x}(t_0)$, we calculate the time function $\mathbf{x}(t)$ for $t_0 \leqslant t \leqslant \tau$, where

---

† With the exception of the controlled source, this example has been worked out in detail in [6].

τ is the time interval whereby the solution is desired. The ability to do this automatically by a computer is unquestionably a major breakthrough in nonlinear network analysis. Unfortunately, this computational breakthrough is not very useful for analyzing new networks whose general behavior is unknown. It is well known that depending on the initial state $\mathbf{x}(t_0)$, a nonlinear network can exhibit widely different phenomena [32]. For example, in the case of an *autonomous* network†, it may be possible to obtain a solution that approaches a constant value or a limit cycle (and hence periodic). In the case of a non-autonomous network under a periodic excitation, it may be possible to obtain a periodic solution at the same frequency, at a harmonic frequency, or at a subharmonic frequency of the excitation. Hence, unless we know what initial states to pick, a solution to the initial-value problem does not tell the whole story. In fact, since the time-interval $[t_0, \tau]$ must by necessity be finite, we may not even know the existence of a periodic solution unless the initial state is chosen near a limit cycle (for otherwise, the transient may not yet reach steady state). It is therefore most important from a practical standpoint to investigate the "global behavior" of a new network. Once the general property is known, then we can apply appropriate computer network analysis programs to find the specific solutions.

The global behavior of an autonomous network can be found by solving for all the *equilibrium states* of the network [1]. The equilibrium states are simply the solutions to the nonlinear equations

$$\dot{\mathbf{x}} = \mathbf{f}(t, \mathbf{x}) = \mathbf{0} \tag{4.105}$$

The solutions to (4.105) can be interpreted as the solution of the resistive network obtained by replacing all inductors by short circuits and all capacitors by open circuits.* If this resistive network contains one or more non-monotonic elements such as tunnel diodes, it is very likely that the network has more than one solution. In fact, all switching circuits are of this nature and must be analyzed as *multivalued networks*.

Since most numerical techniques are either based on the assumption that there is only one solution, or that the iteration would converge toward the desired solution, they cannot be used to analyze multivalued

† A nonlinear network is said to be autonomous if its normal form equation is independent of time. Otherwise, it is non-autonomous.

* This assumes the state variables to be inductor currents and capacitor voltages. Otherwise, the equilibrium states will be a function of the short-circuit inductor current and open-circuit capacitor voltages.

networks. For this type of nonlinear networks, a different approach is needed.

4.5.1 The Iterative Piecewise-Linear Method

In this approach, all nonlinear resistors are assumed to be characterized by piecewise-linear *v-i* curves. Each *segment k of resistor R_j* is completely specified by three pieces of information as shown in Fig. 4.10(a); namely, (1) its conductance ${}_kg_j$, or its resistance ${}_kr_j$; (2) its current intercept ${}_kI_j$, or its voltage intercept ${}_kE_j$; (3) its current interval of definition $[{}_kI_j^-, {}_kI_j^+]$, or its voltage interval of definition $[{}_kE_j^-, {}_kE_j^+]$. In view of the piecewise-linear nature of the resistor *v-i* curves, each resistor *k* may be replaced by its *iterative Norton equivalent circuit* as shown in Fig. 4.10(b), or by its *iterative Thevenin equivalent circuit* as shown in

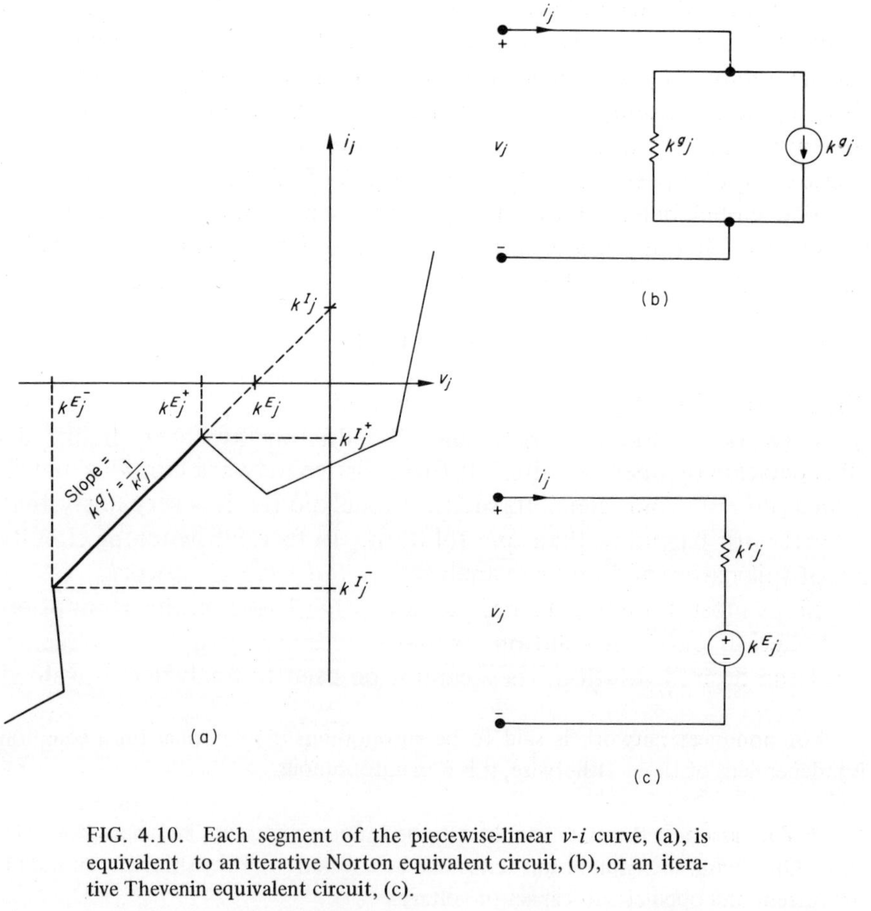

FIG. 4.10. Each segment of the piecewise-linear *v-i* curve, (a), is equivalent to an iterative Norton equivalent circuit, (b), or an iterative Thevenin equivalent circuit, (c).

Fig. 4.10(c). The adjective *iterative* is used to emphasize that the values of the parameters will depend on the particular *segment* being considered during the course of the solution.

There are three basic concepts in the theory of multivalued resistive networks; namely, the *operating point*, the *driving-point characteristic Plot* (DP Plot), and the *transfer-characteristic Plot* (TC Plot). The operating point is, of course, the solution per se. The DP Plot is the locus of voltage and current operating points across a pair of terminals, as the terminal voltage (or current) ranges over all values. It is the analog of the driving-point impedance for linear networks. The TC Plot is the locus of voltage or current operating points across one pair of terminals, as the voltage (or current) across another terminal pair is varied over all values. It is the analog of the transfer-function for linear networks. Whereas there is only one type of DP Plot, there are four distinct types of TC Plots. For multivalued networks, both the DP Plot and the TC Plot may be multivalued.

To solve a multivalued network for its operating points, the iterative piecewise linear method starts by assuming arbitrarily a certain segment combination, thereby obtaining a *linear* network. This network is then solved by any standard techniques and the solution is checked to see whether it agrees with the assumption. If it agrees, the answer is printed out as a legitimate solution. If it disagrees, the answer is discarded. In either case, the procedure is *iterated* by assuming another segment combination. This iterative process continues until all segment combinations are exhausted. It is clear that at the end of the iteration all operating points of the network would be found. Although the method always converges (since only a *finite* number of iterations is involved), the computer time to run through all iterations can be prohibitive if there are many nonlinear resistors. It is desirable therefore to devise algorithms so that not all iterations will be required. Although some special techniques are available, no sufficiently general algorithm is presently known.

In theory, the DP Plot and TC Plot of any network can be found by solving for the operating points while varying the driving-point voltage (or current) over a wide range of values. However, this would be extremely inefficient. A systematic and efficient algorithm for finding the DP Plot and TC Plot of a multivalued network is given in references [33] and [34]. Based on this algorithm, a computer program has been written and will be briefly described as follows:

4.5.2 MECA – Multivalued Electronic Circuit Analysis Program

MECA is a completely automatic computer program written to solve for the operating point, the DP Plot, or the TC Plot of any multivalued resistive network. The network may contain any nonlinear resistor characterized by unicursal piecewise-linear *v-i* curves, independent sources, and controlled sources. Written in FORTRAN IV, MECA is developed at Purdue University under the sponsorship of the U.S. Naval Avionics Facility. The linear analysis program required by the MECA algorithm is the CODED program developed by the U.S. Naval Avionics Facility [35].

The use of MECA requires no programming knowledge. Information is entered by means of three types of cards: (1) control cards, (2) parameter cards, and (3) description cards. The control cards specify the type of solution desired (operating points, DP Plots or TC Plots; any number of DP Plot or TC Plot may be requested) as well as other information needed for processing the circuit. The parameter cards specify the topological information of the network as well as the parameter values of the linear elements and sources. The description cards specify additional information for completely describing the *v-i* curves of the nonlinear resistors.

We will now describe briefly the basic features and operation of the MECA program with the help of the simplified flow chart shown in Fig. 4.11.† The first part of the program reads in the control, parameter, and description cards and stores these pieces of information into the proper arrays. A branch point is reached where the value of DELTA (read from the control card) signifies whether the operating points, or the TC Plots (the DP Plot is interpreted by the program as a special case of the TC Plot) are to be determined. For example, if we set DELTA equal to either "6" or "7", the TC Plot iteration would take place as shown on the chart in detail "B." In the case of TC Plots, we must insert a dummy voltage source (or current source) between the two nodes from which we wish to drive the circuit; namely, the driving-point terminals.

To begin the iteration, the computer is programmed to select an arbitrary segment combination $\sigma \equiv (k_1, k_2, ..., k_m)$ corresponding to each of the m nonlinear resistors. In other words, the parameter values defining segment k_1 of resistor R_1, segment k_2 of resistor R_2, ..., and seg-

† A detailed description of MECA will be given in a forthcoming report and user's manual [36].

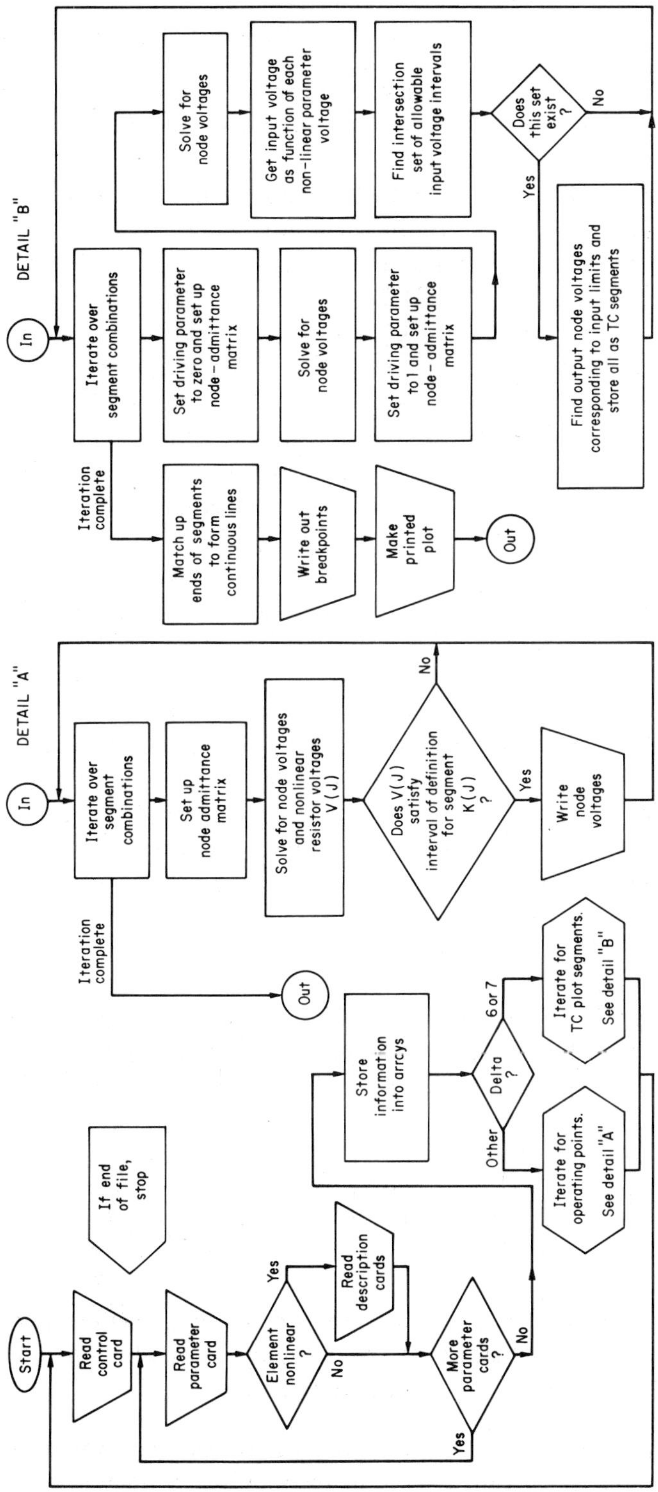

FIG. 4.11. The simplified flow chart of the MECA program.

ment k_m of resistor R_m are assigned in place of the general parameters ${}_kG_j$, ${}_kI_j$, ${}_kI_j^+$, and ${}_kI_j^-$ of the *iterative Norton's equivalent circuit* for each nonlinear resistor (see Fig. 4.10).

If the control card designates that operating points are to be determined, the computer is instructed to obtain the node-admittance matrix of the linear network corresponding to segment combination σ and solve for the node voltages.† See Detail "A". From these node voltages, the branch voltage v_j across *each* of the m nonlinear resistors is calculated and its value is checked to see whether it falls within the interval of definition for segment k_j of resistor R_j. If all m branch voltages meet the test, the initial segment combination σ is legitimate in the sense that an operating point does fall within each of the m assumed segments. The solutions are therefore printed in the output. On the other hand, if even one branch voltage fails the test, the solution is discarded. In any case, the computer selects another segment combination, and the same algorithm is thereby repeated. This procedure is iterated until all segment combinations are exhausted. The final printed output would contain the node voltages corresponding to each legitimate operating point.

If the control card designates the TC Plot (or DP Plot) is to be determined, then the same initial step is taken; namely, the computer selects an initial segment combination σ. Again, because the network is linear under each segment combination, the transfer relationship between the voltage v_j across each nonlinear resistor R_j and the driving-point voltage v is given by

$$v_j = H_j v + K_j, \qquad j = 1, 2, \ldots, m$$

The computer is programmed to calculate the coefficients H_j and K_j by performing two separate operating-point analyses – first with $v = 0$, and then with $v = 1$. The values of K_j and H_j are then obtained from:

$$K_j = v_j \Big]_{v=0} \qquad j = 1, 2, \ldots, m$$

$$H_j = v_j - K_j \Big]_{v=1} \qquad j = 1, 2, \ldots, m$$

With the values of K_j and H_j corresponding to each nonlinear resistor R_j known, the range of the driving-point voltage v for which R_j will remain operating in segment k_j can be calculated from

† It is not necessary to use the node-admittance method of analysis. Our choice is motivated by the existence of the CODED Program referred to earlier. Clearly, other linear programs will serve just as well.

$$v = (v_j - K_j)/H_j \qquad j = 1, 2, \ldots, m$$

by assigning $v_j = {}_{k_j}E_j^+$ and $v_j = {}_{k_j}E_j^-$ – the end points of the interval of definition for segment k_j of resistor R_j. An interval of driving-point voltages is therefore obtained corresponding to each resistor R_j. The common interval – that is, their intersection – gives the interval of the driving-point voltage for segment combination σ.† If this common interval is not empty, the computer is then instructed to calculate the TC Plot corresponding to this segment combination between the specified output voltage (which need not be across a nonlinear resistor) and the driving-point voltage. This is accomplished by a procedure similar to that described above; namely, two separate operating points are evaluated corresponding to $v = 0$ and $v = 1$, thereby obtaining the slope and intercept of the TC Plot. This information, together with the common interval obtained earlier, completely specify one segment of the desired TC Plot. The above procedure is then repeated over all possible segment combinations. Observe that *this algorithm is much more efficient than that of incrementing the values of the driving-point voltages* – a procedure invariably used in all existing programs.

In the case of operating point determination, the printed output of the MECA program consists of any specified set of voltages and currents corresponding to each legitimate operating point. In the case of TC Plot, or DP Plot, the printed output consists of the break points of each segment. These are ordered consecutively – to form a continuous string of segments – in case the TC Plot is a unicursal curve. Otherwise, the TC Plot will consist of several unicursal branches, each branch being referred to as a *curve* in this program. In this case, the breakpoints corresponding to each line are printed in their consecutive order. Finally, the graph of the TC Plot will be printed out as a set of points. The scale to be used in the plotting is done automatically by the computer.

We will now illustrate the application of MECA to the analysis of a well-known multivalued electronic circuit; namely, the Schmitt trigger circuit shown in Fig. 4.12(a). It is well known that the output collector voltage v_0 vs. the input trigger voltage v_{in} exhibits a hysteresis relationship, and is therefore multivalued. Since all existing programs are written under the explicit assumption that the network has a unique solution, they cannot be used to find this TC Plot. This serious limitation is what prompted the development of the MECA program.

† For a more precise derivation of these relationships, the reader is referred to Chua [33].

To apply the MECA program, a model must first be chosen for the transistors. Applying the modeling techniques described in [1], the model shown in Fig. 4.12(b) is constructed. Observe that for better accuracy, the usual collector diode is replaced by a nonlinear resistor whose *v-i* curve is shown in Fig. 4.12(c). The output collector-to-emitter curves of this model are shown in Fig. 4.12(d), and represent a fairly good representation of the actual transistor characteristics. After this model is substituted in place of the transistors, all nodes and elements of the resulting circuit are assigned numbers, and the necessary control, parameter, and description cards are punched. A list of the input information is shown in the upper portion of the print-out sheet shown in Fig. 4.13. The two diodes in the two transistor models are labeled D1 and D2, and the two nonlinear resistors are labeled NL1 and NL2. The middle portion of Fig. 4.13 shows the breakpoints of the desired TC Plot. Since the TC Plot in this case consists of a single unicursal branch,

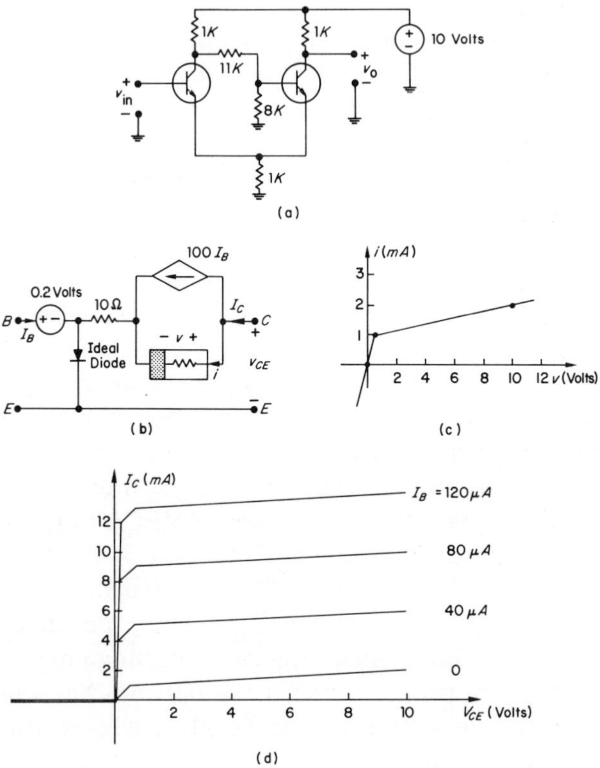

FIG. 4.12. A Schmitt trigger circuit and the transistor model used for the computer analysis.

```
  CASE          DELTA
TRIGGER           6

 PARAMETER      INIT TERM                                  INPUT                INPUT
NUMBER NAME     NODE NODE TYPE INDEX                     VALUE A              VALUE B
   1  RC1         6    2    1                        1.0000000E+03
   2  RC2         6    1    1                        1.0000000E+03
   3  RCB         2    5    1                        1.1000000E+04
   4  RB2         5    0    1                        8.0000000E+03
   5  RF          4    0    1                        1.0000000E+03
   6  E1          0    6   11                        1.0000000E+01
   7   D1        10    4   22                        3.0000000E+00
                           26                        0.             *
                           30                        0.             *     0.             *
                           27                        1.0000000E+00  *
   8   NL1        2    7   22                        3.0000000E+00
                           28                        0.             *     0.             *
                           30                        5.0000000E-01  *     1.0000000E-03  *
                           29                        1.0000000E+01  *     2.0000000E-03  *
   9   BETA1      2    7    7    14                  1.0000000E+02
  10   D2         8    4   22                        3.0000000E+00
                           26                        0.             *
                           30                        0.             *     0.             *
                           27                        1.0000000E+00  *
  11   NL2        1    9   22                        3.0000000E+00
                           28                        0.             *     0.             *
                           30                        5.0000000E-01  *     1.0000000E-03  *
                           29                        1.0000000E+01  *     2.0000000E-03  *
  12   BETA2      1    9    7    16                  1.0000000E+02
  13  VIN         0    3   11                        0.
  14   VCO1       3   10   11                       -2.0000000E-01
  15   RA1       10    7    1                        1.0000000E+01
  16   VCO2       5    8   11                       -2.0000000E-01
  17   RA2        8    9    1                        1.0000000E+01

* INDICATES INPUT DATA FOR NONLINEAR PARAMETERS.  OTHER INPUTS ARE NOMINAL VALUES.

.......................................................................................

NONLINEAR COMPONENTS

VOLTAGE LIMITS FOR SEGMENTS ARE VA AND VB.
CURRENT LIMITS FOR SEGMENTS ARE IA AND IB.
THE POINT (VA,IA) IS NOT IN THE REGION OF DEFINITION UNLESS THE SEGMENT IS A FINITE END SEGMENT.
THE POINT (VB,IB) IS IN THE REGION OF DEFINITION WHENEVER THE POINT IS FINITE.

 PARAMETER   SEG-                                                                                  SHORT CIRCUIT
NUMBER NAME  MENT        VA               VB              IA               IB          CONDUCTANCE      CURRENT
   7   D1     1        - INF.        0.               0.               0.               0.              0.
              2   0.                   + INF.         0.                 + INF.         1.0000000E+00   0.
   8   NL1    1        - INF.        5.0000000E-01      - INF.         1.0000000E-03    2.0000000E-03   0.
              2   5.0000000E-01        + INF.         1.0000000E-03      + INF.         1.0526316E-04   9.4736842E-04
  10   D2     1        - INF.        0.               0.               0.               0.              0.
              2   0.                   + INF.         0.                 + INF.         1.0000000E+00   0.
  11   NL2    1        - INF.        5.0000000E-01      - INF.         1.0000000E-03    2.0000000E-03   0.
              2   5.0000000E-01        + INF.         1.0000000E-03      + INF.         1.0526316E-04   9.4736842E-04

...................................................................................................

TRANSFER CHARACTERISTIC

DRIVING PARAMETER
   NUMBER    NAME    INIT NODE    TERM NODE    INDEPENDENT VARIABLE
     13      VIN         0            3               VOLTAGE

OUTPUT NODE  1

PLOT BREAKPOINTS LISTED BY INDIVIDUAL LINE

   LINE  1
POINT       DRIVING       OUTPUT          DRIVING      OUTPUT          DRIVING      OUTPUT          DRIVING      OUTPUT
 NO.        VOLTAGE       VOLTAGE         VOLTAGE      VOLTAGE         VOLTAGE      VOLTAGE         VOLTAGE      VOLTAGE
 1- 4     1.00000E+01   9.98031E+00     4.81520E+00  9.98236E+00     2.97893E+00  9.98311E+00     3.87377E+00  6.35047E+00
 5- 5     0.            6.35198E+00

* PRECEDES A POINT ON AN INFINITE SEGMENT
...................................................................................................
```

(a)

```
TRANSFER CHARACTERISTIC          INPUT PARAMETER NUMBER 13     OUTPUT NODE  1

ALL SEGMENTS WHICH EXTEND TO EDGE OF GRAPH ARE INFINITE.
BASIC VOLTAGE UNITS ARE VOLTS.    CURRENT UNITS ARE AMPS.

LINE  1              VERTICAL AXIS IS  OUTPUT VOLTAGE  X  1E+01
```

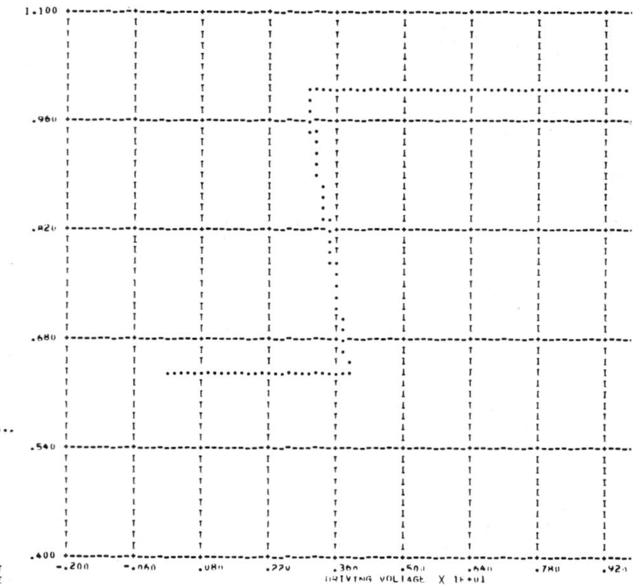

(b)

FIG. 4.13. Actual computer printed output of the Schmitt trigger analysis by MECA.

all breakpoints are printed under Curve 1 in Fig. 4.13. The final portion of Fig. 4.13 shows the printed graph of this TC Plot. Observe that the collector output voltage $v_0(t)$ will jump from 6.4 volts to 10 volts whenever the input voltage $v_{in}(t)$ increases beyond 4 volts. When this voltage is then decreased, the output remains at 10 volts. It will take $v_{in}(t)$ to decrease below 3 volts before the output voltage drops back to 6.4 volts. This hysteresis phenomenon is exactly what is observed in practice.

4.5.3 Global Analysis of Second-Order Large-Signal Electronic Circuits

The iterative piecewise-linear method can be used to solve for all equilibrium states of any multivalued network. The behavior of the trajectories in the vicinity of each equilibrium state can be found by computer solution of the initial-value problems with the initial-conditions chosen near the equilibrium states. For second-order large-signal electronic circuits, the phase plane method is usually more instructive [32]. We will discuss this method for the network shown in Fig. 4.14(a),

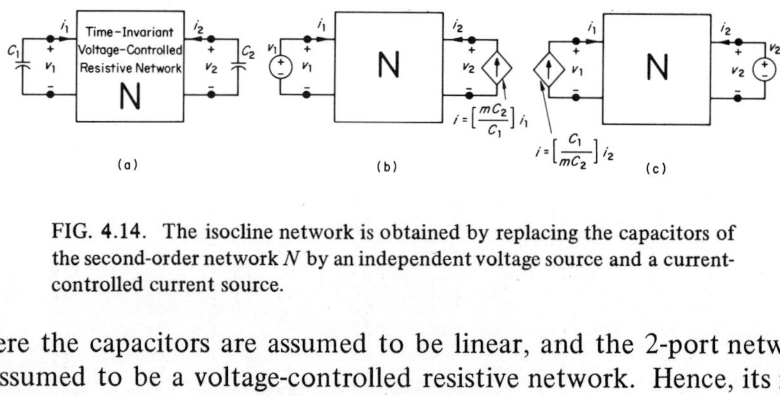

FIG. 4.14. The isocline network is obtained by replacing the capacitors of the second-order network N by an independent voltage source and a current-controlled current source.

where the capacitors are assumed to be linear, and the 2-port network is assumed to be a voltage-controlled resistive network. Hence, its normal form equations can be written as follows:

$$dv_1/dt = f_1(v_1, v_2) \tag{4.106}$$

$$dv_2/dt = f_2(v_1, v_2) \tag{4.107}$$

Dividing (4.107) by (4.106), we obtain

$$dv_2/dv_1 = f_2(v_1, v_2)/f_1(v_1, v_2) \tag{4.108}$$

Equation (4.108) represents the slope of the trajectories in the v_1-v_2 plane – the phase plane. Conversely, by equating (4.108) to a constant m; namely

$$f_2(v_1, v_2) = mf_1(v_1, v_2) \tag{4.109}$$

we obtain the equation of the *isocline* with slope m. If we draw a short segment with slope m – called a director – through each point of this isocline, then any trajectory intersecting the isocline at a point P must be tangent to the director at P. If enough isoclines and their associated directors are drawn, we obtain the so-called *phase portrait*. If the solution is unique, then no isoclines may intersect one another except at equilibrium points where (4.109) is indeterminate. In other words, the intersections of all isoclines automatically give the location of the equilibrium states.

Given any initial state, a trajectory can be easily drawn tangent to the directors. Hence, the behavior of the trajectories in the vicinity of each equilibrium state can be easily sketched. In other words, the phase portrait gives a complete picture of the behavior of the network.

In spite of its utility, very few second-order large-signal electronic circuits have been analyzed by the phase plane method. The reason can be traced to the difficulty in deriving the isocline equation (4.109). In fact, as we indicated earlier, it is impractical to write the normal form equations analytically. Hence, in order to obtain the isocline equations, a means must be found independent of the normal form equations. It is possible to do this if we construct a new resistive network by replacing the capacitors C_1 and C_2 of the network in Fig. 4.14(a) by an independent source and a current-controlled current source as shown in Fig. 4.14(b). Alternately, we may construct the resistive network shown in Fig. 4.14(c). For reasons that will be obvious, each of these networks is called an *isocline network* [37].

Isocline Network Theorem. The isocline (with slope m) of the second-order capacitor network shown in Fig. 4.14(a) is identical to the v_2 vs. v_1 TC Plot of the isocline network shown in Fig. 4.14(b), or the v_1 vs. v_2 TC Plot of the isocline network shown in Fig. 4.14(c).

Proof: From Fig. 4.14(a), we obtain

$$dv_1/dt = -i_1/C_1 \tag{4.110}$$

$$dv_2/dt = -i_2/C_2 \tag{4.111}$$

Dividing (4.111) by (4.110) and equating the result to m, we obtain

$$i_2 = [mC_2/C_1]i_1 \tag{4.112}$$

$$i_1 = [C_1/mC_2]i_2 \tag{4.113}$$

Observe that the constraint imposed by (4.112) is satisfied by the resistive network shown in Fig. 4.14(b). Similarly, the constraint imposed by (4.113) is satisfied by the resistive network shown in Fig. 4.14(c). Since the isocline of the network shown in Fig. 4.14(a) is simply a "relationship" between v_2 and v_1 when (4.112) and (4.113) is satisfied, it is clear that it must coincide with the v_2 vs. v_1 TC Plot of the network shown in Fig. 4.14(b), or the v_1 vs. v_2 TC Plot of the network shown in Fig. 4.14(c). This completes the proof.†

To illustrate the application of this theorem, consider the switching circuit shown in Fig. 4.15(a), where the tunnel diode v_d-i_d curve is shown in Fig. 4.15(b). The isocline network is easily drawn as in Fig. 4.15(c) for each value of m, we solve for the v_2 vs. v_1 TC Plot, and hence

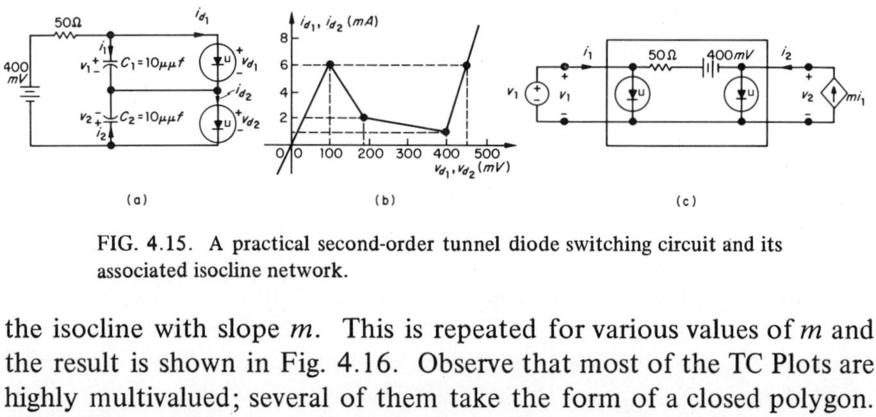

FIG. 4.15. A practical second-order tunnel diode switching circuit and its associated isocline network.

the isocline with slope m. This is repeated for various values of m and the result is shown in Fig. 4.16. Observe that most of the TC Plots are highly multivalued; several of them take the form of a closed polygon. This clearly shows that the iterative piecewise linear method is useful not only for finding multivalued operating points, but also multivalued TC Plots. The reader should notice that all TC Plots in Fig. 4.16 intersect at five points – the equilibrium states. From the three typical trajectories shown in Fig. 4.16, it is clear that the equilibrium states Q_1, Q_3, and Q_5, are stable, whereas the rest are unstable.

† It is easy to derive an analogous theorem corresponding to the case where the energy storage elements consist of two inductors, or one capacitor and one inductor, respectively [37].

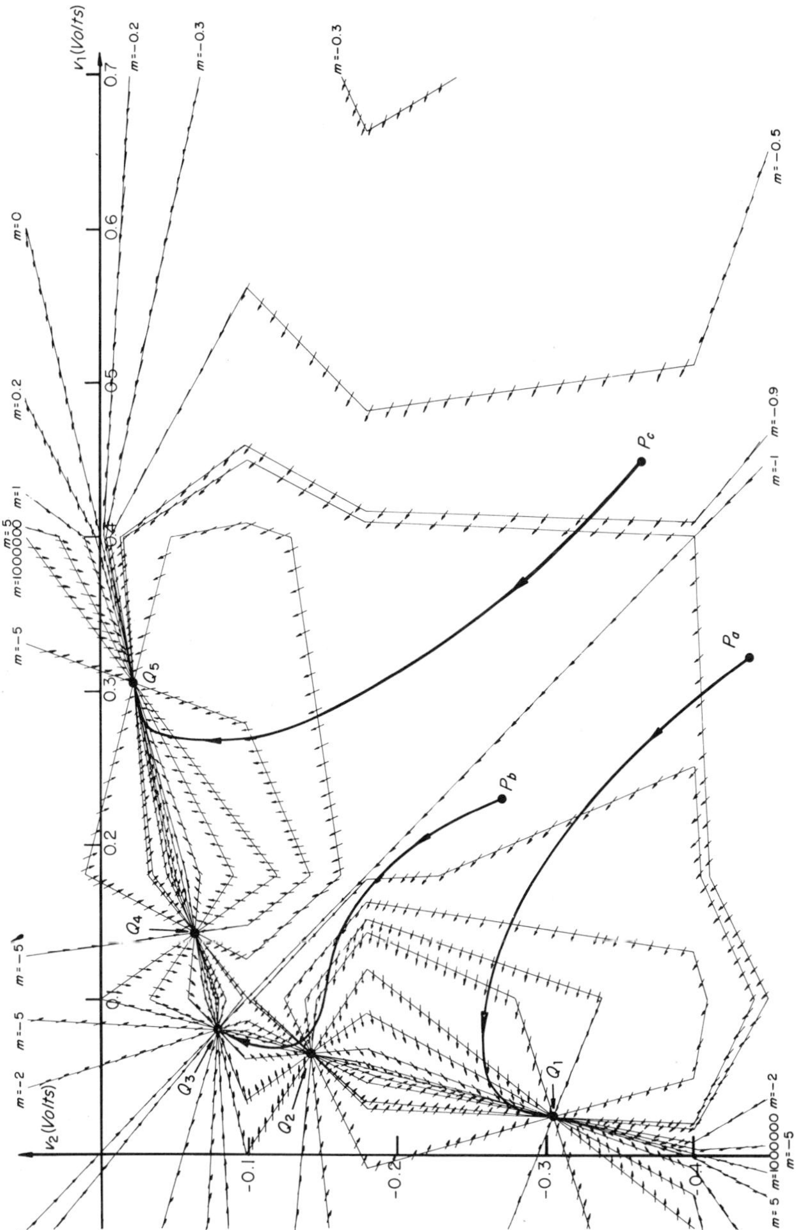

FIG. 4.16. The phase-portrait of the tunnel diode switching circuit is obtained by applying the isocline network theorem.

REFERENCES

1. L. O. Chua, *Introduction to Nonlinear Network Theory*, McGraw-Hill Book Company, Inc., New York, 1969.

2. C. A. Desoer and E. S. Kuh, *Basic Circuit Theory*, McGraw-Hill Book Company, Inc., New York, 1969.

3. T. E. Stern, *Theory of Nonlinear Networks and Systems, An Introduction,* Addison-Wesley Publishing Company, Reading, Massachusetts, 1965.

4. R. K. Brayton and J. K. Moser, "A theory of nonlinear networks," *Quart. Appl. Math.* vol. 12, pp. 1-33, 81-104, April, 1964.

5. C. A. Desoer and J. Katzenelson, "Nonlinear RLC networks," Bell Sys. Tech. J., pp. 161-198, January, 1965.

6. L. O. Chua and R. A. Rohrer, "On the dynamic equations of nonlinear RLC networks," *IEEE Trans. on Circuit Theory*, vol. CT-12, pp. 475-489, December, 1965.

7. T. E. Stern, "On the equations of nonlinear networks," *IEEE Trans. on Circuit Theory*, vol. CT-13, pp. 74-81, March, 1966.

8. E. S. Kuh, "Representation of nonlinear networks," *Proc. NEC,* vol. XXI, 1965.

9. H. N. Ghosh, F. H. DeLa Moneda, and N. R. Dono, "Computer-aided transistor design, characterization, and optimization," *Proc. of the IEEE*, vol. 55, No. 11, November, 1967, pp. 1897-1912.

10. B. D. Roberts, Jr. and C. O. Harbourt, "Computer models of the field-effect transistor," *Proc. of the IEEE*, vol. 55, No. 11, November, 1967, pp. 1921-1929.

11. L. O. Chua and R. A. Rohrer, "Nonlinear network analysis – the parametric approach," *Coordinated Science Laboratory*, Report R-224, University of Illinois, July 1, 1964.

12. T. M. Apostol, *Mathematical Analysis*, Addison-Wesley, Reading, Mass., 1961.

13. A. Ralston, *A First Course in Numerical Analysis*, McGraw-Hill Book Company, Inc., New York, 1965.

14. R. W. Hamming, *Numerical Methods for Scientists and Engineers*, McGraw-Hill Book Company, Inc., New York, 1962.

15. P. Henrici, *Discrete Variable Methods in Ordinary Differential Equations*, John Wiley and Sons, Inc., 1962.

16. P. R. Bryant, "The order of complexity of electrical networks," *Proc. IEE* (London), Monograph 335E, vol. 106 C., pp. 174-188, June, 1959.

17. S. Seshu and M. B. Reed, *Linear Graphs and Electrical Networks*, Addison-Wesley, Reading, Mass., 1961.

18. F. H. Branin, Jr., "The inverse of the incidence matrix of a tree and the formulation of the algebraic-first-order differential equations of an RLC network," *IEEE Trans. on Circuit Theory* (*Correspondence*), vol. CT-10, pp. 543-544, December, 1963.

19. J. Katzenelson, "AEDNET: A simulator for nonlinear networks," *Proc. of the IEEE*, vol. 54, No. 11, November, 1966, pp. 1536-1552.

20. R. Palais, "Natural operations on differential forms," *Trans. of American Mathematical Society*, vol. 92, No. 1, pp. 125-141, July, 1959.

21. H. R. Hall, *Analysis of Nonlinear Networks*, Ph.D. Dissertation, University of California, Berkeley, California, 1967.

22. R. J. Duffin, "Nonlinear networks (I, II, and II b)," *Bull. Amer. Math. Soc.*, vols. 52, 53 and 54, pp. 836-838, 963-971, and 119-127; 1946, 1947, and 1948.

23. A. M. Ostrowski, *Solution of Equations and Systems of Equations*, 2nd ed., Academic Press, New York, 1966.

24. J. F. Traub, *Iterative Methods for the Solution of Equations*, Prentice-Hall, Inc., Englewood Cliffs, N. J.

25. L. Collatz, *Functional Analysis and Numerical Mathematics*, Academic Press, New York, 1966.

26. C. G. Broyden, "A class of methods for solving nonlinear simultaneous equations," *Mathematics of Computation*, vol. 19, pp. 577-593, 1965.

27. F. H. Branin, Jr., "A fast reliable iteration method for dc analysis of nonlinear networks," *Proc. of the IEEE*, vol. 55, No. 11, Nov., 1967, pp. 1819-1826.

28. E. A. Coddington and N. Levinson, *Theory of Ordinary Differential Equations*, McGraw-Hill, New York, 1955.

29. V. V. Nemytskii and V. V. Stepanov, *Qualitative Theory of Differential Equations*, Princeton University Press, Princeton, New Jersey, 1960.

30. A. Wintner, "The non-local existence problem of ordinary differential equations," *Amer. Jour. Math.*, vol. 67, 1945, pp. 277-284.

31. L. O. Chua, "State variable formulation of nonlinear RLC networks in explicit normal form," *Proc. of the IEEE*, vol. 53, No. 2, February, 1965, pp. 206-207.

32. N. Minorsky, *Nonlinear Oscillations,* Van Nostrand, Princeton, New Jersey, 1962.

33. L. O. Chua, "Analysis and synthesis of multivalued nonlinear networks," *IEEE Trans. on Circuit Theory*, vol. CT-15, June 1967, pp. 192-209.

34. L. O. Chua and W. H. Stellhorn, "Computer aided analysis and synthesis of multivalued memoryless networks," Purdue University Report TR-EE-67-8, School of Electrical Engineering, June, 1967.

35. Coded Circuit Analysis, vol. 1, "Instructional manual for dc-ac applications," Technical Report 544, U.S. Naval Avionics Facility, Indianapolis, Indiana.

36. L. O. Chua and P. Medlock, "NECA—Multivalued electronic circuit analysis program," Purdue University Technical Report TR-EE-69-7, School of Electrical Engineering, April 30, 1969.

37. L. O. Chua, "Two new theorems in nonlinear networks," *Proc. of the Eleventh Midwest Symposium on Circuit Theory,* University of Notre Dame, South Bend, Indiana, May 13-14, 1968.

CHAPTER 5

OPTIMIZATION METHODS IN CIRCUIT DESIGN

G. C. Temes, Ampex Corporation

INTRODUCTION

Optimization techniques are needed in the frequent case when the synthesis procedures of classical circuit theory are, for some reason, inapplicable. They are, typically, *iterative* in nature; the network is repeatedly analyzed, its performance compared with the specifications, and, if necessary, the parameters are adjusted *in a systematic way*. Without this last qualification, of course, the analysis-cum-adjustment process degenerates into a computer-aided cut-and-try method, the equivalent of the screwdriver-adjustment technique used by "on-the-bench" designers. As anyone who has attempted this procedure knows, convergence via cut-and-try is hopeless for circuits with, say, seven or more variable parameters. For this reason, the sophisticated analysis methods of the previous chapters should be supplemented by equally clever techniques which enable the designer to improve his initial or intermediate results.

Depending on the requirements, an optimization program may solve only the approximation problem and exit when a suitable transfer function has been obtained, or it may perform the realization only, from a given transfer function, or, finally, it may carry out the whole task, obtaining the circuit directly from the specifications. The flowcharts for these three cases are shown in Fig. 5.1.

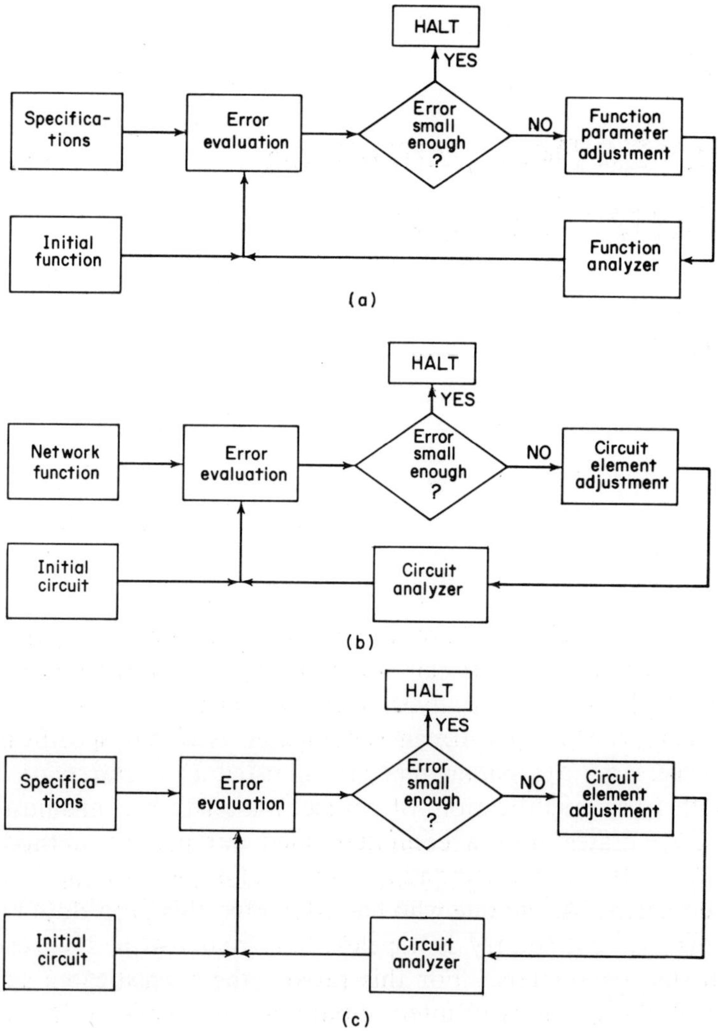

FIG. 5.1. Flow chart for optimization programs. (a) approximation, (b) realization, (c) direct design.

The partial procedures described by the flowcharts of Fig. 5.1(a) and (b) are justified in practice if the rest of the design can be carried out via exact synthesis. In general, *any* part of the design procedure which does not absolutely require the use of iterative techniques, should be done in a noniterative way. The reason behind this is that iterative procedures are notoriously slow: the required computer-time is typically one or two orders of magnitude longer for an iterative design than for a comparable noniterative one. In addition, iterative techniques may not converge at all, and, if they do, the convergence may be to a *local optimum*, whereas most classical synthesis procedures are guaranteed to lead to a *global optimum*.

Any iterative optimization method should, ideally, satisfy the following requirements:

1) Fast execution
2) Ease of programming for a computer
3) Ease of feeding input data, circuit configuration into the program
4) Readily interpreted output
5) Assured convergence to a global optimum
6) Flexibility in accepting a wide variety of circuits and constraints
7) Accuracy of output
8) Insensitivity to round-off errors, ill-conditioned inputs

These properties are, unfortunately, contradictory A reasonable compromise, based on the application and anticipated frequency of use must be found. For a program that is widely and often used, for example, requirements 1), 3), and 4) outweigh requirement 2), and vice versa.

5.1 DEFINITIONS AND NOTATION

The notations of an earlier work [1] will be followed. Accordingly, let E be the error criterion, i.e., a single number which measures the performance of the network; let x be the independent variable of the response function, representing usually frequency and/or time, let $F(x)$ be the desired response and $F^j(x)$ the actual response after the jth iterative adjustment.† Furthermore, let the vector

$$\mathbf{c} = (c_1, c_2, \ldots, c_n)^T \tag{5.1}$$

contain the n adjustable circuit parameters c_i as components; the value

† Here the superscript j is an iteration index, not an exponent.

of $\mathbf{c}$ after the jth iteration will be called $\mathbf{c}^j$ and its increment, obtained in the next iteration, $\Delta\mathbf{c}^j$.

The performance of the network at any single given x is described by the error function†

$$\epsilon(x, \mathbf{c}^j) \triangleq w(x)\,[F(x) - F^j(x, \mathbf{c}^j)] \tag{5.2}$$

where $w(x) \geqslant 0$ is the weighting function. The error criterion E may be obtained from $\epsilon(x)$ several ways. Typically, several sample points x_i are selected in the approximation range of interest $a \leqslant x \leqslant b$ and E is defined via

$$E \triangleq \sum_{(i)} \epsilon_i^2 \tag{5.3}$$

where $\epsilon_i \triangleq \epsilon(x_i, \mathbf{c}^j)$; or

$$E \triangleq \max_{(i)} |\epsilon_i| \,. \tag{5.4}$$

In the former case, minimization of $E(\mathbf{c})$ leads to a least-squares minimum; in the latter, to a min-max or Chebyshev solution of the problem.

In general, $E(\mathbf{c})$ may be expanded into a Taylor series about a given $\mathbf{c}$

$$E(\mathbf{c} + \Delta\mathbf{c}) = E(c_1, c_2, \ldots, c_n) + \sum_{i=1}^{n} \frac{\partial E}{\partial c_i}\Delta c_i + \tag{5.5}$$

$$\frac{1}{2}\sum_{i=1}^{n}\sum_{j=1}^{n} \frac{\partial^2 E}{\partial c_i \partial c_j}\Delta c_i \Delta c_j + \cdots = E(\mathbf{c}) + \nabla\mathbf{E}^T \cdot \Delta\mathbf{c} + \frac{1}{2}\Delta\mathbf{c}^T\,\mathbf{H}\,\Delta\mathbf{c} + \ldots$$

In (5.5),

$$\nabla\mathbf{E} \triangleq \left(\frac{\partial E}{\partial c_1}, \frac{\partial E}{\partial c_2}, \ldots, \frac{\partial E}{\partial c_n}\right)^T$$

is the error gradient, and $\mathbf{H}$, the symmetric $n \times n$ matrix of second partial derivatives, with its i, j element defined as $\partial^2 E/\partial c_i \partial c_j$, is the Hessian matrix.

† The sign $\triangleq$ means "defined as."

Assume now that, at some $\mathbf{c} = \hat{\mathbf{c}}$, E is minimized. Then, for *any* small $\Delta\mathbf{c}$, $E(\mathbf{c})$ of (5.5) is increased. Hence, necessarily $\nabla \mathbf{E} = 0$ and $\mathbf{H}$ must be a positive definite matrix.† If $\mathbf{H}$ is positive definite in some region R of the $\mathbf{c}$-space, then $E(\mathbf{c})$ is called convex in that region. An equivalent condition [2] of convexity is that

$$E[\mathbf{c}_a + \lambda(\mathbf{c}_b - \mathbf{c}_a)] \leqslant E(\mathbf{c}_a) + \lambda[E(\mathbf{c}_b) - E(\mathbf{c}_a)] \tag{5.6}$$

for $0 \leqslant \lambda \leqslant 1$ and for any $\mathbf{c}_a$, $\mathbf{c}_b$ in R. This condition may easily be visualized via Fig. 5.2.

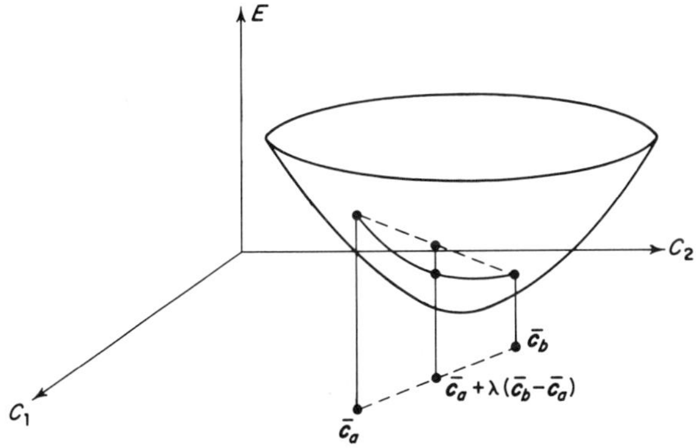

FIG. 5.2. Convex error-surface.

5.2 LINEAR AND LINEARIZED PROGRAMMING

Statement of the Problem

In some design situations E depends linearly on the c_i. For example, if $F^j(x, \mathbf{c}^j)$ is a linear function of the c_h;

$$F^j(x, \mathbf{c}^j) = \sum_{h=1}^{n} c_h f_h(x) \tag{5.7}$$

and a Chebyshev error-criterion is chosen, then, by (5.4)

† In degenerate cases, H may be semidefinite at the minimum. This leads to a much more involved situation, which will not be discussed here.

$$E \triangleq \max_{(i)} \left| w(x_i)\,[F(x_i) - \sum_{h=1}^{n} c_h f_h(x_i)] \right| \tag{5.8}$$

so that at all sample points x_i the inequalities

$$\begin{aligned} E - w(x_i)[F(x_i) - \sum_{h=1}^{n} c_h f_h(x_i)] &\geqslant 0 \\ E + w(x_i)[F(x_i) - \sum_{h=1}^{n} c_h f_h(x_i)] &\geqslant 0 \end{aligned} \tag{5.9}$$

hold. The design aim is then to minimize E subject to the inequalities (5.9). Notice that the inequality relations (5.9) are linear in the unknowns E and all c_h, as postulated.

This problem and others in this category may be stated in the general form of a *linear program (LP)* as follows: find the unknowns $x_1, x_2, \ldots, x_k$ such as to maximize

$$z \triangleq r_1 x_1 + r_2 x_2 + \cdots + r_k x_k \tag{5.10}$$

subject to the conditions:

$$\begin{aligned} a_{11}x_1 + a_{12}x_2 + \cdots + a_{1k}x_k &\leqslant l_1 \\ a_{21}x_1 + a_{22}x_2 + \cdots + a_{2k}x_k &\leqslant l_2 \\ \cdots\cdots\cdots & \qquad m \leqslant k \\ a_{m1}x_1 + a_{m2}x_2 + \cdots + a_{mk}x_k &\leqslant l_k \end{aligned} \tag{5.11}$$

and

$$x_1 \geqslant 0, \qquad x_2 \geqslant 0, \qquad \ldots, \qquad x_k \geqslant 0. \tag{5.12}$$

LP may equivalently be expressed in the matrix form:
Maximize

$$z \triangleq \mathbf{r}^T\mathbf{x} \tag{5.13}$$

subject to

$$\mathbf{A}\mathbf{x} \leqslant \boldsymbol{l} \tag{5.14}$$

and

$$\mathbf{x} \geqslant \mathbf{0}. \tag{5.15}$$

In yet another formulation, LP is:

$$z \triangleq \mathbf{r}^T \mathbf{x} \tag{5.16}$$

subject to

$$\sum_{i=1}^{k} \mathbf{a}_i x_i \leqslant \boldsymbol{l} \tag{5.17}$$

and

$$\mathbf{x} \geqslant 0. \tag{5.18}$$

Usually, z is called the *objective function,* $\boldsymbol{l}$ the *requirement vector,* $\mathbf{r}$ *the price vector* (where r_i is the *price* of x_i) and the $\mathbf{a}_i$ the *activity vectors.* Any vector $\mathbf{x}$ satisfying the inequalities (5.11) – (5.12) is called a *feasible solution*; if it also maximizes z, it will be called the *optimum solution.*

The above given formulation of the optimization problem appears to be unnecessarily restricted and hence overspecialized. In reality, however, a number of simple manipulations may be employed in order to extend the usefulness of the method to a wide variety of problems. Some of these will be listed below.

1. *Unrestricted variables.* If some variable x_i is not restricted to positive values, it may be replaced by two new nonnegative variables x_i^+ and x_i^-, via $x_i \triangleq x_i^+ - x_i^-$, in (5.10) and (5.11). Then, (5.12) will be replaced by

$$\begin{gathered} x_1 \geqslant 0,\ x_2 \geqslant 0,\ \ldots,\ x_{i-1} \geqslant 0,\ x_i^+ \geqslant 0,\ x_i^- \geqslant 0, \\ x_{i+1} \geqslant 0, \ldots,\ x_k \geqslant 0, \end{gathered} \tag{5.19}$$

and, in terms of these new variables, again a standard LP results. Note that in this case in the optimum solution either x_i^+ or x_i^- (or both) will be zero, so x_i will become x_i^+ or $-x_i^-$ (or zero).

2. *Minimization of the objective function.* If a linear form $\mathbf{r}^T\mathbf{x}$ is to be minimized rather than maximized, then replacing $\mathbf{r}$ by $-\mathbf{r}$ results in the requirement of maximizing $(-\mathbf{r}^T\mathbf{x})$. This again is in the form (5.13).

3. *Variables restricted from above.* If, instead of (5.12), some variables x_i are constrained by inequalities of the form

$$x_i \leqslant a_i \tag{5.20}$$

then defining

$$x_i' \triangleq a_i - x_i \tag{5.21}$$

will again result in a standard LP.

4. *Variables restricted from below.* For constraints of the form

$$x_i \geqslant b_i \tag{5.22}$$

the new variables

$$x_i' = x_i - b_i \tag{5.23}$$

may be used.

5. *Maximization of absolute values.* If some of the variables are unrestricted, and their absolute values appear in z

$$z = r_1 x_1 + r_2 x_2 + \cdots + r_i \, | x_i | + \cdots + r_k x_k \tag{5.24}$$

then again using the substitution $x_i \triangleq x_i^+ - x_i^-$ in (5.11) and replacing the ith term in z by $r_i(x_i^+ + x_i^-)$ will result in standard form. The validity of this process is due to the fact, mentioned above, that the optimum value of x_i will be either x_i^+ or $-x_i^-$ or zero.

If *only* the ith term is present in z, the techniques used in deriving (5.9) may be expediently used.

6. *Variables restricted from above and below.* If constraints of the form

$$b_i \leqslant x_i \leqslant a_i \tag{5.25}$$

are present, then introducing

$$x_i' \triangleq \frac{a_i - x_i}{a_i - b_i} \tag{5.26}$$

results in the new constraints

$$x_i' \leqslant 1$$
$$x_i' \geqslant 0. \tag{5.27}$$

These are in standard form.

7. *Equality constraints.* Some (or all) of the inequalities in (5.11) may be replaced by equality constraints of the form

$$a_{j_1}x_1 + a_{j_2}x_2 + \cdots + a_{j_k}x_k = l_j \tag{5.28}$$

This does not impede solution, as shown later in the section.

8. *"Larger than" constraints.* If the jth constraint contains the "$\geqslant$" rather than "$\leqslant$" sign, then replacing all a_{j_1}, a_{j_2}, ..., and also l_j by their negatives leads to standard formulation.

9. *The case $m > k$.* This case can be handled on the basis of the duality theorem, explained in eqs. (5.49) – (5.54), below.

10. *Nonlinear relations.* If either the object function or any of the constraints is nonlinear, Taylor series expansion may be used to linearize all relations. This technique, called *approximation programming*, will be described later on in this section.

The Simplex Method

An efficient solution of the LP problem described in Eqs. (5.13) – (5.15) is possible using the so-called simplex method. To explain the principles of this method, assume initially that

$$\boldsymbol{l} \geqslant 0 \tag{5.29}$$

and that none of relations in (5.14) is an equality. Then, (5.14) – (5.15) represent $m + k$ inequality constraints on the k unknown x_i. In the k-dimensional parameter space these constraints define a feasible (i.e., permissible) polyhedral subspace. A *vertex* is defined as a point in the parameter-space, where exactly k constraints are active, i.e., are satisfied with the equality sign. There are a total of

$$\binom{m+k}{k} = \binom{m+k}{m} = \frac{(m+k)!}{m!k!}$$

vertices, since the k active constraints may be selected from the total of $m + k$ in $\binom{m+k}{k}$ different ways. If the m inactive inequalities are also satisfied (with the $<$ or $>$ sign), then the vertex is a geometric vertex of the permissible polyhedron; it is then called a *feasible vertex*.

The solution of the LP problem lies at one of the feasible vertices. This is intuitively obvious from $k = 2$ or 3. See, e.g., Fig. 5.3,* where

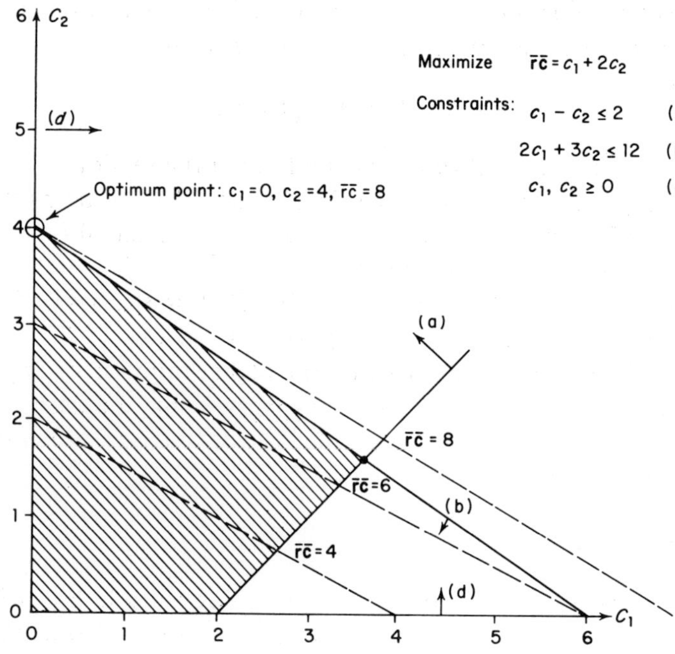

FIG. 5.3. Linear program.*

the simple case $\mathbf{r}^T = (1, 2)$, $\mathbf{a}_1^T = (1, 2)$, $\mathbf{a}_2^T = (-1, 3)$, $\boldsymbol{l}^T = (2, 12)$ is illustrated [1]. There, only 4 of the 6 vertices are feasible† and evidently for any objective function (manifested by dotted parallel lines) the optimum lies at one of them. For the general case, assume that the optimum solution does *not* lie at a feasible vertex. Then only $l < k$ inequalities are active; use these to express any l of the x_i in terms of the others. Thus, we obtain $k - l$ variables x_i, unhindered by active inequalities, upon which z depends linearly. Surely it is possible to change one or more of these x_i by small amounts such as to increase z further, without violating the inactive $m + k - l$ inequalities. Hence the initial solution was not optimal.

† The non-feasible ones are at (0, 6) and (–2, 0).

* In Figs. 5.3 and 5.4, $\mathbf{c} \equiv \mathbf{x}$.

It is clear at this point that by evaluating z at all feasible vertices the optimum solution can be found. Unfortunately, this involves the solution of k linear equations, plus checking the inactive constraints, at all $\binom{m+k}{m}$ vertices. For a moderately complex problem (e.g., $k = 10$, $m = 7$) the number of vertices can easily become huge (in our example, 19,448) ruling out this simplistic approach. Hence, instead of this full evaluation, a gradual movement from one feasible vertex to an *adjacent feasible vertex* is performed. An adjacent vertex, in our context, is one obtained by replacing one of the active constraints by a new one, hitherto inactive. A systematic way of performing this in such a way that z is increased by a large amount in each step is provided by the *simplex method* [3].

For the simplifying assumption (5.29) and in the absence of equality relations in (5.14), the procedure involves the following steps:

1. Add an m element column vector $\mathbf{x}_s$ to the LHS of (5.14) so as to transform it into a system of equalities. Then

$$\mathbf{x}_s \triangleq \boldsymbol{l} - \mathbf{A}\mathbf{x} > 0 \tag{5.30}$$

$$\mathbf{A}\mathbf{x} + \mathbf{x}_s = \mathbf{A}\mathbf{x} + \mathbf{I}_m \mathbf{x}_s = \boldsymbol{l} \tag{5.31}$$

The components of $\mathbf{x}_s$ are called, for obvious reasons, *slack variables.* Defining the $(m + k)$ component vectors $\mathbf{s}$ and $\mathbf{y}$, as well as the m by $(m + k)$ matrix $\mathbf{B}$ via

$$\mathbf{s} \triangleq \begin{pmatrix} \mathbf{r} \\ \mathbf{0} \end{pmatrix}; \quad \mathbf{y} \triangleq \begin{pmatrix} \mathbf{x} \\ \mathbf{x}_s \end{pmatrix}; \quad \mathbf{B} \triangleq (\mathbf{A} \,\vdots\, \mathbf{I}_m) \tag{5.32}$$

the optimization problem becomes:

Maximize: $$z \triangleq \mathbf{s}^T \mathbf{y}$$

subject to: $$\mathbf{B}\mathbf{y} = \boldsymbol{l} \tag{5.33}$$

and $$\mathbf{y} \geqslant 0. \tag{5.34}$$

In terms of $\mathbf{y}$, a value is feasible if (5.33) – (5.34) are satisfied; a vertex is obtained if any k components of $\mathbf{y}$ equal zero, leaving m nonzero components.

2. A feasible vertex is chosen as initial point. This now can be obtained by selecting $\mathbf{x} = \mathbf{0}$, $\mathbf{x}_s = \boldsymbol{l}$, so that

$$\mathbf{y}_0 = \begin{pmatrix} \mathbf{0} \\ \boldsymbol{l} \end{pmatrix}$$

$\mathbf{y}_0$ is certainly a feasible vertex, as substitution into (5.33) – (5.34) shows, although the corresponding objective function $z_0 = 0$ is presumably not optimal (unless all $r_i < 0$). Hence, we proceed to an adjacent vertex.

3. The next vertex should be feasible, adjacent and such that z is increased by a large amount. To reach an adjacent vertex, one of the nonzero components $y_i\,(i > k)$ of $\mathbf{y}_0$ should be made zero and, vice versa, a zero component $y_j\,(j \leqslant k)$ should be given a nonzero value. This involves two decisions. The first one is the choice of y_j. This can be made on the basis that the corresponding increase in z will be $s_j y_j = r_j y_j$. Hence, choosing j such that

$$r_j = \max_{(l)} r_l$$

seems to be advisable.† Next, the value of y_j and the choice of y_i should be determined. Consider (5.33). After y_i is removed and y_j introduced, this equation system will be of the form

$$\begin{aligned}
a_{1j}y_j + y_{k+1} &= l_1 \\
a_{2j}y_j + y_{k+2} &= l_2 \\
\cdots\cdots\cdots \\
a_{i-k,j}y_j + 0 &= l_{i-k} \\
\cdots\cdots\cdots \\
a_{mj}y_j + y_{k+m} &= l_m
\end{aligned} \tag{5.35}$$

From (5.35)

$$y_j = \frac{l_{i-k}}{a_{i-k,j}} \tag{5.36}$$

and, for $h > k$,

† An even better process would be to maximize $r_j y_j$. The value of y_j, however, depends upon i (cf. (5.36), below) and hence this criterion leads to a much more complicated program.

$$y_h = l_{h-k} - a_{h-k,j} y_j = l_{h-k} \left[1 - \frac{a_{h-k,j}}{a_{i-k,j}} \frac{l_{i-k}}{l_{h-k}} \right] . \tag{5.37}$$

Now since (5.34) constrains all y_h to be nonnegative, i must be chosen so that

$$\frac{l_{i-k}}{a_{i-k,j}} = \min_{(h)} \frac{l_{h-k}}{a_{h-k,j}} \tag{5.38}$$

Eq. (5.38) establishes i; then, (5.36) – (5.37) give all m new nonzero y-components. Thus, the next vertex, feasible and with improved z, is obtained.

4. In steps following the initial one, the arithmetic becomes more complicated, but the principle remains exactly the same; given a set of m nonzero components and k zero components, bring in a new variable y_j such as to maximize the anticipated increase of z. Having thus established j, the subscript of the resurrected component, the subscript and value of all new y_h is uniquely determined by the feasibility condition (5.34). The iteration continues until a vertex is found with larger z value than any adjacent one. This is the required optimum.

The initial feasible solution cannot be found by the described method if some of the $l_i < 0$, or if any of the constraints is in the form of an equality. For inequalities with $l_i < 0$, dummy variables (now called *surplus variables*), defined as in (5.30), may be added to the LHS of the relation to convert it into an equality. Now, however, $\mathbf{x} = 0$ no longer represents a feasible solution, since it would lead to $x_{si} = l_i < 0$. Similarly, for equalities in the original system, $\mathbf{x} = 0$ is only feasible if $l_j = 0$.

To be able to derive initial feasible solution for the general case, all inequalities are first transformed into equations. After that, if necessary, some are multiplied by -1, so that all constants on the RHS are nonnegative ($|l_i|$). The resulting equations then are augmented with new variables, and thus brought to the new form

$$\begin{aligned} \mathbf{C}\mathbf{x} + \mathbf{I}_m \mathbf{x}_a &= \mathbf{b} \\ \mathbf{x} \geqslant 0, \qquad \mathbf{x}_a &\geqslant 0 \end{aligned} \tag{5.39}$$

Here, $\mathbf{b} > 0$ is the vector with components $|l_i|$; $\mathbf{x}_a$ represents a vector of m additional variables, called *artificial variables*. Evidently a feasible initial solution is obtained by setting $\mathbf{x} = 0$, $\mathbf{x}_a = \mathbf{b}$.

If slack variables are present, they can form part of $\mathbf{x}_a$; surplus variables, however, have -1 coefficients in (5.39) and hence cannot be included. In equality constraints, only artificial variables enter.†

To illustrate the process, consider the example shown in Fig. 5.4. This is the same as that of Fig. 5.3, except that a new constraint, calling

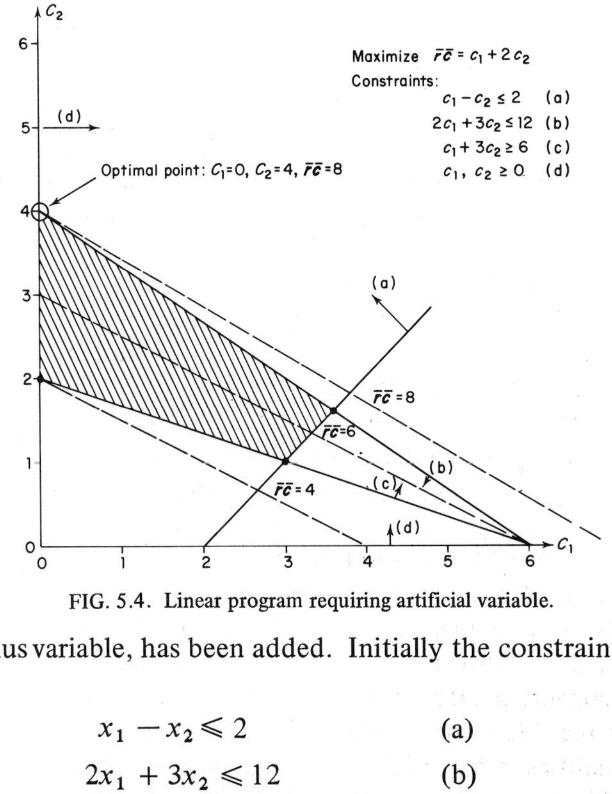

FIG. 5.4. Linear program requiring artificial variable.

for a surplus variable, has been added. Initially the constraints are in the form

$$
\begin{aligned}
x_1 - x_2 &\leqslant 2 && \text{(a)}\\
2x_1 + 3x_2 &\leqslant 12 && \text{(b)}\\
-x_1 - 3x_2 &\leqslant -6 && \text{(c)}
\end{aligned}
\tag{5.40}
$$

Adding slack variables x_3, x_4 as well as surplus variable x_5, and multiplying (c) by -1,

$$
\begin{aligned}
x_1 - x_2 + x_3 &= 2 && \text{(a)}\\
2x_1 + 3x_2 + x_4 &= 12 && \text{(b)}\\
x_1 + 3x_2 - x_5 &= 6 && \text{(c)}
\end{aligned}
\tag{5.41}
$$

Finally, adding the artificial variable x_6 to the LHS of (c)

† It may be argued that for equality-constraints it might be simpler to express some of the x_i in terms of others, thus getting rid both of x_i and the constraint. In reality, however, this would transform the $x_i \geqslant 0$ constraint into a more complicated one, requiring dummy variables again.

$$\begin{pmatrix} 1 & -1 & 0 \\ 2 & 3 & 0 \\ 1 & 3 & -1 \end{pmatrix} \begin{pmatrix} x_1 \\ x_2 \\ x_5 \end{pmatrix} + \begin{pmatrix} 1 & 0 & 0 \\ 0 & 1 & 0 \\ 0 & 0 & 1 \end{pmatrix} \begin{pmatrix} x_3 \\ x_4 \\ x_6 \end{pmatrix} = \begin{pmatrix} 2 \\ 12 \\ 6 \end{pmatrix} \quad (5.42)$$

This is in the form of (5.39). Obviously, a feasible initial vertex is

$$x_1 = x_2 = x_5 = 0; \qquad x_3 = 2, \quad x_4 = 12, \quad x_6 = 6. \quad (5.43)$$

Now the simplex method is applicable for the vector

$$\mathbf{y} \triangleq (x_1, x_2, x_3, x_4, x_5, x_6)^T.$$

The slack and surplus variables are physically meaningful quantities (the margins in the inactive inequalities). They are irrelevant as far as the objective function is concerned; hence, their "price" is zero. The artificial variables, on the other hand, are added to the LHS of valid equations, upsetting them. Hence, in the final solution they must be equal to zero. This can be forced by attributing a large *negative* price to them.† Maximization of z then automatically drives the artificial variable to zero.

The actual algorithm used in the simplex method, based on the described principles, involves an inordinate amount of bookkeeping and elementary manipulations. For this reason, and also since it is widely available either as independent program or as subroutine for computer users, it will not be discussed any further here.

Duality. The important duality theorem of linear programming states that if $\hat{\mathbf{x}}$ is the optimum solution of the (*primal*) LP given in (5.13) – (5.15), then the *dual problem*

Minimize $$z \triangleq \boldsymbol{l}^T \mathbf{y} \quad (5.44)$$

subject to: $$\mathbf{A}^T \mathbf{y} \geqslant \mathbf{r} \quad (5.45)$$

and $$\mathbf{y} \geqslant 0 \quad (5.46)$$

has an optimal solution $\hat{\mathbf{y}}$ satisfying

$$\mathbf{r}^T \hat{\mathbf{x}} = \boldsymbol{l}^T \hat{\mathbf{y}} \quad (5.47)$$

---

† A more sophisticated process, called two-phase method [3], initially attaches −1 price to the artificial variables and zero price to the original variables. This results in a feasible solution for the original problem, which then forms the second phase of the procedure.

i.e. the optimized objective functions are equal. The dual optimal solution $\mathbf{y}$ may be obtained from

$$\hat{\mathbf{y}}^T = \hat{\mathbf{r}}^T \mathbf{B}^{-1} . \tag{5.48}$$

Here, $\hat{\mathbf{r}}$ is the vector of those components of r which correspond to the nonzero components of $\hat{\mathbf{x}}$. Similarly, $\mathbf{B}$ is a matrix whose columns are the activity vectors $\mathbf{a}_i$ (cf. (5.17)) associated with these nonzero components of $\hat{\mathbf{x}}$. Thus, solution of the primal problem immediately supplies the solution of the dual. Since, furthermore, definitions (5.13) – (5.15) on the one hand, and (5.44) – (5.46) on the other are perfectly symmetrical, it is arbitrary which one is considered primal.

The great importance of the duality theorem for circuit designers lies in the fact that the inequality systems occuring in circuit optimization generally consist of a very large number m of inequalities (typically, satisfied at all sample points), while the number of unknowns, k, is relatively small. Since the amount of computations in the simplex method depends mostly on m, it is more efficient to derive and solve the dual LP rather than the physical problem. The required physical solution can then be obtained from (5.48).

For circuit applications, the following more specialized form [3] of the duality theorem is useful. Let $\hat{\mathbf{u}}$ be the solution of the primal LP

Minimize $$z \triangleq \boldsymbol{l}^T \mathbf{v} \tag{5.49}$$

subject to $$\mathbf{A}^T \mathbf{u} \geqslant \mathbf{r}. \tag{5.50}$$

(Note the absence of nonnegativity conditions on $\mathbf{u}$.) Then the dual problem

Maximize $$z \triangleq \mathbf{r}^T \mathbf{v} \tag{5.51}$$

subject to $$\mathbf{A}\mathbf{v} = \boldsymbol{l} \tag{5.52}$$

$$\mathbf{v} \geqslant 0 \tag{5.53}$$

(note here that (5.52) is an *equality*) has a solution $\hat{\mathbf{v}}$ satisfying (5.47). It can be obtained from

$$\hat{\mathbf{v}}^T = \boldsymbol{l}^T \mathbf{C}^{-1} \tag{5.54}$$

where $\boldsymbol{l}$ and $\mathbf{C}$ are defined analogously to $\hat{\mathbf{r}}$ and $\mathbf{B}$ of (5.48).

Chebyshev approximation using LP [4], [5]. Now the problem stated at the beginning of this section (Eqs. (5.7) – (5.9) can be tackled. In order to establish continuity of notation, let

$$
\begin{aligned}
u_h &\triangleq c_h \\
u_k &\triangleq E \\
r_{2i-1} &\triangleq w(x_i)F(x_i) \\
r_{2i} &\triangleq -w(x_i)F(x_i) \\
a_{h,2i-1} &\triangleq w(x_i)f_h(x_i) \\
a_{h,2i} &\triangleq -w(x_i)f_h(x_i) \\
a_{k,2i-1} &\triangleq a_{k,2i} \triangleq 1 \\
l_h &\triangleq 0 \\
l_k &\triangleq 1
\end{aligned}
\tag{5.55}
$$

for

$$
\begin{aligned}
h &= 1, 2, \ldots, k-1\,(\triangleq n) \\
i &= 1, 2, \ldots, \frac{m}{2}\,.
\end{aligned}
\tag{5.56}
$$

Then the LP of (5.49) – (5.50) is obtained. For the reasons explained above, the dual problem

$$\text{Maximize} \qquad z \triangleq \sum_{i=1}^{m/2} w(x_i)F(x_i)(v_{2i-1} - v_{2i}) \tag{5.57}$$

$$
\begin{aligned}
\text{subject to } \sum_{i=1}^{m/2} w(x_i)f_h(x_i)(v_{2i-1} - v_{2i}) &= 0 \qquad h = 1, 2, \ldots, k-1 \\
\sum_{i=1}^{m/2} (v_{2i-1} + v_{2i}) &= 1
\end{aligned}
\tag{5.58}
$$

should preferably be solved
Equations (5.57) – (5.58) are in a suitable form for solution via the simplex method.

It is of interest to note that for the primal problem, at all vertices (and thus at the optimum solution as well) $k = n + 1$ constraints are active. Hence the value E is taken on by $|w[F - \Sigma c_h f_h]|$ at $n + 1$ points.

Especially for filters, a tolerated error $e(x)$ is often prescribed for the approximation. Then the problem becomes

Minimize
$$z \triangleq E \tag{5.59}$$

subject to
$$w(x_i)\left[F(x_i) - \sum_{h=1}^{n} c_h f_h(x_i)\right] \leqslant e(x_i) + E \tag{5.60}$$

$$-w(x_i)\left[F(x_i) - \sum_{h=1}^{n} c_h f_h(x_i)\right] \leqslant e(x_i) + E. \tag{5.61}$$

It is evident that this problem can also be treated using duality and the simplex method

One example of the generalized polynomial approximant (5.7) is the squared modulus of the input/output transfer ratio of a network with given natural mode polynomial $Q(s)$

$$F^j(\omega^2, \mathbf{c}^j) \triangleq \sum_{h=1}^{n} c_h \frac{\omega^{2h-2}}{|Q(j\omega)|^2}. \tag{5.62}$$

Another is the impulse response of such a network

$$F^j(t, \mathbf{c}^j) \triangleq \sum_{h=1}^{n} c_h e^{s_h t} \tag{5.63}$$

where the s_h are the roots of $Q(s)$.

Various realizability conditions, such as the positiveness of $F^j(\omega^2, \mathbf{c}^j)$ or the Fialkow-Gerst condition, are expressible as linear constraints on the c_h, and can thus be incorporated in the LP [5].

For the case of generalized rational functions

$$\sum_h c_h f_h(x) \Big/ \sum_l d_l g_l(x),$$

the process may be extended at the cost of some added complications and a small loss in optimality by phrasing the problem as follows:

Minimize
$$z \triangleq E \tag{5.64}$$

subject to

$$\pm w(x_i)\left[F(x_i) - \frac{\Sigma c_h f_h(x_i)}{\Sigma d_l g_l(x_i)}\right] \leqslant e(x_i) + \frac{E}{\Sigma d_l g_l(x_i)} . \tag{5.65}$$

Clearing fractions results once again in an LP. Alternatively, a series of LP's may be performed, with a modified error function

$$w(x_i)\frac{\Sigma d_l^j g_l(x_i)}{\Sigma d_l^{j-1} g_l(x_i)}\left[F(x_i) - \frac{\Sigma c_h^j f_h(x_i)}{\Sigma d_l^j g_l(x_i)}\right] \tag{5.66}$$

being used in the jth iteration. Convergence is reached when all

$$d_l^j \approx d_l^{j-1} .$$

Approximation programming [6], [7]. Nonlinear optimization problems may be reduced to linear ones by using series expansion. Assume that a nonlinear $F_j(\mathbf{c})$ function is to be used in a Chebyshev optimization, subject to the nonlinear constraints represented by

$$\mathbf{h}(\mathbf{x}, \mathbf{c}) \leqslant \mathbf{0}. \tag{5.67}$$

Linearization is obtained by expressing the unknown value of $\mathbf{c}^{j+1}$ as $\mathbf{c}^j + \Delta\mathbf{c}$, where $\mathbf{c}^j$ is known and then using first-order Taylor-series expansions as in (5.5). This leads to the linear problem

minimize
$$E$$

subject to

$$\pm w(x_i)[F(x_i) - F^j(x_i, \mathbf{c}^j) - \nabla F^{j^T}(x_i, \mathbf{c}^j)\Delta\mathbf{c}] \leqslant E \tag{5.68}$$

also

$$\pm \Delta\mathbf{c} \leqslant \Delta \tag{5.69}$$

and

$$\mathbf{h}(x_i, \mathbf{c}^j) + \mathbf{A}\Delta\mathbf{c} \leqslant \boldsymbol{\delta} \tag{5.70}$$

Here, Δ is a vector of small numbers which bounds $\Delta\mathbf{c}$ and hence prevents it from becoming so large as to make the Taylor-expansions invalid; also,

$$\mathbf{A} \triangleq \left(\frac{\partial h_k}{\partial c_l}\right) \tag{5.71}$$

and, finally, $\boldsymbol{\delta} < \mathbf{0}$ is the vector of small safety margins, needed to offset the inaccuracies of the Taylor expansion.

Equations (5.68) – (5.71) describe an LP. Having solved it for $\Delta\mathbf{c}$, the parameter-vector $\mathbf{c}^j + \Delta\mathbf{c}$ may or may not give a smaller E than $\mathbf{c}^j$. It can, however, be shown [7] that $\Delta\mathbf{c}$ points in a downhill direction on the $E(\mathbf{c})$ surface at the point $\mathbf{c}^j$, i.e., that for a sufficiently small $\lambda > 0$

$$E(\mathbf{c}^j + \lambda\Delta\mathbf{c}) < E(\mathbf{c}^j) \tag{5.72}$$

and also (5.67) is satisfied. Hence, if $\mathbf{c}^j + \Delta\mathbf{c}$ does not decrease E, a linear search in the direction of $\Delta\mathbf{c}$ may be performed and the λ minimizing E, while still satisfying (5.67), chosen. Next, the process is repeated using the new value of $\mathbf{c}$ as a starting point. The iteration continues until the linear search yields $\lambda = 0$, at which time the constrained optimum has been achieved.

5.3 GRADIENT TECHNIQUES

Newton's Method

The conceptually simplest method of matching $F(x)$ by a nonlinear function $F^j(x, \mathbf{c}^j)$ of $\mathbf{c}$ is via Newton's method. Selecting n points x_i and denoting the matrix of partial derivatives

$$j_{kl} \triangleq \frac{\partial F^j(x_k, \mathbf{c}^j)}{\partial c_l}$$

by $\mathbf{J}$ (for *Jacobian* matrix), the correction $\Delta\mathbf{c}$ to $\mathbf{c}^j$ may be obtained from

$$\mathbf{J}\Delta\mathbf{c} = \mathbf{F} - \mathbf{F}^j \tag{5.73}$$

where

$$\mathbf{F} \triangleq [F(x_1), F(x_2), ..., F(x_n)]^T$$
$$\mathbf{F}^j \triangleq [F^j(x_1), F^j(x_2), ..., F^j(x_n)]^T \tag{5.74}$$

This method is simple but often effective. Damping may be applied to improve convergence, by using $\mathbf{c}^{j+1} = \mathbf{c}^j + \lambda\Delta\mathbf{c}$, $0 < \lambda < 1$. The process needs to be repeated, using $\mathbf{c}^{j+1}$ as the starting value, until $\mathbf{F} = \mathbf{F}^j$ within prescribed tolerance.

Disadvantages of the method are the need for the tedious computation† of the n^2 derivatives j_{kl}, the necessity of inverting of $\mathbf{J}$, and the absence of guaranteed convergence. Also, it is equivalent to interpolating $\mathbf{F}$ at n points via $\mathbf{F}^j$; between these points, anything can happen.

Descent Along the Negative Gradient

Let $\mathbf{c}^{j+1}$ be written in the form

$$\mathbf{c}^{j+1} = \mathbf{c}^j + \alpha^j \mathbf{s}^j \tag{5.75}$$

where $\mathbf{s}^j$ is a unit vector. It is well known [1] that the derivative

$$\left.\frac{\partial}{\partial \alpha^j} E(\mathbf{c}^j + \alpha^j \mathbf{s}^j)\right|_{\alpha^j=0}$$

is minimized if

$$\mathbf{s}^j = -\frac{\nabla \mathbf{E}}{|\nabla \mathbf{E}|}\,. \tag{5.76}$$

The choice of α^j can be made several ways. An obvious one is to perform a linear search in the $-\nabla\mathbf{E}$ direction and select α^j such as to minimize E. For such a choice, $\alpha_{\min}$, $\nabla\mathbf{E}(\mathbf{c}^j)$ will obviously be orthogonal to $\nabla\mathbf{E}(\mathbf{c}^{j+1})$. Hence, the process will "zig-zag" toward the minimum. In particular, in 2 dimensions,* $\nabla\mathbf{E}(\mathbf{c}^j)$, $\nabla\mathbf{E}(\mathbf{c}^{j+2})$, ..., will all be parallel with each other and orthogonal to $\nabla\mathbf{E}(\mathbf{c}^{j+1})$, $\nabla\mathbf{E}(\mathbf{c}^{j+3})$, ... (Fig. 5.5). This suggests that choosing $\alpha^j = \lambda\alpha^j_{\min}$ $(0 < \lambda < 1)$ may improve convergence. Also, from time to time choosing

---

† If the j_{kl} are approximated by difference quotients, this involves $n(n + 1)$ evaluation of $F^j(x, \mathbf{c}^j)$.

* This property does not carry over for $n > 2$.

$$\mathbf{s}^j = \frac{\mathbf{c}^j - \mathbf{c}^{j-2}}{|\mathbf{c}^j - \mathbf{c}^{j-2}|}$$

is likely to accelerate the process, as Fig. 5.5 indicates.

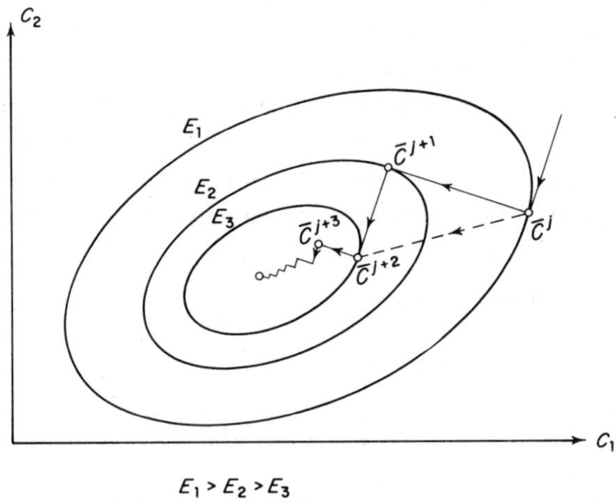

FIG. 5.5. Steepest descent.

If the Hessian matrix **H** is available, α^j may be chosen using (5.5):

$$E\left(\mathbf{c}^j - \alpha^j \frac{\nabla \mathbf{E}}{|\nabla \mathbf{E}|}\right) \cong E(\mathbf{c}^j) - \alpha^j |\nabla \mathbf{E}| + \frac{1}{2} (\alpha^j)^2 \frac{\nabla \mathbf{E}^T \mathbf{H} \nabla \mathbf{E}}{|\nabla \mathbf{E}|^2}$$

$$\frac{\partial E}{\partial \alpha^j} \cong -|\nabla \mathbf{E}| + \alpha^j \frac{\nabla \mathbf{E}^T \mathbf{H} \nabla \mathbf{E}}{|\nabla \mathbf{E}|^2} \stackrel{!}{=} 0 \tag{5.77}$$

$$\alpha^j = \frac{|\nabla \mathbf{E}|^3}{\nabla \mathbf{E}^T \mathbf{H} \nabla \mathbf{E}}; \qquad \nabla \mathbf{c} = -\frac{|\nabla \mathbf{E}|^2}{\nabla \mathbf{E}^T \mathbf{H} \nabla \mathbf{E}} \cdot \nabla \mathbf{E}.$$

Note that for positive definite **H** (convex $E(\mathbf{c})$) $\alpha^j > 0$.

To avoid the tedious computation of **H**, the following manipulation [8] may be used. By (5.5), choosing arbitrarily

$$\Delta \mathbf{c}^0 = -\frac{\nabla \mathbf{E}(\mathbf{c}^j)}{|\nabla \mathbf{E}(\mathbf{c}^j)|}, \tag{5.78}$$

$$E\left(\mathbf{c}^j - \frac{\nabla \mathbf{E}(\mathbf{c}^j)}{|\nabla \mathbf{E}(\mathbf{c}^j)|}\right) = E(\mathbf{c}^j) - |\nabla \mathbf{E}(\mathbf{c}^j)| + \frac{1}{2} \frac{\nabla \mathbf{E}(\mathbf{c}^j)^T \mathbf{H}(\mathbf{c}^j) \nabla \mathbf{E}(\mathbf{c}^j)}{|\nabla \mathbf{E}(\mathbf{c}^j)|^2}.$$

Expressing the last term on the RHS and substituting into (5.77)

$$\Delta \mathbf{c} = \frac{1}{2} \frac{-\nabla \mathbf{E}(\mathbf{c}^j)}{|\nabla \mathbf{E}(\mathbf{c}^j)| - E(\mathbf{c}^j) + E\left(\mathbf{c}^j - \dfrac{\nabla \mathbf{E}(\mathbf{c}^j)}{|\nabla \mathbf{E}(\mathbf{c}^j)|}\right)} \tag{5.79}$$

which does not involve $\mathbf{H}$ anymore. This simple process gave excellent results in numerical problems [8].

If it is known that at the minimum $E(\hat{\mathbf{c}}) \ll E(\mathbf{c}^j)$, then an application of Newton's process in the $\nabla\mathbf{E}$ direction gives [9]

$$\alpha^j = \frac{E(\mathbf{c}^j)}{|\nabla \mathbf{E}|} > 0. \tag{5.80}$$

An interpolation of the $E(\alpha^j)$ by quadratic or cubic polynomials, and calculation of $\alpha^j_{\min}$ as the minimum of this parabola is also often used [9].

A different strategy [10], [11] called *PARTAN* (short for *parallel tangents*) should be mentioned. It is performed in the following steps [11]:

(1) Choose $\mathbf{c}^0$. Let

$$\mathbf{c}^2 \triangleq \mathbf{c}^0 - \alpha^2_{\min} \frac{\nabla \mathbf{E}(\mathbf{c}^0)}{|\nabla \mathbf{E}(\mathbf{c}^0)|}$$

(2) In any later jth iteration cycle, let

$$\begin{aligned} \mathbf{c}^{2j+1} &\triangleq \mathbf{c}^{2j} - \alpha^{2j+1}_{\min} \frac{\nabla \mathbf{E}(\mathbf{c}^{2j})}{|\nabla \mathbf{E}(\mathbf{c}^{2j})|}. \\ \mathbf{c}^{2j+2} &\triangleq \mathbf{c}^{2j+1} - \alpha^{2j+2}_{\min} \frac{\mathbf{c}^{2j+1} - \mathbf{c}^{2j-1}}{|\mathbf{c}^{2j+1} - \mathbf{c}^{2j-1}|}. \end{aligned} \tag{5.81}$$

This method minimizes a true quadratic function $E(\mathbf{c})$ in $j = n$ cycles.

Finally, it is evident from Fig. 5.5 that good scaling (one which makes the contour-lines nearly circular) is very advantageous for fast convergence of the described methods.

Steepest Descent

The term *steepest descent* is somewhat ambiguous, it is often used to denote the descent along the negative gradient, while other writers reserve it for the method discussed next.† By (5.5)

$$\nabla \mathbf{E}(\mathbf{c}^{j+1}) = \nabla \mathbf{E}(\mathbf{c}^{j}) + \mathbf{H}\Delta \mathbf{c} + \cdots \tag{5.82}$$

Hence, if it is desired that $\mathbf{c}^{j+1} \overset{!}{=} \hat{\mathbf{c}}$, the minimum of $E(\mathbf{c})$, then

$$\begin{aligned} \nabla \mathbf{E}(\hat{\mathbf{c}}) &= 0 \cong \nabla \mathbf{E}(\mathbf{c}^{j}) + \mathbf{H}\,\nabla \mathbf{c} \\ \Delta \mathbf{c} &\cong -\mathbf{H}^{-1}(\mathbf{c}^{j})\,\nabla \mathbf{E}(\mathbf{c}^{j})\,. \end{aligned} \tag{5.83}$$

It should be noted that the differences between the optimum increments given by (5.77) and (5.83) are the following: first, the $\Delta\mathbf{c}$ of (5.77) points in the $-\Delta\mathbf{E}$ direction, while that of (5.83) does not; secondly, (5.83) requires the inversion of $\mathbf{H}$, while (5.77) does not. As a result, the process which uses (5.83) requires usually much more computation per iteration cycle, but much fewer such cycles than the descent along the negative gradient. Again, (5.83) may be interpreted as a Newton-type process.

Least pth Approximation [12]

Choosing the error-measure in the form

$$E \triangleq \sum_{i=1}^{m} \epsilon_i^p \qquad m > n \tag{5.84}$$

where $\epsilon_i \triangleq \epsilon(x_i, \mathbf{c})$, p is an even integer and ϵ is defined by (5.2), we get

$$\nabla E_l \triangleq \frac{\partial E}{\partial c_l} = p \sum_{i=1}^{m} \epsilon_i^{p-1} \frac{\partial \epsilon_i}{\partial c_l} \tag{5.85}$$

and the k, l element of $\mathbf{H}$ becomes

$$\begin{aligned} h_{kl} = h_{lk} &\triangleq \frac{\partial^2 E}{\partial c_k \partial c_l} \\ &= p \sum_{i=1}^{m} \left[\frac{\partial \epsilon_i}{\partial c_l}(p-1)\epsilon_i^{p-2}\frac{\partial \epsilon_i}{\partial c_k} + \epsilon_i^{p-1}\frac{\partial^2 \epsilon_i}{\partial c_k \partial c_l} \right]. \end{aligned} \tag{5.86}$$

† This latter is also often called Newton's method.

Hence, using (5.83), the steepest descent method is applicable. It is advantageous, however, to assume that for all i, k, l

$$(p-1)\frac{\partial \epsilon_i}{\partial c_k}\frac{\partial \epsilon_i}{\partial c_l} \gg \epsilon_i \frac{\partial^2 \epsilon_i}{\partial c_k \partial c_l}\,. \tag{5.87}$$

The following reasons can be listed in support of assuming (5.87):

(1) For many practical problems it is valid.

(2) It leads to great saving in computation time.

(3) The resulting iteration procedure is guaranteed to converge to a minimum, rather than saddle-point or maximum, of E.

(4) The resulting process possesses the downhill property.

(5) At convergence, all $\Delta c_k \to 0$, hence both methods – with or without second derivatives – converge to the same $\hat{\mathbf{c}}$.

It should be noted that (5.87) holds better for higher p values. This leads to an improvement, observed in computations, of the convergence of least-4th, least-6th, ... process over that of the least squares. Using the definitions

$$\begin{aligned}
&\mathbf{e} \triangleq (\epsilon_1^{p-1}, \epsilon_2^{p-1}, \ldots, \epsilon_m^{p-1})^T \\
&\mathbf{A} \triangleq (a_{il});\; a_{il} \triangleq -\frac{\partial \epsilon_i}{\partial c_l};\; \mathbf{A} : m \times n \\
&\mathbf{B} \triangleq \operatorname{diag}(\epsilon_i^{p-2}) \qquad i = 1, 2, \ldots, m
\end{aligned} \tag{5.88}$$

by (5.85),

$$\nabla \mathbf{E} = -p\mathbf{A}^T\mathbf{e} \tag{5.89}$$

and by (5.86) and (5.87)

$$\mathbf{H} \cong p(p-1)\mathbf{A}^T\mathbf{B}\mathbf{A}. \tag{5.90}$$

Hence, using (5.83)

$$\Delta \mathbf{c} = (p-1)^{-1}(\mathbf{A}^T\mathbf{B}\mathbf{A})^{-1}\mathbf{A}^T\mathbf{e} \tag{5.91}$$

For $p = 2$, (5.91) reduces to the well-known least-squares equations [9], [13] due to Gauss

$$\Delta \mathbf{c} = (\mathbf{A}^T\mathbf{A})^{-1}\mathbf{A}^T\mathbf{e}. \tag{5.92}$$

The downhill property of the process may easily be proven. Let

$$\mathbf{c}_\lambda \triangleq \mathbf{c}^j + \lambda\Delta\mathbf{c} \qquad 0 < \lambda < 1 \tag{5.93}$$

$$E_\lambda \triangleq E(\mathbf{c}_\lambda).$$

Then, for $\lambda \to 0$, by (5.5), (5.89) and (5.91)

$$\begin{aligned} E_\lambda &\to E(\mathbf{c}^j) + \lambda\, \nabla \mathbf{E}^T \Delta\mathbf{c} = \\ &= E(\mathbf{c}^j) - \frac{\lambda}{p(p-1)} \nabla\mathbf{E}^T(\mathbf{A}^T\mathbf{B}\mathbf{A})^{-1}\nabla\mathbf{E}. \end{aligned} \tag{5.94}$$

But it can easily be shown [12] that $\mathbf{A}^T\mathbf{B}\mathbf{A}$ as well as $(\mathbf{A}^T\mathbf{B}\mathbf{A})^{-1}$ are positive definite symmetric matrices.† Hence $E_\lambda \leqslant E(\mathbf{c}^j)$ as stated. This suggests that using a linear search in the $\Delta\mathbf{c}$ direction in each iteration, E may be decreased every time.

The positive definite character of $\mathbf{A}^T\mathbf{B}\mathbf{A}$ guarantees that $E(\mathbf{c})$ is convex around $\mathbf{c}^j$; hence, a minimum is approached.

In practical applications, the following process worked well. $E(\mathbf{c}^j + \Delta\mathbf{c})$ was computed and compared with $E(\mathbf{c}^j)$. If there was improvement, the new point was accepted. If not, $E(\mathbf{c}^j + \lambda\Delta\mathbf{c})$ was computed. If this was still greater than $E(\mathbf{c}^j)$, $E(\mathbf{c}^j + \lambda^2\Delta\mathbf{c})$ was calculated, etc. The value of λ was about 0.75.

Damping of the least pth process may also be accomplished by penalizing large values of $|\Delta\mathbf{c}|$. This is achieved, if

$$E_k \triangleq E(\mathbf{c}) + k\Delta\mathbf{c}^T\Delta\mathbf{c} \tag{5.95}$$

is minimized. $k > 0$ represents the relative weight attached to the size of $\Delta\mathbf{c}$. Using (5.5) and substituting the values of $\nabla\mathbf{E}$ and $\mathbf{H}$ from (5.89) – (5.90), the design equations

$$\begin{aligned} E_k &\cong E(\mathbf{c}^j) - p\mathbf{e}^T\mathbf{A}\Delta\mathbf{c} + \frac{1}{2}\Delta\mathbf{c}^T[p(p-1)\mathbf{A}^T\mathbf{B}\mathbf{A} + 2k\mathbf{I}_n]\Delta\mathbf{c} \\ \mathbf{H}_k &\cong p(p-1)\mathbf{A}^T\mathbf{B}\mathbf{A} + 2k\mathbf{I}_n \\ \Delta\mathbf{c}_k &= (p-1)^{-1}\left[\mathbf{A}^T\mathbf{B}\mathbf{A} + \frac{2k}{p(p-1)}\mathbf{I}_n\right]^{-1}\mathbf{A}^T\mathbf{e} \end{aligned} \tag{5.96}$$

† Assuming that $\mathbf{A}$ has the full rank n.

are obtained. It is easy to see that the downhill property and assured convergence to a minimum of E_k are unaffected. Also, using $\mathbf{c}_k \triangleq \mathbf{c}^j + \Delta\mathbf{c}_k$,

$$E(\mathbf{c}_k) < E_k(\mathbf{c}_k) < E(\mathbf{c}^j) \tag{5.97}$$

and, using the $\Delta\mathbf{c}$ of (5.91)

$$E_k(\mathbf{c}_k) < E(\mathbf{c}^j + \Delta\mathbf{c}) + k\Delta\mathbf{c}^T\Delta\mathbf{c} < E(\mathbf{c}_k) + k\Delta\mathbf{c}^T\Delta\mathbf{c}. \tag{5.98}$$

Hence

$$\Delta\mathbf{c}_k^T\Delta\mathbf{c}_k < \Delta\mathbf{c}^T\Delta\mathbf{c}. \tag{5.99}$$

Equation (5.97) shows that the quadratic approximant $E(\mathbf{c})$ is decreased in each iteration cycle; (5.99) shows that the change vector $\Delta\mathbf{c}_k$ is smaller than $\Delta\mathbf{c}$.

Since $E_k(\mathbf{c}_k)$ still has the downhill property, a linear search in $\mathbf{c}^j + \lambda\Delta\mathbf{c}_k (0 < \lambda < 1)$ is feasible here as well.

It is evident from (5.96) that the only effect of damping in the actual formulas is that all diagonal terms in $\mathbf{A}^T\mathbf{B}\mathbf{A}$ are incremented by $2k/p(p-1) > 0$. The choice of k is a somewhat unclarified problem. Evidently, for $k \to \infty$, $\Delta\mathbf{c}_k$ turns in the direction of negative gradient, while for $k \to 0$ the $\Delta\mathbf{c}$ of the undamped least pth solution results. For $0 < k < \infty$, an intermediate direction is obtained. A method for choosing k dynamically in each iteration is given by Beale [13]. This essentially involves choosing a small k if the iteration is converging rapidly, and, vice versa, a large k if divergence occurs.†

The described general procedures simplify somewhat for special degrees and/or special approximants $F^j(x, \mathbf{c}^j)$. In particular, if $m = n$, $\mathbf{A}$ becomes a square matrix and (5.91) simplifies to

$$\Delta\mathbf{c} = (p-1)^{-1}(\mathbf{B}\mathbf{A})^{-1}\mathbf{e} \tag{5.100}$$

or, for $p = 2$,

$$\Delta\mathbf{c} = \mathbf{A}^{-1}\mathbf{e} \tag{5.101}$$

a known result [1]. If $m < n$, for $p = 2$ the equation

$$\mathbf{A}\Delta\mathbf{c} = \mathbf{e} \tag{5.102}$$

† Values of $2k/[p(p-1)]$ as small as 1% of the average value of the diagonal elements of $\mathbf{A}^T\mathbf{B}\mathbf{A}$ are usually sufficient to assure convergence.

may be derived, which does not have a unique solution $\Delta \mathbf{c}$. In any case, for practical results $m > 5n$ is usually necessary, making these considerations somewhat academic.

For least squares minimization, two cases deserve special mentioning. If the approximant F^j is a generalized polynomial of the form (5.7), Eq. (5.5) becomes an *exact* formula for $E(\mathbf{c})$ which now is quadratic in $\mathbf{c}$. The matrix $\mathbf{A}$ in this case has, as general element,

$$a_{ih} = w(x_i) f_h(x_i) \tag{5.103}$$

which is independent of $\mathbf{c}$. Formula (5.92) gives therefore the correct answer in one step.

If the approximant is a generalized rational function, as shown in the LHS of (5.65), the modified error-function of (5.66) can be used to linearize the problem in d_l^j and c_h^j. The linearized function may now be optimized in one shot, but the process must be iterated until all

$$d_l^j \approx d_l^{j-1}.$$

The solution of the least pth problem may also be obtained using orthonormalization. This approach avoids the matrix-inversion needed otherwise, but involves instead a comparably tedious sequence of simple manipulations.

The Fletcher-Powell Method [14]

Algorithm. The basic purpose of this powerful method is to simulate the steepest descent equation (5.83) without going through the tedium of computing and inverting $\mathbf{H}$. To achieve this, a matrix $\mathbf{G}^j$ is derived† as an approximation to $\mathbf{H}^{-1}$. Based on the results $\mathbf{c}^j$ and $\nabla E(\mathbf{c}^j)$ obtained in each iteration cycle, $\mathbf{G}^j$ is updated in each cycle until it converges to $\mathbf{H}^{-1}$ at the minimum. Because of the utility and widespread use of this process, it will be discussed in some detail next.

Assume that a "good" approximation $\mathbf{G}^0$ is available to $\mathbf{H}^{-1}(\mathbf{c}^0)$, where $\mathbf{c}^0$ is the starting point of the iteration. $\mathbf{G}^0$ may be obtained by computing and inverting the Hessian matrix at $\mathbf{c}^0$; usually, however, it

† The notations used in this section conform with our earlier notations, rather than with those used in Ref. [14].

is simply taken as the unit matrix $\mathbf{I}_n$. Now, in accordance with (5.83), the direction in which the minimum lies is assumed to be

$$\mathbf{d}^0 \triangleq -\mathbf{G}^0 \nabla^0 \tag{5.104}$$

where $\nabla^0 \triangleq \nabla \mathbf{E}(\mathbf{c}^0)$. Also,

$$\Delta \mathbf{c}^0 = \lambda^0 \mathbf{d}^0 \tag{5.105}$$

where λ^0 is obtained by minimizing $E(\mathbf{c})$ along $\mathbf{d}^0$, starting at $\mathbf{c}^0$. If $E(\mathbf{c})$ is quadratic in $\mathbf{c}$ and $\mathbf{G}^0 = \mathbf{H}^{-1}$ holds, $\lambda^0 = 1$ results. Actually, λ^0 may be different by orders of magnitude.

Next, let $\mathbf{c}^1 \triangleq \mathbf{c}^0 + \Delta \mathbf{c}^0$ and compute $\nabla^1 \triangleq \nabla \mathbf{E}(\mathbf{c}^1)$. Define

$$\mathbf{y}^0 \triangleq \nabla^1 - \nabla^0. \tag{5.106}$$

By (5.82), if a quadratic form $E(\mathbf{c})$ is to be minimized, its Hessian $\mathbf{H}$ should satisfy

$$\mathbf{H}^{-1}(\nabla^1 - \nabla^0) = \mathbf{H}^{-1}\mathbf{y}^0 = \Delta \mathbf{c}^0. \tag{5.107}$$

$\mathbf{G}^1$ is now found such as to satisfy (5.107), i.e.

$$\mathbf{G}^1 \mathbf{y}^0 = \Delta \mathbf{c}^0. \tag{5.108}$$

To create a $\mathbf{G}^1$ satisfying (5.108), let

$$\mathbf{G}^1 \triangleq \mathbf{G}^0 + \mathbf{A}^0 + \mathbf{B}^0. \tag{5.109}$$

Now $\mathbf{A}^0$, $\mathbf{B}^0$ are to be selected such that (5.108) is valid and also $\mathbf{G}^1$ be positive definite and symmetric. This latter condition is needed for $\mathbf{H}$ to be positive definite; it also guarantees convergence. Substituting (5.109) into (5.108), the simple choice

$$\begin{aligned} \mathbf{A}^0 \mathbf{y}^0 &= \Delta \mathbf{c}^0 \\ \mathbf{B}^0 \mathbf{y}^0 &= -\mathbf{G}^0 \mathbf{y}^0 \end{aligned} \tag{5.110}$$

may be made. The first of these equations can easily be satisfied by selecting

$$\mathbf{A}^0 \triangleq \frac{\Delta \mathbf{c}^0 \Delta \mathbf{c}^{0T}}{\Delta \mathbf{c}^{0T} \mathbf{y}^0} \tag{5.111}$$

as substitution directly shows. It is to be realized that, while the denominator of $\mathbf{A}^0$ is a scalar, its numerator, according to the rules of matrix multiplication, is a matrix:

$$\begin{pmatrix} \Delta c_1 \\ \Delta c_2 \\ \cdot \\ \cdot \\ \cdot \\ \Delta c_n \end{pmatrix} (\Delta c_1, \Delta c_2, \ldots, \Delta c_n) = \begin{pmatrix} (\Delta c_1)^2 & \Delta c_1 \Delta c_2 & \ldots & \Delta c_1 \Delta c_n \\ \Delta c_2 \Delta c_1 & (\Delta c_2)^2 & \ldots & \Delta c_2 \Delta c_n \\ \cdot & \cdot & \cdot & \cdot \\ \Delta c_n \Delta c_1 & \Delta c_n \Delta c_2 & \ldots & (\Delta c_n)^2 \end{pmatrix}$$

The rank of this matrix is only 1, since the ith row (or column) may be obtained from the jth via multiplication by $\Delta c_i / \Delta c_j$.

To satisfy the second of Eqs. (5.110), the "obvious" solution $\mathbf{B}^0 = -\mathbf{G}^0$ seems indicated. This would, however, give $\mathbf{G}^1 = \mathbf{A}^0$. Now $\mathbf{A}^0$ is a degenerate matrix, not positive definite and in addition, according to (5.104) and (5.111) would lead to a "new" direction

$$\mathbf{d}^1 = -\mathbf{G}^1 \nabla^1 \overset{?}{=} \frac{\Delta \mathbf{c}^{0T} \nabla^1}{\Delta \mathbf{c}^{0T} \mathbf{y}^0} \Delta \mathbf{c}^0 \tag{5.112}$$

In other words,the direction of $\mathbf{d}$ would remain unchanged, and the process get stuck. For this reason, the next simplest choice

$$\mathbf{B}^0 \triangleq - \frac{(\mathbf{G}^0 \mathbf{y}^0) \mathbf{z}^T}{\mathbf{y}^{0T} \mathbf{z}} \tag{5.113}$$

is made. This again evidently satisfies (5.110) as substitution shows. Next, to insure the symmetry of $\mathbf{B}^0$ and thus $\mathbf{G}^1$, let $\mathbf{z} \triangleq \mathbf{G}^0 \mathbf{y}^0$. Then

$$\mathbf{G}^1 = \mathbf{G}^0 + \frac{\Delta \mathbf{c}^0 \Delta \mathbf{c}^{0T}}{\Delta \mathbf{c}^{0T} \mathbf{y}^0} - \frac{\mathbf{G}^0 \mathbf{y}^0 \mathbf{y}^{0T} \mathbf{G}^0}{\mathbf{y}^{0T} \mathbf{G}^0 \mathbf{y}} . \tag{5.114}$$

This process is now repeated; the formulae can be obtained simply by increasing the superscripts in (5.104) – (5.114). Thus $\mathbf{d}^1$, $\Delta \mathbf{c}^1$, $\mathbf{y}^1$, $\mathbf{A}^1$, $\mathbf{B}^1$, and $\mathbf{G}^2$ are derived, in this order. The iteration continues in this manner, until convergence is achieved.

Convergence. To show convergence, first the downhill property

$$\Delta E \triangleq \lim_{\lambda \to +0} \lambda \mathbf{d}^{i^T} \nabla^i = -\lim_{\lambda \to +0} \lambda \nabla^{i^T} \mathbf{G}^i \nabla^i < 0 \tag{5.115}$$

will be proven. It was initially assumed that $\mathbf{G}^0$ was chosen positive definite (e.g., the unit matrix). By (5.115), then, it is sufficient to show that $\mathbf{G}^{i+1}$ is a positive definite (p.d.) matrix, provided $\mathbf{G}^i$ is. Denoting† $\mathbf{p} \triangleq (\mathbf{G}^i)^{1/2}\mathbf{c}$, $\mathbf{q} \triangleq (\mathbf{G}^i)^{1/2}\mathbf{y}^i$, for any arbitrary nonzero $\mathbf{c}$ by (5.114)

$$\mathbf{c}^T \mathbf{G}^{i+1} \mathbf{c} = \frac{|\mathbf{p}|^2 |\mathbf{q}|^2 - (\mathbf{p}^T \mathbf{q})^2}{|\mathbf{q}|^2} + \frac{(\mathbf{c}^T \Delta \mathbf{c}^i)^2}{\Delta \mathbf{c}^{i^T} \mathbf{y}^i} . \tag{5.116}$$

For a p.d. $\mathbf{G}^{i+1}$ this needs to be positive. The first term evidently is *nonnegative*; so is the second, if $\Delta \mathbf{c}^{iT} \mathbf{y}^i$ is. But, by (5.104) – (5.106)

$$\Delta \mathbf{c}^{i^T} \mathbf{y}^i \triangleq \Delta \mathbf{c}^{i^T} (\nabla^{i+1} - \nabla^i) = \Delta \mathbf{c}^{i^T} \nabla^{i+1} + \lambda^i \nabla^{i^T} \mathbf{G}^i \nabla^i . \tag{5.117}$$

Here,

$$\Delta \mathbf{c}^{iT} \nabla^{i+1} \triangleq 0 \tag{5.118}$$

since the size of $\Delta \mathbf{c}^i$ was derived by finding a point along $\mathbf{d}^i$ such that E was minimum, thereby assuring that ∇^{i+1} has no component in the $\Delta \mathbf{c}^i$ direction. By hypothesis, $\nabla^{iT} \mathbf{G}^i \nabla^i$ is of course positive. To complete the proof, it is evident that the first term in (5.116) is only zero* if $\mathbf{c} = k\mathbf{y}$, while the second only if $\mathbf{c}^T \Delta \mathbf{c}^i = 0$. But $\Delta \mathbf{c}^{i^T} \mathbf{y}$ is positive and hence $\mathbf{c}$ cannot satisfy these conditions simultaneously.

A significant aspect of the proof is that it relies on $\Delta \mathbf{c}^{iT} \mathbf{y}^i$ being nonnegative. This means that the linear search establishing the value of λ^i and thus $\Delta \mathbf{c}^i$ should be accurate, or else a safety margin can be provided by choosing λ^i slightly larger than the presumed minimizing parameter value.

Next, it can be shown that the method is quadratically convergent in the sense that it finds the minimum of a quadratic form such as that of (5.5) in exactly n iterations. The proof, while a straightforward induction process, is lengthy [14]. Hence, only its outline will be given here. First, it is proven that

$$\mathbf{G}^j \mathbf{H} \Delta \mathbf{c}^i = \Delta \mathbf{c}^i \qquad 0 \leqslant i < j \leqslant n \tag{5.119}$$

---

† $(\mathbf{G}^i)^{1/2}$ exists, since $\mathbf{G}^i$ is by hypothesis p.d.

* Since $(\mathbf{G}^i)^{-1/2}$ exists.

i.e., that all $\Delta \mathbf{c}^i$ are eigenvectors of $\mathbf{G}^j\mathbf{H}$ for $i < j$, with unity eigenvalues. Next, it is shown that

$$\Delta \mathbf{c}^{j^T} \mathbf{H} \Delta \mathbf{c}^i = 0 \qquad 0 \leqslant i < j < n \tag{5.120}$$

i.e., that all $\Delta \mathbf{c}^i$ are conjugate with respect to $\mathbf{H}$. This latter property implies that the $\Delta \mathbf{c}^i$ are linearly independent. For, if they are not,

$$\mathbf{b} = \sum_{i=0}^{j} \beta_i \Delta \mathbf{c}^i = 0 \tag{5.121}$$

with some β_i and $\Delta \mathbf{c}^i$ nonzero. Then, using (5.120),

$$\mathbf{b}^T\mathbf{H}\mathbf{b} = \sum_{i=0}^{j} \beta_i^2 (\Delta \mathbf{c}^{i^T} \mathbf{H} \Delta \mathbf{c}^i) = 0 \tag{5.121}$$

which is impossible, since $\mathbf{H}$ is p.d. The linear independence of the $\Delta \mathbf{c}^i$ together with (5.119) implies that

$$\mathbf{G}^n \mathbf{H} = \mathbf{I}_n \tag{5.122}$$

For, defining the nonsingular matrix

$$\mathbf{S} \triangleq (\Delta \mathbf{c}^0, \Delta \mathbf{c}^1, \ldots, \Delta \mathbf{c}^{n-1}) \tag{5.123}$$

by (5.119)

$$\mathbf{G}^n \mathbf{H} \mathbf{S} = \mathbf{S}, \tag{5.124}$$

which implies (5.122). Thus $\mathbf{G}^n$ is the inverse of the Hessian and therefore (5.83) gives directly the optimizing value

$$\Delta \mathbf{c}^n = -\mathbf{H}^{-1} \nabla^n = -\mathbf{G}^n \nabla^n \tag{5.125}$$

It is of interest to note that as $\mathbf{G}^j \to \mathbf{H}^{-1}$,

$$\sum_{i=0}^{j} \mathbf{A}^i \to \mathbf{H}^{-1} \quad \text{and} \quad \sum_{i=0}^{j} \mathbf{B}^i \to -\mathbf{G}^0 = -\mathbf{I}^n.$$

Practical considerations. In applications, the process had the typical convergence characteristic indicated in Fig. 5.6. Here, region (a) corres-

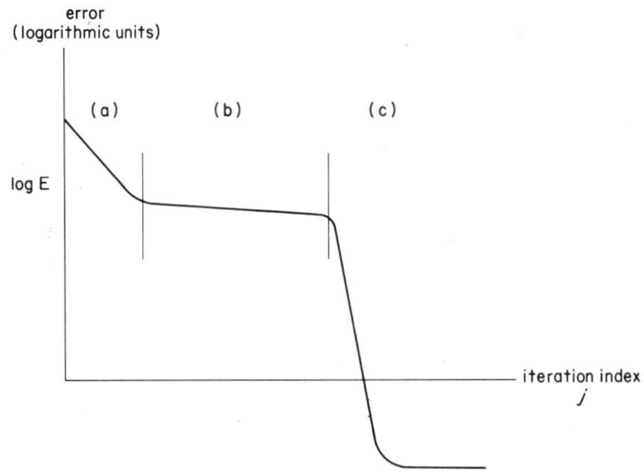

FIG. 5.6. Convergence characteristic for Fletcher-Powell optimization.

ponds to a steepest descent, since $\mathbf{G}^i \sim \mathbf{I}_n$ in that region. In (b), the steepest descent flattens out but $\mathbf{G}^i \to \mathbf{H}^{-1}$. When $\mathbf{G}^i$ approximates the local Hessian well, quadratic convergence sets in (region (c)). In tests, the process converged in $n \leqslant i \leqslant 20n$ iterations.

As the iteration converges, $\mathbf{d}^i \to \Delta\mathbf{c}^i \to -\mathbf{H}^{-1}\nabla \to 0$. Hence, $|\mathbf{d}^i|$ or $\max_{(k)} |\mathbf{d}_k^i|$ gives a good convergence criterion. It is advisable, however, to check $|\Delta\mathbf{c}^i|$ as well, and, in any case, perform at least n iterations, regardless of the size of $|\Delta\mathbf{c}^i|$ and $|\mathbf{d}^i|$.

In practical applications, it was found that the speed of convergence and also the p.d. character of $\mathbf{G}^i$ depended strongly on the accuracy with which λ^i was found. This is theoretically justified, since the proofs of downhill property, p.d. Hessian and quadratic convergence all need (5.118) to be valid. For this reason, the linear search for λ^i is the most critical part of the process; all failures experienced in using the method could be traced back to it. The eigenvalue λ^n is 1, as shown by (5.125); however, λ^0 may be many orders of magnitude different.

Broyden's Method

A valuable generalization of the Fletcher-Powell method was provided by Broyden [15]. The purpose is to solve n nonlinear equations in n unknowns

$$\mathbf{f}(\mathbf{c}) = 0 \tag{5.126}$$

where both $\mathbf{f}$ and $\mathbf{c}$ are n-dimensional vectors. Evidently, if $\mathbf{f} \equiv \nabla\mathbf{E}$, this problem is equivalent to finding a stationary point of $E(\mathbf{c})$. However,

(5.126) is less restricted, since the Jacobian **J** with components

$$j_{kl} \triangleq \frac{\partial f_k}{\partial c_l}$$

replaces the Hessian **H**, and **J** is in general neither symmetric nor p.d.

Using Newton's method to find an approximate solution of (5.126), one has to iterate by repeatedly calculating

$$\begin{aligned} \mathbf{f}(\mathbf{c}^i) + \mathbf{J}^i \Delta \mathbf{c}^i &\cong 0 \\ \Delta \mathbf{c}^i &\cong -\mathbf{J}^{i^{-1}} \mathbf{f}(\mathbf{c}^i). \end{aligned} \tag{5.127}$$

To avoid repeated computation of the n^2 elements j_{kl} and the inversion of **J**, let the approximating matrix

$$\mathbf{G}^i \cong \mathbf{J}^{i^{-1}} \tag{5.128}$$

be used. Also, let

$$\begin{aligned} \mathbf{d}^i &\triangleq -\mathbf{G}^i \mathbf{f}(\mathbf{c}^i) \\ \Delta \mathbf{c}^i &\triangleq \lambda^i \mathbf{d}^i \end{aligned} \tag{5.129}$$

as in the Fletcher-Powell method. The scalar λ^i may be selected so as to minimize or at least significantly reduce $|\mathbf{f}(\mathbf{c}^i + \lambda^i \mathbf{d}^i)|$. To update $\mathbf{G}^i$, the relation

$$\frac{d\mathbf{f}}{d\lambda^i} = \mathbf{J}^i \mathbf{d}^i \tag{5.130}$$

which follows by indirect differentiation, may be used. Let the derivative $d\mathbf{f}/d\lambda^i$ be approximated by a difference-quotient:

$$\frac{d\mathbf{f}}{d\lambda^i} \cong \frac{1}{\Delta\lambda^i} \left[\mathbf{f}(\mathbf{c}^{i+1}) - \mathbf{f}(\mathbf{c}^{i+1} - \Delta\lambda^i \mathbf{d}^i)\right], \tag{5.131}$$

where $\Delta\lambda^i$ is an appropriately selected small scalar, and define

$$\mathbf{y}^i \triangleq \mathbf{f}(\mathbf{c}^{i+1}) - \mathbf{f}(\mathbf{c}^{i+1} - \Delta\lambda^i \mathbf{d}^i). \tag{5.132}$$

Then, from (5.127) – (5.132), $\mathbf{G}^{i+1}$ satisfies

$$\mathbf{G}^{i+1}\mathbf{y}^i = \Delta\lambda^i \mathbf{d}^i. \tag{5.133}$$

The close relation between (5.129) and (5.133) on one hand, and the Fletcher-Powell equations (5.104) and (5.108) on the other is obvious.

Equation (5.133) has been derived using information about the variation of $\mathbf{f}$ in the direction $\mathbf{d}^i$. To make the choice of $\mathbf{G}^{i+1}$ unique, it will be postulated that the predicted change in $\mathbf{f}$ in a direction $\mathbf{e}^i$, orthogonal to $\mathbf{d}^i$, should be the same when calculated from either $\mathbf{G}^i$ or $\mathbf{G}^{i+1}$.

$$\mathbf{J}^i\mathbf{e}^i \cong \mathbf{G}^{i^{-1}}\mathbf{e}^i = \mathbf{G}^{i+1^{-1}}\mathbf{e}^i \cong \mathbf{J}^{i+1}\mathbf{e}^i. \tag{5.134}$$

It may be verified† by substitution that the unique $\mathbf{G}^{i+1}$ satisfying (5.133) and (5.134) is

$$\mathbf{G}^{i+1} = \mathbf{G}^i \frac{(\Delta\lambda^i\mathbf{d}^i - \mathbf{G}^i\mathbf{y}^i)\mathbf{d}^{i^T}\mathbf{G}^i}{\mathbf{d}^{i^T}\mathbf{G}^i\mathbf{y}^i}. \tag{5.135}$$

The computation proceeds by calculating (analytically or numerically) the inverse Jacobian $\mathbf{G}^0$ at the starting point. Next, the equation (5.129) is used to get $\mathbf{d}^0$ and $\Delta\mathbf{c}^0$. Now $\mathbf{f}(\mathbf{c}^1) = \mathbf{f}(\mathbf{c}^0 + \Delta\mathbf{c}^0)$ is obtained. If $|\mathbf{f}(\mathbf{c}^1)|$ is too large, then, via (5.132), $\mathbf{y}^0$ may be obtained. Finally, (5.135) yields $\mathbf{G}^1$ and the process may be repeated. The reader is referred to Broyden's paper [15] for practical hints in choosing $\Delta\lambda^i$ and for a discussion of convergence properties. Note that convergence is not guaranteed, although steps may be taken to prevent divergence.

One-Dimensional Search

Almost all gradient techniques involve a search for a minimum of $E(\mathbf{c})$ in a given direction $\Delta\mathbf{c}$ (i.e., for given ratios $\Delta c_i/\Delta c_j, i \neq j$). Some optimization methods, the so-called multidimensional search techniques [1], consist entirely of sequences of such directional, one-dimensional searches. In a typical gradient optimization such as one using the Fletcher-Powell method, easily half or more of the computer time

---

† Householder's formula

$$(\mathbf{J} + \mathbf{x}\mathbf{d}^T)^{-1} = \mathbf{J}^{-1} - \frac{\mathbf{J}^{-1}\mathbf{x}\mathbf{d}^T\mathbf{J}^{-1}}{1 + \mathbf{d}^T\mathbf{J}^{-1}\mathbf{x}}$$

is needed in this process.

needed is spent on directional search.† This part of the computation therefore must be organized carefully, to make it fast, efficient and, as far as possible, immune against pathological error-surfaces. Several types of search techniques are available; [10], [1] the only one discussed here is a little less efficient (by about 17%) but much easier to program than the fastest of them, the so-called *Fibonacci-search* [10].

The problem is to find λ such that $E(\mathbf{c} + \lambda\Delta\mathbf{c})$ is minimized. It will be assumed that $\mathbf{c}$, $\Delta\mathbf{c}$ are given and that $\Delta\mathbf{c}$ points in a downhill direction; also, that the desired minimum λ_m is in the range $(0, \lambda_u)$ where $E(\lambda)$ is unimodal. Finally, no *a priori* knowledge will be assumed about the whereabouts of λ_m in the interval $(0, \lambda_u)$

The basic idea utilized in both the Fibonacci search and the Golden Section search to be discussed here is that each evaluated $\lambda, E(\lambda)$ pair is utilized more than once. Refer to Fig. 5.7, showing the jth and the

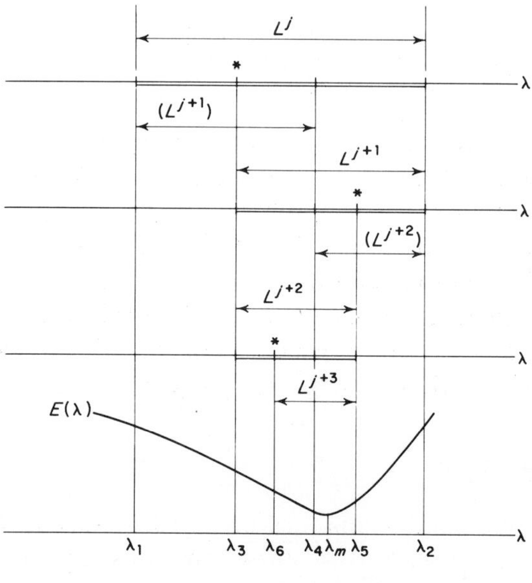

FIG. 5.7. Linear search.

$(j + 1)$st step of the process. Assume that λ_m is already known to be in the L^j range, between λ_1 and λ_2. Evaluating now $E(\lambda)$ at two internal points λ_3 and λ_4 yields $E(\lambda_3) > (\lambda_4)$. (The asterisk indicates the internal point with the larger E value.) Hence, λ_m must lie between λ_3 and λ_2 in L^{j+1}. Next, however, only one additional evaluation is

† Also called *linear search*.

needed to narrow the range even more, to L^{j+2}, since $E(\lambda_4)$ is already known. Hence, finding $E(\lambda_5) > E(\lambda_4)$ establishes L^{j+2}, between λ_3 and λ_5. Picking λ_6, from the value of $E(\lambda_6) > E(\lambda_5)$ it can be established that $\lambda_6 < \lambda_m < \lambda_5$, Proceeding in such manner, the location of λ_m may be found to the required accuracy, performing only one new calculation of $E(\lambda)$ for each cycle.

Lacking *a priori* knowledge about the whereabouts of λ_m, it seems logical to make the following assumptions on the process:

(1) the internal points in any interval should be symmetrically located; e.g., in L^j, $\lambda_3 - \lambda_1 = \lambda_2 - \lambda_4$, or in L^{j+1}, $\lambda_4 - \lambda_3 = \lambda_2 - \lambda_5$.

(2) the relative reduction in the size of the interval, r, should be the same in each cycle.

Because of condition (1),

$$\begin{gathered} \cdots \\ L^j = L^{j+1} + L^{j+2} \\ L^{j+2} + L^{j+2} + L^{j+3} \\ \cdots \end{gathered} \tag{5.136}$$

Because of condition (2),

$$\cdots = \frac{L^j}{L^{j+1}} = \frac{L^{j+1}}{L^{j+2}} = \frac{L^{j+2}}{L^{j+3}} = \cdots \triangleq r. \tag{5.137}$$

From these equations

$$\frac{L^j}{L^{j+1}} = r = \frac{L^{j+1} + L^{j+2}}{L^{j+1}} = 1 + \frac{1}{r} \tag{5.138}$$

or

$$\begin{gathered} r^2 - r - 1 = 0, \\ r = \frac{1 + \sqrt{5}}{2} \cong 1.618. \end{gathered} \tag{5.139}$$

Hence, the internal points in the initial interval $(0, \lambda_u)$.should be chosen at

$$\begin{gathered} \lambda_1 = \lambda_u (1 - \frac{1}{r}) = (2 - r)\lambda_u \\ \lambda_2 = \frac{\lambda_u}{r} = r\lambda_1 \ ; \end{gathered} \tag{5.140}$$

This is the so-called *Golden Section* of the interval. From then on, the new internal point should be picked symmetrically with respect to the one already established in the interval. After k such evaluations, by (5.137), the interval is

$$L^k = \lambda_u r^{1-k}. \tag{5.141}$$

Once this is less than the required accuracy of λ_m, the search may be terminated. Alternatively, the search may be abandoned earlier and polynomial interpolation [14], [1] can be used to find λ_m accurately. The attraction of this latter method is that polynomial interpolation converges rapidly (at least quadratically) in the vicinity of λ_m, while the search is only linearly convergent, since $L^j \propto L^{j-1}$.

So far it was assumed that λ_u is given. Actually, in most cases only the direction $\Delta\mathbf{c}$ is known, and λ_u must be found in the course of the search. An efficient technique for this consists of choosing a proper step-size λ_0 and then computing $E(\lambda_i)$ for

$$\lambda_i \triangleq \sum_{j=0}^{i} r^j \lambda_0 \qquad i = 0, 1, 2, \ldots \tag{5.142}$$

until $E(\lambda_i) \geqslant E(0)$ for some $\lambda_1 \triangleq \lambda_u$. The advantage of this method is that at least the last 3 of the $E(\lambda_i)$ are reusable in the subsequent Golden Section search.

5.4 CONSTRAINED OPTIMIZATION METHOD

Constraints

One valuable aspect of circuit optimization is the possibility of incorporating constraints on the parameters into the design process. Such constraints are, in general, frequency-dependent as exemplified by a constrained Q-factor $\omega L/r$ for a coil; or they may simply specify the permissible or practical range of a parameter

$$0 \leqslant l_i \leqslant c_i \leqslant u_i \tag{5.143}$$

where l_i, u_i are respectively the lower and upper limit on c_i. Alternatively, they may constrain the ratios of parameters

$$0 \leqslant l_{ij} \leqslant \frac{c_i}{c_j} \leqslant u_{ij} \tag{5.144}$$

as e.g. in the case of circuits containing piezoelectric resonators where the range of capacitance-ratios within a resonator, as well as the ratio of inductances of different resonators, is prescribed. In any case, all constraints may be put in the general form

$$g_j(x, \mathbf{c}) \geqslant 0, \tag{5.145}$$

where sometimes only the $>$ or only the $=$ sign is present. For example, (5.144) may be rewritten as

$$(c_i - c_j l_{ij})\,(c_j u_{ij} - c_i) \geqslant 0 \tag{5.146}$$

if $c_i, c_j > 0$.

Parameter-vectors $\mathbf{c}$ satisfying all existing constraints occupy a subspace of the whole parameter-space. This is called the *feasible region* and will be denoted by $\mathbf{R}$. $\mathbf{R}$ will be called *convex*, if the entire line segment joining any two of its points is inside $\mathbf{R}$. A convex ($\mathbf{R}_1$) and a non-convex ($\mathbf{R}_2$) region are shown in Fig. 5.8. Mathematically, if $\mathbf{c}_a, \mathbf{c}_b \epsilon \mathbf{R}$, then also $\mathbf{c}_a + \lambda(\mathbf{c}_b - \mathbf{c}_a) \epsilon R$ for any $0 \leqslant \lambda \leqslant 1$.

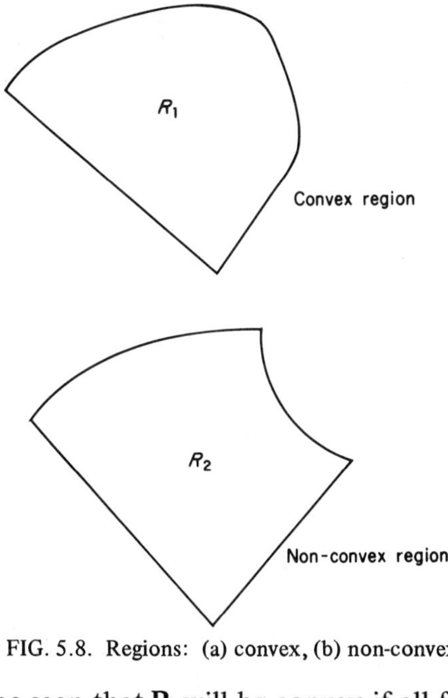

FIG. 5.8. Regions: (a) convex, (b) non-convex.

It can easily be seen that $\mathbf{R}$ will be convex if all functions $-g_j(\mathbf{c})$ are convex, i.e., if all $g_j(\mathbf{c})$ are concave. Here, convexity of the function $g_j(\mathbf{c})$ is defined as that of $E(\mathbf{c})$ in (5.6); concavity by replacing $\leqslant$ by $\geqslant$.

If a convex $E(\mathbf{c})$ is to be minimized over a convex R, the problem is that of *convex program*. While this seldom occurs in circuit design, it may be valid over limited regions of the $\mathbf{c}$ space. Its prime importance lies in the easily visualized fact that in a convex program any minimum is unique and thus global. A very thorough discussion of constraints and convex programming may be found in Refs. [2], [6] and [11].

Evidently, for some special constraints and/or optimization techniques, the constraints do not cause undue difficulties. That is, if the $g_j(\mathbf{c})$ are linear, or may be linearized, they may be incorporated into a linear program. If search methods are used, constraints may be accommodated by adding a large "penalty" to E each time a constraint is violated. In any case, it is possible to ignore the constraints in the optimization, except for checking their validity periodically. If a parameter (or parameter-ratio) violates one of its bounds, it is made equal to its extreme value permitted by the bound and "frozen" there for a number of iterations, or until convergence is achieved. After convergence a linear search may be performed in the feasible range of this parameter to achieve further improvement.

Transformation of Variables

The most straightforward exact way to enforce the constraints in the optimization is to change the variables. E.g., (5.143) may be assured by replacing c_i using

$$c_i = l_i + (u_i - l_i)\sin^2 d_i \tag{5.147}$$

and carrying out the optimization in terms of d_i. Evidently, any resulting real d_i will lead to a satisfactory c_i. A catalog of such transformations [16], along with the corresponding constraints, is given in Table 5.1.

The situation becomes more complicated if a constraint involves two or more of the c_i. If the constraints are

$$l_i \leqslant c_i \leqslant c_2 \leqslant \cdots \leqslant c_i \leqslant c_{i+1} \leqslant \cdots \tag{5.148}$$

the transformation of variables via

$$\begin{aligned} c_1 &= l_1 + d_1^2 \\ c_2 &= l_1 + d_1^2 + d_2^2 \\ &\bullet\ \bullet\ \bullet\ \bullet\ \bullet\ \bullet\ \bullet\ \bullet \\ c_i &= l_1 + \sum_{j=1}^{i} d_j^2 \\ &\bullet\ \bullet\ \bullet\ \bullet\ \bullet\ \bullet\ \bullet\ \bullet \end{aligned} \tag{5.149}$$

TABLE 5.1

Constraint	*Transformation*
$c_i \geqslant 0$	$c_i = d_i^2$
	$c_i = \exp d_i$
	$c_i = \lvert d_i \rvert$
$0 \leqslant c_i \leqslant u_i$	$c_i = u_i \sin^2 d_i$
	$c_i = u_i \dfrac{\exp d_i}{\exp d_i + 1}$
$c_i \geqslant l_i$	$c_i = l_i + d_1^2$
	$c_i = l_i + \exp d_i$
	$c_i = l_i + \lvert d_i \rvert$
$l_i \leqslant c_i \leqslant u_i$	$c_i = l_i + (u_i - l_i) \sin^2 d_i$
	$c_i = l_i + (u_i - l_i) \dfrac{\exp d_i}{\exp d_i + 1}$
	$c_i = \dfrac{l_i + u_i}{2} + \dfrac{u_i - l_i}{2} \sin d_i$

is usable. In other cases, a multi-stage process, requring much ingenuity, may be needed [16].

Penalty Functions

Constraints may be enforced by adding, arbitrarily, a new function $F(\mathbf{c})$ to $E(\mathbf{c})$ prior to minimization. The value of $F(\mathbf{c})$ should be large when one or more constraints are violated; small or zero otherwise. $F(\mathbf{c})$ is usually written in the form

$$F(\mathbf{c}) = r \sum_{(j)} G[g_j(\mathbf{c})], \tag{5.150}$$

where $r > 0$ is a constant. The functional $G(g_j)$ represents the punishment for violating the constraint $g_j \geqslant 0$; it is frequently called *penalty-function*. Evidently $G(y)$ must be a decreasing function of y; it is also easy to show [11] that if $G(y)$ is convex, then for a concave $g_j(\mathbf{c})$, $G[g_j(\mathbf{c})]$ is also convex. Hence

$$E + r \sum_{(j)} G(\mathbf{g}_j)$$

is also a convex function of $\mathbf{c}$. The simplest choice of $G(y)$ is the linear spline function $|y| - y$. For this choice, the modified error-function is

$$E_r(\mathbf{c}) = E(\mathbf{c}) + r \sum_{(j)} (|g_j(\mathbf{c})| - g_j(\mathbf{c})). \tag{5.151}$$

Minimizing this function starting from a point $\mathbf{c}^0$ outside of $\mathbf{R}$ results in a final vector $\hat{\mathbf{c}}$ which is close to the solution of the constrained problem. The choice of r is critical; if it is too small, the constraints will not be enforced; if too large, inaccuracies will prevent convergence. It is possible to repeat the calculations for increasing r values, choosing the solution $\hat{\mathbf{c}}^k$ of the kth cycle as the starting vector of the $(k+1)$st. Even more generally, both r and p may be changed (r increased, p decreased) in a penalty function defined by

$$r \sum_{(j)} (|g_j| - g_j)^p.$$

These simple techniques gave excellent results for relatively well-conditioned functions involving not more than five variables.

The Fiacco-McCormick Method†

A much better explored technique is that due to Carrol, Fiacco and McCormick. [17], [18] Here, $G(y) \triangleq 1/y$ for $y > 0$, $G(y) \to \infty$ for $y \leqslant 0$, and the optimization proceeds in the following steps.

(1) A feasible point $\mathbf{c}^0 \in \mathbf{R}$ is chosen.

(2) For a well selected r^1,

$$E + r^1 \sum_{(j)} \frac{1}{g_j(\mathbf{c})}$$

is minimized, giving $\hat{\mathbf{c}}^1$ as a result.

(3) Starting from $\mathbf{c}^1$,

$$E + r^2 \sum_{(j)} \frac{1}{g_j(\mathbf{c})}$$

is minimized for $r^2 < r^1$, giving $\hat{\mathbf{c}}$.

(4) Step (3) is repeated for a decreasing sequence of r^k, until the constrained minimum of E is reached.

---

† This method is also called SUMT, which is short for "Sequential Unconstrained Minimization Technique."

For a convex program, it may be shown that the constrained minimum is indeed approached as $r^k \to 0$. In addition, for any two subsequent cycle s ($\mathbf{k} > 1$).

$$E(\hat{\mathbf{c}}^{k+1}) \leqslant E(\hat{\mathbf{c}}^k)$$

$$\sum_{(j)} \frac{1}{g_j(\hat{\mathbf{c}}^{k+1})} \geqslant \sum_{j} \frac{1}{g_j(\hat{\mathbf{c}}^k)} \tag{5.152}$$

$$E(\hat{\mathbf{c}}^{k+1}) + r^{k+1} \sum_{(j)} \frac{1}{g_j(\hat{\mathbf{c}}^{k+1})} \leqslant E(\hat{\mathbf{c}}^k) + r^k \sum_{(j)} \frac{1}{g_j(\hat{\mathbf{c}}^k)}$$

and, for any cycle (k > 1),

$$E(\hat{\mathbf{c}}^k) - r^k \sum_{(j)} \frac{1}{g_j(\hat{\mathbf{c}}^k)} \leqslant \min_{\mathbf{c} \in \mathbf{R}} E(\mathbf{c}) \leqslant E(\hat{\mathbf{c}}^k) \tag{5.153}$$

Hence, by the second of Eqs. (5.152), the minimum is being approached from the inside of **R** moving toward the boundary, while E and

$$E + r \sum_{(j)} \frac{1}{g_j}$$

is monotonically decreasing. These relations are, however, invalid for the first cycle, since both $\mathbf{c}^0$ and r^1 are arbitrarily chosen.

The initial steps (1) and (2) are critical: if r^1 is chosen too large, **c** will be in the middle of **R** and a large number of iterations will be needed to approach the boundary; if r^1 is too small, the trajectory $\mathbf{c}^k$ crawls along the boundary which again leads to lengthy calculation. Two procedures are given in Ref. [18] for choosing r^1 properly.

If several variables and constraints are present, choosing $\mathbf{c}^0$ in **R** may not be a trivial problem. A possible method to find $\mathbf{c}^0$ is to carry out the (unconstrained) minimization of

$$\sum_{(j)} (|g_j| - g_j)^P .$$

If this is positive, $\mathbf{R} \equiv 0$. A "bootstrap" method [11] consists of picking a $\mathbf{c}$ arbitrarily and then grouping the satisfied constraints g_j, $j \in J$, and the violated constraints g_i, $i \in I$. Next,

$$-\sum_{i \in I} g_i(\mathbf{c})$$

is minimized, e.g., via the Fiacco-McCormick method, subject to $g_j(\mathbf{c}) \geqslant 0, j \in J$. This will result in satisfying some additional constraints, previously violated, at the minimum $\mathbf{c}^1$. The process is now repeated from $\mathbf{c}^1$. An alternative bootstrap method, satisfying the constraints one by one, is described in Ref. [17]. Equality-constraints of the form $g_i(\mathbf{c}) = 0$ may be incorporated into the error-function via terms

$$(r^k)^{-1/2} \sum_{(i)} [g_i(\mathbf{c})]^2.$$

As r^k decreases, these enforce the convergence of all $g_i(\mathbf{c})$ to zero.

Other penalty function techniques. The Fiacco-McCormick SUMT† method has several computational disadvantages. These include the arbitrariness in choosing the r^k; the possibility of a large increase from $E(\mathbf{c}^0)$ to $E(\hat{\mathbf{c}}^1)$; and, importantly, the steep "constraint-walls" introduced by the $1/y$ type penalty-function. These make the gradient methods used in the minimization inaccurate. Finally, a feasible solution $\mathbf{c}^0$ is needed to start the iteration.

Recognizing those shortcomings, several other penalty-function methods have been introduced by Fiacco, McCormick and others. A different penalty-function $G(y) \triangleq \log 1/y$ has been used successfully [21]. This has the advantage of more suitable constraint walls, and also gives the bounding

$$E(\hat{\mathbf{c}}^k) - mr^k \leqslant \min_{\mathbf{c} \in R} E(\mathbf{c}) \leqslant E(\hat{\mathbf{c}}^k). \tag{5.154}$$

Here, m is the number of constraints. Eq. (5.154) connects the accuracy of the kth approximation explicitly with r^k; this makes it possible to choose r^1 in such a way as to meet a prescribed tolerance on min E.

Two variations of their SUMT method have been described by the originators. In the first one, [19] the slack-parameters

$$t_i = g_i(x) \geqslant 0 \tag{5.155}$$

† See footnote on p. 232.

are introduced and used to convert all inequality constraints into equalities. Then the problem is to minimize

$$E(\mathbf{c}) + \frac{1}{r^k} \sum_{(j)} [g_i(\mathbf{c}) - t_i]^2,$$

subject to $t_i \geqslant 0$. The advantages claimed are (1) it approaches the optimum from outside of **R** and hence requires no effort to obtain $\mathbf{c}^0$; (2) it can be used even if the *interior* of **R** is empty; (3) it requires less computation per iteration.

The other method minimizes, in the kth cycle,

$$[E(\mathbf{c}^k) - E(\mathbf{c})]^{-1} + \sum_{(j)} 1/g_j(\mathbf{c}),$$

over **c** values satisfying $E(\mathbf{c}) < E(\mathbf{c}^k)$. This procedure turns out to be equivalent to the SUMT for the special choice $r_k \equiv E(\mathbf{c}^k) - E(\mathbf{c}^{k+1})$. This choice turns out to be such as to decrease $E(\mathbf{c})$ from the very first iteration. It also disposes of any arbitrariness in the process.

While all these methods claim advantages over SUMT, insufficient computing experience prevents an objective comparison.

5.5 CIRCUIT ORIENTED OPTIMIZATION TECHNIQUES

Constrained Optimization of Filters

Most network optimization problems may be formulated as constrained linear or nonlinear programs. By choosing a least pth error-criterion, gradient or search-techniques are directly applicable to design in either the time and/or the frequency domain. In the design of filters, however, Chebyshev-approximation is usually more desirable. To obtain this in the case of linear $E(\mathbf{c})$ and $g_j(\mathbf{c})$, the methods of Sec. 5.2 may be used. In the more usual situation, the transfer function A (e.g., the loss) is a nonlinear function of **c**, and it has to satisfy requirements of the form [22]

$$\begin{aligned} &A(\omega, \mathbf{c}) \leqslant A_p(\omega) \qquad \omega \in PB_j, \qquad j = 1, 2, \ldots \\ &A(\omega, \mathbf{c}) \geqslant A_s(\omega) \qquad \omega \in SB_k, \qquad k = 1, 2, \ldots \end{aligned} \tag{5.156}$$

where the PB_j are the passbands, the SB_k the stopbands, while $A_p(\omega)$ and $A_s(\omega)$ are the required passband ripple and guaranteed stopband loss, respectively. In addition, constraints of the form given in (5.143) – (5.146) may bound the c_i. To bring the problem into the standard form tractable by the methods of Sec. 5.4, a nonlinear program is formulated as follows: [22]

Minimize $E(\mathbf{c})$, subject to

$$\begin{aligned} A(\omega, \mathbf{c}) - Ew(\omega) &\leqslant A_p(\omega), \qquad && \omega \in PB_j \\ A(\omega, \mathbf{c}) + Ew(\omega) &\geqslant A_s(\omega), && \omega \in SB_k \end{aligned} \tag{5.157}$$

and subject to constraints, if any, on the c_i. Here, $w(\omega)$ is a (nonnegative) weight function. Evidently, the specifications are surpassed by the margin $-Ew(\omega)$. If in the solution $E \leqslant 0$, the specifications are met; otherwise, they are violated by the amount $+Ew(\omega)$.

The solution, using the Fiacco-McCormick SUMT-method, consists of the repeated unconstrained minimization of

$$\begin{aligned} E + r^k \Bigg\{ &\sum_{(j)} \sum_{\omega_h \epsilon PB_j} \Big[-A(\omega_h, \mathbf{c}) + Ew(\omega_h) + A_p(\omega_h)\Big]^{-1} + \\ &+ \sum_{(k)} \sum_{\omega_h \epsilon SB_k} \Big[A(\omega_h, \mathbf{c}) + Ew(\omega_h) - A_s(\omega_h)\Big]^{-1} + \\ &+ \sum_{(i)} \Big[(c_i - l_i)(u_i - c_i)\Big]^{-1} + \sum_{(i)} \sum_{(i)} \Big[(c_i - l_{ij}c_j)(u_{ij}c_j - c_i)\Big]^{-1} \Bigg\} \end{aligned} \tag{5.158}$$

for a decreasing sequence of $r^k \to 0$. Here, the ω_h are the sample points.

For filters consisting of many cascaded sections, $A(\omega)$ should be written in terms of the chain-matrices of the individual sections [22]. Then, each circuit parameter affects only the elements of one chain-matrix. This simplifies the calculation of the derivatives needed in the optimization.

Constrained Optimization of Equalizers

For equalizers, the usual Chebyshev-error criterion leads to the specifications†

$$\left| w(\omega)\left[A(\omega, \mathrm{c}) - A_{\mathrm{spec}}(\omega)\right]\right| \leqslant \epsilon_{\max}, \qquad \omega \in PB \tag{5.159}$$

where $w(\omega)$ is the weight, $A_{\mathrm{spec}}(\omega)$ the specified loss-response and $\epsilon_{\max}$ the tolerance in the approximation. (Here $\epsilon_{\max}$ is merely a number, e.g., 0.1 db, independent of ω.) Introducing the error-function

$$\epsilon(\omega, \mathrm{c}) \triangleq w(\omega)[A(\omega, \mathrm{c}) - A_{\mathrm{spec}}(\omega)], \tag{5.160}$$

as well as the margin $-E$ via

$$\epsilon_{\max} - |\epsilon| \geqslant -E \tag{5.161}$$

the problem may be written

minimize E, subject to

$$\left.\begin{aligned} \epsilon_{\max} - \epsilon(\omega, \mathrm{c}) + E &\geqslant 0 \\ \epsilon_{\max} + \epsilon(\omega, \mathrm{c}) + E &\geqslant 0 \end{aligned}\right. \qquad \omega \in PB \tag{5.162}$$

as well as all constraints upon the c_i.

The SUMT solution therefore requires the repeated minimization of

$$\begin{aligned} E + r^k \Big\{ &\sum_{\omega_h \in PB} \left[(\epsilon_{\max} - \epsilon(\omega, \mathrm{c}) + E)^{-1} + (\epsilon_{\max} + \epsilon(\omega, \mathrm{c}) + E)^{-1}\right] + \\ &+ \sum_{(i)} \left[(c_i - l_i)(u_i - c_i)\right]^{-1} + \sum_{(i)} \sum_{(j)} \left[(c_i - l_{ij}c_j)(u_{ij}c_j - c_i)\right]^{-1} \Big\}. \end{aligned} \tag{5.163}$$

As before, $E < 0$ indicates a margin, $E > 0$ a violation of the specifications by an amount $|E|$.

---

† The problem here is described, for definiteness, for loss-equalizers. Exactly analog formulation is, of course, valid for phase-equalizers or for equalization in the time-domain [23].

Iterative Chebyshev Approximation; the Remez Algorithm

Some iterative techniques for obtaining Chebyshev approximation via $F^j(x, \mathbf{c})$ which is linear (or linearized) in $\mathbf{c}$ were described in Sec. 5.2. An alternative method, found especially suitable for solving the approximation problem in either frequency or time domain, will be discussed here. This method, originally developed by and named after Remez, achieves an equal-ripple error-function by gradual equalization of the error-extrema. The process is illustrated (for 3 parameters) in Fig. 5.9.

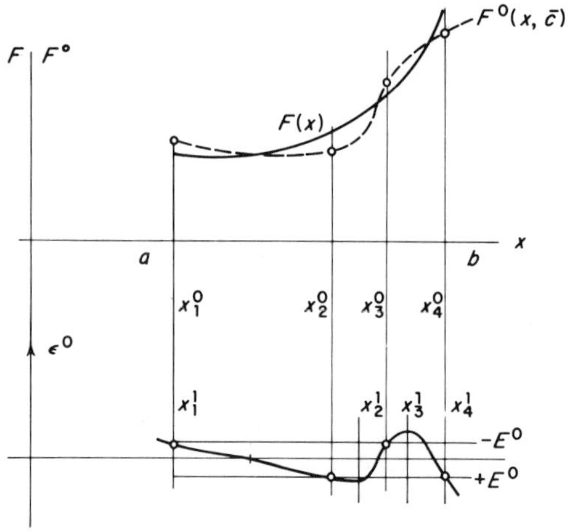

FIG. 5.9. The Remez algorithm.

First, $n + 1$ independent variable values $x_1^0, x_2^0, \ldots, x_{n+1}^0$ are chosen. Next, $F^0(x, \mathbf{c})$ is obtained by forcing the error ϵ (given in (5.2), above) to take on equal but alternating values at the x_k^0, i.e., by solving, for $k = 1, 2, \ldots, n + 1$

$$\epsilon^0(x_k^0) \triangleq w(x_k^0)\,[F(x_k^0) - F^0(x_k^0, \mathbf{c}^0)] = (-1)^k E^0 \qquad (5.164)$$

In (5.164), E^0 is one of the unknowns, $c_1^0, c_2^0, \ldots, c_n^0$ the others. In principle, at least, from the $n + 1$ equations c^0 and E^0 can be found, and thus $F^0(x, \mathbf{c}^0)$ established. Next, the $n + 1$ alternating extrema of $\epsilon^0(x)$ are found and chosen as the new abscissa-set $[x_k^1]$. Now, Eqs. (5.164) are solved again (with all superscripts changed to 1) for $\mathbf{c}^1$, E^1. In the normal course of iteration, E^j keeps increasing with j until finally it represents the amplitude of the resulting equal-ripple error-function.

For the class of problems where $\epsilon(x)$ satisfies the conditions

(a) if the difference of two error-functions has more than $n - 1$ zeros, then they are identical;
(b) any error-function is uniquely and continuously determined by its values at n coordinates;

the convergence of the Remez method to a unique Chebyshev approximant $F^j(x, \mathbf{c})$ is assured. The most important representative of the class is the case of polynomial approximant

$$F^j(x, \mathbf{c}) = \sum_{i=0}^{n-1} c_i x^i. \tag{5.165}$$

For this simplest problem, explicit formulae may easily be obtained for the c_i^j and E^j, by noting that now (5.164) is of the form

$$\mathbf{F} - \mathbf{X}_{n-1}\mathbf{c} = \boldsymbol{\epsilon}$$

where

$$\mathbf{F} \triangleq (F(x_1), F(x_2), \ldots, F(x_{n+1}))^T$$

$$\mathbf{X}_{n-1} \triangleq \begin{pmatrix} 1 & x_1 & x_1^2 & \cdots & x_1^{n-1} \\ 1 & x_2 & x_2^2 & \cdots & x_2^{n-1} \\ \cdot & \cdot & \cdot & \cdot & \cdot \\ 1 & x_{n+1} & x_{n+1}^2 & \cdots & x_{n+1}^{n-1} \end{pmatrix} \tag{5.166}$$

$$\boldsymbol{\epsilon} \triangleq E\left(-\frac{1}{w(x_1)}, +\frac{1}{w(x_2)}, \ldots, \frac{(-1)^{n+1}}{w(x_{n+1})}\right)^T$$

and that the minors are Vandermonde determinants.

Rational and quasirational approximants. [1], [24] In solving the approximation problem in circuit design, the approximant is either a rational expression in $x \triangleq \omega^2$, or a simple functional of a rational function. Hence, in the most general case,

$$F^j(x, \mathbf{c}) = \phi\left[\frac{P_{n-m-1}(x)}{Q_m(x)}\right], \tag{5.167}$$

where $P(x)$, $Q(x)$ are polynomials of degrees indicated by their subscripts. Now (5.164) gives directly

$$(\mathbf{X}_{n-m-1} \,\vdots\, \mathbf{G}\mathbf{X}_m)\mathbf{c} = 0 \tag{5.168}$$

where the symbol (⋮) indicates that the matrix may be partitioned along the dotted line. $\mathbf{X}_{n-m-1}$, $\mathbf{X}_m$ are defined similarly to $\mathbf{X}_{n-1}$, and G is a diagonal matrix with its (k, k) component given by

$$g_{kk} = -\Phi^{-1}\left[F(x_k) - (-1)^k \frac{E}{w(x_k)}\right]. \qquad (5.169)$$

Finally, $\mathbf{c}$ is the $n + 1$ element vector composed of the coefficients of P and Q. Evidently, any one of the c_k may be chosen arbitrarily, leaving n free parameters in $\mathbf{c}$.

Equation (5.168) is linear in $\mathbf{c}$; it is, however, usually very nonlinear in E which enters via the g_{kk}. Theoretically, (5.168) may be reduced to the single nonlinear equation

$$\det(\mathbf{X}_{n-m-1} \vdots \mathbf{G}\mathbf{X}_m) = 0 \qquad (5.170)$$

in the single variable E. In most network problems, however, (5.170) turns out to be an $m + 1$ order polynomial equation in E or in a simple function $f(E)$ [24]. This has $m + 1$ roots, all real, of which only one can lead to continuous $\epsilon(x)$, but often there is no clear indication which one. For this reason, as well as for computational simplicity, iterative methods [1], [24] are often employed to find E and $\mathbf{c}$ from (5.168).

General approximant. For a general function $F^j(x, \mathbf{c})$, linearization of (5.164) gives [1]

$$\mathbf{F} - \mathbf{F}^j - \mathbf{J}\Delta\mathbf{c} = \boldsymbol{\epsilon} \qquad (5.171)$$

where $\mathbf{F}^j$ has components $F^j(x_k)$ and where $\mathbf{J}$ is the Jacobian with its (k, l) component equal to $\partial F^j(x_k)/\partial c_l$. Equation (5.171) is linear in $\Delta\mathbf{c}$ and E. Having obtained $\Delta\mathbf{c}$, the improved $\mathbf{c}^{j+1}$ may be obtained as $\mathbf{c}^j + \Delta\mathbf{c}$, or, if damping is needed, as

$$\mathbf{c}^j + \lambda\Delta\mathbf{c},\ 0 < \lambda \leqslant 1.$$

Choosing λ sufficiently small, convergence has invariably been obtained in practical problems.

The Remez algorithm, originally developed for polynomial approximation and subsequently widely used to obtain rational function approximants for computer subprograms, proved to be a very useful tool for the Chebyshev design of equalizers, filters and delay lines.

Coefficient Matching

If the coefficients of the rational network function are known, the circuit realization in a given configuration may be obtained via the method of coefficient matching. [15], [1] Let the desired function be given by

$$T_s(s) = \frac{\sum_{j=0}^{n} a_{j+m+1} s^j}{\sum_{i=0}^{m} a_i s^i}. \tag{5.172}$$

For a network of proper configuration and complexity the network function has the same form

$$T(s) = \frac{\sum_{j=0}^{n} b_{j+m+1}(\mathbf{c}) s^j}{\sum_{i=0}^{m} b_i(\mathbf{c}) s^i} \qquad (5.173$$

but the coefficients $b_i(\mathbf{c})$ will not in general equal the a_i. The b_i are functions of the element value vector **c**, and may be calculated using topological formulas. Since the $b_i(\mathbf{c})$ are nonlinear functions of **c**, containing products of the c_i, iterative techniques must be used for finding the **c** such that $T(s) = T_s(s)$.

For canonical networks, where the number of variable c_i's is one less than the number of the b_i, a Newton method described by

$$b_i + \sum_{(l)} \frac{\partial b_i}{\partial c_l} \Delta c_l = K a_i + a_i \Delta K \qquad i = 1, 2, \ldots \tag{5.174}$$

may be used to find Δc_l and ΔK. Here, K is a constant multiplier needed, since in general only a proportionality $b_i = K a_i$ should be achieved between a_i and b_i. In general, the numbers of b_i and c_l can be quite different, either because of a non-canonical (e.g., active) circuit, or because only a part of the network function needs to be matched, as in the case of dominant poles. Then, an error-criterion of the form

$$E \triangleq \sum_{(i)} \left[Ka_i - b_i(\mathbf{c})\right]^p \qquad (p \text{ even}) \qquad (5.175)$$

may be defined and the method of least pth approximation or gradient techniques used to achieve the only acceptable solution, namely $E = 0$.

The main advantage of the coefficient-matching method is its versatility. It can take any lumped linear time-invariant circuit, active or passive, stable or unstable, as initial approximant and usually converge to the required stable solution quite rapidly. It requires, however, *a priori* solution of the approximation problem (e.g., via the Remez method) and also explicit expressions, obtained using linear graph theory or conventional circuit analysis for all $b_i(\mathbf{c})$ functions.

5.6 DESIGN EXAMPLE

To illustrate the use of optimization techniques in a practical design problem, the synthesis of a simple active RC equalizer [12] will next be described. The specified gain response, in the 1 MHz $\leqslant f \leqslant$ 2 MHz range. is given by

$$G^{(\text{db})} = 5 + 5f^{(\text{MHz})}$$

(Fig. 5.10). The assumed circuit configuration uses an operational amplifier and five passive elements (Fig. 5.11). With ideal components, the gain-response is given by

$$G_c^{(\text{db})} \triangleq 20 \log_{10} \left| \frac{V_2}{V_1} \right| = 20 \log_{10} |RY| =$$

$$= 10 \log_{10} \left[\frac{R^2}{R_1^2} \frac{(1 - \omega^2 R_1 C_1 R_2 C_2)^2 + \omega^2 (R_1 C_1 + R_1 C_2 + R_2 C_2)^2}{1 + (\omega R_2 C_2)^2} \right]$$

where Y is the series admittance built from R_1, C_1, R_2 and C_2. Evidently R serves merely as a scaling factor and hence $R = 1$ may be chosen without restricting generality.

To illustrate various processes, this circuit will be optimized using different criteria and techniques. Let us first define

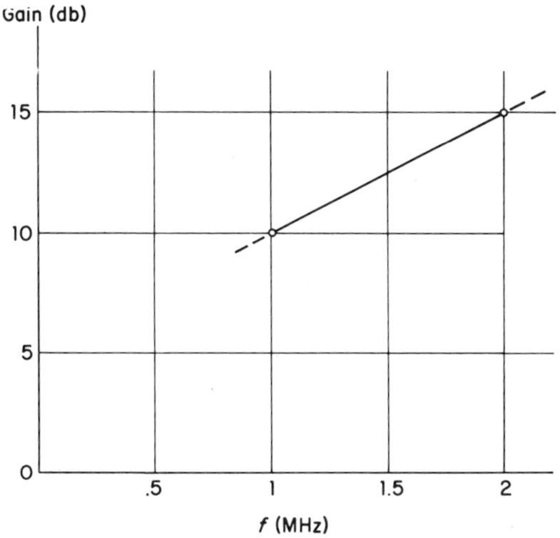

FIG. 5.10. Specified gain characteristics.

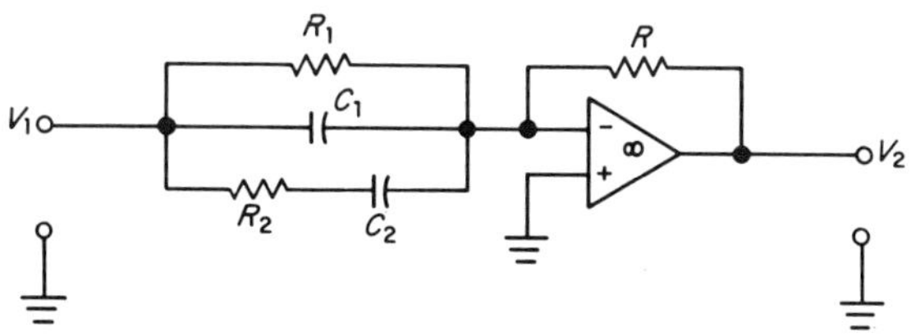

FIG. 5.11. Active equalizer circuit.

$$\mathbf{c} \triangleq (C_1, C_2, R_1, R_2)^T, \qquad x \triangleq f, \qquad F \triangleq G^{(\text{db})},$$

$$F^j(\mathbf{c}^j) \triangleq G_c^{(\text{db})}(C_1^j, C_2^j, R_1^j, R_2^j);$$

also, let

$$E^j \triangleq \sum_{i=1}^{m} [F(x_i) - F^j(x_i, \mathbf{c}^j)]^p$$

be selected as error-criterion, for p even. Then the least pth approximation techniques described in Sec. 5.3 are directly applicable. Let initially $\mathbf{c}^0 = (1, 1, 1, 1)^T$; 21 sample points are chosen ($m = 21$) in steps of

0.05 from 1 to 2. For the first run, $p = 10$ is selected. The error-function for the initial approximation is shown in Fig. 5.12. The correspon-

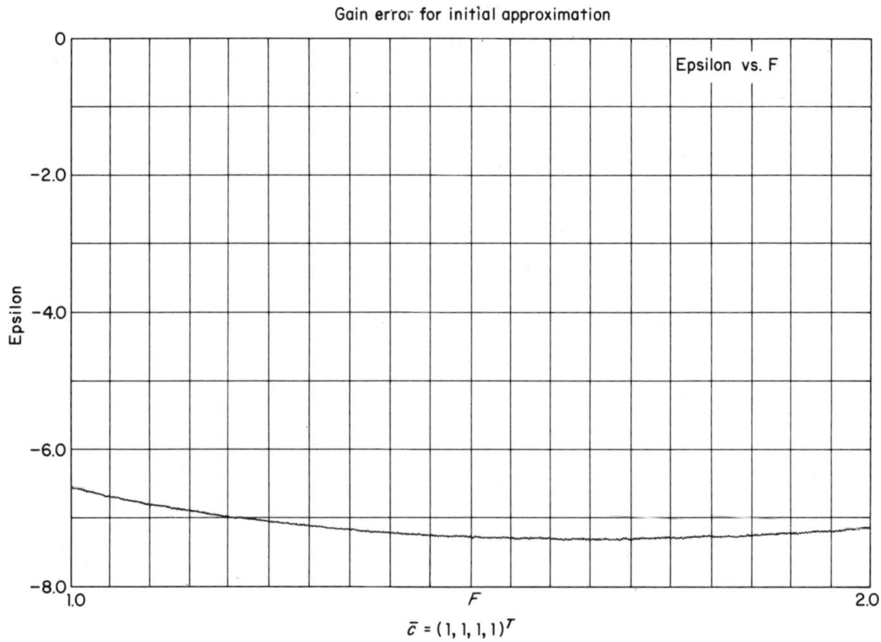

FIG. 5.12. Gain error for initial approximation.

ding value is $E^0 \cong 7.3 \times 10^9$. The decrease of E^j as a function of j in the course of the iteration is shown in Fig. 5.13, in logarithmic units. The total change is by about a factor of 10^{-17}. The final element vector is $\mathbf{c}_a = (0.41949, 0.03121, 0.6161, 39.884)^T$. The corresponding error curve is shown in Fig. 5.14.

Next, the same calculation was performed from the initial-vector $\mathbf{c}^0 = (0.5, 0.5, 0.5, 0.5)^T$. The iteration now converged to the final element vector $\mathbf{c}_b = (0.419514, 1.1408, 0.8445, 2.286)^T$. With the exception of c_1, all components have changed drastically. The final error curve shows a quality of approximation comparable to that of Fig. 5.14. The two E values were within 20% of each other.

Next, the $p \triangleq 2$, i.e., least squares approximation was found, using $\mathbf{c}_a$ as initial value. The result was a final vector $\mathbf{c}_c = (0.41631, 0.0644, 0.5950, 10009)^T$.

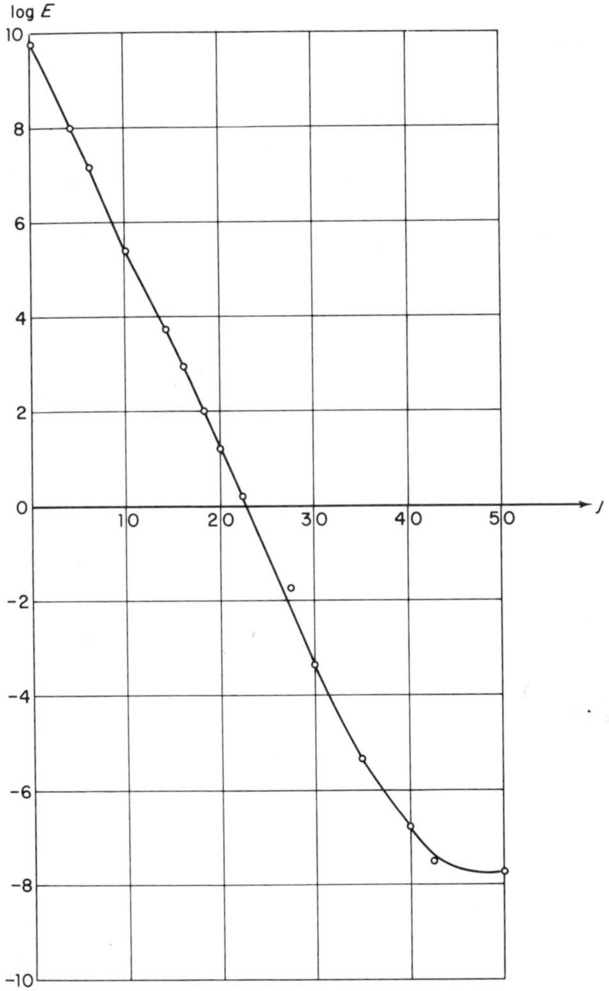

FIG. 5.13. Error criterion vs. iteration number.

As $\mathbf{c}_a$ and especially $\mathbf{c}_c$ suggests, C_2 and R_2 can be left out of the circuit without serious impairment of performance. This was borne out by computing the error-curve for $\mathbf{c}_a$ but with $1/R_2 = C_2 = 0$.

For further comparison, the same optimization problem was worked out using the Fletcher-Powell method. Now the ideal voltage gain $|V_2/V_1|$ was chosen as F, the actual voltage gain as F^j, and

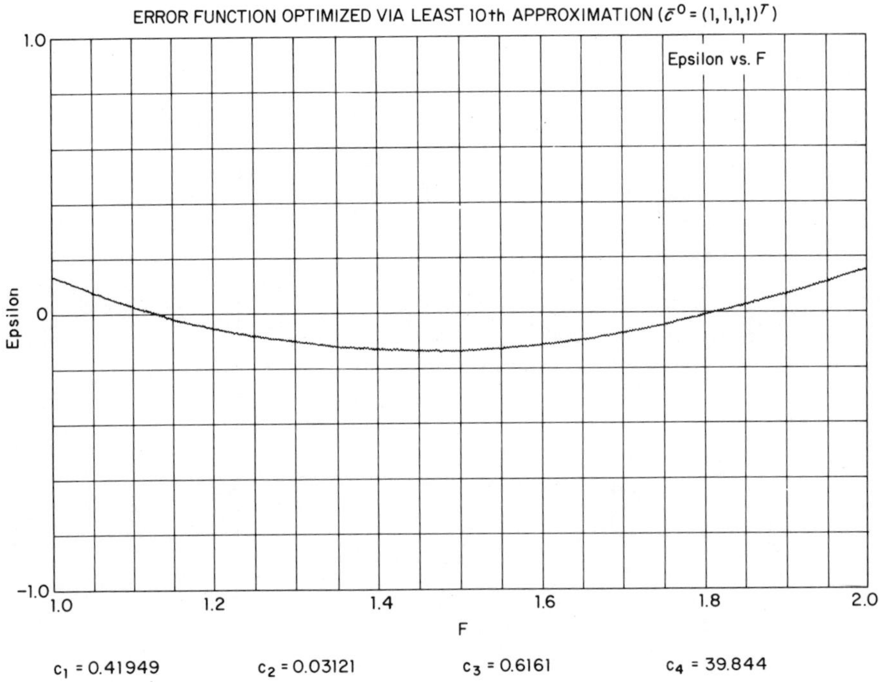

FIG. 5.14. Error function optimized via least 10th approximation ($c^0 = (1, 1, 1, 1)^T$).

$$E \triangleq \sum_{i=1}^{m} [F(x_i) - F^j(x_i, c^j)]^2.$$

Again, various starting points were chosen. Starting with

$$\mathbf{c}^0 = \left(\frac{0.5}{2\pi}, \frac{0.5}{2\pi}, 0.5, 0.5\right)^T,$$

the iteration "converged" to a local minimum, where (for $m = 50$), $E \cong 5.06$ was achieved. The error-curve, accordingly, was none too good; the element-vector is $\mathbf{c}_d = 0.16016, 0.15969, 0.3282, 0.25006)^T$. Starting with $\mathbf{c}^0 = (1/2\pi, 1/2\pi, 1, 1)^T$, the vector $\mathbf{c}_e = (0.30155, 0.15590, 0.81785, 0.31239)^T$ is obtained, giving $E \cong 1.5 \times 10^{-4}$. Finally, for the initial values $\mathbf{c}^0 = (1.5/2\pi, 1.5/2\pi, 1.5, 1.5)^T$, the final values $\mathbf{c}_f = (0.42766, 0.34110, 1.1844, 1.5541)^T$, $E \cong 2.08 \times 10^{-2}$.

The best result was obtained for the initial choice $\mathbf{c}^0 = (1, 1, 1, 1)$; $\mathbf{c}_g = (0.42258, 2.2629, 1.4877, 1.26814)^T$. The corresponding error-curve is shown in Fig. 5.15.

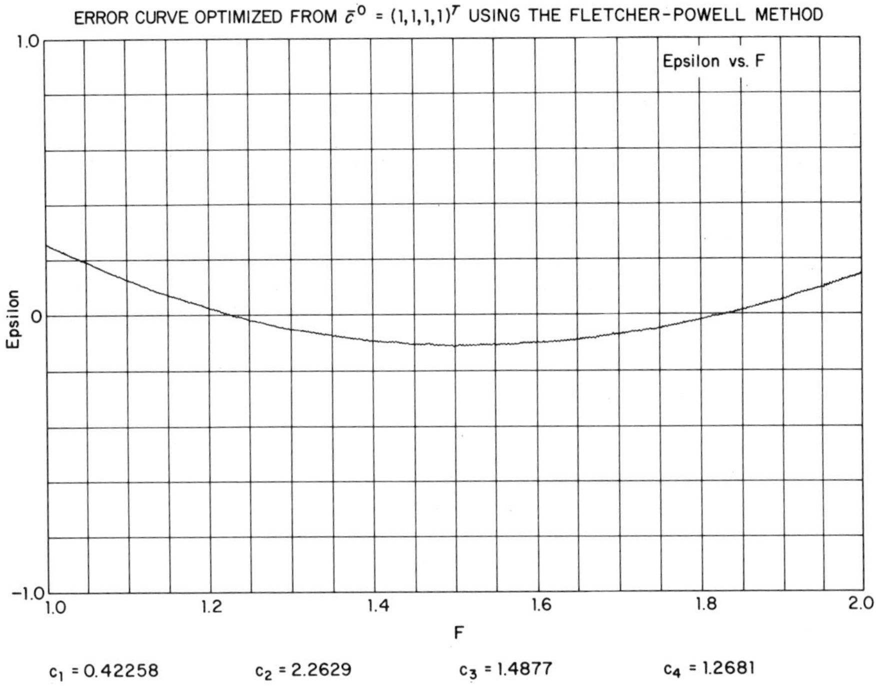

FIG. 5.15. Error curve optimized from $\mathbf{c}^0 = (1, 1, 1, 1)^T$, using the Fletcher-Powell method.

Evidently, for each starting point the program found a "nearby" local minimum, and it is purely a matter of luck (or intuition) in choosing the initial values, how good the ultimate result becomes.†

REFERENCES*

1. G. C. Temes, D. A. Calahan, "Computer-aided network optimization, the state of the art," *Proc. IEEE*, vol. 55, pp. 1832-1863, November, 1967.

2. G. Zoutendijk, *Methods of Feasible Directions*, D. Van Nostrand Co., Inc., New York, 1960.

† The writer is indebted to Dr. D. Zai and students of Stanford University class EE305 for obtaining the above, and about 20 additional local optima of this problem.

* The philosophy in compiling this bibliography was to reference tutorial descriptions of topics, rather than their original (often difficult) expositions, whenever this was possible.

3. For an excellent summary and further references, see e.g. *Linear Programming* by Lloyd Rosenberg in *Digital Computer User's Handbook*, ed. M. Klerer and G. A. Korn, McGraw-Hill Book Co., New York, 1967.

4. G. A. Matthews, F. Davis, J. McKee, R. Cavin and G. Sibbles, "Approximating transfer functions by linear programming," *Proc. First Allerton Conf.*, November, 1963.

5. A. A. Lanne, "Some approximation problems in electrical circuit theory," *Telecommunications* (transl. from Russian), No. 8, August, 1965.

6. E. M. L. Beale, "Nonlinear programming" in Klerer and Korn, *op. cit.*

7. Y. Ishizaki, H. Wakazuki and H. Watanabe, "Some considerations on min-max approximation problem for network design," *Proc. Third Allerton Conf.*, October, 1965.

8. R. A. Rohrer, personal communication.

9. M. R. Aaron, "The use of least squares in system design," *IRE Trans. on Circuit Theory*, vol. CT-3, December, 1956.

10. D. J. Wilde, *Optimum Seeking Methods*, Prentice-Hall, Inc., 1964.

11. P. Wolfe, "Methods of nonlinear programming," in *Nonlinear Programming*, ed. J. Abadie, John Wiley & Sons, Inc., 1967.

12. G. C. Temes and D. Y. F. Zai, "Least *p*th approximation: theory and application," to be published in the *IEEE Trans. on Circuit Theory*, 1969.

13. E. M. L. Beale, "Numerical methods," in Abadie, *op. cit.*

14. R. Fletcher and M. J. D. Powell, "A rapidly convergent descent method for minimization," *Brit. Computer J.*, vol. 6, pp. 163-168, 1963.

15. C. G. Broyden, "A class of methods for solving nonlinear simultaneous equations," *Math. of Comp.*, vol. 19, pp. 577-593; 1965.

16. M. J. Box, "A comparison of several current optimization methods, and the use of transformations in constrained problems," *Computer J.*, vol. 9, pp. 67-77; May, 1966.

17. A. V. Fiacco and G. P. McCormick, "The sequential unconstrained minimization technique for nonlinear programming, a primal-dual method," *Management Science*, vol. 10, no. 2, pp. 360-366; January, 1964.

18. ——— and ———, "Computational algorithm for the SUMT for nonlinear programming," *ibid*, vol. 10, no. 4, pp. 601, 617; July, 1964.

19 ———— and ———— , "The slacked unconstrained minimization technique for convex programming," *SIAM J. Appl. Math.*, vol. 15, no. 3, pp. 505-515; May, 1967.

20. ———— and ———— , "The sequential unconstrained minimization technique (SUMT) without parameters," *Operations Research*, vol. 15, no. 5, pp. 820-827; 1967.

21. F. A. Lootsma, "Logarithmic programming: a method of solving nonlinear programming problems," *Philips Res. Repts.*, vol. 22, pp. 329-344; 1967.

22. A. D. Waren and L. S. Lasdon, "Practical filter design using mathematical optimization," *Proc. 3rd Allerton Conf. on Circuit and System Theory*, pp. 677-689; October, 1965.

23. T. A. Athanassopoulos, J. D. Schoeffler and A. D. Waren, "Time domain synthesis by nonlinear programming," *Proc. 4th Allerton Conf. on Circuit and System Theory*, 1966.

24. G. C. Temes, J. A. C. Bingham, "Iterative Chebyshev approximation technique for network synthesis," *IEEE Trans. on Circuit Theory*, vol. 14. no. 1; March, 1967.

25. D. A. Calahan, "Computer design of linear frequency-selective networks," *Proc. IEEE*, vol. 53, pp. 1701-1706; November, 1965.

CHAPTER 6

FILTER DESIGN IN TRANSFORMED FREQUENCY VARIABLE

G. C. TEMES, Ampex Corporation

INTRODUCTION

Recent developments in filter theory will be summarized in this chapter. The prime motivation for these new techniques was to make the filter design procedure suitable for digital computer programming. To achieve this, human decisions during the synthesis should be eliminated, the number of unknowns minimized and, last but not least, the *accuracy* problem inherent in the insertion loss synthesis procedure of highly selective filters should be alleviated as much as possible. It turns out that all these can be achieved by various mappings of the complex frequency-variable s. Hence, this chapter will be devoted to the use of these mappings in filter design.

For convenience, the main steps of insertion loss filter design will be recapitulated next. For a double-terminated reactance two-port (Fig.

6.1), the design process usually starts with the solution of the *approximation problem.* This involves finding either the *transducer function*:

$$H(s) \triangleq \frac{1}{2}\sqrt{\frac{R_2}{R_1}}\,\frac{E_g}{E_2} = \sqrt{\frac{P_{\text{max}}}{-E_2 I_2}}; \quad |H(j\omega)|^2 = \frac{P_{\text{max}}}{P_2} \tag{6.1}$$

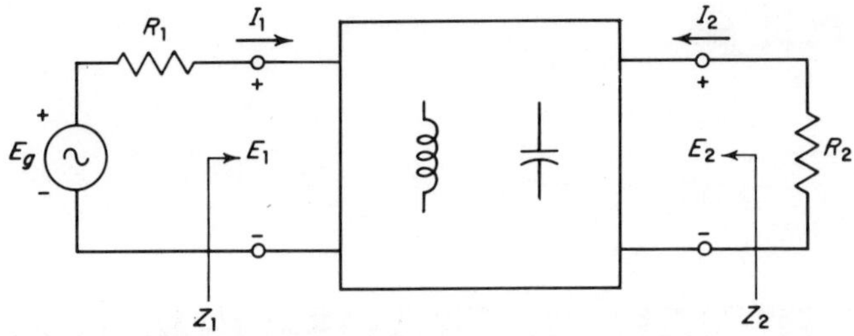

FIG. 6.1. Double-terminated reactance two-port.

where $P_{\text{max}} = E_g^2/4R_1$ is the maximum generator-power and P_2 the power in the load R_2; or the *characteristic function* given by

$$K(s) \triangleq \rho_1 H = \frac{R_1 - Z_1}{R_1 + Z_1}\,H$$
$$|K(j\omega)|^2 = \frac{P_r}{P_2}. \tag{6.2}$$

Here, ρ_1 is the primary reflection factor, and P_r the "reflected power," $P_{\text{max}} - P_2$. Obviously,

$$|H(j\omega)|^2 = |K(j\omega)|^2 + 1 \geqslant 1 \tag{6.3}$$

so the polynomials $E(s)$, $F(s)$, $P(s)$ defined by

$$H(s) = \frac{E(s)}{P(s)}; \qquad K(s) = \frac{F(s)}{P(s)} \tag{6.4}$$

satisfy

$$E(s)E(-s) = F(s)F(-s) + P(s)P(-s). \tag{6.5}$$

The zeros of E, F, and P are the natural modes, reflection zeros and attenuation poles, respectively.

Typically, if only the loss response of the filter is of interest, the approximation problem is solved by finding $K(s)$; if the phase and/or time-response is also prescribed, $H(s)$ is found first. In either case, via (6.5), the missing polynomial $E(s)$ or $F(s)$ may be found from the other two, by factorization.

Having found $H(s)$ and $K(s)$, the driving-point immittance parameters of the two-port may be found from

$$
\begin{aligned}
z_{11} &= R_1 \frac{H_e - K_e}{H_0 + K_0} \\
z_{22} &= R_2 \frac{H_e + K_e}{H_0 + K_0} \\
y_{11} &= \frac{1}{R_1} \frac{H_e + K_e}{H_0 - K_0} \\
y_{22} &= \frac{1}{R_2} \frac{H_e - K_e}{H_0 - K_0}
\end{aligned}
\tag{6.6}
$$

From these, the element values may be obtained by expanding one of the immittances into a ladder network. Often two immittances, one from the primary and one from the secondary side, are expanded, and then the resulting element-values compared to obtain an indication of the accuracy. The details of this process may be found in tutorial papers [1] or books [2], [3] on filter synthesis.

6.1 EQUAL-RIPPLE PASSBAND LOWPASS FILTERS

Frequency Transformation, Design Immittances

In many cases, the maximum amount of passband loss A_p is specified. If only the loss (and not the phase- or time-response) is of interest, the optimum way to satisfy this requirement is to allow the loss to oscillate between 0 and A_p as many times as possible. This leads to the characteristic function indicated in Fig. 6.2. There, the ordinate is the squared modulus of the characteristic function $K(s)$,

$$|K|^2 = |H|^2 - 1 = 10^{A/10} - 1 \tag{6.7}$$

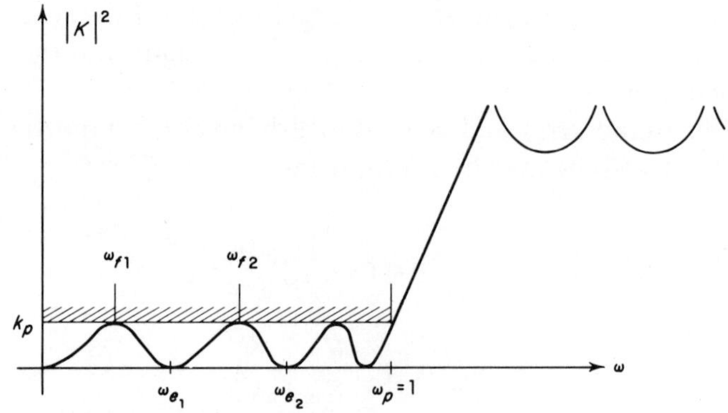

FIG. 6.2. Equal-ripple passband characteristic function for a lowpass filter.

where H is the transmission factor and A the loss.† Note that the vertical scale is distorted; also, that the general shape of the curve remains unchanged if A rather than $|K|^2$ is the ordinate.

It will next be assumed that the passband loss ripple A_p and thus

$$k_p \triangleq 10^{Ap/10} - 1 \tag{6.8}$$

are prescribed, and so is the degree of the network

$$\begin{aligned} n &\triangleq \text{no. of natural modes} \\ &= \text{no. of loss poles,} \end{aligned} \tag{6.9}$$

as well as the number m and location of the finite $j\omega$-axis loss-poles. The problem is to find $K(s)$ and $H(s)$ from these data, such that the equiripple passband response of Fig. 6.2 is realized.

Consider $|K|^2$. It must be of the form

$$|K|^2 = \omega^2 \frac{\prod_{i=1}^{k} \left(\omega^2 - \omega_{e_i}^2\right)^2}{\text{denominator}} \tag{6.10}$$

† A is in decibels.

for the optimum design [4]. Similarly,

$$k_p - |K|^2 = (1-\omega^2)\,\frac{\prod_{j=1}^{k}\left(\omega^2-\omega_{f_j}^2\right)^2}{\text{denominator}} \tag{6.11}$$

where the (unknown) denominators in (6.10) and (6.11) are identical, except perhaps for a constant factor. If there are loss-poles at $\omega \to \infty$, then the factor equals unity, as a consideration of (6.10), (6.11) and (6.12) (below) shows. By (6.10) and (6.11),

$$(1-\omega^2)\,\frac{\prod_{(j)}}{\text{den.}} = k_p - \frac{\omega^2 \prod_{(i)}}{\text{den.}} \tag{6.12}$$

and

$$\text{den.} = \frac{1}{k_p}\left[\omega^2 \prod_{(i)} + (1-\omega^2)\prod_{(j)}\right] \tag{6.13}$$

so that, by (6.8) and (6.10),

$$|K|^2 = \left(10^{A_p/10}-1\right)\frac{\omega^2 \prod_{(i)}\left(\omega^2-\omega_{e_i}^2\right)^2}{\omega^2 \prod_{(i)}\left(\omega^2-\omega_{e_i}^2\right)^2 - (\omega^2-1)\prod_{(j)}\left(\omega^2-\omega_{f_j}^2\right)^2} \tag{6.14}$$

This expression satisfies the equal-ripple passband criterion for any choice of ω_{e_i}, ω_f. It does not, however, exhibit the prescribed loss-poles. The task, then, is to find ω_{e_i}, ω_{f_j}, such that the denominator of $|K|^2$ has double roots at the prescribed poles ω_k and that ω_{e_i}, ω_{f_j} are real and satisfy

$$0 < \omega_{f_1} < \omega_{e_1} < \omega_{f_2} < \ldots . < \omega_{f_l} < \omega_{e_k} < 1 \tag{6.15}$$

This is equivalent to requiring that

$$F(\omega) = \sqrt{\frac{\omega^2-1}{\omega^2}}\;\frac{\prod_{(j)}\left(\omega^2-\omega_{f_j}^2\right)}{\prod_{(i)}\left(\omega^2-\omega_{e_i}^2\right)} \tag{6.16}$$

which has a pole at $\omega = 0$ and zero at $\omega = 1$, behave like a reactance-function for $0 < \omega < 1$, i.e. be monotone increasing with interlaced zeros and poles. Consider now the transformation

$$z = \sqrt{\frac{\omega^2 - 1}{\omega^2}} = \sqrt{1 + s^{-2}} \qquad \text{Re}\, z \leqslant 0 \tag{6.17}$$

which transforms the $0 \leqslant \omega \leqslant 1$ passband into the imaginary z-axis (jy-axis) and the rest of the $j\omega$-axis into the negative real z-axis ($-x$-axis). Eq. (6.17) transforms $F(\omega)$ into

$$\Phi(z) = z \frac{\prod_{j=1}^{k} \dfrac{z^2 - z_{f_j}^2}{1 - z_{f_j}^2}}{\prod_{i=1}^{k} \dfrac{z^2 - z_{e_i}^2}{1 - z_{e_i}^2}} = Cz \frac{\prod_{j=1}^{k} \left(z^2 - z_{f_j}^2\right)}{\prod_{i=1}^{k} \left(z^2 - z_{e_i}^2\right)} \tag{6.18}$$

and thus makes this job much easier. The transformation is illustrated in Fig. 6.3.

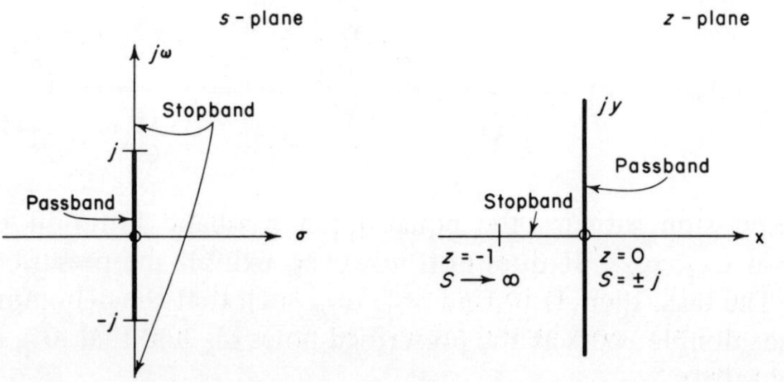

FIG. 6.3. Frequency-transformation for equal-ripple passband lowpass filters.

Defining the even polynomials

$$\begin{aligned} e(z^2) &\triangleq \prod_{i=1}^{k} \frac{z^2 - z_{e_i}^2}{1 - z_{e_i}^2} \\ f(z^2) &\triangleq \prod_{j=1}^{k} \frac{z^2 - z_{f_j}^2}{1 - z_{f_j}^2} \end{aligned} \tag{6.19}$$

the function

$$\Phi(jy) = jy \frac{f(-y^2)}{e(-y^2)} \tag{6.20}$$

needs to be a reactance-function in jy. This will happen if and only if $e + zf$ is a Hurwitz-polynomial or an anti-Hurwitz polynomial. Noting that $e + zf$ is a polynomial containing the (transformed) loss-poles and eq. (6.17), it is evident that this condition is automatically satisfied.

Hence, the following procedure will accomplish the required task:

1) The loss-poles are transformed into the z-plane via (6.17). Note that poles at $\omega \to \infty$ must also be included; they transform into $z = -1$, with original multiplicity. The transformed loss-poles are z_k.
2) From the z_k

$$e + zf \triangleq \prod_{k=1}^{n} (z_k - z) \tag{6.21}$$

is calculated.
3) The required $|K|^2$ is by (6.14) and (6.19),

$$|K|^2 = (10^{A_p/10} - 1) \left. \frac{e^2(z^2)}{e^2(z^2) - z^2 f^2(z^2)} \right|_{z^2 = 1 + s^{-2}} \tag{6.22}$$

4) The natural modes are the zeros of $H(s)$. Hence, by (6.7) and (6.22), they are zeros of [3]

$$10^{A_p/10} e^2 - z^2 f^2 = (10^{A_p/20} e + zf)(10^{A_p/20} e - zf) \tag{6.23}$$

(Here, $e + zf$ is Hurwitz and thus e/zf is a reactance function. Therefore, Ce/zf is also a reactance and so $Ce + zf$ is Hurwitz. (6.17) shows that the zeros of the *first* factor on the RHS of (6.23) are the transformed natural modes.)

The retransformation of the natural modes can be accomplished, factor by factor, via the equations [5]

$$\begin{aligned} z + a &\to s + (a^2 - 1)^{-1/2} \\ z^2 + bz + c &\to s^2 + ds + e \end{aligned} \tag{6.24}$$

where

$$d^2 = \frac{2}{\sqrt{(1+c)^2 - b^2}} - \frac{2(1+c) - b^2}{(1+c)^2 - b^2}$$

$$e = [(1+c)^2 - b^2]^{-1/2} \tag{6.25}$$

Eqs. (6.24) and (6.25) can easily be derived from (6.17).

5) From $|K|$ and

$$|H|^2 = |K|^2 + 1 = \left. \frac{10^{A_p/10} e^2 - z^2 f^2}{e^2 - z^2 f^2} \right|_{z^2 = 1 + s^{-2}} \tag{6.26}$$

as well as the natural modes, the design impedances can be obtained.

It should be noted that generally it is not advantageous to transform $|K|^2$ and $|H|^2$ into the s-plane, as the above steps do. Both the numerical accuracy and speed of computations are enhanced by calculating the design impedances as well as the element values in the z-plane. To accomplish this,

$$10^{A_p/10} e^2 - z^2 f^2$$

is factorized (using z^2 as variable); factors corresponding to real roots are replaced by their (s-plane) Hurwitz-factors according to

$$z^2 - k_0 \to \sqrt{z^2 - 1} + \sqrt{k_0 - 1}$$

and complex root-factors via

$$z^4 + k_i z^2 + l_i \to z^2 + \sqrt{\frac{(4l_i - k_i^2)(z^2 - 1)}{2\sqrt{1 + k_i + l_i} + 2 + k_i}} + \frac{k_i + l_i}{\sqrt{1 + k_i + l_i} + 1}$$

Next, these Hurwitz-factors are multiplied to give $A(z^2) + \sqrt{z^2 - 1} \times B(z^2)$, where $A(z^2)$, $B(z^2)$ are polynomials in z^2. Then, the driving-point immittances may be obtained from

$$\frac{z_{11}}{R_1} = \frac{z_{22}}{R_2} = \frac{\sqrt{z^2-1}\,B(z^2)}{A(z^2)+te(z^2)}$$
$$y_{11}R_1 = y_{22}R_2 = \frac{\sqrt{z^2-1}\,B(z^2)}{A(z^2)-te(z^2)} \tag{6.27}$$

if n is odd (symmetric filters), and from

$$\frac{z_{11}}{R_1} = y_{22}R_2 = \frac{A(z^2)-te(z^2)}{\sqrt{z^2-1}\,B(z^2)}$$
$$\frac{z_{22}}{R_2} = y_{11}R_1 = \frac{A(z^2)+te(z^2)}{\sqrt{z^2-1}\,B(z^2)} \tag{6.28}$$

if n is even (antimetric filters). Here, the abbreviation $t \triangleq \sqrt{10^{A_p/10}-1}$ has been used. Note that $A - te$ contains a factor $z^2 - 1$, which makes it necessary to cancel $\sqrt{z^2-1}$ in some of the expressions before realization.

It is easy to show that the above procedure is the exact z-plane equivalent of finding $E(s)$ from given $F(s)$, $P(s)$ via (6.5), and then subsequently calculating the z_{ii}, y_{ii} from (6.6). The design immittance for a single-terminated network may be found by substituting $e(z^2) \equiv 0$ into (6.27) or (6.28)

Calculation of the loss-poles. In practical situations, the loss-poles are not given. Instead the required minimum stopband loss $A_s(\omega)$ is specified and n as well as the loss-poles ω_k must be obtained from this and also from the given A_p and ω_p. The procedure for finding ω_k from this information is discussed next.

Assuming a realistic $A_s > 20$db, $|K|^2 > 99 \gg 1$, so that in the transformed stop-band $-1 \leqslant z \leqslant 0$,

$$\begin{aligned} A_s &= 10\log_{10}(1+|K|^2) \cong 10\log_{10}|K|^2 = \\ &= 10\log_{10}(10^{A_p/10}-1) + 10\log_{10}\frac{e^2}{e^2-z^2f^2}, \end{aligned} \tag{6.29}$$

where (6.22) was used. But

$$\frac{e^2}{e^2 - z^2 f^2} = \frac{1}{2} + \frac{1}{4}\,\frac{e + zf}{e - zf} + \frac{1}{4}\,\frac{e - zf}{e + zf} \gg 1 \tag{6.30}$$

since $(10^{A_p/10} - 1)$ is of order 1 or less. Of the three terms in (6.30), only

$$\frac{1}{4}\,\frac{e - zf}{e + zf}$$

can be large, since the zeros of $e - zf$ are in the mirror image of the stopband, in $0 \leqslant z \leqslant 1$. Therefore,

$$A_s \cong 10 \log_{10} \frac{10^{A_p/10} - 1}{4} + 10 \log_{10} \frac{e - zf}{e + zf}. \tag{6.31}$$

It should be noted that in all approximations, $|K|^2 + 1$ was decreased by neglecting positive terms. Hence, these approximations actually introduce a safety margin (although extremely small one) into the computation.

Let now A_s be expressed in terms of the transformed loss-poles. With

$$e + zf = \prod_{k=1}^{n} (z_k - z) = \prod_{k=1}^{n} (e^{\gamma_k} - e^{\gamma}) \tag{6.32}$$

we get

$$A_s(\gamma) \cong A_c + \sum_{k=1}^{\frac{n - n_\infty}{2}} 20 \log_{10} \coth \frac{|\gamma - \gamma_k|}{2} + 10\, n_\infty \log_{10} \coth \frac{|\gamma|}{2}. \tag{6.33}$$

Here,

$$A_c \triangleq 10 \log_{10} \frac{10^{A_p/10} - 1}{4} \tag{6.34}$$

and

$$\gamma = -\log_e(-x) = -\frac{1}{2}\log_e(1-\omega^{-2}); \tag{6.35}$$

also n_∞ is the number of loss-poles at $\omega \to \infty$, i.e., at $z_\infty = -1$ and $\gamma_\infty = 0$. It should be noticed that the mapping of (6.17) transforms a pole $j\omega_k$ as well as its conjugate $-j\omega_k$ into a double pole at z_k. Hence the upper limit and constant factor of the Σ in (6.33). A_s needs to be larger than the specified loss A_{spec}.

Now since log coth $|\gamma - \gamma_k|/2$ is an even function in only $\gamma - \gamma_k$, $A_s(\gamma)$ can be constructed as a sum of template-curves, which are identical except for horizontal position. By cutting the 10 $\log_{10}$ coth $\gamma/2$ curve (for $\gamma > 0$) as well as the 20 $\log_{10}$ coth $\gamma/2$ out of clear plastic, a graphical procedure can be used to find the proper location of the γ_k. The procedure starts with the plotting of $A_{spec}(\gamma)$ along a linear γ-scale. Next, $A_c + 10n_\infty \log_{10}$ coth $|\gamma|/2$ is graphically subtracted from $A_{spec}(\gamma)$. (Note that $A_c < 0$ in all practical cases.) Finally, the template curve 20 $\log_{10}$ coth $|\gamma|/2$ is drawn and shifted along the γ-axis until

$$\sum_{k=1}^{\frac{n-n_\infty}{2}} 20\log_{10}\coth\frac{|\gamma-\gamma_k|}{2} > A_{spec} - A_c - 10n_\infty\log_{10}\coth\frac{|\gamma|}{2} \tag{6.36}$$

is achieved, with a more or less constant safety margin. The corresponding pole frequencies are, by (6.35)

$$\omega_k = \pm\,(1 - e^{2\gamma_k})^{-1/2} \tag{6.37}$$

The template-curve is shown in Fig. 6.4.

The procedure becomes quite messy for $n - n_\infty > 6$. Then it is mandatory, and in any case it is preferable, to carry out the location of the loss-poles numerically, using a computer. The iterative algorithm performing this calculation automatically [6] is carried out in the following steps:

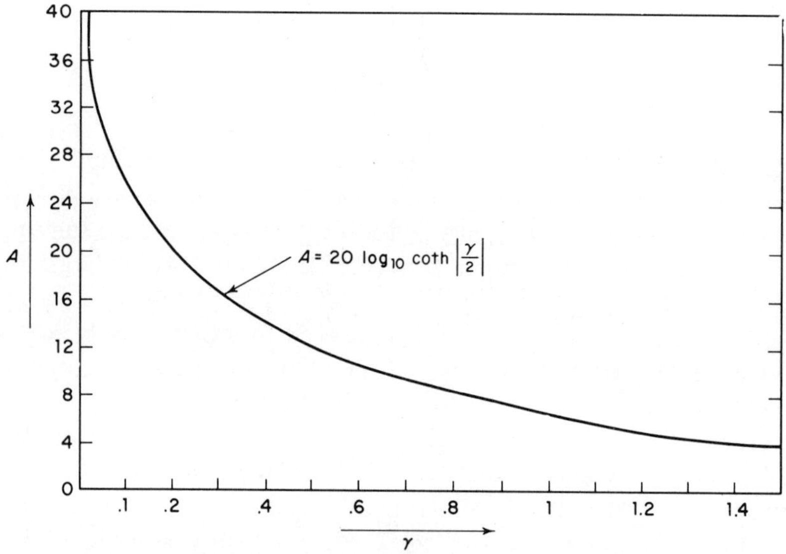

FIG. 6.4. Template-curve for equal-ripple passband filters.

1) Assuming that n_∞ is given, a trial value for n is calculated from

$$n = \frac{0.23\,\overline{A}_s + 2.78 - \log_e(e^{0.23A_p} - 1)}{\log_e \coth \frac{\gamma_s}{4} + 0.693} \tag{6.38}$$

where $\gamma_s = -\log_e\sqrt{1-\omega_s^{-2}}$ is the transformed stop-band limit and

$$\overline{A}_s \triangleq \frac{1}{\gamma_s}\int_0^{\gamma_s} A_s(\gamma)\,d\gamma \tag{6.39}$$

is the mean stopband loss, in decibels.

2) The largest γ_k is initially located at 0.95 γ_s; the others are distributed equidistantly. Let these values be

$$\gamma_k^{(0)} \quad (k = 1, 2, \ldots, \frac{n - n_\infty}{2}).$$

3) Let the specified stopband loss be $A_s(\gamma)$. Then locate the $\frac{n-n_\infty}{2}$ + 1 local minima $\mu_j^{(0)}$ of the loss margin

$$\Delta(\gamma, \gamma_k^{(0)}) \triangleq A_s(\gamma) - A_c - 10n_\infty \log_{10} \coth \frac{|\gamma|}{2} - \sum_{k=1}^{\frac{n-n_\infty}{2}} 20 \log_{10} \coth \frac{|\gamma - \gamma_k^{(0)}|}{2}. \qquad (6.40)$$

4) Change the $\gamma_k^{(0)}$ to new values $\gamma_k^{(1)}$ such that $\Delta(\gamma)$ takes on the same value $\Delta^{(1)}$ at all μ_j; i.e., solve

$$\Delta(\mu_j, \gamma_k^{(1)}) = \Delta^{(1)} \qquad j = 1, 2, \ldots, \frac{n - n_\infty}{2} + 1 \qquad (6.41)$$

for all $\gamma_k^{(1)}$, $k = 1, 2, \ldots, \frac{n - n_\infty}{2}$, and $\Delta^{(1)}$.

5) Locate the new minima $\mu_j^{(1)}$ for the new approximation and repeat step (4). Iterate until the $\gamma_k^{(i)}$ do not change anymore. If the final $\Delta^{(i)}$ is positive, a valid solution is obtained; otherwise, n must be increased and the process repeated.

In practice, an effective way to solve (6.41) was to linearize it, by expanding the last term on the RHS of (6.40) into a first-order Taylor-polynomial in terms of

$$\Delta\gamma_k^{(i+1)} = \gamma_k^{(i+1)} - \gamma_k^{(i)} . \qquad (6.42)$$

Then (6.41) becomes a system of $\frac{n - n_\infty}{2} + 1$ linear equations in the $\frac{n - n_\infty}{2}$ unknown $\Delta\gamma_k^{(i+1)}$ and in $\Delta^{(i+1)}$. The resulting program gave uniformly excellent results over several years of use.

Ladder expansion. The steps involved in carrying out a ladder expansion in the s-plane are concisely summarized in Fig. 3 of Ref. [1]. Identical operations may be carried out in the s-plane, and a similar table can be compiled. Here, however, lowpass and bandpass ladders must be treated separately, because the z-mappings used in the two cases are different. Table 6.1 sums up the necessary steps for lowpass ladder expansion. It should be noted that $N_1(z^2)$, $N_2(z^2)$ and $D_2(z^2)$ are polynomials (not rational functions) in z^2.

The preceding considerations show that the transformation (6.17) of the frequency-variable is a useful tool for simplifying the computations for lowpass filters with equal-ripple passbands. Fortunately, it has other valuable properties as well. For a sharp filter ($\omega_s/\omega_p - 1 \ll 1$), most critical frequencies: natural modes, reflection zeros, loss poles as well as the zeros and poles of the driving-point immittances tend to crowd around the $j\omega_p$ point in the s-plane. This makes the necessary solution of polynomial equations very ill-conditioned; it also makes the coefficients in the numerators and denominators of the z_{ii}, y_{jj} very sensitive to the accuracy of the preceding calculations. Finally, the calculation of element values via ladder-expansion necessitates the subtraction of nearly equal large numbers. The transformation (6.17) maps the $j\omega_p$ point of the origin, thereby increasing the relative distance between neighboring points near $j\omega_p$. As a result, the necessary number of significant figures needed in the calculations is drastically reduced. For s-plane calculations, this number is of the order n; carrying out all calculations in the z-plane, this is reduced to about $n/3$.

6.2 EQUAL-RIPPLE PASSBAND BANDPASS FILTERS

Frequency Transformation, Design Immittances

For frequency-asymmetric bandpass filter with equal-ripple passband, similar considerations hold. The required response is illustrated schematically in Fig. 6.5.

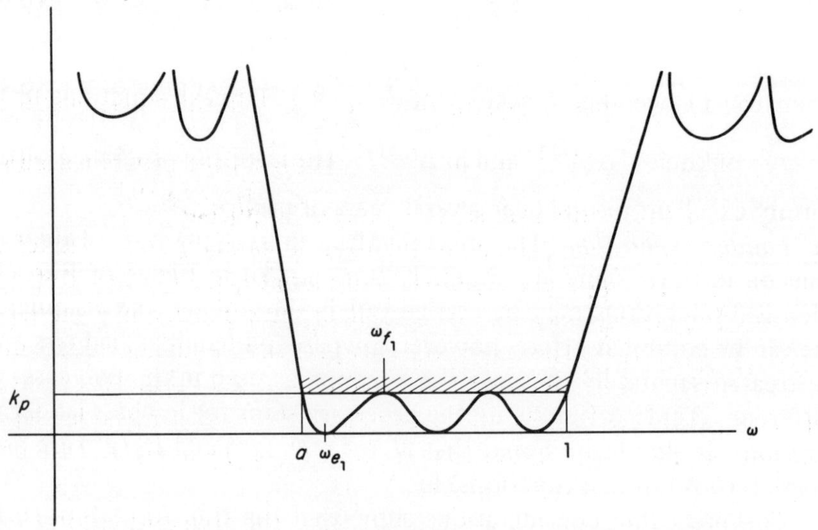

FIG. 6.5. Equal-ripple passband characteristic function for a bandpass filter.

By an argument similar to that presented in the case of lowpass filters, one is led to consider the complex frequency-transformation.

$$z = \sqrt{\frac{\omega^2 - 1}{\omega^2 - a^2}} = \sqrt{\frac{s^2 + 1}{s^2 + a^2}} \qquad \text{Re } z \leqslant 0 \tag{6.43}$$

The properties of the transformation are shown in Fig. 6.6.

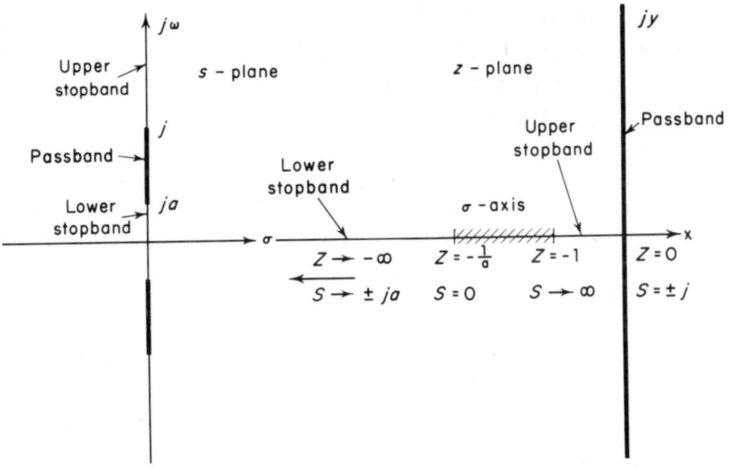

FIG. 6.6. Frequency-transformation for equal-ripple passband bandpass filters.

(It should be noted that for $a \to 0$ the lowpass mapping of eq. (6.17) and Fig. 6.3 is obtained.) For an equal ripple passband, again (6.21) and (6.22) result. However, if the retransformation of $|K|^2$ is carried out into the s-plane via (6.22) and (6.43), the result is, by

$$z_k^2 - z^2 = \frac{1 - a^2}{s_k^2 + a^2} \frac{s^2 - s_k^2}{s^2 + a^2}, \tag{6.44}$$

$$|K|^2 = \text{Const.} \frac{(s^2 + a^2)^n \left[e(z)^2\right]_{z^2 = \frac{s^2+1}{s^2+a^2}}}{\prod_{k=1}^{n-n_\infty} (s^2 - s_k^2)}. \tag{6.45}$$

Evidently the numerator of $K(s)$ will only be rational and hence $K(s)$ realizable, if

$$n = n_0 + n_\infty + 2m \tag{6.46}$$

is *even*. In (6.46), n_0, n_∞ and $2m$ are the numbers of poles at 0, ∞ and finite nonzero frequencies, respectively. Hence, $n_0 + n_\infty$ must be even.

Once again, the advantages of completing the design in the z-plane are great. The gain in numerical accuracy is especially substantial for highly selective narrowband filters [7], [8], [9]. In other cases, if n is not too high, the circuit can be synthesized in the s-plane. The natural frequencies are the inverse transforms of the roots of $10^{A_p/20}e + zf$. A linear factor transforms back as

$$z + \alpha \rightarrow s + \left(\frac{1 - a^2\alpha^2}{\alpha^2 - 1}\right)^{1/2} \tag{6.47}$$

while a quadratic according to [5]

$$z^2 + bz + c \rightarrow s^2 + \frac{b(1 - a^2)\sqrt{4c - b^2}}{\sqrt{R(2\sqrt{RT} + S)}}\, s + \sqrt{\frac{T}{R}} \tag{6.48}$$

where

$$\begin{aligned} R &= (1 + c)^2 - b^2 \\ S &= 2(1 + c)(1 + ca^2) - b^2(1 + a^2) \\ T &= (1 + ca^2)^2 - b^2a^2 \end{aligned} \tag{6.49}$$

A good initial choice can be obtained [9] for the roots of $10^{A_p/20}e + zf$, to be refined by the root-solving program.

The z-plane design may be carried out in the following steps: First, the polynomial

$$10^{A_p/10}e^2 - z^2f^2$$

is factorized in terms of z^2. The degree is even; hence, only factors of the form $z^4 + k_iz^2 + l_i$ need to be considered. Such a factor is then replaced by its s-plane Hurwitz factor $a_i z^2 + b_i[(1 - z^2 a^2)(z^2 - 1)]^{1/2} + c_i$. Here, the constants may be obtained via

$$a_i = \frac{1 + a^2(1 + k_i)}{d_i + a^2 e_i}$$

$$b_i = \sqrt{\frac{4l_i - k_i^2}{2d_i e_i + f_i}} \tag{6.50}$$

$$c_i = \frac{k_i + l_i(1 + a^2)}{d_i + e_i}$$

where

$$d_i = \sqrt{1 + k_i a^2 + l_i a^4}$$

$$e_i = \sqrt{1 + k_i + l_i} \tag{6.51}$$

$$f_i = k_i(1 + a^2) + 2(1 + a^2 l_i).$$

Next, the Hurwitz-factors are multiplied together and also multiplied by $10^{A_p/20}$, to obtain the z-plane equivalent of $E(s)$: $A(z^2) + B(z^2)[(1 - z^2 a^2)(z^2 - 1)]^{1/2}$. The design immittances may now be obtained from

$$\frac{z_{11}}{R_1} = \frac{z_{22}}{R_2} = \frac{B(z^2)\sqrt{(1 - z^2 a^2)(z^2 - 1)}}{A(z^2) + te(z^2)}$$

$$y_{11}R_1 = y_{22}R_2 = \frac{B(z^2)\sqrt{(1 - z^2 a^2)(z^2 - 1)}}{A(z^2) - te(z^2)} \tag{6.52}$$

if n_0 and n_∞ are both odd (symmetric twoport); from

$$\frac{z_{11}}{R_1} = y_{22}R_2 = \frac{A(z^2) - te(z^2)}{B(z^2)\sqrt{(1 - z^2 a^2)(z^2 - 1)}}$$

$$\frac{z_{22}}{R_2} = y_{11}R_1 = \frac{A(z^2) + te(z^2)}{B(z^2)\sqrt{(1 - z^2 a^2)(z^2 - 1)}} \tag{6.53}$$

if n_0, n_∞ are both even (antimetric twoport). Here again the notation,

$$t \triangleq \sqrt{10^{A_p/10} - 1}$$

was used. The polynomial $A - te$ contains the factor $(1 - z^2a^2)(z^2 - 1)$; hence, the square-root of this factor must be cancelled in all immittances containing $(A - te)$. The design-immittance for single-loaded two-ports may again be found by substituting $e(z^2) = 0$ into (6.52) – (6.53).

Ladder Expansion, Bandpass Filters. For the design of bandpass filters, only the first eight configurations of Fig. 3 in Ref. [1] will be considered. The calculations are summarized in Table 6.2. As before, it should be noted that N_1, N_2, N_3, D_2, D_3 are polynomials (not rational functions) in z^2.

Calculation of the Loss-Poles, Bandpass Filters. In the bandpass case a derivation closely analogous to that given for lowpass filters yields the following results. The stopband loss is very well approximated by the expression

$$A_s \cong A_c + 10 \log_{10} \frac{e - zf}{e + zf} = A_c + 10 \sum_{i=1}^{n} \log_{10} \frac{z_i + z}{z_i - z}, \tag{6.54}$$

as in the lowpass case. Defining now,

$$\gamma = \log_e(-\sqrt{a}\, z) = \log_e\left(+\sqrt{a\frac{1 - \omega^2}{a^2 - \omega^2}}\right) \tag{6.55}$$

the relation

$$\begin{aligned} A_s(\gamma) \cong A_c &+ \frac{n_0}{2} 20 \log_{10} \coth \frac{|\gamma - \gamma_0|}{2} + \\ &+ \sum_{k=1}^{\frac{n-n_0-n_\infty}{2}} 20 \log_{10} \coth \frac{|\gamma - \gamma_k|}{2} + \\ &+ \frac{n_\infty}{2} 20 \log_{10} \coth \frac{|\gamma - \gamma_\infty|}{2} \end{aligned} \tag{6.56}$$

No.	Circuit	Immittance before pole-removal	Element values	Immittance after pole-removal	Notation	Next possible removal No.
1	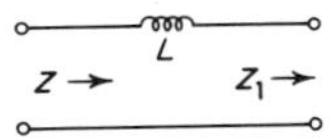	$Z(z)=\dfrac{N(z^2)}{\sqrt{z^2-1}\;D(z^2)}$	$L=\dfrac{N(1)}{D(1)}$	$Z_1(z)=\dfrac{\sqrt{z^2-1}\;N_1(z^2)}{D(z^2)}$	$N_1(z^2)\triangleq\dfrac{N(z^2)-\dfrac{N(1)}{D(1)}D(z^2)}{z^2-1}$	2 or 4
2	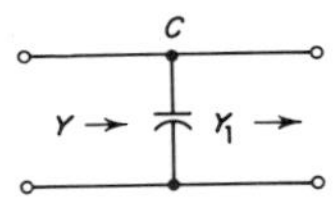	$Y(z)=\dfrac{N(z^2)}{\sqrt{z^2-1}\;D(z^2)}$	$C=\dfrac{N(1)}{D(1)}$	$Y_1(z)=\dfrac{\sqrt{z^2-1}\;N_1(z^2)}{D(z^2)}$	as in 1	1 or 3
3	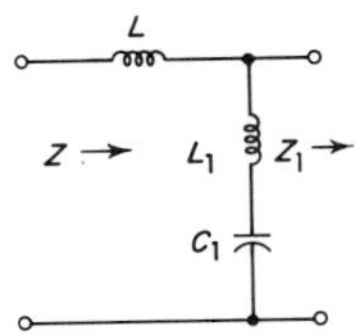 $1<\omega_\infty<\infty$ $0<z_\infty^2<1$	$Z(z)=\dfrac{N(z^2)}{\sqrt{z^2-1}\;D(z^2)}$	$L=\dfrac{N(z_\infty^2)}{D(z_\infty^2)}$ $C_1=\dfrac{D(z_\infty^2)}{N_2(z_\infty^2)}$ $L_1=\dfrac{1-z_\infty^2}{C_1}$	$Z_1(z)=\dfrac{N_2(z^2)}{\sqrt{z^2-1}\;D_2(z^2)}$	$z_\infty^2\triangleq 1-\omega_\infty^{-2}$ $N_2(z^2)=\dfrac{N(z)^2-\dfrac{N(z_\infty^2)}{D(z_\infty^2)}D(z^2)}{z^2-z_\infty^2}$ $D_2(z^2)\triangleq\dfrac{-\dfrac{D(z_\infty^2)}{N_2(z_\infty^2)}N_2(z^2)+D(z^2)}{z^2-z_\infty^2}$	1 or 3
4	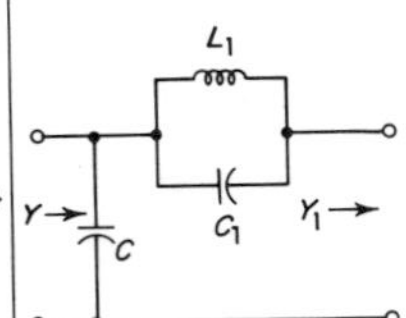 $1<\omega_\infty<\infty$ $0<z_\infty^2<1$	$Y(z)=\dfrac{N(z^2)}{\sqrt{z^2-1}\;D(z^2)}$	$C=\dfrac{N(z_\infty^2)}{D(z_\infty^2)}$ $L_1=\dfrac{D(z_\infty^2)}{N_2(z_\infty^2)}$ $C_1=\dfrac{1-z_\infty^2}{L_1}$	$Y_1(z)=\dfrac{N_2(z^2)}{\sqrt{z^2-1}\;D_2(z^2)}$	as in 3	2 or 4

TABLE 6.1

No.	Circuit	Immittance before pole-removal	Element values	Immittance after pole-removal	Notation	Next possible removal No.
1	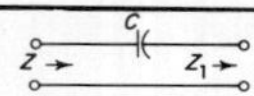	$Z=\sqrt{\frac{z^2-1}{1-a^2z^2}}\,\frac{N(z^2)}{D(z^2)}$	$C=\frac{D(a^{-2})}{N(a^{-2})}$	$Z_1=\sqrt{(z^2-1)(1-a^2z^2)}\,\frac{N_1(z^2)}{D(z^2)}$	$N_1(z^2)\triangleq\frac{N(z^2)-\frac{N(a^{-2})}{D(a^{-2})}}{1-a^2z^2}$	not 1 or 5
2	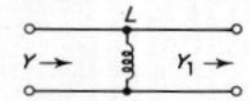	$Y=\sqrt{\frac{z^2-1}{1-a^2z^2}}\,\frac{N(z^2)}{D(z^2)}$	$L=\frac{D(a^{-2})}{N(a^{-2})}$	$Y_1=\sqrt{(z^2-1)(1-a^2z^2)}\,\frac{N_1(z^2)}{D(z^2)}$	as in 1	not 2 or 6
3	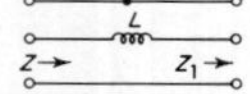	$Z=\sqrt{\frac{1-a^2z^2}{z^2-1}}\,\frac{N(z^2)}{D(z^2)}$	$L=\frac{N(1)}{D(1)}$	$Z_1=\sqrt{(z^2-1)(1-a^2z^2)}\,\frac{N_2(z^2)}{D(z^2)}$	$N_2(z^2)\triangleq\frac{N(z^2)-\frac{N(1)}{D(1)}D(z^2)}{z^2-1}$	not 3 or 7
4	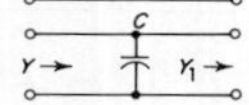	$Y=\sqrt{\frac{1-a^2z^2}{z^2-1}}\,\frac{N(z^2)}{D(z^2)}$	$C=\frac{N(1)}{D(1)}$	$Y_1=\sqrt{(z^2-1)(1-a^2z^2)}\,\frac{N_2(z^2)}{D(z^2)}$	as in 3	not 4 or 8
5	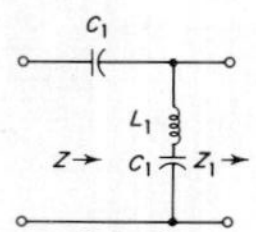 $0<\omega_\infty<a$ $\frac{1}{a^2}<z_\infty^2<\infty$	$Z=\sqrt{\frac{z^2-1}{1-a^2z^2}}\,\frac{N(z^2)}{D(z^2)}$	$C=\frac{D(z_\infty^2)}{N(z_\infty^2)}$ $L_1=\frac{(z_\infty^2-1)^2}{k(a^2-1)}$ $C_1=\frac{k(1-a^2)}{(1-a^2z_\infty^2)(1-z_\infty^2)}$	$Z_1=\sqrt{\frac{z^2-1}{1-a^2z^2}}\,\frac{N_3(z^2)}{D_2(z^2)}$	$z_\infty^2\triangleq\frac{\omega_\infty^2-1}{\omega_\infty^2-a^2}$ $N_3(z^2)\triangleq\frac{N(z^2)-\frac{N(z_\infty^2)}{D(z_\infty^2)}D(z^2)}{z^2-z_\infty^2}$ $k\triangleq\frac{D(z_\infty^2)}{N_3(z_\infty^2)}$ $D_2(z^2)\triangleq\frac{D(z^2)-k\frac{z^2-1}{z_\infty^2-1}N_3(z^2)}{z^2-z_\infty^2}$	not 2 or 6
6	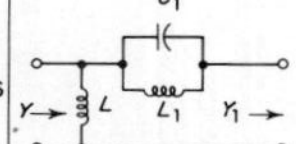 $1<\omega_\infty<\infty$ $0<z_\infty^2<1$	$Y=\sqrt{\frac{z^2-1}{1-a^2z^2}}\,\frac{N(z^2)}{D(z^2)}$	$L=\frac{D(z_\infty^2)}{N(z_\infty^2)}$ $C_1=\frac{(z_\infty^2-1)^2}{k(a^2-1)}$ $L_1=\frac{k(1-a^2)}{(1-a^2z_\infty^2)(1-z_\infty^2)}$	$Y_1=\sqrt{\frac{z^2-1}{1-a^2z^2}}\,\frac{N_3(z^2)}{D_2(z^2)}$	as in 5	not 1 or 5
7	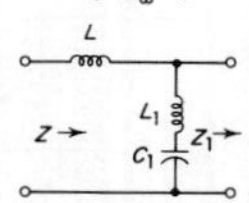 $1<\omega_\infty<\infty$ $0<z_\infty^2<1$	$Z=\sqrt{\frac{1-a^2z^2}{z^2-1}}\,\frac{N(z^2)}{D(z^2)}$	$L=\frac{N(z_\infty^2)}{D(z_\infty^2)}$ $L_1=\frac{(1-z_\infty^2)(a^2z_\infty^2-1)}{k(a^2-1)}$ $C_1=\frac{k(1-a^2)}{(1-a^2z_\infty^2)^2}$	$Z_1=\sqrt{\frac{a^2z^2-1}{1-z^2}}\,\frac{N_3(z^2)}{D_3(z^2)}$	as in 5; also $D_3(z^2)\triangleq\frac{D(z^2)-k\frac{a^2z^2-1}{a^2z_\infty^2-1}N_3(z^2)}{z^2-z_\infty^2}$	not 4 or 8
8	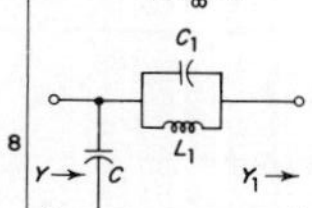	$Y=\sqrt{\frac{1-a^2z^2}{z^2-1}}\,\frac{N(z^2)}{D(z^2)}$	$C=\frac{N(z_\infty^2)}{D(z_\infty^2)}$ $C_1=\frac{(1-z_\infty^2)(a_\infty^2z^2-1)}{k(a^2-1)}$ $L_1=\frac{k(1-a^2)}{(1-a^2z_\infty^2)^2}$	$Y_1=\sqrt{\frac{a^2z^2-1}{1-z^2}}\,\frac{N_3(z^2)}{D_3(z^2)}$	as in 5 and 7	not 3 or 7

TABLE 6.2

results. In (6.56), A_c is given by (6.34), and

$$\gamma_\infty = -\gamma_0 = \log_e(-\sqrt{a}\, z_\infty) = -\log_e(-\sqrt{a}\, z_0) = \frac{1}{2}\log_e a. \qquad (6.57)$$

Evidently, the template curve is the same as for lowpass networks. However, the pole-locating procedure is more involved. The γ-axis has to be drawn in both positive and negative directions and the specified $A_s(\omega)$ transformed into $A_s(\gamma)$. This is illustrated for a simple specification in Fig. 6.7. Notice that the relative location of the passbands has

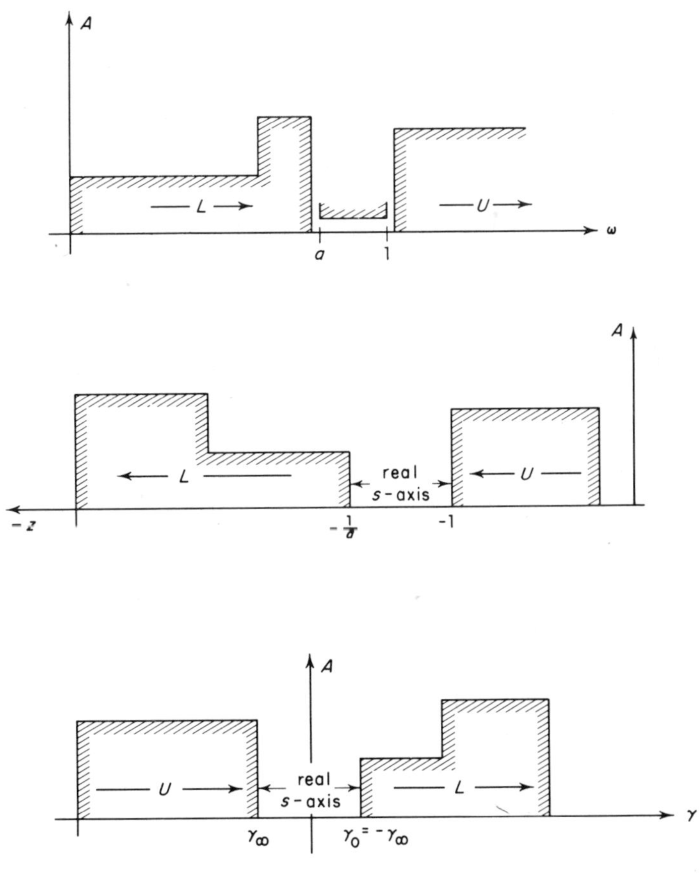

FIG. 6.7. Transformation of the stopbands.

been changed, although within a band increasing ω and increasing γ correspond to each other. Having plotted $A_{spec}(\gamma)$, next

$$\left[A_{spec} - A_c - 10n_0 \log_{10} \coth \frac{|\gamma - \gamma_0|}{2} - 10n_\infty \log_{10} \coth \frac{|\gamma - \gamma_\infty|}{2}\right]$$

is constructed and compared with the sum of template-curves

$$\sum_{k=1}^{\frac{n-n_0-n_\infty}{2}} 20 \log_{10} \coth \frac{|\gamma - \gamma_k|}{2}.$$

The number and location of the γ_k is changed until the requirements are met with a proper safety margin. Note that, for a ladder realization, n_0, $n_\infty > 0$ and also no loss-pole shall be located in the $\gamma_\infty < \gamma < \gamma_0$ region, which corresponds to the real s — (σ) axis. Exceptions to the last rule will be discussed next.

The automatic pole-placing routine described in connection with lowpass filters has been used to carry out these manipulations on a computer [6].

Parametric Filters

In most practical cases, the number of coils in the circuit is to be kept to a minimum, since they are lossy, nonlinear, bulky and expensive. The minimum number of coils is determined by the required degree of the network, i.e., by the necessary number of natural modes given by

$$n = n_L + n_c - (\nu_L + \nu_c + l_L + l_c) \tag{6.58}$$

In (6.58), n_L is the number of coils, ν_L the number of inductive cutsets, l_L the number of inductive tiesets. n_c, ν_c, l_c are the corresponding quantities for the C's. Assume that only capacitors are initially present.† Then each coil connected into the circuit can raise the degree by 2: by increasing n_L and by decreasing l_c or ν_c. Similarly, each resistor can increase n by 1, by decreasing l_c or ν_c. The maximum degree, then, is

$$n_{max} = 2n_L + 2 \tag{6.59}$$

† Then $n = 0$, initially.

for a double-terminated LC twoport. The optimum utilization of the terminations then evidently excludes configurations of the type of Fig. 6.8.a,b; permits those of Fig. 6.8.c,d. Note that the optimum n is necessarily *even*.

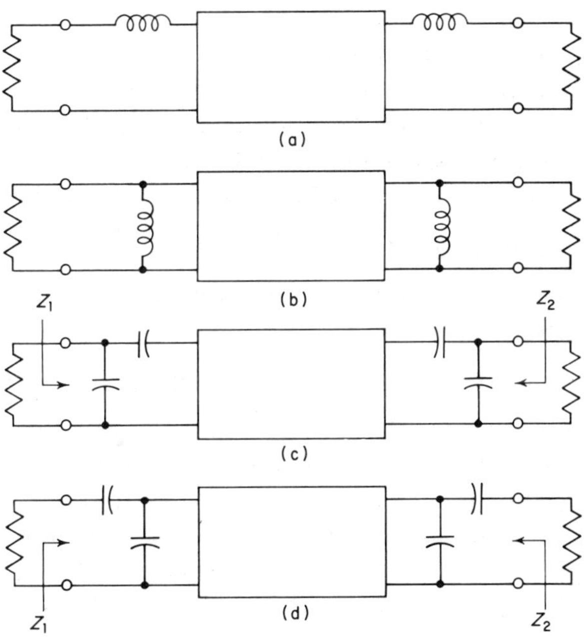

FIG. 6.8. Bandpass filter configurations.

It can be seen that for such networks, the driving-point impedances z_1, z_2 are ∞ at $\omega = 0$, 0 for $\omega \to \infty$. Hence, the reflection coefficients [1]

$$\rho_{1,2} = \frac{R_{1,2} - Z_{1,2}}{R_{1,2} + Z_{1,2}} \to \begin{cases} -1, \text{ if } s \to 0 \\ +1, \text{ if } s \to \infty \end{cases}. \tag{6.60}$$

The zeros of ρ_1, ρ_2 are the zeros of $K(s)$, $K(-s)$. Their poles are the natural modes of the system, so that the denominator polynomial (common to ρ_1, ρ_2) has only positive coefficients. Hence, along the positive real s-axis there must be a reflection zero† to account for the sign-change of the ρ_i along that axis [10].

† Or an *odd* number of reflection zeros.

If the filter is to be symmetrical, $K(s)$ must be odd and hence reflection zeros must be symmetrically located along both half σ-axes; i.e., the numerator $F(s)$ of

$$K(s) = \frac{F(s)}{P(s)} \tag{6.61}$$

must contain a factor of the form $s^2 - b^2$.

For a symmetric minimum inductance bandpass filter, the $K(s)$ is odd. Since there can be no reflection zero at $s = 0$, then it follows that $P(s)$ is odd, $F(s)$ is even, n_0 and n_∞ are both odd while n is even, and $F(s)$ contains a factor $(s^2 - b^2)$. Such a network can be realized using only $\frac{n}{2} - 1$ coils; it will be called a *symmetric parametric filter*, since b can be varied within wide limits without significantly affecting the response and acts thus as a free parameter.

If the required degree is odd, the number of coils is at least $\frac{n-1}{2}$. Then, one termination of the network can be of the type shown in Fig. 6.8.a,b. Only one reflection factor will have a real axis zero along the positive real σ-axis, the other will have a negative real axis zero *and* a negative real axis pole so that its signs at $s = 0$ and ∞ will be the same. Such a filter will be called an *asymmetric parametric filter*.

It should be noted that the real reflection zero (RRZ) does not contribute to the passband flatness of the response and is thus a necessary evil. Hence, although an odd number of RRZ can be used, usually only one exists.

Finally, it should be clear that any circuit built exclusively from capacitors and piezoelectric resonators is, of necessity, a symmetric parametric filter since these elements behave as capacitors at both 0 and ∞ frequencies and hence (6.60) applies to both ports.

Creating Real Reflection Zeros

From Eqs. (6.22) and (6.30) it should be clear that $|K|^2$ has a double zero whenever

$$h(z) \triangleq \frac{e - zf}{e + zf} = -1. \tag{6.62}$$

On this basis, several methods have been developed for creating RRZ's [1], [8], [10]. Only two will be discussed here, since all others have a tendency to create low-attenuation "holes" in the stopband.

A method for creating RRZ's for *symmetric parametric filters* is illustrated in Fig. 6.9.

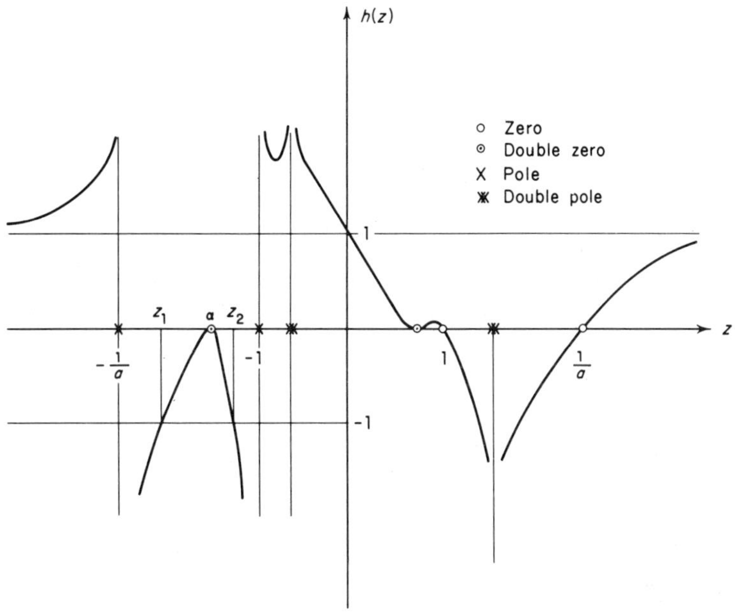

FIG. 6.9. Real reflection zeros for symmetric parametric filters.

It is clear from (6.30) that a double pole of $|K|^2$ has been created at $\pm\alpha$, and also double zeros at $\pm z_1$ and $\pm z_2$. The resulting factors can be exceedingly well approximated by

$$\left[\frac{(z^2 - z_1^2)(z^2 - z_2^2)}{(z^2 - \alpha^2)^2}\right]^2 \cong (z^2 - \beta^2)^2. \tag{6.63}$$

To make the resulting small distortion symmetric in the passband, in addition to matching at $z \to \infty$, the factors should match at $z = 0$. This gives

$$\beta = \frac{z_1 z_2}{\alpha}. \tag{6.64}$$

The choice of α is discussed in Ref. [11]. Generally, choosing

$$\alpha = -a^{-1/2} \tag{6.65}$$

is optimum or nearly so. Then $\gamma_\alpha = 0$. Hence, the template-method can be used by placing a negative template curve $-20 \log_{10} \coth |\gamma|/2$ at the origin of the γ-plane. Having thus found the γ_k, $|K|_0^2$ can be obtained via $e + zf$, which now has a double zero at $z_\alpha = \alpha$. $|K|_0^2$ has a double pole at z_α and double zeros at $\pm z_1$, $\pm z_2$. Now z_1, z_2 can be calculated and the final $|K|^2$ obtained by using (6.63) for replacing

$$\left[\frac{(z^2 - z_1^2)(z^2 - z_2^2)}{z^2 - \alpha^2}\right]^2 \qquad \text{by } (z^2 - \beta^2)^2.$$

A simpler, but somewhat less accurate process is also available [11].

For *asymmetric parametric filters*, a factor $(s + b)$ (or $(s - b)$) is needed. This corresponds to a factor $z^2 - \beta^2$. An approximating method [10] to obtain this is described next. It consists of putting a *single* zero of $e + zf$ at α. Then (Fig. 6.10) a reflection zero is created at z_1. Since $\alpha \sim z_1$, to a good approximation

$$\frac{(z^2 - z_1^2)^2}{z^2 - \alpha^2} \cong z^2 - \beta^2. \tag{6.66}$$

Equating at $z = 0$

$$\beta = \frac{z_1^2}{\alpha}. \tag{6.67}$$

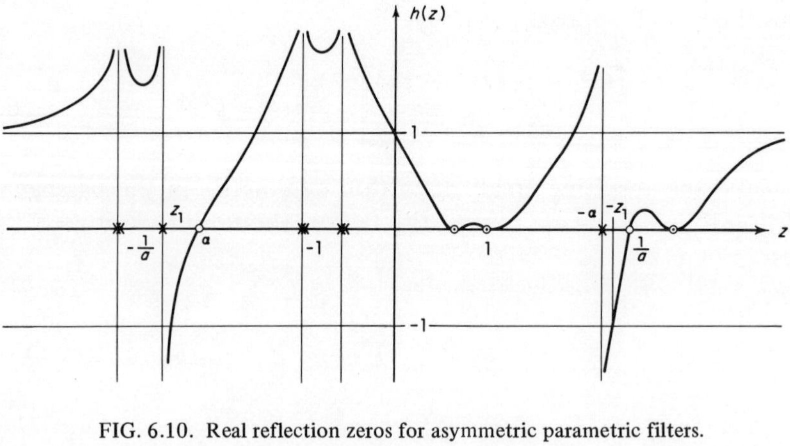

FIG. 6.10. Real reflection zeros for asymmetric parametric filters.

The template method is again applicable, with a negative template curve $-10 \log_{10} \coth |\gamma - \gamma_\alpha|/2$ at γ_α. Again, $\gamma_\alpha = 0$ is a good choice.

The multiplicity of poles at $\omega = 0, \infty (z = -1/a, -1)$ is evidently crucial in both schemes. For asymmetric filters, both $n_0 = 1, n_\infty = 2$, and $n_0 = 2, n_\infty = 1$ lead to realizable $|K|^2$. The simplified method of Ref. [11] is also applicable. Having found $|K|^2$, the design proceeds in the same fashion as for "conventional" bandpass filters.

6.3 FILTERS WITH MAXIMALLY FLAT PASSBANDS

Lowpass Filters

Design Immittances. In some applications, the monotone passband loss response and superior phase and transient characteristic of maximally flat passband filters† outweigh the advantage that equal-ripple passband filters have over them in selectivity. In contrast to Chebyshev passband networks, the characteristic function for Butterworth-filters can be written in the s-plane directly. If the required loss at the passband limit is A_p, the required modulus of the characteristic function is as shown in Fig. 6.11, and its algebraic form is, by inspection,

$$|K|^2 = (10^{A_p/10} - 1)\, \omega^{2n_\infty} \prod_{i=1}^{\frac{n-n_\infty}{2}} \left(\omega^2 \frac{\omega_i^2 - 1}{\omega^2 - \omega_i^2}\right)^2. \tag{6.68}$$

Here too, however, the z-transformation is expedient, to improve numerical accuracy and to simplify the determination of the loss-poles. Accordingly, (6.17) is used to get

$$|K|^2 = \frac{t'^2}{e^2 - z^2 f^2} \tag{6.69}$$

Here, $e(z^2)$ and $f(z^2)$ are defined as before (cf (6.21)) and

$$t'^2 = (10^{A_p/10} - 1) \prod_{k=1}^{n} z_k^2 . \tag{6.70}$$

† Often called Butterworth filters.

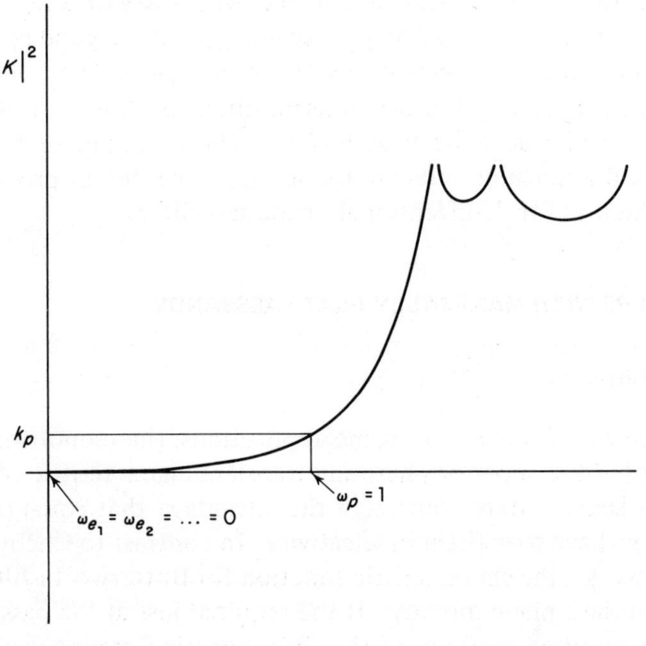

FIG. 6.11. Maximally flat passband characteristic function for a lowpass filter.

Then the transformed natural modes of the network are, by (6.5), roots of the polynomial $e^2 - z^2 f^2 + t'^2$. If this polynomial is factorized, therefore, and each root-factor $z^2 - k_0$ or $z^4 + k_i z^2 + l_i$ is replaced according to the lowpass formulae given for the equal-ripple passband case, then the transformed natural mode polynomial

$$E(z^2) = A(z^2) + \sqrt{z^2 - 1}\, B(z^2) \tag{6.71}$$

may be obtained as the product of the transformed root-factors. Having thus obtained A and B, the design immittances are given for odd n, by,

$$\begin{aligned} \frac{z_{11}}{R_1} = \frac{z_{22}}{R_2} &= \frac{B(z^2)\sqrt{z^2 - 1}}{A(z^2) - t'} \\ y_{11}R_1 = y_{22}R_2 &= \frac{B(z^2)\sqrt{z^2 - 1}}{A(z^2) - t'} \end{aligned} \tag{6.72}$$

For even n,

$$\frac{z_{11}}{R_1} = y_{22}R_2 = \frac{A(z^2) + t'}{B(z^2)\sqrt{z^2 - 1}}$$
$$\frac{z_{22}}{R_2} = y_{11}R_1 = \frac{A(z^2) + t'}{B(z^2)\sqrt{z^2 - 1}} \tag{6.73}$$

From the design immittances, the ladder-expansion method summarized in Table 6.1 can be used to find the element-values.

Locating the Loss-Poles for Maximally Flat Passband [12]: *Lowpass Filters.* Using the transformation (6.35), the $|K|^2$ of (6.69) may be written in the form

$$|K|^2 = \frac{10^{A_p/10} - 1}{(1 - z^2)^{n_\infty} \prod_{i=1}^{\frac{n-n_\infty}{2}} \left(1 - \frac{z^2}{z_i^2}\right)^2} = \frac{10^{A_p/10} - 1}{(1 - e^{2\gamma})^{n_\infty} \prod_{i=1}^{\frac{n-n_\infty}{2}} [1 - e^{2(\gamma-\gamma_i)}]^2} \tag{6.74}$$

in the stopband. Hence, using (6.34),

$$A_s(\gamma) = 10 \log_{10} (1 + |K|^2) \cong 10 \log_{10} |K|^2 \cong$$
$$\cong A_c + 6 - \sum_{i=1}^{\frac{n-n_\infty}{2}} 20 \log_{10} |1 - e^{2(\gamma-\gamma_i)}| \tag{6.75}$$
$$- 10n_\infty \log_{10} |1 - e^{2\gamma}|$$

is obtained. Equation (6.75) indicates that the template-curves to be used here are given by $-10 \log_{10} |1 - e^{2\gamma}|$ and $-20 \log_{10} |1 - e^{2\gamma}|$. These curves are shown in Fig. 6.12. It should be noted that the curves are no longer symmetric; also, that the loss goes negative for $\gamma - \gamma_i > 0.5 \log_e 2$.

The automatic pole-placing program, described for equal-ripple passband lowpass filters, is applicable with minor modifications to these networks as well.

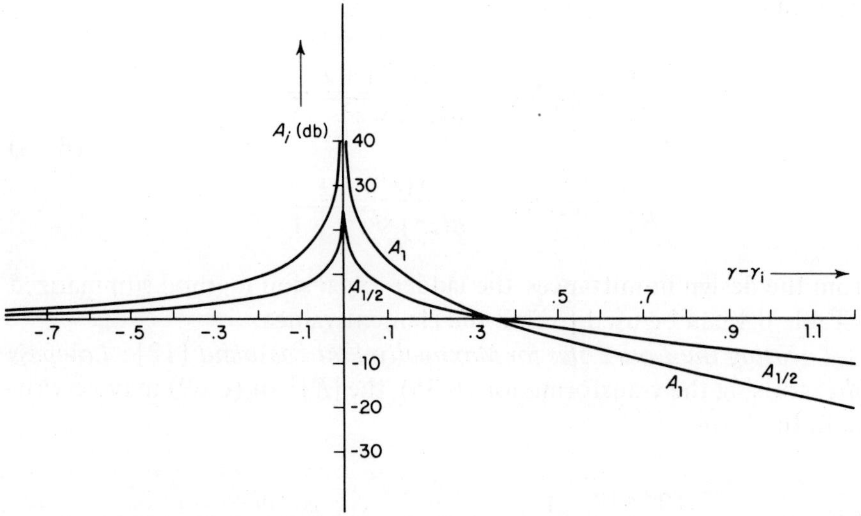

FIG. 6.12. Template curves for maximally flat passband filters.

Bandpass Filters

Design Immittances. The squared modulus of the characteristic function of a conventional maximally flat bandpass filter is evidently of the form

$$|K|^2 = (10^{A_p/10} - 1)\left(\frac{\omega_p}{\omega}\right)^{2n_0}\left(\frac{\omega^2 - \omega_c^2}{\omega_p^2 - \omega_c^2}\right)^{n_0+n_\infty} \times \prod_{i=1}^{\frac{n-n_0-n_\infty}{2}}\left(\frac{\omega^2 - \omega_c^2}{\omega_p^2 - \omega_c^2}\,\frac{\omega_p^2 - \omega_i^2}{\omega^2 - \omega_i^2}\right)^2. \tag{6.76}$$

Here, ω_c is the center-band zero-loss frequency, also ω_p may be either one of the two passband limits, both of which satisfy $A(\omega_p) = A_p$. Again, n as well as $n_0 + n_\infty$ must be even. Normalizing to the upper passband limit $\omega_p = 1$ and introducing z^2 via (6.43), eq. (6.76) becomes simply

$$|K|^2 = (10^{A_p/10} - 1)\frac{\prod_{i=1}^{n} z_i^2}{z_c^{2n}}\,\frac{(z^2 - z_c^2)^n}{\prod_{i=1}^{n}(z^2 - z_i^2)}. \tag{6.77}$$

Here, the loss poles at zero ($z_0^2 = a^{-2}$) and at infinite frequency ($z_\infty^2 = 1$) must be included in $\prod_{(i)}$ with their proper multiplicity. Since at both passband limits ($z = 0, \infty$), $|K|^2 = (10^{A_p/10} - 1)$ must hold,

$$\prod_{i=1}^{n} z_i = (-1)^{n/2} z_c^n \tag{6.78}$$

is necessary. If the passband limits are given but ω_c is not, the proper z_c satisfying (6.78) may be obtained using a simple iterative technique [12]. Note that z_c is in the transformed passband and is therefore imaginary. From $|K|^2$, the transformed natural modes can be obtained by factorizing

$$t^2(z^2 - z_c^2)^n + \prod_{i=1}^{n} (z^2 - z_i^2). \tag{6.79}$$

All factors are of the form $z^4 + k_i z^2 + l_i$, since n must be even. Replacing each factor by its s-plane Hurwitz factor (see eqs. (6.50) – (6.51)) and multiplying the resulting expression by $10^{A_p/10}$

$$A(z^2) + B(z^2)\sqrt{(1 - z^2a^2)(z^2 - 1)}$$

is obtained. The rest of the design proceeds as for equal-ripple passband filters; here, however, $t(z^2 - z_c^2)^{n/2}$ replaces $te(z^2)$ in the design immittances of (6.52) – (6.53).

Parametric maximally flat passband filters may be designed, e.g., by a simple approximating procedure [11]. This consists of including a factor

$$\frac{(z^2 - \alpha^2)^2}{(z^2 - 1)(z^2 - a^{-2})}$$

in the $|K|^2$ of (6.77), if a symmetric parametric filter is required; and a factor $(z^2 - \alpha^2)/(z^2 - a^{-2})$ or $(z^2 - \alpha^2)/(z^2 - 1)$ for an asymmetric parametric filter. Choosing α properly, the loss remains A_p at both passband limits. The multiple reflection zero at ω_0, of course, also remains unaffected. Having found $|K|^2$, the design immittances may be found in a process analogous to the ones previously described.

Table 6.2 can be used to obtain the element-values from the design immittances.

Finding the Loss-Poles for Maximally Flat Passband Bandpass Filters [12]. For these filters, the loss-poles may be more expediently found using the transformation

$$\gamma \triangleq \frac{1}{2} \log_e \frac{\omega^2 - \omega_p^2}{\omega^2 - \omega_c^2} \tag{6.81}$$

rather than via (6.55). Introducing (6.81) into (6.76), the simplified formula

$$|K|^2 = \frac{10^{A_p/10} - 1}{[1 - e^{2(\gamma-\gamma_0)}]^{n_0} \, [1 - e^{2\gamma}]^{n_\infty} \prod_{i=1}^{\frac{n-n_0-n_\infty}{2}} [1 - e^{2(\gamma-\gamma_i)}]^2} \tag{6.82}$$

is obtained. Therefore, in the stopband,

$$\begin{aligned} A_s(\gamma) &\cong 10 \log_{10} |K|^2 \cong A_c + 6 - 10 n_0 \log_{10} |1 - e^{2(\gamma-\gamma_0)}| \\ &- 10 n_\infty \log_{10} |1 - e^{2\gamma}| - 20 \sum_{i=1}^{\frac{n-n_0-n_\infty}{2}} \log_{10} |1 - e^{2(\gamma-\gamma_i)}| \end{aligned} \tag{6.83}$$

is obtained. Here,

$$\gamma_0 = \gamma(\omega = 0) = \log_e \frac{\omega_p}{\omega_c}. \tag{6.84}$$

Once again, the loss may be built up from the template curves of Fig. 6.12. Hence, either graphical procedures or, preferably, the automatic pole-placing program described for Chebyshev-passband filters, may be used to find the loss-poles for prescribed minimum stopband attenuation. Care must be taken, however, to satisfy (6.78) along with the stopband requirements. This may be accomplished, if necessary, using iterative techniques [12].

6.4 FILTERS WITH EQUAL-RIPPLE STOPBANDS [13]

Lowpass Filters

Frequency Transformation. For many applications, noise or unwanted signal with uniform spectrum must be suppressed. If the minimum stopband attenuation is thus a constant, the optimum transfer function is one which has an equal-ripple stopband response [4]. In addition to the stopband specifications, there are always restrictions on the passband loss or phase characteristics, on the transient response, or on several of these simultaneously.

The techniques described in this chapter may be used to such filters, with a minimum amount of computation. This minimum effort is assured by manipulating the smallest possible number of parameters in the course of the iterative portion of the calculations.

The required response is schematically shown in Fig. 6.13. Comparison with Fig. 6.2 shows that in both figures the ω_{fi} (ω_{ej}) are the frequencies where specifications are met with zero (maximum) safety margin. As in the case of Fig. 6.2, the squared modulus of the characteristic function shown in Fig. 6.13 may be written directly from the diagram:

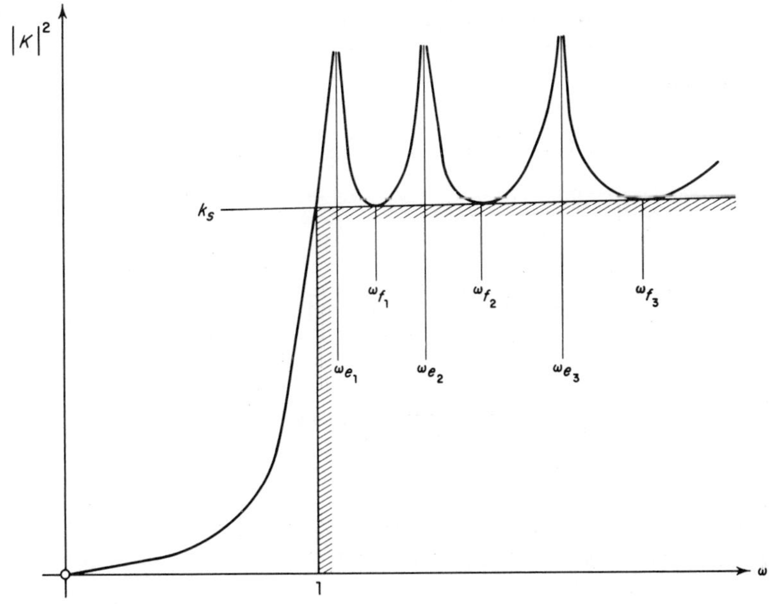

FIG. 6.13. Lowpass filter characteristic with equal-ripple stopband.

$$|K|^2 = k_s \left\{ 1 + (\omega^2 - 1) \left[\frac{\prod_{i=1}^{\frac{n-n_\infty}{2}} \left(1 - \frac{\omega^2}{\omega_{fi}^2}\right)}{\prod_{j=1}^{\frac{n-n_\infty}{2}} \left(1 - \frac{\omega^2}{\omega_{ej}^2}\right)} \right]^2 \right\} \tag{6.85}$$

where

$$k_s = 10^{A_s/10} - 1 \tag{6.86}$$

and A_s is the guaranteed minimum stopband loss. By (6.3),

$$|H|^2 = |K|^2 + 1 = 10^{A_s/10} \left\{ 1 + (1 - 10^{-A_s/10})(\omega^2 - 1) \left[\frac{\prod_{i=1}^{\frac{n-n_\infty}{2}} \left(1 - \frac{\omega^2}{\omega_{fi}^2}\right)}{\prod_{j=1}^{\frac{n-n_\infty}{2}} \left(1 - \frac{\omega^2}{\omega_{ej}^2}\right)} \right] \right\} \tag{6.87}$$

Again, it is desired that all ω_{fi}, ω_{ej} be real and be located interlaced in the $1 < \omega < \infty$ range. This may easily be achieved by defining the transformed variable

$$z \triangleq \sqrt{1 - \omega^2} = \sqrt{1 + s^2}; \qquad \operatorname{Re} z \geqslant 0 \tag{6.88}$$

and even polynomials

$$e(z^2) \triangleq \prod_{(j)} \left(1 - \frac{\omega^2}{\omega_{ej}^2}\right) = \prod_{(j)} \frac{z^2 - z_{ej}^2}{1 - z_{ej}^2} \tag{6.89}$$

$$f(z^2) \triangleq \sqrt{1 - 10^{-A_s/10}} \prod_{(i)} \frac{z^2 - z_{fi}^2}{1 - z_{fi}^2} \tag{6.90}$$

The mapping (6.88) transforms the passband to the $[\sqrt{1 - \omega_p^2}, 1]$ range of the real z-axis; it transforms the stopband to the whole imaginary z-axis (Fig. 6.14).

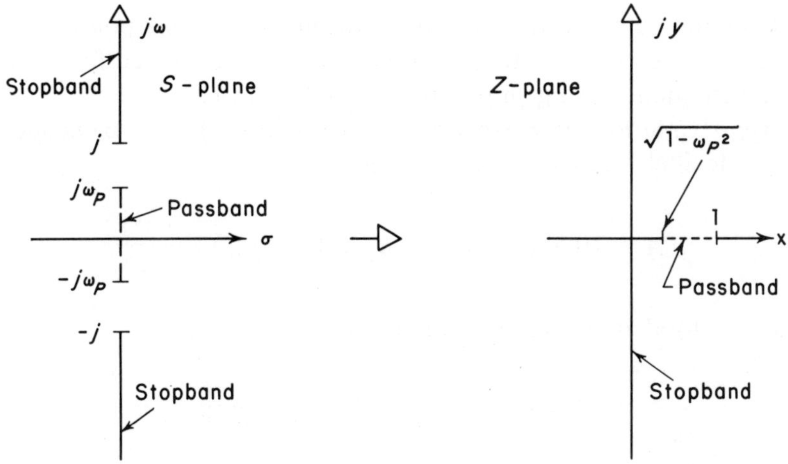

FIG. 6.14. Frequency-transformation for equal-ripple stopband lowpass filters.

Using (6.89) and (6.90), eq. (6.87) may be written in the form

$$|H|^2 = 10^{A_s/10}\left[1-\left(\frac{zf}{e}\right)^2\right] = 10^{A_s/10}\,\frac{e^2 - z^2 f^2}{e^2}, \qquad (6.91)$$

an equation which bears close resemblance to (6.22). Obviously, the right half-plane roots of $e^2 - z^2 f^2$ are the transformed natural modes (cf. (6.88)). If these are chosen as the roots of $e + zf$, then $-z[f(z^2)]/[e(z^2)]$ is a reactance function and hence the ω_{fi}, ω_{ej} have the required locations. Hence, for any desired natural modes s_i, finding their transforms z_i via (6.88), defining

$$e + zf \triangleq \prod_{i=1}^{n} (z_i - z) \qquad (6.92)$$

and then choosing $|H|^2$ in the form (6.91) assures simultaneously the required natural modes as well as equal-ripple stopband with prescribed A_s and real ω_{ei}.

Approximation of a Desired Phase-Response. Since all loss-poles are on the $j\omega$-axis, the phase is uniquely defined by the natural modes. Hence, the roots of existing polynomial approximations for maximally flat or equal-ripple delay may be used to find, via (6.88) and (6.92), the $e + zf$ polynomial. Such lowpass filters have been tabulated for maximally flat delay [14].

It should be noted that for odd n, the number of loss-poles at $\omega \to \infty$ is one, for even n, $n_\infty = 0$. A double pole may be created for even n at infinite frequency using an iterative procedure [14].

Approximation of Passband Loss-Response. In the passband, the loss (in decibels) may be written in the form

$$A = 10 \log_{10} |H|^2 = A_s + 10 \log_{10} \left(1 - \frac{z^2 f^2}{e^2}\right). \tag{6.93}$$

To a very good approximation [13],

$$A \cong A_s + 6.02 - 10 \log_{10} \frac{e - zf}{e + zf}. \tag{6.94}$$

Either (6.93) or (6.94) may be used to find $e + zf$ for a desired passband loss. In particular, if a maximally flat approximation is needed to a desired response given by the expansion

$$A(\omega^2) = a_0 + a_1 \omega^2 + a_2 \omega^4 + \ldots \tag{6.95}$$

around $\omega = 0$, then the design proceeds in the following steps [13]:

(1) Let $b_0 \triangleq 10^{a_0/10}$. Find $b_1, b_2, \ldots, b_{n-1}$ from

$$b_k = \frac{\log_e 10}{10k} \sum_{j=1}^{k} j a_j b_{k-j}. \tag{6.96}$$

(2) Let $c_0 \triangleq 1 - 10^{(a_0 - A_s)/10}$. Find

$$c_k = c_{k-1} - 10^{-A_s/10} b_k; \qquad k = 1, 2, \ldots, n-1. \tag{6.97}$$

(3) Let $d_0 = \sqrt{c_0}$. Find

$$d_j = \frac{1}{d_0}\left[\frac{c_j}{2} - \frac{1}{j} \sum_{k=1}^{j-1} k d_k d_{j-k}\right]; \qquad j = 1, 2, \ldots, n-1 \tag{6.98}$$

(4) Let, for n odd

$$e + zf \triangleq \sum_{k=0}^{\frac{n-1}{2}} e_k z^{2k} + z \sum_{j=0}^{\frac{n-1}{2}} f_j z^{2j} \tag{6.99}$$

and choose $f_{(n-1)/2} = 1$. Then all e_k and f_j may be obtained from the n linear equations

$$(-1)^r \sum_{j=r}^{\frac{n-1}{2}} \binom{j}{r} f_j = \sum_{k=0}^{r} \left[(-1)^k \sum_{l=k}^{\frac{n-1}{2}} \binom{l}{k} e_l \right] d_{r-k}\,; \qquad r = 0, 1, \ldots, n-1. \tag{6.100}$$

This process is fast, accurate, and avoids the solution of any nonlinear equations.

As an application of the process, Figs. 6.15 and 6.16 show the loss- and time-response of a filter with Gaussian passband and equal-ripple stopband, obtained by the described procedure. The resulting network has evidently attractive selectivity and transient characteristics.

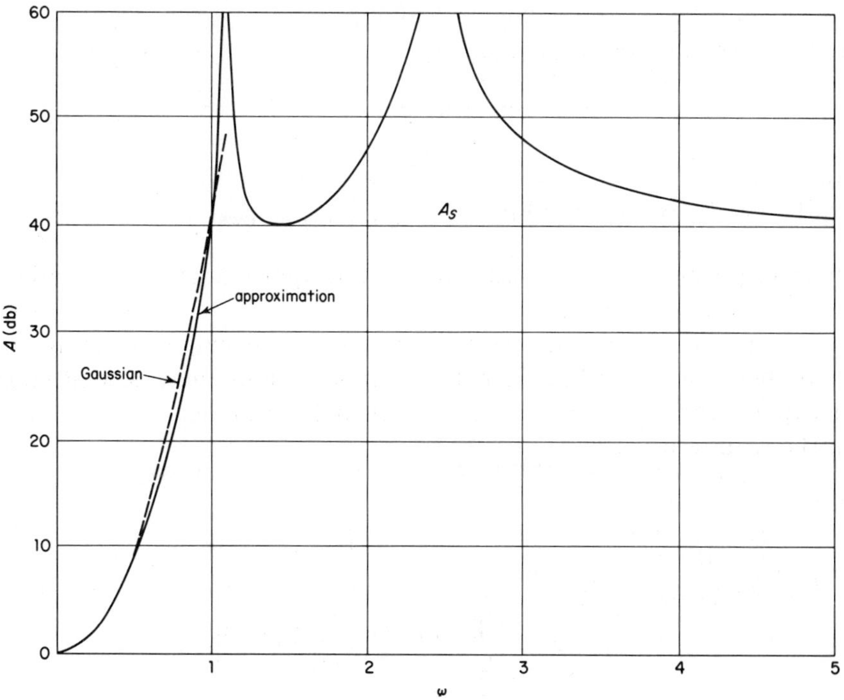

FIG. 6.15. Loss-response of quasi-Gaussian filter.

Chebyshev passband loss approximation may also be obtained to a prescribed smooth response. In particular, the linearized Remez method, discussed in Chapter 5, has been used successfully to obtain an equal-

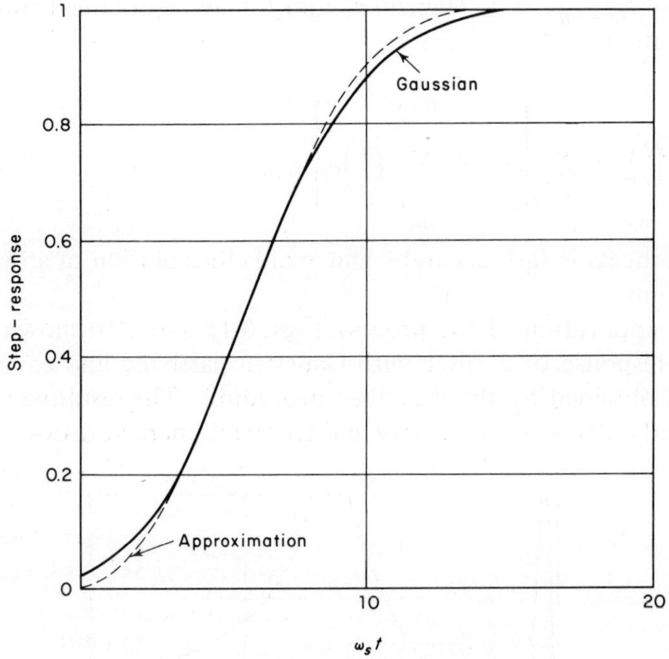

FIG. 6.16. Time-response of quasi-Gaussian filter.

ripple stopband loss and an also equal-ripple approximation to a variety of passband loss-responses.

Approximation of the Time-Response. One major advantage of Chebyshev stopband filters is that their time-response may be written in an explicit form, in terms of only their natural modes. By (6.1) and (6.91), with reference to Fig. 6.1,

$$H \triangleq \sqrt{\frac{R_2}{R_1}}\,\frac{E_g(s)}{2E_2(s)} = 10^{A_s/20}\,\frac{\prod_{i=1}^{n}(s-s_i)}{\left[\text{Even part of } \prod_{i=1}^{n}(z_i-z)\right]_{z^2=1+s^2}}\,. \tag{6.101}$$

Hence, the response to a DC generator voltage of $2\sqrt{R_1/R_2}$ volts ("step-response") is

$$e_2(t) = \mathcal{L}^{-1}\left[\frac{1}{sH(s)}\right] = 1 + 10^{-A_s/20}\sum_{i=1}^{n}\left(\frac{z_i}{s_i}\prod_{\substack{k=1\\k\neq i}}^{n}\frac{z_i+z_k}{s_i-s_k}\right)e^{s_i t}\,, \tag{6.102}$$

and the corresponding impulse-response is

$$h(t) \triangleq \frac{de_2(t)}{dt} = 10^{-A_s/20} \sum_{i=1}^{n} \left(z_i \prod_{\substack{k=1 \\ k \neq i}}^{n} \frac{z_i + z_k}{s_i - s_k} \right) e^{s_i t} . \quad (6.103)$$

For a loss-pole at $\omega \to \infty$, $e_2(0) = 0$ and hence

$$\sum_{i=1}^{n} \left(\frac{z_i}{s_i} \prod_{\substack{k=1 \\ k \neq 1}}^{n} \frac{z_i + z_k}{s_i - s_k} \right) = -10^{-A_s/20} \quad (6.104)$$

From (6.102) – (6.104), the natural modes may be obtained using the maximally flat, equal-ripple or least-pth approximation techniques discussed in Chapter 5. For example, linearized Remez-techniques have been used successfully [13] to obtain equal-ripple approximation to the ideal step-response. Fig. 6.17 shows the initial approximation (an elliptic filter) and the final results after seven Remez-iterations. Families of such filters have been tabulated [15] using Newton-techniques to find all n natural modes and $(n - 1)/2$ loss poles from the simultaneous time- and frequency-domain specifications.

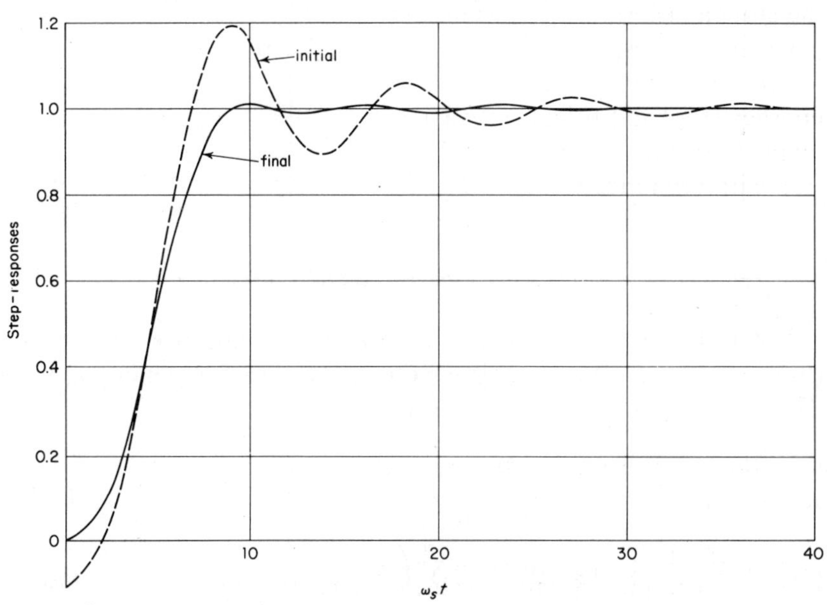

FIG. 6.17. Equal-ripple step-response.

Bandpass Filters

The required loss-response of an equal-ripple stopband bandpass filter is shown in Fig. 6.18. By an argument similar to that used for low-pass filters, the transducer function now is given by [13].

$$|H|^2 = 10^{A_s/10}\left\{1 + C(\omega^2 - \omega_s^2)(\omega^2 - \omega_{-s}^2)\left[\frac{\prod_{i=1}^{(n/2)-1}(\omega^2 - \omega_{fi}^2)}{\omega^{n_0}\prod_{j=1}^{m}(\omega^2 - \omega_{ej}^2)}\right]^2\right\} \tag{6.105}$$

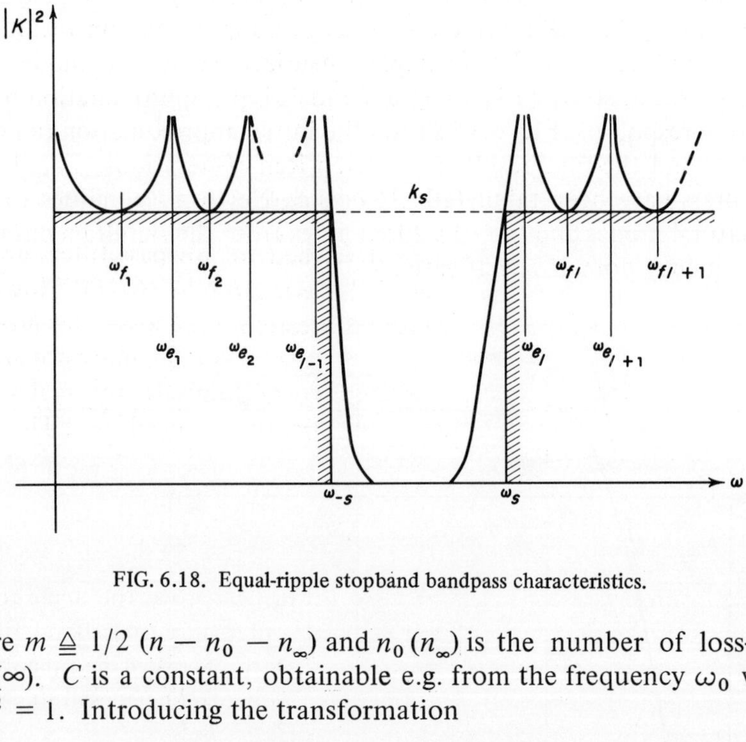

FIG. 6.18. Equal-ripple stopband bandpass characteristics.

where $m \triangleq 1/2\,(n - n_0 - n_\infty)$ and $n_0\,(n_\infty)$ is the number of loss-poles at $0(\infty)$. C is a constant, obtainable e.g. from the frequency ω_0 where $|H|^2 = 1$. Introducing the transformation

$$z^2 = \frac{\omega_s^2 - \omega^2}{\omega^2 - \omega_{-s}^2}\,; \qquad \operatorname{Re} z \geqslant 0, \tag{6.106}$$

the $|H|^2$ of eq. (6.105) may again be brought into the form shown in (6.91), if n is even. The transformation (6.106) maps both stopbands as well as the real s-axis onto the imaginary z-axis; it maps the passband on a portion of the positive real z-axis. The properties of the mapping are illustrated in Fig. 6.19.

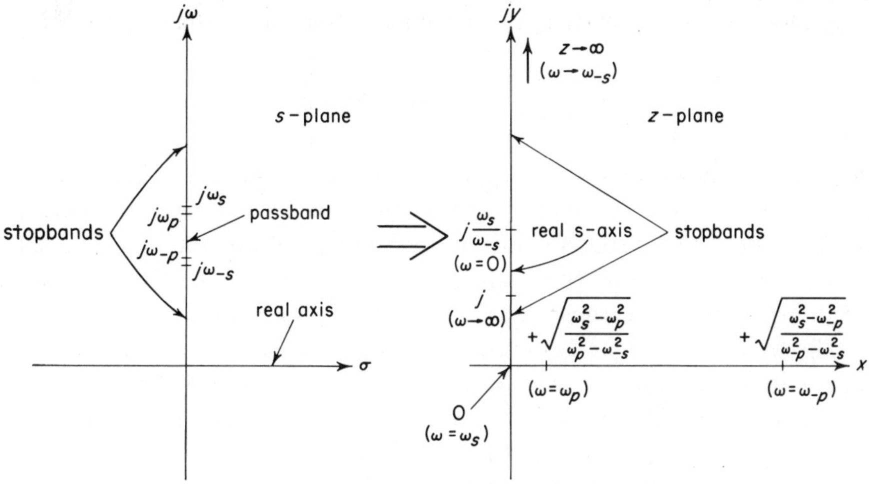

FIG. 6.19. Frequency transformation for equal-ripple stopband bandpass filter.

The approximation techniques described for lowpass filters are applicable here too, since $|H|^2$ is in both cases given by (6.91). However, if a passive ladder network realization is desired, then loss-poles must be obtained at $z_0 = \pm j\ \omega_s/\omega_{-s}$ and $z_\infty = \pm j$; also, loss-poles must not appear between these values, since this range corresponds to the real s-axis. These conditions constrain the approximation procedure. The constraints are absent for active realization.

Accuracy

The mappings (6.88) and (6.106) do not improve the accuracy of computations as do their counterparts of equal-ripple passband filters. In fact, the natural modes tend to crowd near the transformed passband; for lowpass filters especially near $z = 1$. This effect necessitates using double precision arithmetic for about $n \geqslant 10$.

It seems likely that using the transformation $z^2 = (1 + s^2)^{-1}$ leads to better numerical conditioning for equal-ripple passband lowpass filters, while retaining all other advantages of the transformed variable design procedure. In any case, for high degree, it may be worth while to revert to the accuracy-enhancing z-transformation of (6.17) or (6.43) after $|H|^2$ has been obtained, and complete the realization in terms of that variable.

6.5 MÖBIUS-TRANSFORMATIONS

Definitions

A Möbius-transformation is defined as a bilinear change of variable:

$$y = \frac{ax + b}{cx + d} \qquad (ad - bc \neq 0) \tag{6.107}$$

where x, y may be real or complex variables. In circuit applications, x and y are usually the frequency variables, original and transformed, raised to the first or second power. Evidently, all previously discussed transformations such as those given in (6.17), (6.43), (6.106), etc. are within this class, but these represent very special values of a, b, c, d.

The mapping of (6.107) contains three independent constants, e.g. a/c, b/c and d/c. Hence, the desired correspondence of three arbitrary frequencies x_i to their arbitrary images y_i may be enforced by calculating these constants from the three correspondences.

In the following, some special forms of (6.107) will be discussed, along with their applications in circuit design.

Lowpass to Lowpass Transformations

Transformation without Increase of Degree [1], [16]. In some cases, a transformation of the form

$$\Omega^2 = k \frac{\omega^2 - \omega_0^2}{\omega^2 - \omega_\infty^2} . \tag{6.108}$$

is required to obtain a network function with more favorable characteristics. (Here, and throughout this section, Ω represents the physical frequency, whereas ω is the frequency-variable of the "model" filter function from which the transfer function of the physical circuit is obtained via transformation.)

Consider, for example, an antimetric filter (Fig. 6.20 (a)) with equal-ripple passband and stopband. Using classical design techniques, the loss-response shown in Fig. 6.20 (b) results. The loss does not have a pole for $\omega \to \infty$, which precludes ladder-realization. Furthermore, the ratio of terminations, $r \triangleq R_2/R_1$, is related to the zero-frequency loss $A(0)$ by

$$A(0) = 20 \log_{10} \left(\frac{r + 1}{2\sqrt{r}}\right) \tag{6.109}$$

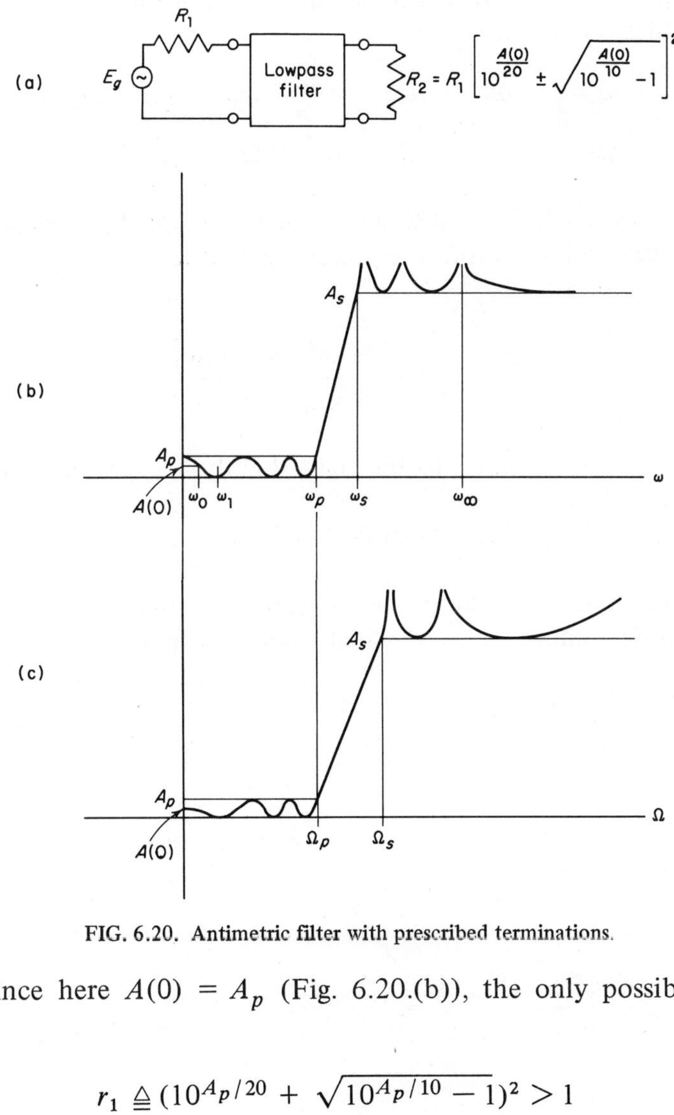

FIG. 6.20. Antimetric filter with prescribed terminations.

Now since here $A(0) = A_p$ (Fig. 6.20.(b)), the only possible values of r are

$$r_1 \triangleq (10^{A_p/20} + \sqrt{10^{A_p/10} - 1})^2 > 1$$
$$r_2 \triangleq (10^{A_p/20} - \sqrt{10^{A_p/10} - 1})^2 = \frac{1}{r_1} < 1. \tag{6.110}$$

If any other termination ratio is desired, the characteristics must be modified. For any such r, a DC loss $A(0)$ can be obtained from (6.109). For a prescribed value r, and a corresponding $A(0)$, a flat loss $A(0) - A_p$ may be included in the loss-characteristics if $r > r_1$ or $r < r_2$; otherwise, $A(0) < A_p$ and frequency-transformation is needed.

As Fig. 6.20 illustrates, the set of correspondences between frequency-values should ideally be

$$\begin{array}{llllll} \omega = & \omega_0 & \omega_p & \omega_s & \omega_\infty \\ \Omega = & 0 & \Omega_p = \omega_p & \Omega_s = \omega_s & \infty . \end{array} \tag{6.111}$$

This set represents four conditions on the three available parameters to (6.108). Dropping therefore the condition that ω_s remain unchanged, the transformation

$$\Omega^2 = \omega_p^2 \frac{\omega_p^2 - \omega_\infty^2}{\omega_p^2 - \omega_0^2} \frac{\omega^2 - \omega_0^2}{\omega^2 - \omega_\infty^2} \tag{6.112}$$

is obtained. Now ω_s shifts to the right slightly; its new value may be obtained from (6.112).

For a desired $R_1 = R_2$, $A(0) = 0$ and $\omega_0 = \omega_1$ should be chosen (Fig. 6.20).

A slight variation of the process is applicable to polynomial filters. These have loss-poles at $\omega \to \infty$ which should be preserved (Fig. 6.21).

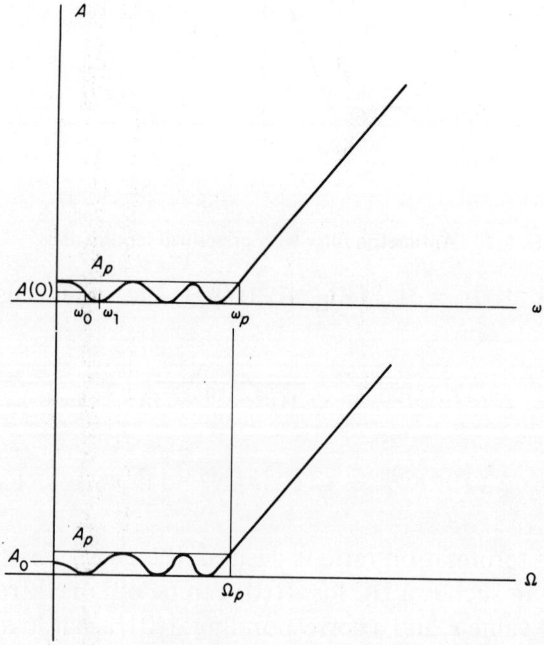

FIG. 6.21. Transformation of polynomial filter characteristics.

Hence, the proper transformation is obtained by setting $\omega_\infty \to \infty$ in (6.112):

$$\Omega^2 = \omega_p^2 \frac{\omega^2 - \omega_0^2}{\omega_p^2 - \omega_0^2} \tag{6.113}$$

It should be noted that these transformations can easily be combined with the z-transformation of (6.17).

Another application is to the matching of two unequal resistors R_1 and R_2 in the $\Omega_1 \leqslant \Omega \leqslant \Omega_2$ frequency-band, using a lowpass filter (Fig. 6.22.(a)). Due to the DC mismatch, a loss $A(0) = 20 \log_{10} (R_1 + R_2)/2$

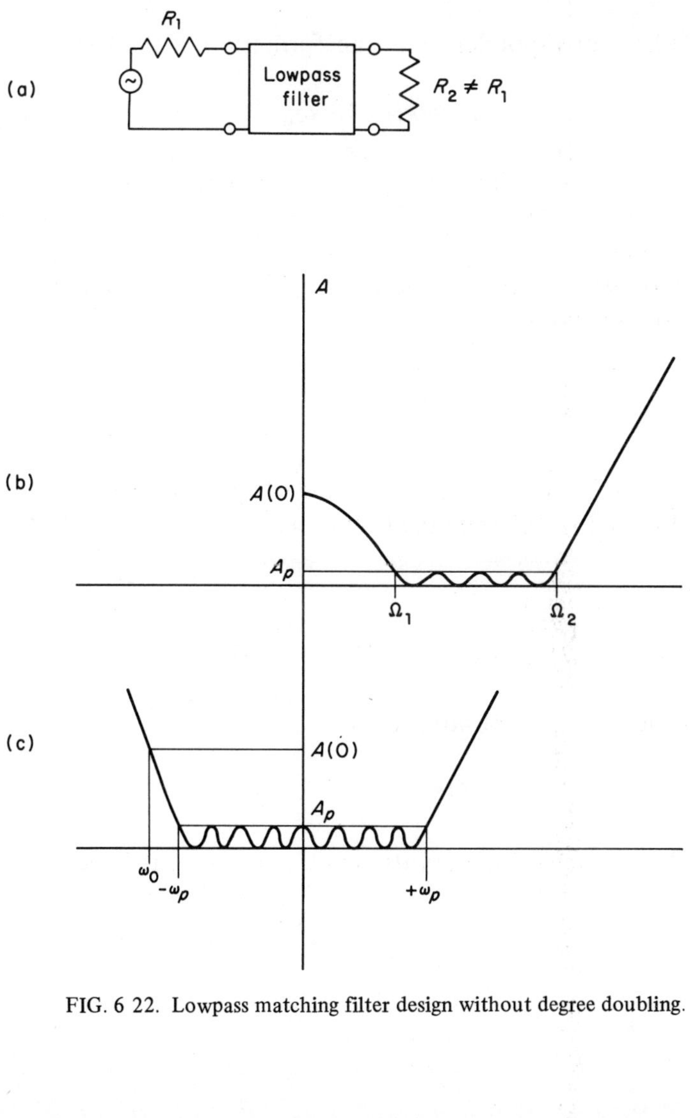

FIG. 6 22. Lowpass matching filter design without degree doubling.

$\sqrt{R_1 R_2}$ exists at zero frequency. Furthermore, if a reflection factor ρ is permissible in the passband interval $[\Omega_1, \Omega_2]$, a passband ripple $A_p = -10 \log_{10}(1 - \rho^2)$ or less is needed. Fig. 6.22 illustrates how these requirements may be met by frequency-transformation from a "model," an equal-ripple passband polynomial filter. For realizability, it is desirable to preserve the attenuation poles at $\omega \to \infty$; hence, the corresponding frequencies are

$$\begin{array}{lcccc} \omega^2 = & \omega_0^2 & \omega_p^2 & 0 & \infty \\ \Omega^2 = & 0 & \Omega_1^2 & \Omega_2^2 & \infty . \end{array} \tag{6.114}$$

Due to the last correspondence, a transformation of the form

$$\Omega^2 = k(\omega^2 - \omega_0^2). \tag{6.115}$$

is required. For the important cases of Chebyshev and elliptic filters, nomographs are available [16] for finding the degree and parameters of the model filter.

Transformation Doubling the Degree [17], [19]. Next, some applications of the mapping

$$\Omega^2 = k \frac{\omega - \omega_0}{\omega - \omega_\infty} \tag{6.116}$$

will be described. If again the matching of the two unequal resistors is required (Fig. 6.23), the specifications become

$$\begin{array}{lcccc} \omega = & \omega_0 & -\omega_p & +\omega_p & \infty \\ \Omega = & 0 & \pm\Omega_1 & \pm\Omega_2 & \infty . \end{array} \tag{6.117}$$

To satisfy the last correspondence,

$$\Omega^2 = k(\omega - \omega_0) \tag{6.118}$$

is required; to satisfy the conditions at $\omega = \pm\omega_p$, it is necessary that

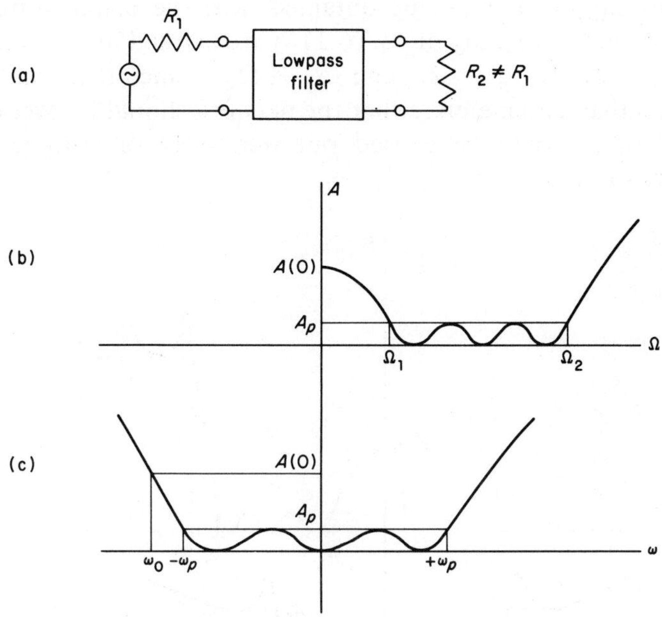

FIG. 6.23. Lowpass matching filter design using degree doubling.

$$\Omega^2 = \frac{\Omega_2^2 - \Omega_1^2}{2\omega_p}\,\omega + \frac{\Omega_1^2 + \Omega_2^2}{2}$$

$$\omega_0 = -\omega_p\,\frac{\Omega_1^2 + \Omega_2^2}{\Omega_2^2 - \Omega_1^2}\,. \tag{6.119}$$

If the model chosen is a Chebyshev-filter [18] (Fig. 6.23), it should have *at most* $A_p = -10\log_{10}(1 - \rho^2)$ passband ripple in $[-\omega_p, \omega_p]$ and *exactly* $A(0)$ loss at $\pm\omega_0$. The necessary degree of such a filter is

$$n \geqslant \frac{\log_{10}\,[(10^{A(0)/10} + 1)\,(\rho^{-2} - 1)]}{2\log\dfrac{\omega_0}{\omega_p}} \tag{6.120}$$

and its design is routine [1].

The same task may be accomplished using elliptic filters [19], [20]. Then, in addition to obtaining the desired matching, a substantial amount of suppression can be obtained near the matched band (Fig. 6.24). Parameters ω_∞, ω_0 of eq. (6.116) are marked in Fig. 6.24; in addition, $\omega_s \rightarrow \Omega_s$, $\omega_{-p} \rightarrow \Omega_1$, and $\omega_p \rightarrow \Omega_2$. Since there are more requirements than parameters, either the network should be over-designed or the design should be carried out iteratively or with the aid of nomographs [21].

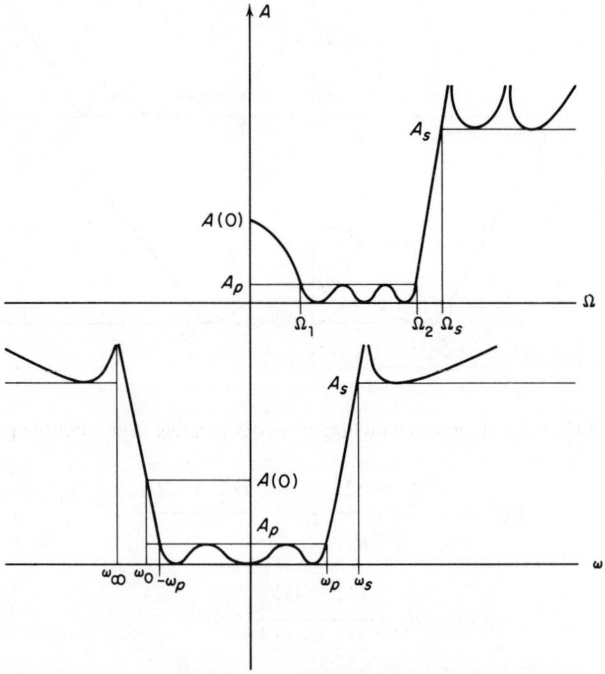

FIG. 6.24. Impedance-matching using an elliptic filter.

An important application of the transformation (6.116) is to the solution of the approximation problem for systems with high degree, of even order [19]. This application is illustrated in Fig. 6.25. Now the correspondence between frequencies is

$$\begin{array}{lllll} \omega = & \omega_0 & \omega_p & \omega_s & \omega_\infty \\ \Omega = & 0 & \Omega_p & \Omega_s & \infty \end{array} \tag{6.121}$$

Fig. 6.25 (b) shows how to visualize this transformation [19]; Fig. 6.26 shows an application for general stopband, equal-ripple passband filters. More complicated patterns are also possible.

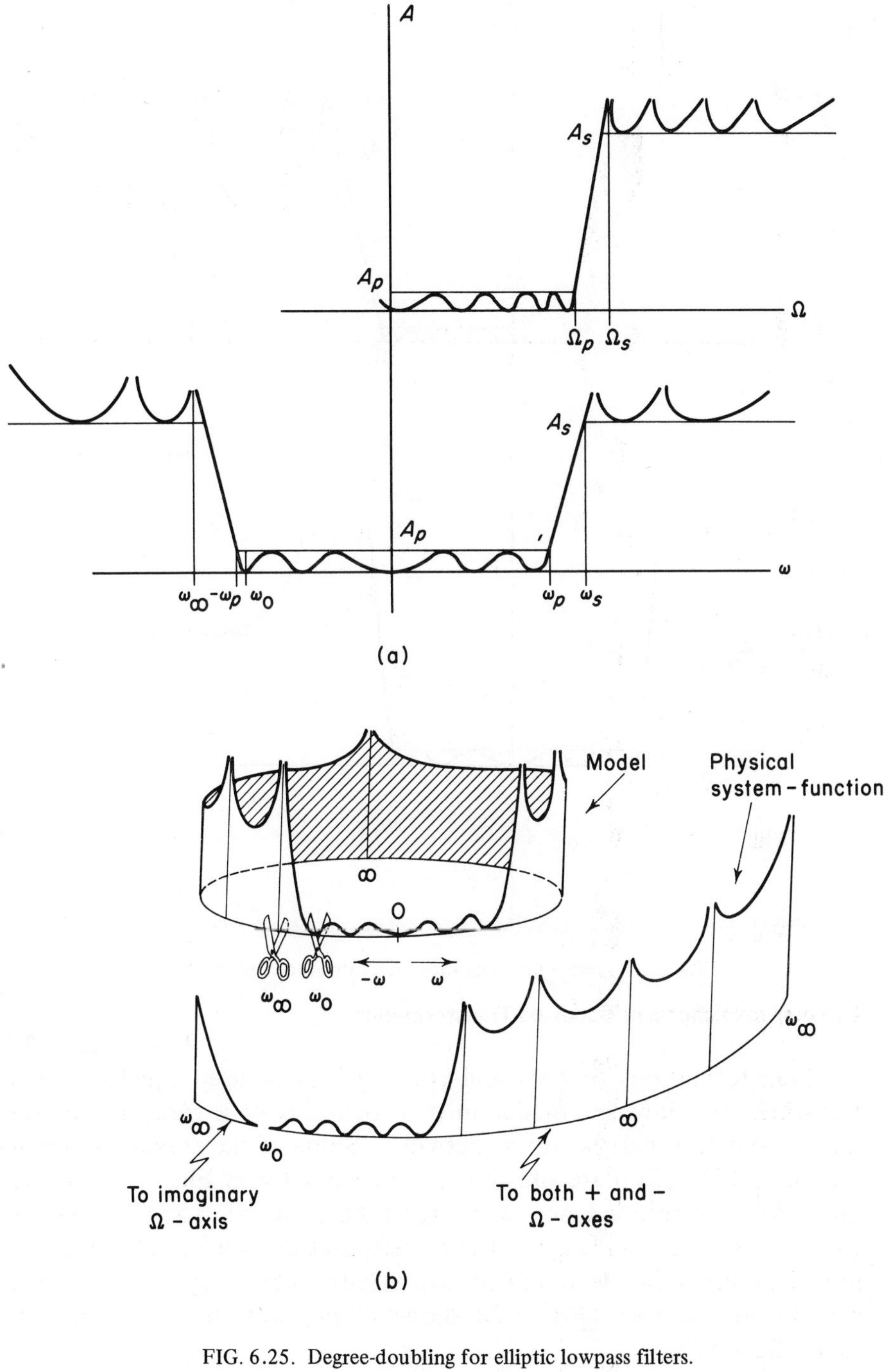

FIG. 6.25. Degree-doubling for elliptic lowpass filters.

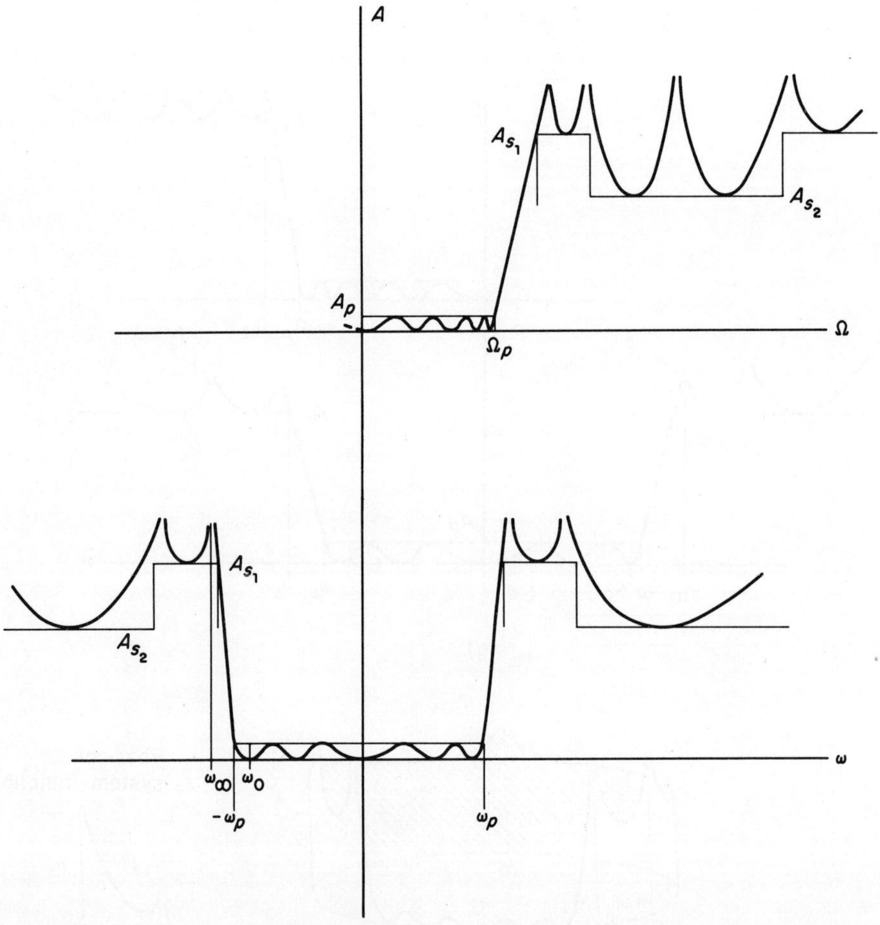

FIG. 6.26. Degree-doubling for general stopband lowpass filters.

Lowpass to Highpass or Bandpass Transformation

Transformations (6.108) and (6.116) may be used equally well to transform the lowpass model into a high-pass or a general (not frequency-symmetrical) bandpass network. Some of the possible transformations [17], [21] are illustrated schematically in Fig. 6.27. By requiring the correspondence of desired points on the ω and Ω axes, the parameters (k, ω_0 and ω_∞) of the transformation as well as those of the model function can be obtained in much the same way as for the lowpass-to-lowpass case. Fig. 6.28 shows a lowpass-to-bandpass mapping as an illustration.

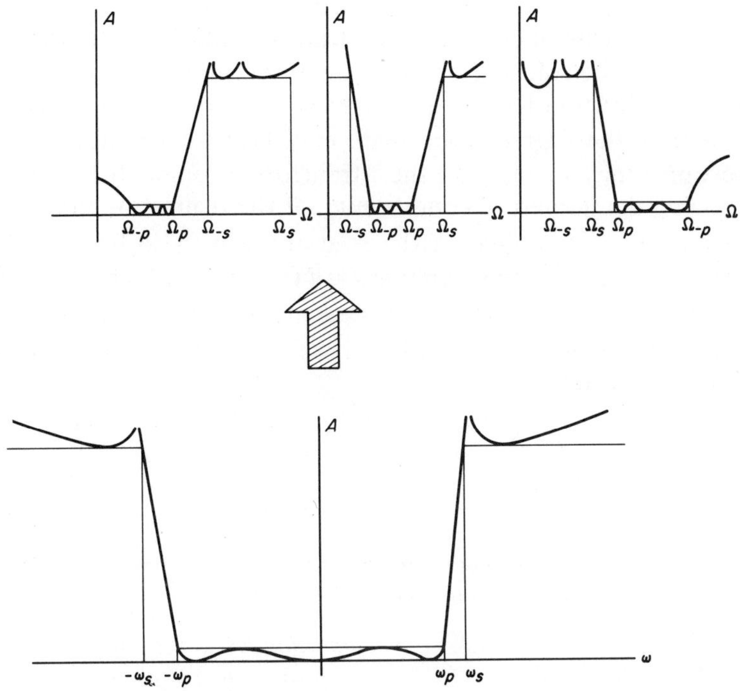

FIG. 6.27. Lowpass-to-highpass and lowpass-to-bandpass transformation.

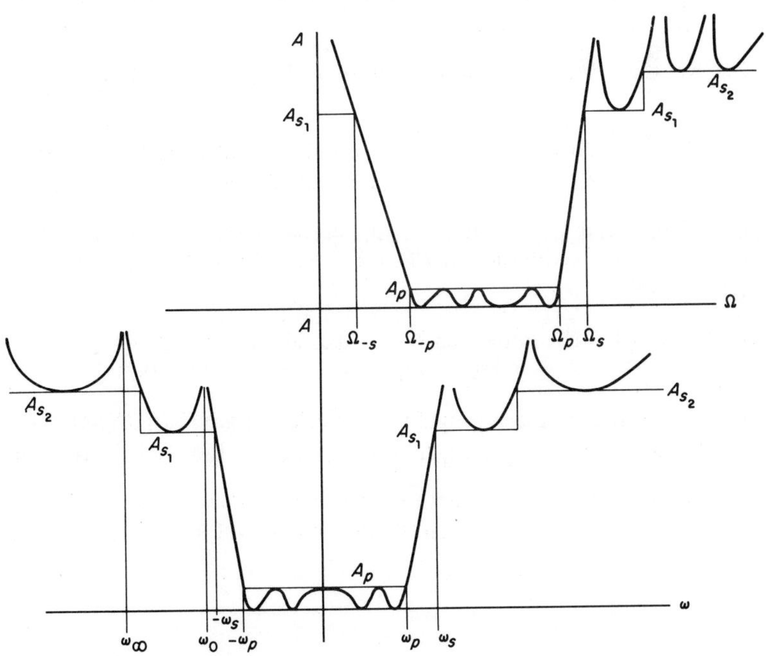

FIG. 6.28. Lowpass-to-bandpass transformation.

It must be remembered that the transformations discussed in this section can aid the solution of only the approximation problem. The subsequent realization has to be performed in terms of the physical frequency and, if the degree is too high, may lead to accuracy-problems. Then recourse may be made to the z-transformation of (6.17) or (6.43) for improving the numerical conditioning of the realization process.

Other rational mappings of the s-variable have also been used. Of particular importance is the mapping $s = k(z - 1/z)$, which transforms a Taylor-series approximation of a prescribed characteristic into a Chebyshev polynomial series approximation [22]. This, however, is amply described in the literature.

REFERENCES

1. R. Saal and E. Ulbrich, "On the design of filters by synthesis," *IRE Trans. on Circuit Theory*, vol. CT-5, pp. 284-327;Dec. 1958

2. W. Cauer, *Synthesis of Linear Communication Networks*, McGraw-Hill Book Co., New York, 1958.

3. J. K. Skwirzynsky, *Design Theory and Data for Electrical Filters*, D. Van Nostrand Co. Ltd., London, 1965

4. N. J. Achieser, *Theory of Approximation*, F. Ungar Publishing Co., New York, 1956.

5. H. J. Orchard, "Insertion transfer functions for filter networks," Post Office Res. Sta. Rept. No. 13134, London, England; June 1951.

6. B. R. Smith and G. C. Temes, "An iterative approximation procedure for automatic filter synthesis," *IEEE Trans. on Circuit Theory*, vol. CT-12, pp. 107-112; March 1965.

7. G. Szentirmai, "Theoretical basis of a digital computer program package for filter synthesis," *Proc. First Allerton Conf.*, Urbana, Ill., 1963.

8. ———, "A filter synthesis program," in *System Analysis by Digital Computer*, ed. F. F. Kuo and J. F. Kaiser, John Wiley & Sons, Inc., New York, 1966.

9. J. A. C. Bingham, "A new method of solving the accuracy problem in filter design," *IEEE Trans. on Circuit Theory*, vol. CT-11, pp. 327-341;Sept. 1964.

10. H. Watanabe, "Approximation theory for filter networks," *IRE Trans. on Circuit Theory*, vol. CT-9, pp. 341-356; Sept. 1961.

11. J. A. C. Bingham, "The approximation problem for both conventional and parametric bandpass filters," *IEEE Trans. on Circuit Theory*, vol. CT-11, pp. 408-410; Sept. 1964.

12. G. C. Temes, "Synthesis of filters with maximally flat passband and arbitrary stopband loss," *Canadian Electr. Engng.*, pp. 29-33; Feb. 1964.

13. G. C. Temes and M. Gyi, "Design of filters with arbitrary passband and Chebyshev stopband attenuation," *IEEE Internat. Conv. Rec.*, Pt. 5, pp. 2-12; March 1967.

14. K. H. Feistel and R. Unbehauen, "Tiefpässe mit Tschebyscheff – Charakter der Betriebsdämpfung im Sperrbereich und maximal geebneter Laufzeit," *Frequenz*, vol. 19, pp. 265-282; August 1965.

15. J. Jess and H. W. Schüssler, "On the design of pulse-forming networks," *IEEE Trans. on Circuit Theory*, vol. CT-12, No. 3; September 1965.

16. E. Christian and E. Eisenmann, "Broadband matching by lowpass transformations," *Proc. Fourth Allerton Conf.*, Urbana, Ill., 1966.

17. W. Cauer, *Synthesis of Linear Communication Networks*, McGraw-Hill Book Co., Inc., New York, 1958; pp. 307-315.

18. G. Szentirmai, "Bandpass matching filter in the form of polynomial lowpass filter," *IEEE Trans. on Circuit Theory*, vol. CT-11, pp. 177-178; March 1964.

19. J. Zdunek, "Generation of filter functions from a given model," *Proc. IEEE*, vol. 110, No. 2; February 1963.

20. K. Antreich and E. Gleissner, "On some applications of the Zdunek transformation in network theory," *NTZ-Comm. J.*, 1967, No. 3, pp. 116-120; March 1967.

21. E. Christian and E. Eisenmann, *Filter Design Tables and Graphs*, John Wiley & Sons, Inc., New York; 1966.

22. S. Darlington, "Network synthesis using Tchebycheff polynomial series," *BSTJ*, vol. 31, pp. 613-665; July 1952.

CHAPTER 7

DEVICE MODELING FOR CIRCUIT ANALYSIS PROGRAMS

M. E. DANIEL, Sandia Laboratories, Albuquerque

INTRODUCTION

The present availability of general purpose circuit analysis programs enables the engineer to use the computer as a design and analysis tool. It is now possible to analyze a circuit in a relatively short time with a precision far exceeding that obtained using manual analysis methods.

Although considerable effort is being expended throughout industry and at various universities in writing general purpose programs, the lack of adequate descriptions for active, nonlinear devices is a serious handicap. The description of the electrical behavior of an active device, including mathematical equations, topology, and methods for obtaining equation coefficients and element values will be referred to as the device model.

The accuracy of the analysis for any system is a direct function of the accuracy of the models used by the engineer to characterize his design. Unfortunately, many of the circuit analysis programs require the user to generate models for the active elements. Consequently, many times the models used are carry-overs from hand analysis methods (piecewise linear approximations, etc.).

The computer can rapidly and accurately solve systems of equations (either linear or nonlinear) which would not be feasible using hand calculations. It is then possible to use accurate, nonlinear, complex models which account for first and second order physical effects. Using these models, it is possible to effectively utilize the computer capabilities to more accurately predict circuit performance with very little additional expenditure of time.

7.1 WHAT CONSTITUTES A MODEL

Types of Models

One type of model is primarily concerned with the device physics. The equations describing the device contain such quantities as base width, junction contact area, doping levels, mobilities, etc. The models are usually constructed as sets of simultaneous partial differential equations which attempt to solve, for example, the charge distribution or the continuity equation in the junction region for one or two spatial dimensions. This microscopic approach rapidly exhausts the capabilities of even the largest and fastest digital computers. The analysis of a circuit containing more than one device is computationally not feasible when using models of this type.

A second type of model is not primarily concerned with the device physics. The quantities of interest are electrical characteristics which can be observed at the external terminals (i.e., a "black-box" approach). The burden of modeling this type of behavior consists of empirically finding a mathematical expression to represent each electrical characteristic.

This chapter will approach device modeling with a compromise of the two above methods, with emphasis toward the latter. Physical constants will be used whenever possible, but if a characteristic cannot be related physically, an empirical description will be used.

7.2 TRANSISTOR-DIODE MODELS

Small-Signal Models

This is the type of model with which most engineers are familiar. It consists of finding an operating point and assuming that all responses are linear about that point. Therefore, this model is valid over only a limited operating range, namely where the response characteristic is linear. An example of a static *V-I* characteristic for an ordinary diode is shown in Fig. 7.2-1.

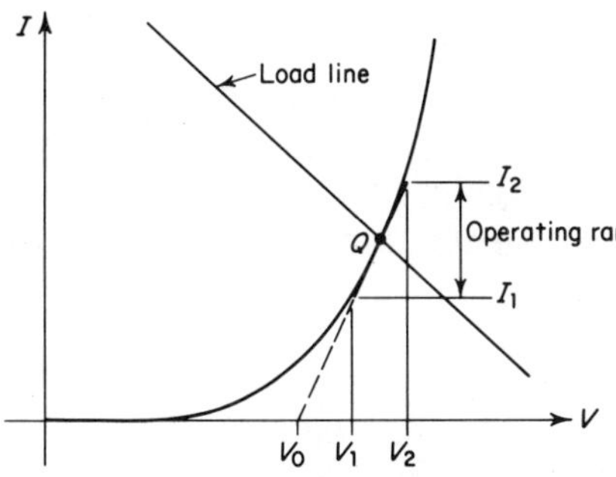

FIG. 7.2-1. Small-Signal Operating Region for an Ordinary Diode.

As long as the current stays in the range of $I_1 - I_2$, the model for D-C operation is given by the circuit of Fig. 7.2-2. The value of R is given by the inverse of the slope of the tangent to the characteristic at point Q.

$$R = \frac{V_2 - V_1}{I_2 - I_1} \tag{7.2-1}$$

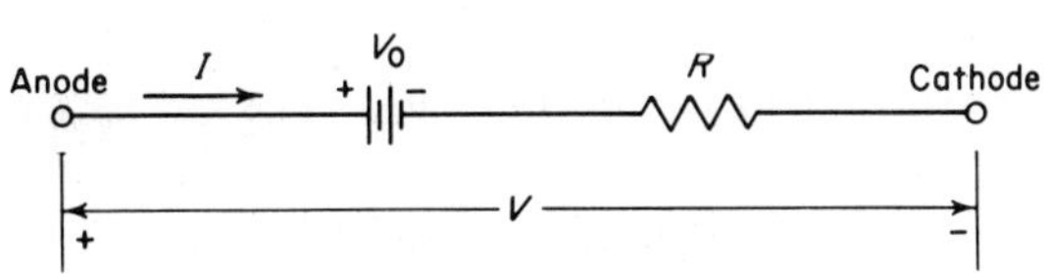

FIG. 7.2-2. Small Signal Diode Model.

V_0 is the intercept of the tangent with the V axis.

A similar approach is used to approximate the response of a junction transistor.

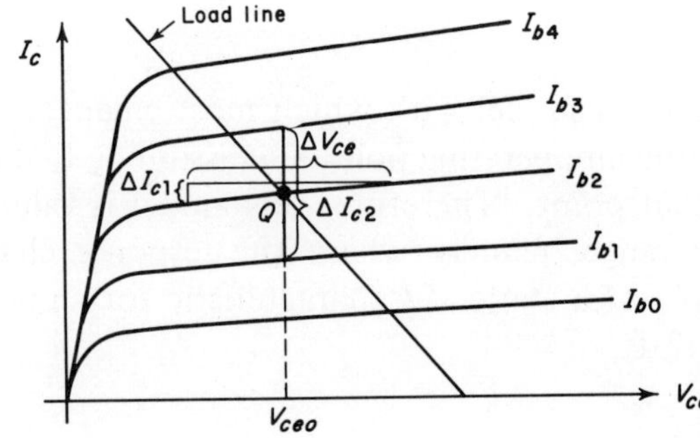

FIG. 7.2-3. Transistor Collector Characteristics (Common Emitter).

One of the better known small-signal equivalent circuits uses the h-parameters. The transistor and its common-emitter equivalent circuit are shown in Fig. 7.2-4.

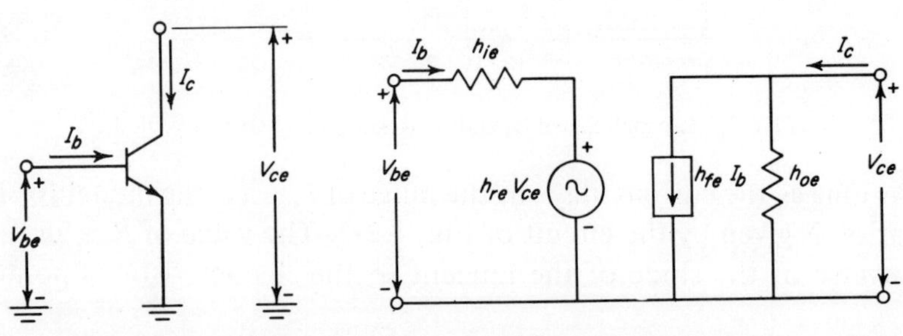

FIG. 7.2-4. Transistor h-Parameter Equivalent Circuit.

The equations which describe the operation of the transistor are

$$\begin{aligned} V_{be} &= h_{ie} I_b + h_{re} V_{ce} \\ I_c &= h_{fe} I_b + h_{oe} V_{ce} \end{aligned} \tag{7.2-2}$$

The hybrid parameters can be written explicitly from the two above equations.

$$h_{ie} = \text{short-circuit input resistance} = \left.\frac{V_{be}}{I_b}\right|_{V_{ce}=0}$$

$$h_{re} = \text{open-circuit reverse voltage transfer ratio} = \left.\frac{V_{be}}{V_{ce}}\right|_{I_b=0}$$

$$h_{fe} = S\text{-}C \text{ forward current transfer ratio} = \left.\frac{I_c}{I_b}\right|_{V_{ce}=0}$$

$$h_{oe} = O\text{-}C \text{ output conductance} = \left.\frac{I_c}{V_{ce}}\right|_{I_b=0}$$

The parameters also can be found from the characteristic curves if linearity is assumed around the operating point, Q (See Fig. 7.2-3).

$$h_{fe} = \left.\frac{\Delta I_{c2}}{I_{b3} - I_{b1}}\right|_{V_{ce}=V_{ceo}} \qquad h_{oe} = \left.\frac{\Delta I_{c1}}{\Delta V_{ce}}\right|_{I_b=I_{b2}}$$

NOTE: Only h_{fe} and h_{oe} can be obtained from Fig. 7.2-3. We must also have another set of curves, such as I_b versus V_{be} with I_c constant, to get h_{ie} and h_{re}.

Another familiar small-signal equivalent circuit, probably more widely used in computer analysis work than the h-parameter equivalent, because it is valid over a wider frequency range, is the hybrid-pi model [1]. It describes the incremental behavior of a transistor for all frequencies at which the device has appreciable current gain. Although it appears fairly complicated in Fig. 7.2-5, it is relatively simple to use for most applications, because it is seldom necessary to consider the effect of all of the circuit elements in any single calculation.

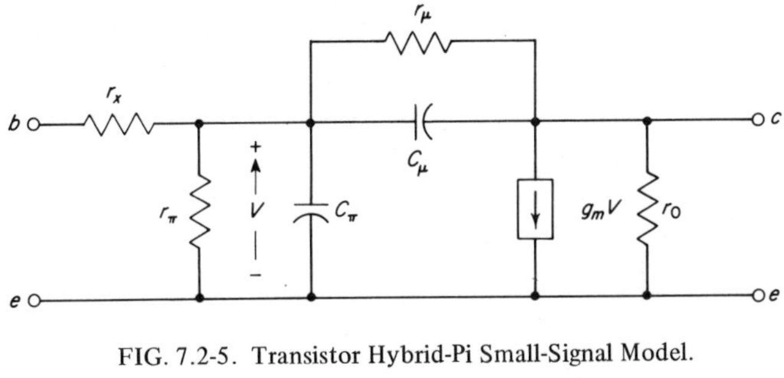

FIG. 7.2-5. Transistor Hybrid-Pi Small-Signal Model.

Chapter three of the book by Searle, et al [1], is devoted entirely to the simplified models for various frequency ranges and to the determination of the hybrid-pi model parameters. The element values are easily found by using the h-parameter equations, (7.2-2), and the y-parameter description, which is as follows.

$$
\begin{aligned}
I_b &= y_{ie} V_{be} + y_{re} V_{ce} \\
I_c &= y_{fe} V_{be} + y_{oe} V_{oe}
\end{aligned}
\tag{7.2-3}
$$

The h-parameters may, of course, be related to the y-parameters by making the transformation of variables required to change (7.2-2) to (7.2-3). The measurements required for determining the hybrid-pi element values are summarized in Table 7.1.

Table 7.1

Hybrid-Pi Model Parameters [1]

Hybrid-Pi Element	*Relationship to Measured Quantities*
g_m	$g_m = \frac{q}{kT} \lvert I_c \rvert$
r_π	$r_\pi = \frac{1}{g_m} h_{fe} \Big\rvert_{1\text{ kHz}}$
r_x (low frequency)	$r_x = h_{ie} \Big\rvert_{1\text{ kHz}} - r_\pi$
r_x(high frequency)	$g_x = \text{Re}\big[y_{ie} \big\rvert \omega \gg g_x/(C_\pi + C_\mu)$
r_μ	$r_\mu \cong \frac{r_\pi}{h_{re} \rvert_{1\text{ kHz}}}$
r_O	$r_O \cong \frac{1}{y_{oe} \rvert_{1\text{ kHz}}}$
C_μ	$C_\mu \cong C_{ob} - C_{bc} - C_{ce}$ (C_{ob} = common-base output capacitance)
C_π	$C_\pi = \frac{g_m}{2\pi f_t} - C_\mu$ ($f_t \cong$ frequency at which $h_{fe} = 1$)

Other important relationships:

$$f_t = \frac{g_m}{2\pi(C_\pi + C_\mu)}$$

$$f_\beta = \frac{1}{2\pi r_\pi (C_\pi + C_\mu)} = \text{3db point on } \lvert h_{fe} \rvert \text{ vs. } f \text{ curve}$$

In general, small-signal models are simple and accurate over a limited range. The circuit elements and equation coefficients are easy to measure, or extract from measured data. They are very useful for frequency analyses and time-domain analyses where transfer functions are used.

Piecewise Linear Models

The piecewise linear approximation to a nonlinear function is used frequently in manual analysis. It consists of several linear segments (or small-signal approximations) each valid over a certain limited range. As more regions are used, there is a corresponding increase in the accuracy. If the boundaries are judiciously chosen, a fairly accurate large-signal model can be obtained.

A switching diode can be represented as shown in Fig. 7.2-6.

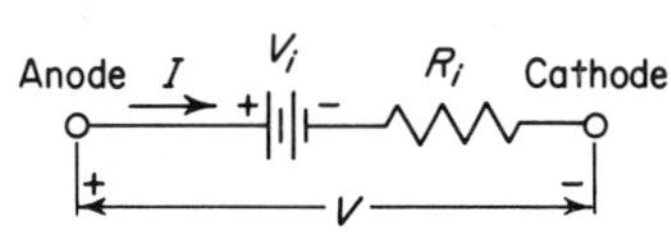

$V \leq 0;\ V_i = 0,\ R_i = R_3$

$0 \leq V \leq V_0;\ V_i = V_1,\ R_i = R_1$

$V \geq V_0;\ V_i = V_2,\ R_i = R_2$

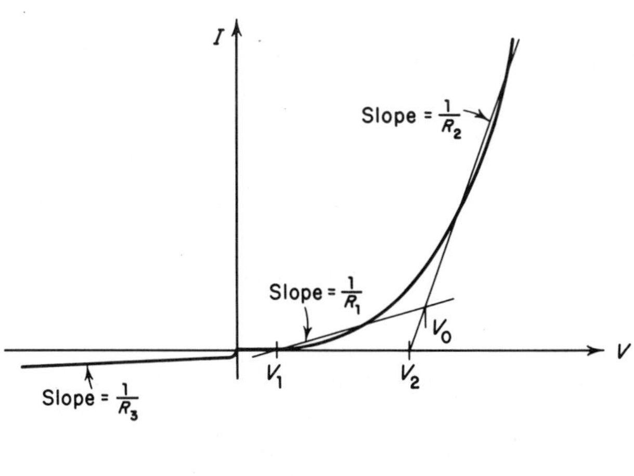

FIG. 7.2-6 Piecewise Linear Approximation for Diode

The same advantages and disadvantages apply to the piecewise linear model as for a small-signal representation, although operation of the device is defined for a wider range of values of V. A piecewise linear model can be made to approximate the actual behavior quite accurately by using several regions. However, as more regions are added, more constants must be provided and the awkwardness involved in determining the correct region of operation and the correct set of constants may prove to be a drawback.

Large-Signal Models

For use in general computer circuit analysis the models discussed above are rather specialized and limited. Ideally, a model should be frequency independent, valid for any current or voltage excitation within the device ratings, and have a fixed topology which provides a one-to-one correspondence with physical processes. It should consist of passive elements (R, L, C, M-mutual inductance), preferably of fixed value; current sources (J) and voltage sources (E), whose mathematical equations are "well-behaved."

A large-signal model is one which can be used over a wide range of operation. By its very nature, it is the best suited model for use in general computer circuit analysis. A number of such models have been proposed for junction transistors; three of which are most widely known and used. They are the Ebers-Moll model [2, 3], the Beaufoy-Sparkes charge-control model [4], and the Linvill lumped model [5]. A comparison of the three models has been made by Hamilton, Lindholm, and Narud [6]. Some of their research and conclusions will be used in the following discussion of the advantages and disadvantages of certain large-signal models for computer circuit analysis and design. A further discussion of device modeling is given by Sparkes [7].

The Ebers-Moll model describes device behavior in terms of easily measured terminal quantities. The "black-box" behavior is expressed by relatively simple mathematical equations. Even though the topology is such that each element represents one physical feature of the actual device, the Ebers-Moll approach provides very little insight into the physics of operation.

The charge-control model is quite similar topologically to the Ebers-Moll model. However, some contact is established with physical processes by describing the generated currents in terms of the minority carrier charge. Although the charge-control model is analytically somewhat more complex, the model parameters can be extracted from terminal measurements and there is a one-to-one correspondence with the Ebers-Moll parameters.

The Linvill model is the most general of the three. It consists of lumped network-like elements (called diffusance, storance, and combinance) which represent physical processes within the device. Thus, a one-to-one correspondence exists between each element and a physical process. However, the values of the lumped elements are not directly measurable and are not expressed in terms of terminal variables.

For use in computer programs for circuit analysis and design, the first two models appear to be most suitable. The primary objective of such programs, and hence of device models to be used in them, is to aid in designing "better" circuits. Normally, the designer is more interested in obtaining a desired response for a given excitation of a circuit, so the terminal behavior of a device is more important to him than the internal physical processes.

7.3 FORMULATION OF GENERAL DEVICE MODELS

There are several steps involved in developing a model for any family of semiconductor devices. Some of the steps have been mentioned in the discussion of existing diode and transistor models. An outline of the development techniques is given in this section. Some of the steps will be discussed in more detail in subsequent sections.

(1) An understanding of the physical processes which produce the external electrical characterisitics is desirable. Although it is possible to use a purely "black-box" approach, it is helpful to know which phenomenon is the result of tunneling, avalanche of the junction, stored charge, conductivity modulation, etc.

(2) The manufacturing process used in making a given device may be pertinent to the model. For example, a 2Nxxx transistor of planar construction will probably exhibit different transient behavior than a transistor having the same 2N number although of annular construction. The result may need to be two different models or two different sets of model parameters, one for each manufacturer or manufacturing process.

(3) A major portion of any model is the set of mathematical equations which describe the physical operation. They may be derived in various ways, but two criteria which must be met are (a) an adequate de-

scription of device behavior, and (b) compatability with the language of the computer.

(4) The model topology is the interconnection of the discrete components (R, L, C, M, J, and E) making up the network which approximates the device. Although not imperative, it is usually desirable to have one element or set of elements to represent each physical phenomenon. (For example, one resistor for bulk resistance and one for leakage resistance; one capacitor to represent transition capacitance and another for diffusion capacitance; a current source for tunnel current, another for minority carrier current, and a third current generator to represent conductivity modulation.)

(5) The development of a model involves a determination of certain defining parameters. These parameters must be obtained from knowledge of the device physics or from operating characteristics. If the operating characteristics are not supplied by the manufacturer, then measurements must be made by the user. Sometimes it is more difficult to determine what measurements to make and how to make them than it is to formulate the equations and the topology.

(6) Once the model has been formulated, the device parameters must be obtained. Usually it is necessary to extract these parameters from the data measured in step 5. The data reduction involves curve-fitting data points or other analytical manipulations to get the parameters explicitly.

7.4 MATHEMATICAL EQUATIONS

Origin of the Equations

The basis for the mathematical equations of a physical semiconductor model are the assumptions of current flow in semiconductor material consisting of drift and diffusion components, continuity of holes and electrons, and the relationship between net space charge and electric field. The equations for hole and electron current density due to drift and diffusion are, respectively:

$$j_p = e\mu_p pE - eD_p \nabla p$$

$$j_n = e\mu_n nE + eD_n \nabla n$$

where e = electronic charge, μ = mobility, D = diffusion constant, E = electric field, p = density of holes, n = density of electrons, ∇p = hole density gradient, ∇n = electron density gradient.

The continuity equation for hole minority carriers in an n-type semiconductor is

$$\frac{\partial p}{\partial t} = \frac{p_n - p}{\tau_p} - \frac{1}{e} \nabla \cdot j_p \, .$$

Similarly, for electron minority carriers in a p-type semiconductor,

$$\frac{\partial n}{\partial t} = \frac{n_p - n}{\tau_n} + \frac{1}{e} \nabla \cdot j_n .$$

where τ_p = lifetime of holes in an n-type semiconductor, τ_n = lifetime of electrons in a p-type semiconductor, p_n and n_p are equilibrium carrier densities.

Such equations are useful when knowledge of the internal behavior of the device is important. It is also possible, by making certain simplifying assumptions, to solve these equations and relate external voltages and currents to internal parameters.

For the purposes of computer aided analysis it is more instructive to relate external behavior to measurable quantities. For example, terminal current and voltage of p-n junction diodes are related by the equation

$$I = I_s(e^{\frac{qV}{MkT}} - 1) \tag{7.4-1}$$

where q/kT = 38.97 at room temperature of 25 degrees C. I_s is the reverse saturation current and M is a constant which defines the curvature of the V-I characteristic. In actuality, V in (7.4-1) is the voltage across an ideal p-n junction and is less than the terminal voltage of a practical diode by a voltage drop IR_b, where R_b is the bulk resistance of the semiconductor material.

Equation (7.4-1) describes terminal behavior in terms of physical constants. Another type of equation is purely empirical, such as approximating the common-emitter current gain of a transistor by a polynomial of degree n.

$$\beta = B_0 + B_1 V_{be} + B_2 (V_{be})^2 + \cdots + B_n (V_{be})^n \tag{7.4-2}$$

The values of B_0, ..., B_n to give a "best" approximation of β to the actual gain are found by curve-fitting (7.4-2) to a set of data points β vs. V_{be}.

Constraints on Equations

The mathematical equations can be nearly any function of known voltages or currents as long as certain conditions are satisfied. Among the conditions are continuity of both the function and the first derivative of the function.

A continuous function is necessary to avoid ambiguity in defining a dependent variable. That is, for any value of an independent variable, there must be one, and only one, value for a function of that variable. This constraint may seem obvious and trivial, but is still sometimes violated in defining a model.

The condition of a continuous first derivative prohibits the use of piecewise linear models. However, piecewise nonlinear models may be used as long as equality of the first derivatives, as well as equality of the functions, is preserved at the boundary. For example, see Fig. 7.4-1.

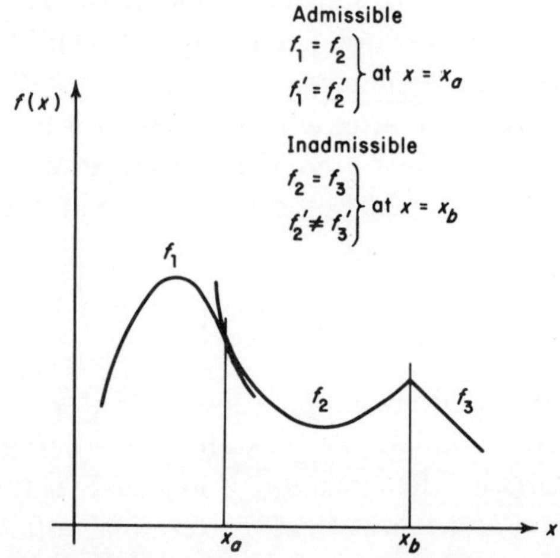

FIG. 7.4-1. Example of Analytical Functions.

The justification for requiring continuity of the first derivative is found in the methods used by various programs for internally formulating the network equations. In virtually all large general purpose programs, a state-variable formulation technique is used. The state equations are found from a linear graph of the complete network being analyzed. If the network contains a circuit of voltage sources and capacitors or a cut-set of current sources and inductors, then a tree and co-tree from which the state equations can be formulated may not exist; unless the first derivatives of current and voltage sources are known. For details of the theory involved in formulating state equations for electrical networks see the referenced paper by Blackwell and Grigsby [8].

An additional constraint on the mathematical equations is that they must have the same form as the network equations used in the general program. For example, in finding a steady-state solution used in the SCEPTRE program [9], a current source cannot be a function of the voltage across it except for the special case of the diode equation (7.4-1). This constraint only means that when putting a model into a general program, one should be aware of any unique features of that program.

7.5 MEASUREMENT TECHNIQUES AND EVALUATION OF MODEL PARAMETERS

Use of Specification Sheet Information

Once the topology and equations have been formulated it becomes necessary to find the model parameters. Ideally, they can be read directly or extracted from the device specification sheet. Unfortunately, this cannot always be done. Most data sheets do not contain sufficient information to define the entire model. For example, specifying the capacitance of a diode at one reverse voltage is not sufficient to define transient operation of the diode circuit.

The information given on specification sheets differs by manufacturer. The tendency is to give data which makes the device performance appear the best, whether it is pertinent to normal application or not. However, until designers agree upon some standardized models, manufacturers cannot be expected to expend the time and money to provide all possible data which might be needed for a new model.

Additional Measurements Necessary

The crux of the matter is that if a designer needs a new model or parameters for an existing model, he must make additional measurements. Some examples of finding parameters for diodes and transistors will be used for illustration.

The most commonly used diode model is the modification of the original Ebers-Moll configuration shown in Fig. 7.5-1. An evaluation of this model was performed and is documented in Reference 10.

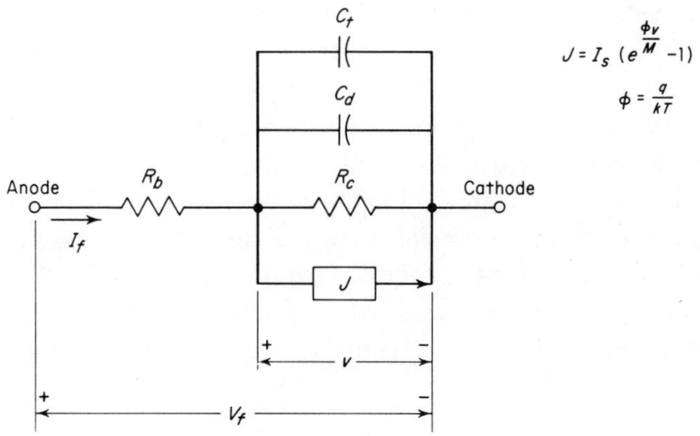

FIG. 7.5-1. Ebers-Moll Diode Model.

For D-C operation, the network equations are

$$V_f = v + I_f R_b \tag{7.5-1}$$

$$I_f = J + \frac{v}{R_c}\,. \tag{7.5-2}$$

A plot of the static $V_f - I_f$ characteristic for positive I_f is shown in Fig. 7.5-2. The current is plotted on a logarithmic scale to illustrate that over a wide range in the forward operating region, $V_f \approx v$.

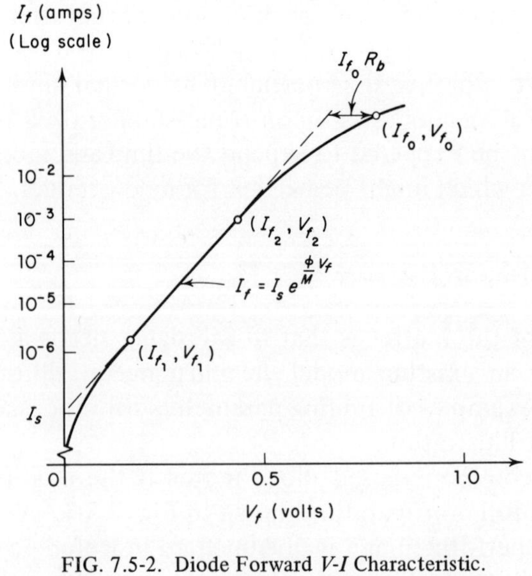

FIG. 7.5-2. Diode Forward V-I Characteristic.

One method for finding the unknown model parameters (I_s, M, R_b) is to curve-fit data points on the forward diode characteristic to (7.5-1) and (7.5-2), where I_f is assumed to equal J. If care is exercised in obtaining the data, fairly close approximations will result.

Another approach to finding the parameters involves the same assumptions but not the curve-fitting. Choose two points on the linear portion of the curve in Fig. 7.5-2, call them (I_{f_1}, V_{f_1}) and (I_{f_2}, V_{f_2}). In the range between the two points we have

$$I_f = I_s e^{\frac{\phi}{M} V_f} . \tag{7.5-3}$$

Taking the logarithm of both sides gives

$$\mathrm{Ln}(I_f) = \mathrm{Ln}(I_s) + \frac{\phi}{M} V_f . \tag{7.5-4}$$

Substituting (I_{f_1}, V_{f_1}) and (I_{f_2}, V_{f_2}) into (7.5-4) and solving for M gives

$$M = \frac{\phi(V_{f_2} - V_{f_1})}{\mathrm{Ln}\,(I_{f_2}/I_{f_1})} . \tag{7.5-5}$$

Substituting either data point and the value of M found in (7.5-5) into (7.5-3),

$$I_s = \frac{I_{f_1}}{e^{\phi/M\, V_{f_1}}} = \frac{I_{f_2}}{e^{\phi/M\, V_{f_2}}} . \tag{7.5-6}$$

The parameter R_b is found from the measured characteristic at (I_{fo}, V_{fo}). The difference between the linear extrapolation and the measured point is the drop due to R_b. Solution of (7.5-1) results in

$$R_b = \frac{V_{fo} - v_{fo}}{I_{fo}}$$

where

$$v_{fo} = \frac{M}{\phi} \mathrm{Ln}\left(\frac{I_{fo} + I_s}{I_s}\right) \approx \frac{M}{\phi} \mathrm{Ln}\left(\frac{I_{fo}}{I_s}\right) .$$

The ohmic leakage resistance, R_c, can be found from one measured point (I_{f_3}, V_{f_3}) in the reverse (cutoff) region shown in Fig. 7.5-3.

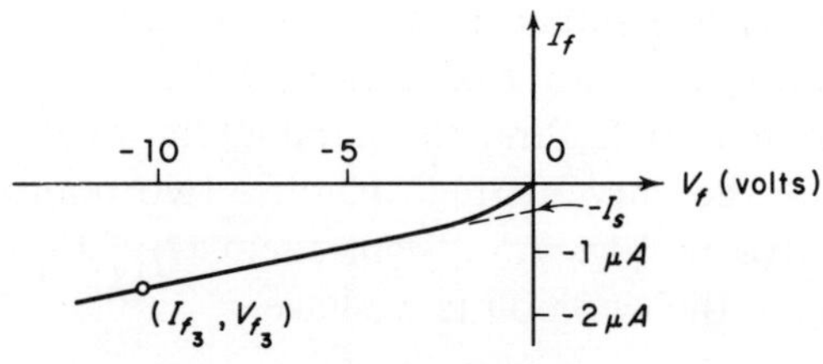

FIG. 7.5-3. Diode Reverse *V-I* Characteristic.

Assuming a linear reverse characteristic, one approximation is

$$R_c = \frac{V_{f3}}{I_{f3}} .$$

A somewhat more accurate expression, as can be seen from Fig. 7.5-3, is

$$R_e = \frac{V_{f3}}{I_{f_3} + I_s}$$

where I_s is given by (7.5-6).

Another description necessary for semiconductor device models which is not provided on most data sheets is the transition capacitance of a *P-N* junction. (This is the junction transition capacitance, C_t, in Fig. 7.5-1). The analytical expression normally used to describe C_t is

$$C_t = \frac{K}{(V_z - v)^N} \tag{7.5-7}$$

We must then find the constants K, N, and R_c from measurements of capacitance vs. V_f.

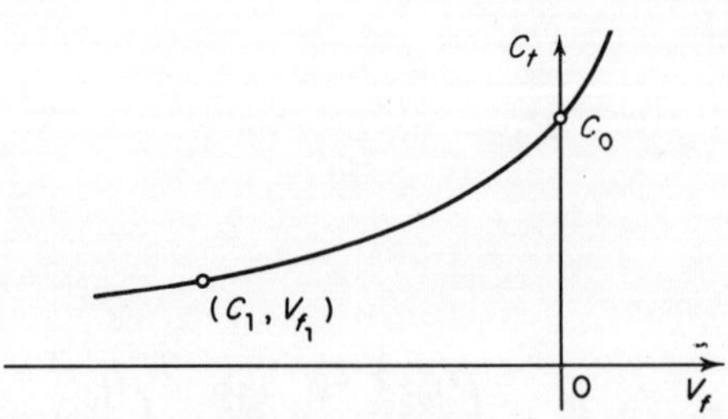

FIG. 7.5-4. Junction Transition Capacitance Characteristic.

At measured values of $V_f \leqslant 0$, $v \approx V_f$ in (7.5-7). Several curve-fitting programs are available [11] for extracting the constants from data points (C_i, V_{fi}). Typical values are $N = 0.3 - 0.5$, $V_z = 0.6 - 1.0$, and K is proportional to C_t (namely, $K = C_t$ when $V_z - v = 1$).

From a practical standpoint, when using (7.5-7) to describe capacitance in a circuit analysis program, care must be taken that $V_z - v$ doesn't approach zero too closely. To avoid the problem, V_z is assumed to be 0.9 — 1.0 volts. A second observation is that for $V_f \gtrsim 0.3$ volts the total capacitance (which becomes very difficult to measure) is the sum of transition capacitance and diffusion capacitance.

Based on the above arguments, it seems that a sufficiently accurate set of parameters can be found as follows. Assume $V_z = 1.0$ volt. Then for $V_f = 0$, $C_t = C_0 = K$. This leaves only N unknown, which can be found from one data point (C_1, V_{f_1}).

$$C_1 = \frac{C_0}{(1 - V_{f_1})^N} \tag{7.5-8}$$

Solving (7.5-8) for N gives

$$N = \frac{\mathrm{Ln}\left(\dfrac{C_0}{C_1}\right)}{\mathrm{Ln}\,(1 - V_{f_1})}\ .$$

Most of the measurements required to obtain the above parameters can be made directly on automatic semiconductor testing equipment. The data points on the static characteristics are measured under DC or pulsed conditions. The capacitance measurements can be made at a fixed frequency on an automatic read-out bridge unless the conductance gets too high (for $V_f > 0$), in which case a double balance (both capacitance and conductance) bridge can sometimes be used.

Some parameters must be found indirectly. An example is the diffusion capacitance, C_d of Fig. 7.5-1. It is defined by

$$C_d = \frac{\emptyset\,(J + I_s)}{2\pi MF} \tag{7.5-9}$$

The parameter F is proportional to the diode cut-off frequency and can be evaluated using Kuno's method [12]. In essence, the method involves rapidly switching a diode from a forward conducting state to a reverse bias as indicated in Fig. 7.5-5.

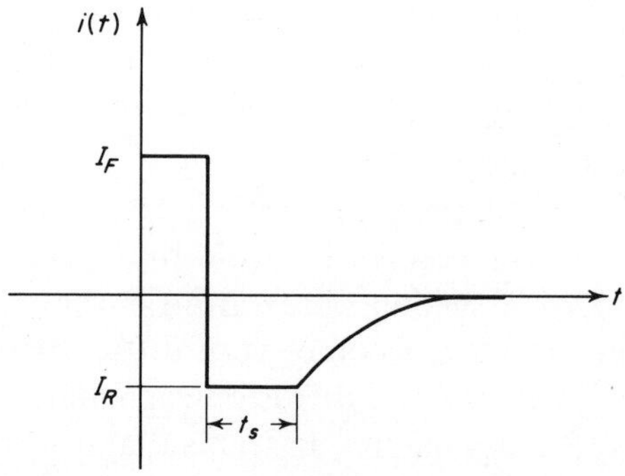

FIG. 7.5-5. Diode Switching Characteristic.

The storage time, t_s, is the time required to sweep the minority carriers out of the junction. (We are not particularly interested in the time it takes for the current to decay from I_R down to $.1I_R$, which also includes recombination effects.) If t_s is found for various ratios of I_F/I_R and a plot is made as in Fig. 7.5-6 the cutoff frequency can be found.

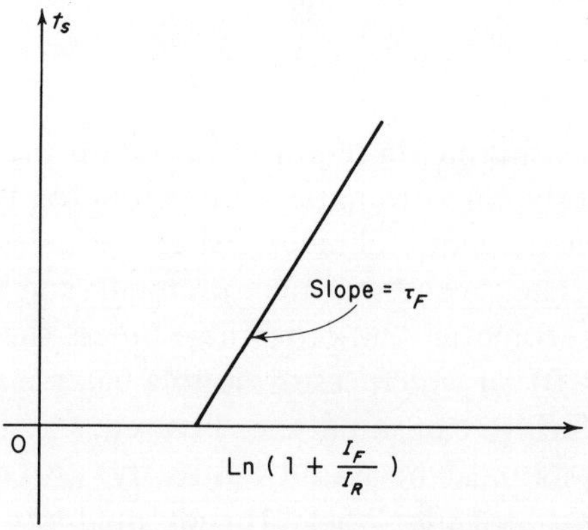

FIG. 7.5-6. Diode Junction Storage Time.

The cutoff frequency is given by

$$F \approx \frac{1}{2\pi\tau_F} .$$

Some unique measurement and data reduction problems are also encountered in finding models and parameters for other devices, such as tunnel diodes, unijunction transistors, etc. Some of these problems will be discussed in the next section along with a presentation of the particular models.

7.6 MODELS FOR OTHER SEMICONDUCTOR DEVICES

Zener/Avalanche Diode Model

Mathematical Equations – The static *V-I* characteristic of an avalanche diode is ideally represented by the curve of Fig. 7.6-1. For voltages more positive than V_b, the device behaves like an ordinary semiconductor diode. When the reverse bias reaches a value of V_b, voltage regulation is achieved. Ideally, after breakdown, for any reverse current greater than I_b, the voltage drop across the diode remains constant at V_b.

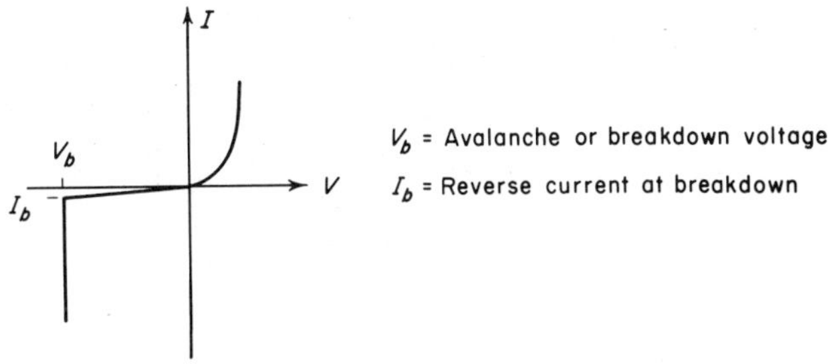

FIG. 7.6-1. Ideal Avalanche Diode Characteristic.

In an actual physical device, breakdown does not suddenly occur at some V_b; there remains a finite resistance across the semiconductor element. The actual static characteristic will appear more like that shown in Fig. 7.6-2. The voltage drop due to the semiconductor bulk resistance is exaggerated to illustrate its effect. The model must include a mathematical description of this curve so that for a given value of voltage, the value of current is known, or vice versa.

Since the avalanche diode is nearly identical to an ordinary diode for operating voltages greater than V_b, it should be possible to modify the Ebers-Moll model (Fig. 7.5-1) to include the avalanche phenomenon.

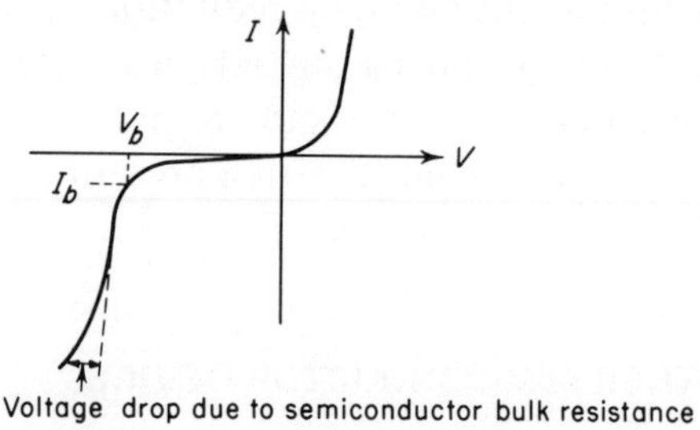

FIG. 7.6-2. Actual Avalanche Diode Characteristic.

First, it is observed that a current source must be inserted to account for the large current flow at V_b. An expression which describes this phenomenon is of the form

$$I_2 = -e^{A(V_b - B'v)} \tag{7.6-1}$$

where $B' = B + 1$ and $|B| \leqslant 0.1$. The value of A can vary from 10 to 100. The constants A and B are determined from the V-I characteristic curve, V_b is the breakdown voltage, and v is the junction voltage.

The postulated static (D-C) model is shown in Fig. 7.6-3.

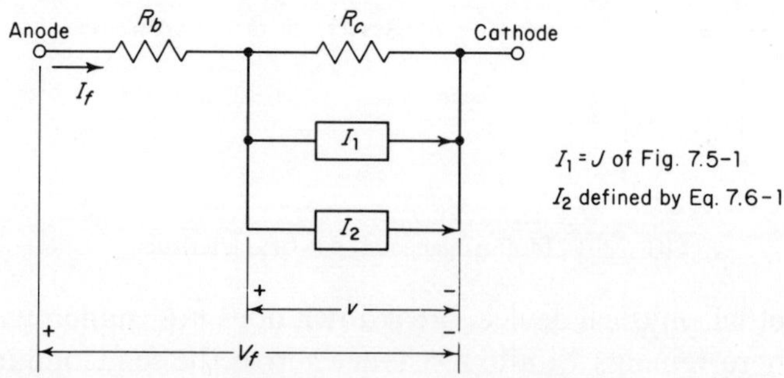

FIG. 7.6-3. Avalanche Diode Static Model.

The static circuit equations are:

$$V_f = v + I_f R_b$$

$$I_f = I_1 + I_2 + \frac{v}{R_c}$$

When v is large and positive, I_2 is effectively zero and I_f is approximately equal to I_1. When v is large and negative, I_1 is effectively zero. With v negative and $|v|$ less than $|V_b|/2, I_2$ is also effectively zero, allowing the leakage current to be approximated by v/R_c. As v is decreased toward V_b, I_2 becomes the dominant current term. The value of B determines when the current begins to break away from its leakage value (i.e., as $|B|$ decreases, the "knee" of the characteristic curve becomes sharper). The value of A determines how rapidly the current increases negatively once breakdown has been achieved (i.e., as A increases, the voltage regulation in the Zener or avalanche region improves).

For transient operation, the model must account for the capacitance across the junction. When the junction voltage is negative, the primary component of capacitance is the transition capacitance, C_t. Since normal operation is in the reverse and avalanche regions, it is assumed that the total capacitance, including the header capacitance (between leads and case), are lumped together into a C_t. The complete postulated model is given in Fig. 7.6-4.

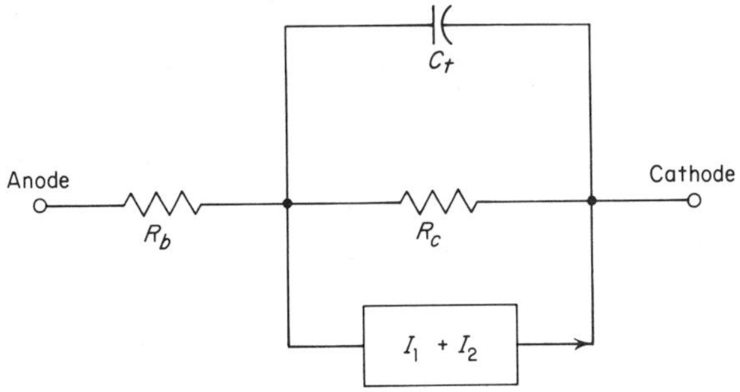

FIG. 7.6-4. Complete Avalanche Diode Model Topology.

The mathematical expression for C_t as a function of junction voltage will be developed after a discussion of the measurements and data reduction techniques.

Measurements and Data Reduction – Sufficient data points must be taken over the entire static characteristic to adequately describe it for curve-fitting by computer. Measurements of capacitance versus reverse voltage must be made at a fixed frequency, for example 1 MHz.

Measurements of junction storage times using Kuno's [12] method can be made. However, only gross estimates of diffusion capacitance are possible from the data. Due to the normal region of operation and the relatively small magnitudes of diffusion capacitance involved, neglecting this capacitance term does not seem to adversely affect the model.

A graph of Ln (I) versus V theoretically should appear as in Fig. 7.6-5. The value of R_b is not constant, but rather has one value in the avalanche region and another value in the forward region. The bulk resistance has been designated R_{b_1} for the forward region and R_{b_2} in breakdown operation. These values generally differ by one to two orders of magnitude.

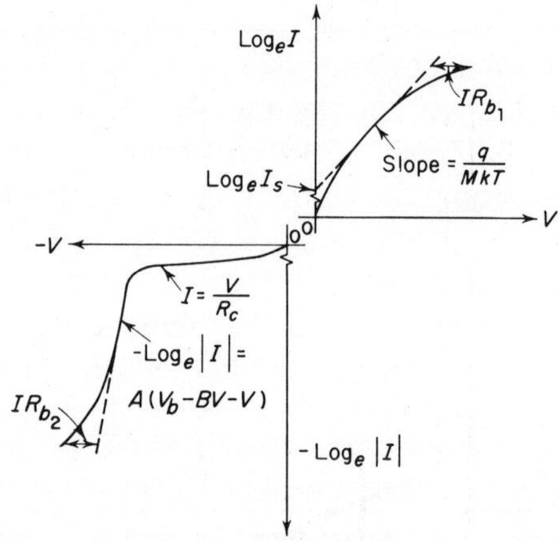

FIG. 7.6-5. Typical Theoretical Static V-I Characteristic.

However, the experimental data does not, in general, agree with the theoretical equations that have been developed. The plot of some typical experimental data is shown in Fig. 7.6-6.

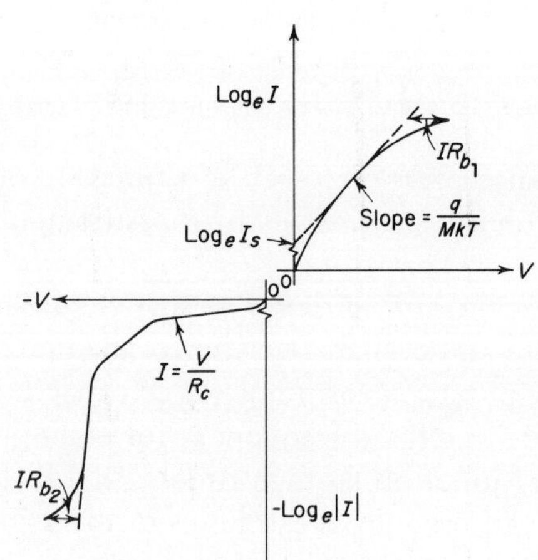

FIG. 7.6-6. Typical Experimental Static V-I Characteristic.

There are two linear regions on the reverse portion of the graph. It appears that the current source which represents the avalanche characteristic should consist of two separate exponential functions instead of only one. Therefore, the equation for the current source is modified to

$$I = I_s(e^{qv/MkT} - 1) - e^{A(V_b - B'v)} - e^{CV_b - Dv} \, . \tag{7.6-2}$$

The model coefficients are related by (7.6-2) and the following equations:

$$I_f = I + \frac{v}{R_c} \tag{7.6-3}$$

$$V_f = v + I_f R_b \, . \tag{7.6-4}$$

The values of I_s, M, and R_b are extracted from the forward V-I curve. R_c is found from the portion of the reverse characteristic where $|v| < .5|V_b|$. Curve-fitting of the data points in the avalanche region produces A, B, C, and D.

A typical curve of junction capacitance versus reverse bias is shown in Fig. 7.6-7.

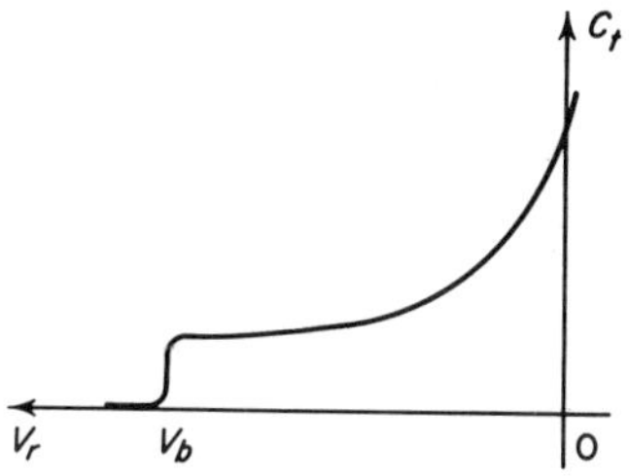

FIG. 7.6-7. Theoretical Avalanche Diode Capacitance Characteristic.

The upper portion of these curves ($V > V_b$) can be curve-fitted to the expression for capacitance of an ordinary diode,

$$C = \frac{K}{(V_z - v)^N} \, . \tag{7.5-7}$$

Equation (7.5-7) is valid only for V greater than V_b (or $|V|$ less than $|V_b|$). After V_b is reached, the capacitance decreases to a small, rather constant, value which is approximately the header capacitance. Equation (7.5-7) can be modified to account for this sudden drop in capacitance as follows:

$$C = C_t + C_h$$

$$C_t = \frac{K}{(V_z - v)^N} \quad \frac{1}{e^{k_1(V_b - v)} + 1} \tag{7.6-5}$$

$$C_h = \text{constant header capacitance}$$

$$K_1 = \text{constant}\,.$$

The voltage at which the capacitance decreases sharply to C_h gives a good definition of the junction breakdown voltage, V_b. Device data sheets normally specify a V_b at some arbitrary I_b; with I_b chosen to make that particular diode look good. It is especially difficult to define an exact point of breakdown from the V-I characteristic of "soft" diodes (i.e., the low voltage diodes which do not have a sharp knee).

The V_b defined from capacitance measurements is a much better reference for the exact point of avalanche breakdown. It seems that this method of specifying V_b could be used to give component users a more uniform definition of junction breakdown.

Tunnel Diode Model

Physical Characteristics – The static voltage-current characteristic curve of a typical tunnel diode is shown in Fig. 7.6-8 along with some of the important design points.

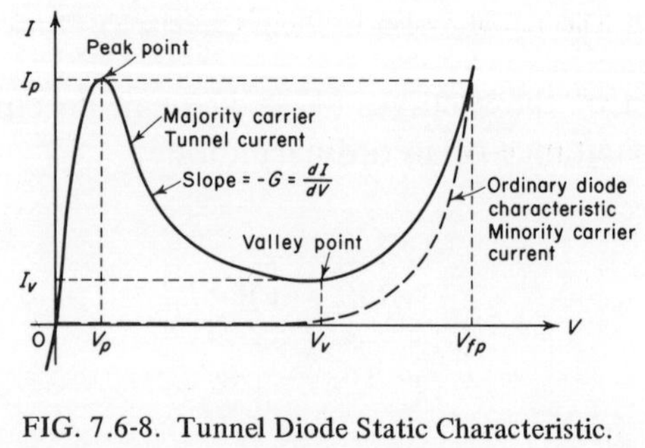

FIG. 7.6-8. Tunnel Diode Static Characteristic.

The unique properties of tunnel diodes which are most often utilized are its negative resistance and its very rapid switching from V_p to a point in the forward conduction region (at a voltage something less than V_{fp}). For voltages greater than V_{fp}, the tunnel diode follows the ordinary diode characteristic.

Most tunnel diode circuits are designed around the peak point and valley point. Any tunnel diode model should accurately approximate these points as well as modeling other important properties. It is relatively easy to measure the peak point, valley point, and the maximum slope in the negative conductance region ($-G$). Measurement of current at voltages greater than V_{fp} is also quite easily accomplished with automatic test equipment.

There is very little resistance to current flow when reverse bias is applied to the junction; i.e., with reverse bias, the tunnel diode is nearly a short circuit. Although models developed in this section are valid for negative bias, it is felt that accurate modeling of reverse behavior is normally of little importance in tunnel diode applications.

Model Topology – The topology assumed to represent a tunnel diode with lumped components is shown in Fig. 7.6-9.

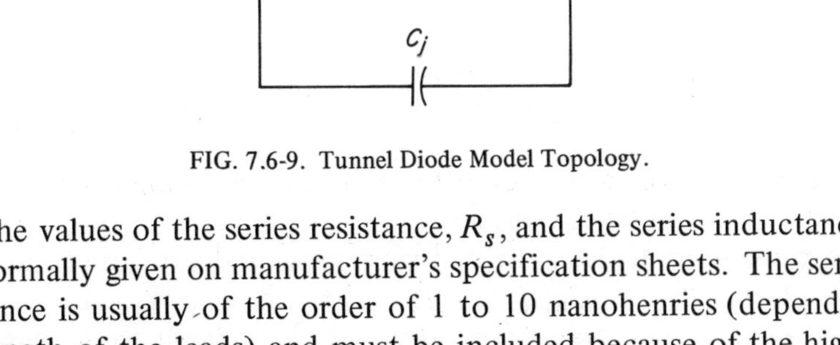

FIG. 7.6-9. Tunnel Diode Model Topology.

The values of the series resistance, R_s, and the series inductance, L_s, are normally given on manufacturer's specification sheets. The series inductance is usually of the order of 1 to 10 nanohenries (depending on the length of the leads) and must be included because of the high normal operating frequencies.

The junction capacitance, C_j, is approximated by the same equation as that used for capacitance of an ordinary diode. Methods for measuring C_j in the forward region and for extracting coefficients of the defining equation from the measured data are discussed later.

The current generator, I, approximates the static V-I characteristic shown in Fig. 7.6-8. Development of a mathematical model includes finding an expression for current as a function of junction voltage, v. It must be possible to obtain any unknown coefficients in the defining equations from external measurements on the device. Methods for extracting the coefficients from measured data are also developed and discussed.

Defining Equations – Several different methods for modeling (I) have been developed. The first one, as postulated in Reference 12, approximates the entire characteristic by one function.

$$I(v) = \text{A}ve^{-av} + B\left(e^{b_1 v} - e^{-b_2 v}\right) + C(e^{cv} - 1) \qquad (7.6\text{-}6)$$

Equation 7.6-6 represents the sum of the tunneling current, the diffusion current, and the excess current, respectively; $I(v) = I_{\text{tun}} + I_{\text{ex}} + I_{\text{diff}}$. The parameters which must be found from external measurements of the diode are A, a, B, b_1, b_2, C and c.

A second method of representing the V-I characteristic is to divide it into several zones and fit the curve in each zone to some mathematical function. One type of function is a polynomial of the form

$$I(v) = A_0 + A_1 v + A_2 v^2 + \ldots, + A_n v^n. \qquad (7.6\text{-}7)$$

Assume the characteristic is divided into k zones and fit to a polynomial $I_j(v)$ of degree n_j in the jth zone.

The polynomials $I_i(v)$ and $I_j(v)$, $(i \neq j;\ i, j \leqslant n)$, representing adjacent zones i and j, are constrained at the boundary voltage (V_b), so that

$$I_i(V_b) = I_j(V_b) \text{ and } I_i'(V_b) = I_j'(V_b), \qquad I_i'(V_b) = \left.\frac{dI_i}{dv}\right|_{v=V_b}$$

This constraint satisfies the condition that all describing equations and their first derivatives be continuous.

The parameters to be extracted from measured data are $A_0, A_1, \ldots, A_{n_1}, B_0, B_1, \ldots, B_{n_2}, \ldots, J_0, J_1, \ldots, J_{n_j}, \ldots, K_0, K_1, \ldots, K_{n_k}$.

Measurements and Data Reduction – It is difficult to measure and record the entire V-I characteristic, including the negative resistance region. Using special low-inductance test fixtures with a CRT display converter, the entire characteristic can be recorded for tunnel diodes with an I_p up to 10 mA [11]. Higher peak-current devices generally cannot be stabilized in the negative region. However, the peak point, valley point, and points in the forward conduction region can be determined.

Although no capacitance measurements can be made in the negative resistance region, sufficient data points are available for $V < V_p$ and $V > V_v$ to determine that a typical C_j versus V curve has the general shape shown in Fig. 7.6-10.

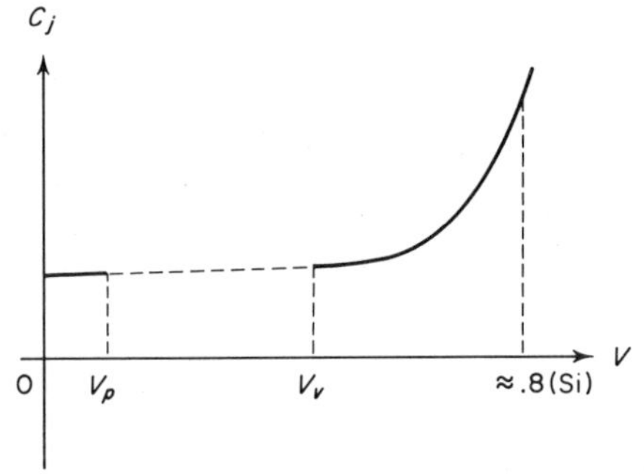

FIG. 7.6-10. Tunnel Diode Capacitance Characteristic.

The primary problem in developing a tunnel diode model is to find methods for extracting coefficients of the defining equations from the measured data. The equations, methods for finding the coefficients, and some of the advantages and disadvantages of each are described below.

Method 1 – Approximation of the *V-I* characteristic by polynomials. Data points taken from tunnel diode curves are curve-fitted to least squares polynomials of specifiable degree. The curve may be divided into as many zones as desired, and in each zone, the degree of the fitting polynomial is chosen arbitrarily.

The curve-fitting is done by starting at the most positive voltage and working back toward the origin. The polynomial for each zone after the first is constrained to match the value and first derivative at a specified point of the previous zone. The zones may overlap, but the constrained point need not lie within the overlap. The polynomial for the last zone is constrained on both ends so that it matches the previous polynomial and also goes through the origin. No attempt is made to fit any data points in the reverse region.

A good choice appears to be division of the curve into three zones, with boundary points at $V_p/3$ and $2V_p$. Using polynomials of maximum degree five, the measured data can be fit to an accuracy of ±3%. Of course, even closer fits can be achieved by using higher-degree polynomials, but the number of coefficients can become prohibitive.

The advantages of this method are that polynomial functions are "well-behaved" and relatively easy to manipulate; polynomial coefficients are easy to extract; and an extremely close fit can be achieved. The main disadvantage is that the entire characteristic curve must be known. As discussed previously, this is always difficult and, for higher current diodes ($I > 10$ mA), virtually impossible to obtain.

Method 2 – Approximation of the *V-I* characteristic by (7.6-6). The entire *V-I* characteristic could be fitted to (7.6-6), but this is not always feasible. First, the entire curve would have to be known as in Method 1. Second, the solution of one equation in seven unknowns can be tedious at best. Although approximations and assumptions can be made to reduce the number of unknowns, a computer search routine to do the job could be relatively inefficient and inaccurate.

It seems that, with what is known about the unique tunnel diode curve and some assumptions, a reasonably good approximation can be made without knowing the entire characteristic. A procedure for obtaining the equation coefficients with only a limited amount of information will now be developed.

Measure four points: the peak point, $[V_p, I_p]$, the valley point, $[V_v, I_v]$, and two points in the forward conduction region, $[V_f, I_f]$ and $[V_m, I_m]$, where $V_m > V_f > V_{fp}$. The last two points are normally chosen so that I_m is near the maximum rated current of the device and I_f is about $1.5I_p$. The value of $-G$ is also needed; either it can be taken from most diode specification sheets or it can easily be measured.

Given

$$I(v) = \text{Av}e^{-av} + B\left(e^{b_1 v} - e^{-b_2 v}\right) + C(e^{cv} - 1). \qquad (7.6\text{-}6)$$

For $v > V_{fp}$, the last term is dominant, so it can be assumed that

$$I(v) = Ce^{cv}, \qquad v > V_{fp}.$$

Using the data points in the forward conduction region, (7.6-7)

$$c = \frac{\text{Ln}\,\dfrac{I_m}{I_f}}{V_m - V_f}. \qquad (7.6\text{-}8)$$

$$C = \frac{I_f}{e^{cV_f} - 1} = \frac{I_m}{e^{cV_m} - 1}. \qquad (7.6\text{-}9)$$

Since the peak point is a relative maximum and the valley point is a relative minimum, the first derivative (slope) of $I(v)$ must be zero at these points. For $v \leqslant V_p$, it can be shown that the diffusion current is negligible $[C(e^{cv} - 1) \approx 0]$. When $v \cong V_v$, none of the three current expressions is normally negligible, although I_{ex} is the majority term. Therefore,

$$I_p = AV_p e^{-aV_p} + B(e^{b_1 V_p} - e^{-b_2 V_p}) \tag{7.6-10}$$

$$I'(V_p) = 0 = Ae^{-aV_p}(1 - aV_p) + B\,(b_1 e^{b_1 V_p} + b_2 e^{-b_2 V_p}) \tag{7.6-11}$$

$$I_v = AV_v e^{-aV_v} + B(e^{b_1 V_v} - e^{-b_2 V_v}) + C(e^{cV_v} - 1) \tag{7.6-12}$$

$$I'(V_v) = 0 = Ae^{-aV_v}(1 - aV_v) + + B(b_1 e^{b_1 V_v} + b_2 e^{-b_2 V_v}) + cCe^{cV_v}\,. \tag{7.6-13}$$

This gives four equations in five unknowns. One more item of information is known, the maximum negative conductance. Since $-G$ is the maximum slope of the negative conductance region, it is at the point of inflection, V_g, of $I(v)$. However, since V_g is not known, some approximations must be made. The diffusion current has very little effect and the excess current is nearly linear in this region. Therefore, assume that the inflection point of Ave^{-av} is also the inflection point of $I(v)$.

$$\frac{d^2}{dv^2}\left[\text{Ave}^{-av}\right] = (av - 2)aAe^{-av} = 0.$$

Since a, A, and e^{-av} are not zero, this implies that $(av - 2) = 0$ and

$$V_g = \frac{2}{a}\,.$$

$$I'(V_g) = -G = Ae^{-aV_g}(1 - aV_g) + B(b_1 e^{b_1 V_g} + b_2 e^{-b_2 V_g}). \tag{7.6-14}$$

Equations (7.6-10) through (7.6-14) constitute a system of equations which must be solved to obtain A, a, B, b_1, and b_2. A method of iteration can be used to obtain the solution as follows:

Assume initial values for B, b_1, b_2. Using these values, define

$$K_1 = B\left(e^{b_1 V_p} - e^{-b_2 V_p}\right)$$

$$K_2 = B\left(b_1 e^{b_1 V_p} + b_2 e^{-b_2 V_p}\right)$$

Then,

$$a = \frac{1}{V_p} + \frac{K_2}{(I_p - K_1)} \tag{7.6-15}$$

$$A = \frac{I_p - K_1}{V_p e^{-aV_p}} \tag{7.6-16}$$

The values of (A) and (a) found above are substituted into (7.6-12, 13, 14).

Define

$$K_3 = I_v - AV_v e^{-aV_v} - Ce^{cV_v} + 1$$

$$K_4 = (aV_v - 1)Ae^{-aV_v} - cCe^{cV_v}$$

$$K_5 = (aV_g - 1)Ae^{-aV_g} - G.$$

In the expression for K_5, V_g is known since $V_g = 2/a$. We now have

$$\frac{b_1 e^{b_1 V_v} + b_2 e^{-b_2 V_v}}{e^{b_1 V_v} - e^{-b_2 V_v}} = \frac{K_4}{K_3} = K_6 \tag{7.6-17}$$

$$\frac{b_1 e^{b_1 V_g} + b_2 e^{-b_2 V_g}}{e^{b_1 V_v} - e^{-b_2 V_v}} = \frac{K_5}{K_3} = K_7. \tag{7.6-18}$$

The last two equations provide a definition of b_1 and b_2. Several iteration procedures for solving such equations exist, such as the Newton-Raphson method, which provides a quadratic convergence to the solution (b_1, b_2). See Reference 10 for a discussion of some methods which may be used.

After b_1 and b_2 have been calculated, a new value of B is found from

$$B = \frac{K_3}{e^{b_1 V_v} - e^{-b_2 V_v}} . \tag{7.6-19}$$

The new B, b_1, and b_2 found are substituted into (7.6-15) and (7.6-16) to calculate better values for A and a. The entire iteration is continued until the parameter values converge within a predetermined accuracy.

The advantages for this method are: (a) the need for only specification sheet information plus two easy measurements in the forward region, and (b) the fit is exact at the two most critical points, peak point and valley point. The disadvantages are that a good fit to the entire characteristic is not always possible and that the five equations relating A, a, B, b_1, and b_2 are difficult to solve; the iteration sometimes doesn't converge. In fact, the iteration for this method usually will not converge for germanium devices.

Unijunction Transistors

Physical Characteristics [14] – The unijunction symbol and nomenclature along with the well-known static emitter characteristic curves are shown in Fig. 7.6-11.

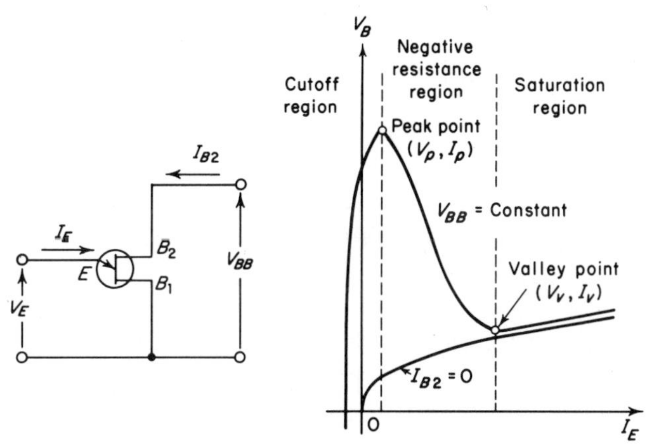

FIG. 7.6-11. Unijunction Symbol and Emitter Characteristics.

The peak point is defined by the expression

$$V_p = v + \eta V_{BB} \tag{7.6-20}$$

where ν is defined by the $I_{B2} = 0$ curve of Fig. 7.6-11. The quantity η, called the intrinsic standoff ratio, will be discussed later. When I_E becomes greater than I_p, the process of conductivity modulation occurs in the semiconductor material between emitter and base-one, causing the negative resistance characteristic. As I_E is increased well beyond I_p there is also a slight conductivity modulation of the $B1 - B2$ region resulting in an increased I_{B2}.

A model which will adequately describe the static behavior of a unijunction transistor will be formulated. The topology is shown in Fig. 7.6-12.

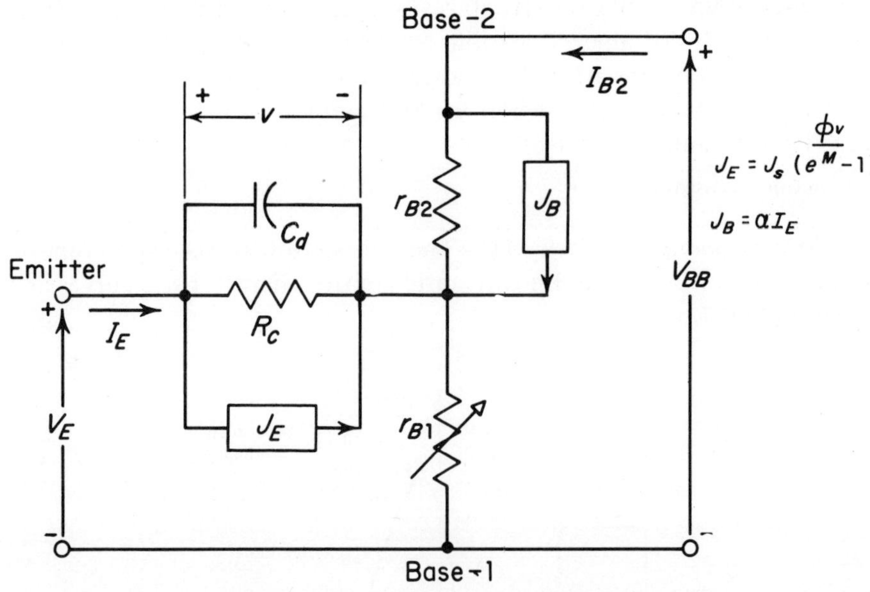

FIG. 7.6-12. Unijunction Model Topology.

The independent variables are V_{BB} (the applied bias) and I_E; since the static characteristic is a voltage function which is single-valued in I_E.

The terminal equations are

$$V_E = \nu + (I_E + I_{B2})r_{B1} \tag{7.6-21}$$

$$V_{BB} = V_{rB1} + V_{rB2}$$

$$= (I_E + I_{B2})r_{B1} + (I_{B2} - J_B)r_{B2}$$

Therefore

$$I_{B2} = \frac{V_{BB} + J_B r_{B2} - I_E r_{B1}}{r_{B1} + r_{B2}}$$

$$= \frac{V_{BB} + I_E(\alpha r_{B2} - r_{B1})}{r_{B1} + r_{B2}} \, . \qquad (7.6\text{-}22)$$

When $I_E = 0$, the resistance between $B1$ and $B2$ (which equals $r_{B1} + r_{B2}$) is designated r_{BB}, the interbase resistance. At the peak point, r_{BB} is divided between r_{B1} and r_{B2} according to the intrinsic standoff ratio. From (7.6-20),

$$V_p - v = I_{B2} r_{B1} = \eta V_{BB} = \eta I_{B2} r_{BB} \, .$$

Therefore

$$r_{B1} = \eta r_{BB} \, . \qquad (7.6\text{-}23)$$

Consequently

$$r_{B2} = r_{BB} - r_{B1} = (1 - \eta) r_{BB} \, . \qquad (7.6\text{-}24)$$

The quantities η and r_{BB} are both dependent on V_{BB}. To a reasonable first approximation, they both show a linear variation. If a reference voltage $V_{BB} = VK$ is chosen ($VK = 10$ volts is usually a good choice), then η and r_{BB} can be written

$$\eta = \eta_{VK} - a(V_{BB} - VK) \qquad (7.6\text{-}25)$$

$$r_{BB} = r_{BB_{VK}} + b(V_{BB} - VK) \, . \qquad (7.6\text{-}26)$$

The constants (model parameters) then are η_{VK} and $r_{BB\,VK}$, (the values at $V_{BB} = VK$), and (a) and (b), normally of the order of .005 and 30, respectively, when r_{BB} is expressed in ohms.

The base-one resistance, r_{B1}, given by (7.6-23) is valid only for very low I_E. Due to the conductivity modulation, for $I_E > I_p$ the base-one resistance decreases logarithmically with I_E. A good empirical approximation is

$$r_{B1} = \frac{r_{B10}}{1\left(+\dfrac{I_E}{I_{E1}}\right)^N} \tag{7.6-27}$$

where $r_{B10} = \eta r_{BB}$, I_{E1} and N are parameters to be determined. Typical values for N and I_{E1} and 0.55 and $5\eta A$, respectively. They are easily found from two measured data points in the saturation region; for example, ($V_{BB} = 10v, I_E = 20$ mA) and ($V_{BB} = 10v, I_E = 50$ mA).

It is assumed that r_{B2} is constant and the conductivity modulation of the $B1 - B2$ region is modeled by the current source J_B. If I_{B2} is measured at one of the above data points (e.g., $V_{BB} = 10v, I_E = 50$ mA), α can be found from (7.6-22).

$$\alpha = \frac{1}{I_E}\left[I_{B2} - \frac{V_{BB} - (I_E + I_{B2})r_{B1}}{r_{B2}}\right]. \tag{7.6-28}$$

The diffusion capacitance, C_d, is found from measurements of stored charge between emitter and base-one, with the emitter forward biased. The stored charge, Q_s, is approximately linear with forward current [15].

$$Q_s = \tau_p\,(I_E + I_s) \tag{7.6-29}$$

where τ_p is the recombination lifetime of the holes in the n-type base-one, I_E is the forward current, and I_s is the junction saturation current. The diffusion capacitance is then related to the stored charge by

$$C_d = \frac{qQ_s}{MkT} = \frac{q(I_E + I_s)\tau_p}{MkT} \tag{7.6-30}$$

The charge stored in a *UJT* can become quite large, e.g., at a forward current of 10 ma., Q_s is about 600 picocoulombs for a 2N2646 and about 950 picocoulombs for a 2N491, giving $C_d = .015\ \mu f$ and $.024\ \mu f$, respectively.

Some measurements of junction capacitance were made on a bridge and the maximum value measured between any pair of terminals was less than 5 pf. When using a *UJT* in most applications (oscillators, trigger circuits, etc.), the junction capacitance is negligible compared to the diffusion capacitance and the values of capacitors in the external circuit. Therefore, C_d is the only A.C. parameter included in the model.

The only remaining parameter in the model is R_c, the leakage resistance. It can be found by applying a reverse bias from $B2$ to E with $B1$ open. The leakage current is normally designated I_{E0} and R_c is the applied reverse bias divided by I_{E0}. (This calculation for R_c assumes $J_E = 0$ for $v < 0$. Actually, $R_c = \text{bias}/(I_{E0} - I_s)$.)

The model postulated above presumes a fixed operating temperature. Nearly all of the model parameters are quite susceptible to temperature variation. Additional information on effects of temperature change are available on most data sheets and application notes.

Field-Effect Transistors

Although field-effect transistors have been used for some time, and are being used extensively at present, there is no one model that will describe all such devices. One reason is that the physical processes causing various modes of operation differ with the type of FET. There are two main types, Junction (JFET) and Insulated Gate (IGFET), sometimes called the "Metal-Oxide Semiconductor (MOS) transistor. This would suggest the use of at least two different models.

A small-signal equivalent circuit [16] for a JFET is shown in Fig. 7.6-13. This model will provide a reasonable approximation of FET characteristics for low frequencies of operation. The capacitors C_{GD} and C_{GS} are dependent on the applied voltages. The value of g_{DS} is the slope of the output characteristics in the pinch-off region. Values r_{DB} and r_{SB} are the bulk resistances of the semiconductor path from the channel edges to the drain and source ohmic contacts, respectively.

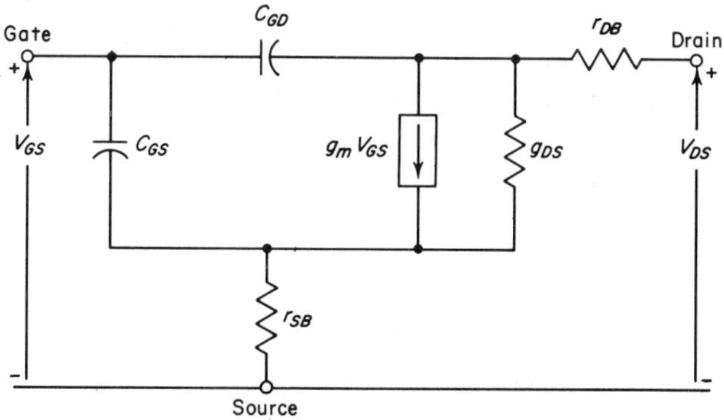

FIG. 7.6-13. Small Signal JFET Model.

A piecewise linear model for a FET has been formulated by Roberts and Harbourt [17]. It is designed for circuit analysis using the ECAP program [18]. However, as indicated in Section 6.1, piecewise linear models have some deficiencies when applied to general circuit analysis programs. A large-signal model for an IGFET, shown in Fig. 7.6-14, has been developed by Hodges and Shichman [19] and has been tested on CIRPAC [20] – a Bell Telephone Laboratories program.

Silicon Controlled Rectifiers

Since the *SCR* is a four-layer, three-junction device with only three leads, it is difficult to describe the internal behavior. The two-transistor analogue [21] shown in Fig. 7.6-13 provides a good intuitive feet for the operation, but it is difficult to obtain the transistor parameters since *B*1 and *C*2 are not accessible.

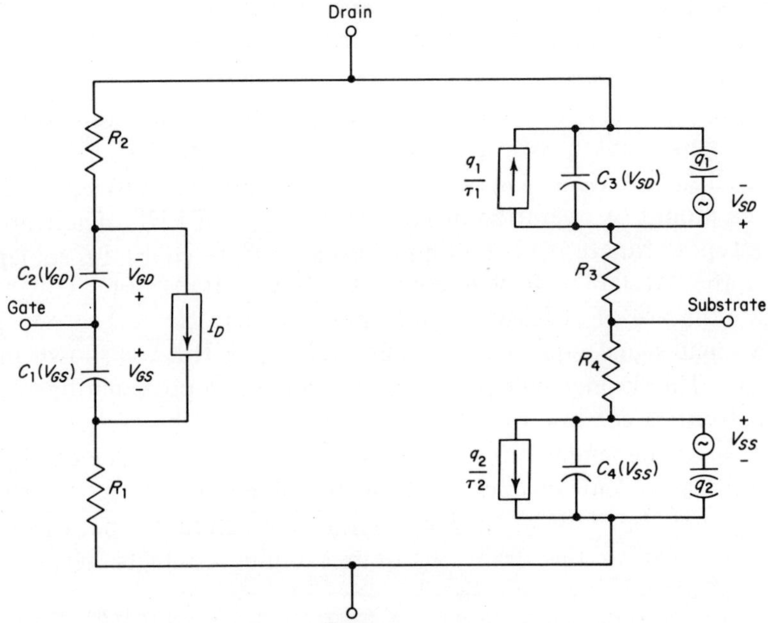

I_D = complex function of V_{CS}, V_{GD}, V_{SD}, V_{SS}, and internal device physics.
R_1, R_2 = series resistance between S and D and the active channel.
R_3, R_4 = bulk spreading resistance.
C_1, C_2 = total gate capacitance, equal for symmetrical geometry.
C_3, C_4 = diode depletion layer capacitance.
q_1, q_2 = diode stored charge.
τ_1, τ_2 = storage time constant.

FIG. 7.6.14. Equivalent Circuit for IGFET.

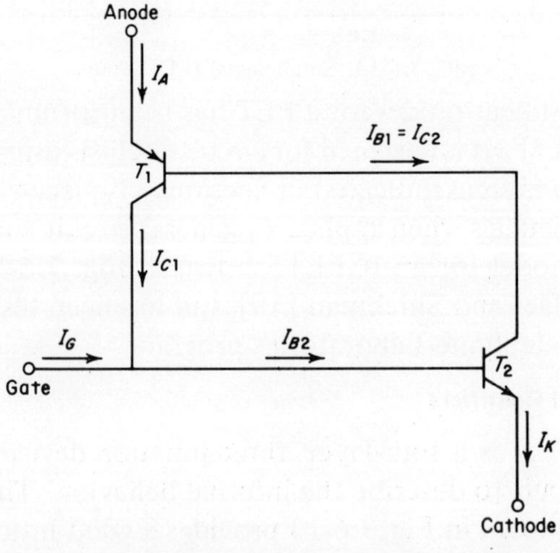

FIG. 7.6.15. Two-Transistor Analogue of Silicon Controlled Rectifier.

A model based on the two-transistor representation has been developed by Steele [22]. He requires that all parameters assigned to the model be obtainable in a simple manner from the specification sheet describing the SCR. It appears, however, that some parameter values are not readily obtainable and, hence, are assumed or fairly well known from experience.

The two-transistor approximation is rather inefficient from the standpoint of computing time required. It should be possible to represent an SCR by empirically relating the internal behavior to terminal characteristics even though the physical processes are somewhat more complex than for some of the devices discussed previously.

7.7 SOLUTION ACCURACY AND MODEL COMPLEXITY

Accuracy of Model

There has been considerable discussion regarding how accurate a model must be; that is, how closely must the terminal characteristics of the mathematical model approximate the actual characteristics of a given device. Probably of greatest importance is the fact that several devices, each with the same 1Nxxx or 2Nxxx number may have certain characteristics which vary over a wide range. It does no good to find a parameter for a given 2NABC to within ± 5% of its measured value if (because of variation in doping levels, wafer thickness, diffusion depth, etc., etc.) the same parameter for a second 2NABC varies from the first device by 40%. By the same token, curve-fitting a set of data points to within ± 1% of their measured values is a waste of computer time if the measured data is accurate only within ± 5%.

The most promising approach to solve the first problem seems to be obtaining "typical" parameters along with their statistical distributions. Probabilistic circuit analysis techniques (Monte Carlo methods, etc.) are then used to find typical and worst-case performances of complete circuits and systems.

Complexity of Model

It should be obvious by this time that the more accurate a model must be, the more elements must be used and/or the more complicated the mathematical equations must be. The addition of either elements or mathematical terms increases the computer time required for solution. The additional mathematical complexity also increases the possibilities for instabilities and non-convergence of solutions. As indicated, though, the increase in solution accuracy may well be worth the extra time and expense required. These are questions which must be answered by the circuit analyst or designer. Some possible answers to the inevitable questions are suggested below.

Trade-Offs of Accuracy vs. Complexity

The model user must look at the application; or the type of circuit in which the device is used. For example, if a tunnel diode is to be used for switching purposes, only the peak point and valley point need to be modeled exactly. On the other hand, if the tunnel diode is used in an oscillator circuit where operation is around a bias point in the negative resistance region, then the negative resistance characteristic may need to be approximated quite precisely.

The desired response of a circuit is obviously more sensitive to changes in values of some elements than others. The same is true for semiconductor models. A small change in one parameter may result in a large variation in the model response. All parameters can be checked for sensitivity by taking the partial derivative of the response with respect to the parameter. However, judicious choices of the parameters to be checked can reduce the computation time considerably. The most sensitive parameters then must be modeled precisely while somewhat more gross approximations may suffice for the others.

The final decision for how precise a particular model must be rests with the user, taking into account the application and the type of analysis desired. It is normally easier to begin with a more detailed model and simplify it (by linearizing elements, by ignoring second-order effects, by omitting some terms in the equations, etc.) than to start with a simple model and expand it.

The models outlined in this chapter are not considered optimum or ideal for every application. They are intended only to be used as starting points for aiding the analysis and design of networks containing such devices.

REFERENCES

1. C. L. Searle, et al., *Elementary Circuit Properties of Transistors*, SEEG, vol. 3, New York: John Wiley & Sons, Inc., 1964. Chapter 3.

2. J. J. Ebers and J. L. Moll, "Large-signal behavior of junction transistors," *Proc. IRE*, vol. 42, pp. 1761-1772, December 1954.

3. J. L. Moll, "Large-signal transient response of junction transistors," *Proc. IRE*, vol. 42, pp. 1773-1784, December 1954.

4. R. Beaufoy and J. J. Sparkes, "The junction transistor as a charge-controlled device," *A. T. E. J.*, vol. 13, pp. 310-327, October 1957.

5. J. G. Linvill, "Lumped models of transistors and diodes," *Proc. IRE*, vol. 46, pp. 1141-1152, June 1958.

6. D. J. Hamilton, F. A. Lindholm, and J. A. Narud, "Comparison of large-signal models for junction transistors," *Proc. IEEE*, vol. 52, pp. 239-248, March 1964.

7. J. J. Sparkes, "Device Modeling," *IEEE Trans. on Electron Devices*, vol. ED-14, pp. 229-232, May 1967.

8. W. A. Blackwell and L. L. Grigsby, "A simple formulation procedure for state-space models of electric networks," *J. Franklin Institute*, vol. 281, pp. 486-498, June 1966.

9. H. W. Mathers, S. R. Sedore, and J. R. Sents, "Automated digital computer program for determining responses of electronic circuits to transient nuclear radiation (SCEPTRE)," IBM Corp., Electronic Systems Center, AFWL-TR-67-124, April 1968.

10. M. E. Daniel and C. H. Purdue, "Computer model specifications for nonlinear devices," *Proc. 1967 Symposium on Reliability*, pp. 162-166, January 1967.

11. M. E. Daniel, "Development of mathematical models of semiconductor devices for computer-aided circuit analysis," *Proc. IEEE*, vol. 55, pp. 1913-1920, November 1967.

12. H. J. Kuno, "Analysis and characterization of p-n junction diode switching," *IEEE Trans. on Electron Devices*, vol. ED-11, pp. 8-14, January 1964.

13. The Boeing Company, "Analytical methods and fundamental parameters for predicting responses of electronic circuits to transient nuclear radiation, with application to hardened circuit design," Tech. Report No. AFWL TR-65-105, p. 4, July 1965.

14. T. B. Bergersen, "Theory and characteristics of the unijunction transistor," Motorola Application Note AN-293, April 1967.

15. M. E. Daniel, "Unijunction transistor model for computer circuit analysis," *IEEE Journal of Solid-State Circuits* (to be published March 1969).

16. L. J. Sevin, Jr., *Field Effect Transistors*, New York: McGraw-Hill Book Company, 1965, p. 41.

17. B. D. Roberts, Jr. and C. O. Harbourt, "Computer models of the field-effect transistor," *Proc. IEEE*, vol. 55, pp. 1921-1929, November 1967.

18. *1620 Electronic Circuit Analysis Program (ECAP) User's Manual.* White Plains, N. Y.: IBM Corporation, 1965.

19. D. A. Hodges and H. Shichman, "Modeling and simulation of insulated-gate field-effect transistor switching circuits," *IEEE Journal of Solid-State Circuits*, vol. SC-3, no. 3, pp. 285-289, September 1968.

20. H. Shichman, "CIRPAC user's manual," Bell Telephone Laboratories Tech. Memo, MM67-2612-1, July 1967.

21. F. E. Gentry, et al., *Semiconductor Controlled Rectifiers*. Englewood Cliffs, N. J.: Prentice-Hall, Inc., 1964.

22. E. L. Steele, "A silicon-controlled rectifier model for computerized circuit analysis," *Proc. National Electronics Conference*, October 1967.

CHAPTER 8

USING ANALYSIS PROGRAMS IN CIRCUIT DESIGN

W. G. MAGNUSON, JR., Lawrence Radiation Laboratory, Livermore
W. J. WALSH, University of California, Berkeley

INTRODUCTION

Analysis, coupled with experimental investigation, is an important step in designing electronic circuits. Fortunately, with today's circuit analysis programs, analyzing relatively complex networks is not as formidable as in the past. Ideally, in computer automated design, the user should be able to input functional performance specifications directly to the program and have as the output a design which fulfills the requirements and even produce perhaps, the layout and integrated circuit masks.

Unfortunately, there does not exist a well defined algorithm for going directly from specification to circuit realization. A designer must propose a network configuration which he thinks (hopes) will meet the specifications and then analyze the proposed design either analytically or/and experimentally. However, electrical network analysis is well-defined and several good computer programs exist today. This chapter discusses five programs which are available and representative of the current capabilities in network analysis programs.

The material in this chapter is organized in an abbreviated user's manual fashion for each of the five programs.

The linear analysis programs (CALAHAN [1], POTTLE [2], and ECAP [3]) are presented together in one section and the large-signal programs (CIRCUS [4] and SCEPTRE [5]) in another section.† The general features of each program are presented, and reasons given for choosing one program rather than another based on individual analysis needs and computer facilities.

One goal in each subsection is to give the user an idea of which programs are best for a given problem, and under what conditions a specific program should or should not be used. The casual reader and nonuser of these programs could well believe that a practical, easy, and final solution to the network analysis problem has been found in computer programs. This, of course, is not true, and greater understanding of the limitations and applications of the various approaches to the network analysis problem is needed. Since most user manuals ignore these limitations this chapter points out some of these limitations. An attempt is made to help prevent the disillusionment that comes to a potential user of computer-based network analysis programs when he encounters the limitations. This disillusionment often obscures the real value of the analysis approach.

The final section of this chapter discusses the system aspects of computer-based analysis programs. The objective is to show how the decision of which program to use depends on:

1. The purpose of the analysis,
2. the machine to be used,
3. the user's experience with computers,
4. the programming support available, and
5. the mode of operation (time-sharing, batch-processing).

Finally, it should be mentioned that many of the comments regarding the analysis programs described in this chapter reflect their present status. The techniques, both numerical and programming, used in formulating the programs are constantly improving and the programs themselves are still being developed to take advantage of these improvements. This chapter reflects then, the programs as they exist today. Certainly many of the restrictions mentioned in the sections following will be eliminated or at least lessened in revised and new programs.

---

† The "Linear Network Analysis Digital Computer Program" and "State-Space Analysis Program" are called CALAHAN and POTTLE respectively in this chapter.

8.1 LINEAR ANALYSIS PROGRAMS

This section describes three general purpose linear network analysis programs: CALAHAN, developed by D. A. CALAHAN, University of Michigan, Ann Arbor, Michigan; POTTLE, developed by C. POTTLE, Cornell University, Ithaca, New York; and ECAP, developed by IBM Corporation, Poughkeepsie, New York.

These programs are representative of the available linear network analysis programs and yet their differences are significant enough to justify a separate discussion of each. As a group, they present two types of network analysis schemes: *Type I* – whose object is to obtain the network function as the ratio of polynomials is represented by the CALAHAN and POTTLE programs; *Type II* – whose object is a numerical determination of a network function is represented by ECAP.

In the following subsections, CALAHAN, POTTLE, and ECAP are discussed from the viewpoints of: their mathematical approach, available features, and computational aspects.

8.1.1 CALAHAN and POTTLE, Basic Formulation

The major difference between the CALAHAN and POTTLE programs is the mathematical formulation used to find the network function as the ratio of polynomials. The CALAHAN program uses a topological approach to obtain the network function, whereas the POTTLE program uses a state-variable approach.

Topological approach – The topological methods used in CALAHAN are the standard techniques used for a passive network described as a two-port [6]. The basic analysis problem of solving the node voltage equations by matrix methods is transformed by the Binet-Cauchy theorem to the topological and combinatorial problem of finding the trees and associated n-tree products of a linear graph. For active networks these comments still apply, except that the n-tree products now have a sign associated with them. The CALAHAN program uses the two-graph approach to determine the sign of the tree products [7] (see also Chapter 2).

Implications of the topological approach – There are a number of advantages and disadvantages to the topological approach. The primary advantage is that the network function, when calculated, will be essentially exact because little or no round-off error is associated with the technique. The principal disadvantage is that large networks have many admittance products to be assembled, thus the computation time can become excessive. A practical network may have more than several hundred thousand trees, and with the present algorithm, a prohibitive amount of computer time is used to find these trees. The upper limit on the size of the network is approximately 15-20 nodes and 20-30 branches. See Chapter 2 for a figure depicting the growth in solution time as the network size increases. All variations of the topological approach have this fundamental limitation of the tree-finding problem or its equivalent in the flow-graph technique, the nth-order loop-reduction problem.

State-variable approach – The state variable techniques used in POTTLE are the standard methods as described in the literature [8,9]. Basically, the program accepts an active/passive network description and generates the state equations which minimally describe the network. The program then solves the transformed state equations for the network function in terms of the complex frequency variable s. To calculate the network function, the program finds the functional matrix inverse of the transformed state transition matrix $(sI\text{-}A)$. At the present time, this inverse is found by the Souriau-Frame algorithm which is very fast but suffers from severe round-off problems.

Implications of the state-variable approach – The primary advantage of this approach is that the mathematical description of the network now has a dimension equal to the number of state variables. This in turn implies a computation time which depends not on the size of the network but rather on the number of state variables. In many instances the state variable method is two or three orders of magnitude faster than the topological method. However, a disadvantage of the approach is its numerical inaccuracy when computing the functional matrix inverse. The round-off problem encountered, can, in most instances, cause the answer to be completely wrong. This is not an insoluble problem, and ways around it are discussed in section 8.3 and in Chapter 3.

8.1.2 Features Available with CALAHAN and POTTLE Programs

The CALAHAN and POTTLE programs offer the user essentially the same output options and have similar input features. The programs will analyze a linear network composed of resistors, capacitors, inductors and active devices. The programs calculate the network function, find the poles and zeros, calculate the frequency response, and permit transient response calculations. The features of each program are described briefly, and although repetitious, will point out differences in the programs.

CALAHAN features – CALAHAN is a FORTRAN II linear network analysis program which calculates one network function per computer run. The network, described as a two-port, may contain resistors, capacitors, inductors, and active elements. The active elements must be described as voltage-controlled current sources. A special feature of the CALAHAN program is its parameter variation capability. This option allows one element to vary and the network function is recalculated for each increment of this element.

Input requirements and output features for CALAHAN – Each element must be described (one per card) by *two node numbers*, an *element type*, and a *value*. The user selects the two-port function to be calculated and supplies bounds on the frequency and time response (if desired). If an element is varied, these values must be supplied. A piecewise linear description of the input waveform is necessary if the transient response is calculated. For each computer run, the CALAHAN program calculates one network function and its poles and zeros of the function. This is repeated for each value of the parameter variation (if any), In addition, the frequency response (magnitude, phase, and delay) is calculated over the given frequency range. All of the output is in tabular form but can be readily converted to graphical form if the user desires.

The circuit in Fig. 8.1 will be used to illustrate the CALAHAN input format. The input data format for the CALAHAN program is shown in Fig. 8.2a. The network being analyzed (Fig. 8.1a) is a broadband amplifier using transistors for isolation and a tunnel diode for gain. The first line (Fig. 8.2a) indicates that the circuit will be a topological input description. The second line gives the number of elements, sources, and nodes, the input node pair, the output node pair, the transfer function desired, the output desired (pole-zeros and a time response in this example), the element to be varied, and finally the number of values the variable element is to assume. The next ten lines are a topological description of the network. Following this description is a line containing the four values for the variable element. The next two lines describe the input waveform which, in this case, is a step input of unity. The last card is simply an end-of-job indicator.

POTTLE features – The POTTLE program is a FORTRAN 63/FORTRAN II linear network analysis program which allows a number of output function calculations per computer run. The network may contain

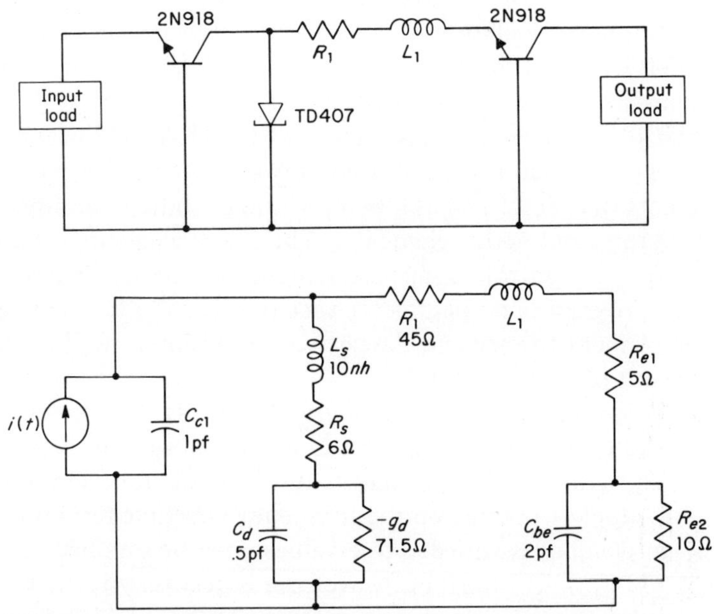

FIG. 8.1. (a) Broadband tunnel diode amplifier circuit using common-base transistors for isolation. (b) Equivalent circuit for computer analysis.

resistors, capacitors, inductors and active elements. The active elements must be described by any of the following controlled sources: voltage-controlled voltage sources, voltage-controlled current sources, current-controlled voltage sources, and current-controlled current sources.

Input requirements and output features for POTTLE – The input requirements for POTTLE are essentially the same as for CALAHAN: a topological description where each element is described by a type, a pair of node numbers, and a value; and an indication of the network functions chosen for calculation. Figure 8.2b shows the POTTLE input for the network of Fig. 8.1. The first card is a title card and following that are ten cards describing network topology. The values for the elements are placed in columns 21 through 40. Columns 17-18 are used to define output ports and column 20 to define controlled sources. A card with a blank in column 1 terminates the network description and on the same card (columns 21-40 and 41-60) frequency and impedance scaling factors are given. Finally the 02 indicates the transfer function and roots are being requested as output and −1 signifies the end of job. The output features of POTTLE are the same as the CALAHAN program except that POTTLE allows several network functions to be calculated simul-

```
01    (CALAHAN)
10    00    07    02    01    07    01    03    03    10    04
02    01    C 1.
02    03    L .0001
03    04    R .006
04    01    C .5
04    01    R -.0715
02    05    R .045
06    07    R .005
07    01    C 2.
07    01    R .01
05    06    L .003
.003        .004        .005        .006
01
1.          0.          1.          5.          .25
00
```

(a)

```
TUNNEL DIODE, COMMON BASE AMPLIFIER (POTTLE)
IG          00 01
L1          01 03                         1.0E-04
R1          03 04              .006
R2          04 00              .5
C1          04 00              -.0715
R3          01 02              .045
L2          02 05                         1.0E+03
R4          05 06                         5.0E-03
R5          06 00  +V          2.
C2          06 00              .01
                               1.               1.
02
-1
```

(b)

FIG. 8.2. (a) Input statements representing FIG. 8.1(b) for the CALAHAN analysis program. (b) Fig. 8.1(b) represented for analysis by the POTTLE program.

taneously whereas CALAHAN permits one only. The network functions are given as ratios of polynomials and poles and zeros are calculated. Frequency response and transient response output can easily be added to the POTTLE program. In fact, the frequency and time response routines from CALAHAN can be linked to POTTLE with little difficulty.

8.1.3 ECAP

The ECAP program is considerably different from the CALAHAN and POTTLE programs primarily because it calculates the network function in numerical form, rather than as a polynomial. From a programming viewpoint, the differences go further than just the form of the solution. ECAP is essentially three separate programs – dc, ac, and transient – all linked by means of an input language. In addition, the ECAP transient program is more complete then either CALAHAN or POTTLE, especially in its use of piecewise-linear elements.

The following sections describe the ECAP program, its basic formulation, its input language, input and output features, and make some general comments on the more important features of the program.

Basic formulation – ECAP's mathematical formulation is described in the literature, and is briefly discussed in the ECAP user's manual. The program assembles the topological, loop impedance, and nodal admittance matrices. These matrices are in turn used in some part of the dc, ac, or transient phase of the program. ECAP's use of the matrices differs from POTTLE in that the elements of ECAP's matrices are numerical while those of POTTLE are functional. Because of this, ECAP has no way of calculating the poles and zeros, but can easily calculate the frequency response.

Input structure of ECAP – ECAP is a general linear network analysis program which performs three analyses – dc, ac, and transient. Each analysis phase of the program permits a number of different solutions. DC analysis calculates nominal solution, partial derivatives, sensitivity coefficients, worst case solution, and standard deviations. AC analysis calculates the nominal solution and permits parameter modifications. Transient analysis calculates the transient response and equilibrium solution.

These three analysis programs have as their common means of communication a special language – the ECAP language. Briefly described, the language consists of six classifications: data, solution control, output specification, command, comment, and system control. Each is described below.

1. *Data* – Data cards contain a code letter (B, T, M, S) when a Branch, Transistor, Mutual inductance, or Switch is being described on the data card. In addition to the code letter, node numbers and an element type and value are also supplied. These cards also contain parameter variation, voltage, and current supply information.

2. *Solution control* – These cards indicate the type of solution desired such as: worst case, standard deviation, sensitivity, etc. This card assures the user that these quantities will be calculated.
3. *Output specification* – These cards indicate which information will be printed as output. The specifications include: element currents, branch voltages, sensitivities, and partial derivatives, etc. The output specification assures the user that those quantities are given as output.
4. *Command* – These cards indicate which analysis (ac, dc, or transient) is desired; is used to modify parameters; and form the beginning and the end of the data.
5. *Comment* – These cards allow the user to insert comments for his own information.
6. *System control* – These cards enable the user to have the output information put on a line printer, console typewriter, or punched cards as he desires

Figure 8.3a presents a network example in order to illustrate the ECAP circuit input data format (see Fig. 8.3b).

ECAP dc analysis – The ECAP dc analysis calculates the dc steady-state solution for a linear circuit composed of resistors, fixed voltage sources, fixed current sources, and dependent current sources. The program performs the following types of calculations: nominal solution, parameter modifications, partial derivatives, and sensitivity coefficients, and worst case analysis. A number of output options are available, these include: node voltages, branch voltages, element voltages, branch currents, element currents, and element power losses.

ECAP ac analysis – The ECAP ac analysis program calculates the ac steady-state solution at a given frequency for a linear circuit composed of resistors, capacitors, inductors, fixed (rms) voltage and current sources, dependent current sources, and coupled mutual inductors. The program calculates the nominal solution and allows parameter modification. The output options are identical with the dc analysis output options.

ECAP transient analysis – The ECAP transient analysis program calculates the transient and equilibrium responses of a linear or piecewise-linear network. If the network is nonlinear, the nonlinear elements must be described in a piecewise-linear fashion by means of switches. A switch simply controls the region of linear approxmiation which is appropriate for a given phase of the solution. The linear portion of the network must be composed of resistors, capacitors, inductors, fixed or

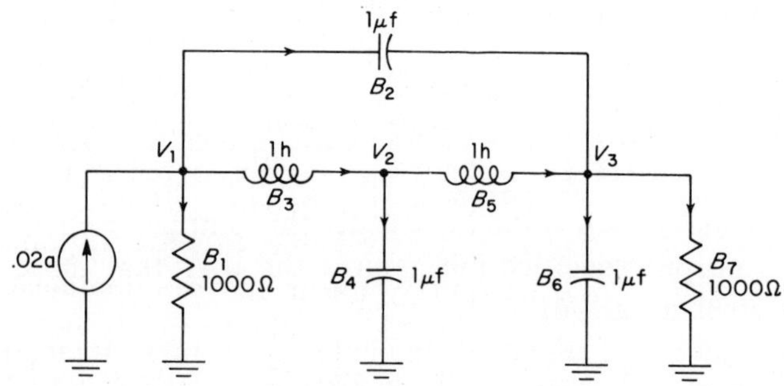

(a)

```
C     BRIDGED-TEE CIRCUIT
C
C
      TRANSIENT ANALYSIS
C
C
B1    N(1,0), R=1000, 1=.02
B2    N(1,3), C=1E-6
B3    N(1,2), L=1
B4    N(2,0), C=1E-6
B5    N(2,3), L=1
B6    N(3,0), C=1E-6
B7    N(3,0), R=1000
      TIME STEP=.01E-3
      OUTPUT INTERVAL=50
      FINISH TIME=2E-3
      PRINT, VOLTAGES
      EXECUTE
```

(b)

FIG. 8.3. (a) Bridged-Tee circuit to illustrate the ECAP analysis program input. (b) Input statements for ECAP calling for a transient analysis.

time-dependent voltage and current sources, dependent current sources, and switches. The output options available include: node voltages, element currents, and switch actuation times.

Comments on the ECAP program – The ECAP program has been widely used and found to be generally quite helpful. The dc analysis is more than adequate as long as the user does not wish to have an exact analysis performed. That is, with the active devices described with linear elements – specifically, using dependent current sources. The ac analysis, as previously discussed, is useful when frequency response is desired. In fact, it will give the frequency response of every element and element parameter, if desired, in one computer run. The transient response program is also useful, but suffers from the time constant problem (see Chapter 3) – as do all numerical integration programs. Although the program permits piecewise-linear elements in the network, at times the ordeal of arranging the proper combinations of switches makes the procedure of dubious worth. Ideally one would prefer to use a nonlinear program (as discussed in section 8.2) rather than ECAP transient. However, if a nonlinear program is not available, then ECAP transient serves to fill the gap between the purely linear and the nonlinear networks. Many users have found that, for their particular needs, only one or two parts of ECAP are of interest. In this case (as will be discussed in section 8.3) the program can be easily decomposed into three separate programs which often has many advantages both in convenience and in computer availability.

8.2 LARGE-SIGNAL ANALYSIS PROGRAMS

This section describes two general-purpose, large-signal network analysis programs: CIRCUS, developed by Dickhaut, Milliman, Massena, and Mong at The Boeing Company, Seattle, Washington; and SCEPTRE, developed by Mathers, Sedore, and Sents at International Business Machines Corporation, Owego, New York. Three subsections are used to describe the programs. Their basic formulation is discussed in order to better understand both the program features and program limitations. Next, the features available in each program are given and finally some of these features are illustrated in the third section through both specific program input language descriptions and an analysis example. It is intended that a description of the features and input language for these programs form a brief user's manual for CIRCUS and SCEPTRE.

CIRCUS and SCEPTRE are just two of several large-signal network analysis programs available to users today. Other programs that might have been included are NET-1, TAG, General Network Analysis Program, STRAP, and TRAC, to mention a few [10, 11]. CIRCUS and SCEPTRE were chosen because they are readily available and represent the present state-of-the-art in terms of program performance.

8.2.1 Basic Formulation

The mathematical formulation of CIRCUS and SCEPTRE is based on the state-variable approach, widely used in major circuit analysis programs. The formulation procedure is described, quite briefly, in the following.

We are generally interested in analyzing networks which may contain semiconductor devices such as transistors, diodes, SCR's, FET, etc. When each of these devices is represented with an equivalent circuit a network consisting of both linear and nonlinear RLC elements and generators is created. It is possible to use equivalent circuits to simulate these devices such that the resulting representation consists entirely of two-terminal elements.

First a particular tree of this network is selected (a tree is a connected subgraph of a network which includes every node but forms no closed loops). When the tree branches are chosen from the circuit elements in the order: voltage sources, capacitors, resistors, inductors, current sources, and an appropriate choice of network variables is made, a set of first-order differential equations can be written down. One set of variables that yields this set of equations is the voltages across the capacitive branches and the currents through the inductive chords of the particular tree. These variables are called the state variables, and together with all inputs, are sufficient to determine all other system quantities at any instant in time.

This formulation is essentially Bryant's method of expressing the state variables and has become standard in most general-purpose network analysis programs [12]. Both CIRCUS and SCEPTRE make use of this method as modified by Wilson and Massena [13].

The solution procedure is shown schematically in Fig. 8.4. A description of methods used in establishing the solution equations and procedure for solving these equations is given in Chapter 3. A repetitive evaluation of the transient solution equations and subsequent integration of the state-variable derivatives produces the time-history of the network response. At each integration time step, several algebraic matrix equations must be evaluated.

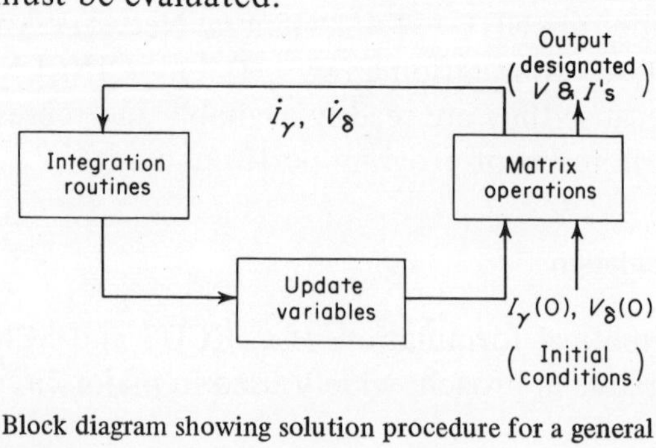

FIG. 8.4 Block diagram showing solution procedure for a general network analysis program.

The formulation of a general network analysis computer program can be divided into two phases: (1) obtaining a set of simultaneous ordinary differential equations that describe a general network, and (2) establishing a numerical procedure to solve the network equations. The block diagram of Fig. 8.4 illustrates the two primary factors that determine the amount of computer time that will be used for a transient solution. The number of solution steps in the solution will depend on the network time constants and the integration routine used. The amount of time per step will depend on the efficiency of the solution of the algebraic equations and to a lesser degree also on the integration method used [14].

Network Size and Running Time – In solving the algebraic network equations, several matrix operations are required, in particular: addition, subtraction, transposition, inversion, and multiplication.

Each of these operations is performed numerous times at each integration time step. The time required to complete each matrix operation is a function of the matrix size. Figure 8.5 depicts how the amount of computer time required for each operation varies as a function of matrix size or order for rather simple, unsophisticated matrix programs, written in FORTRAN, and run on a CDC 3600 computer. Timing for the operations was obtained from an internal clock on the computer and repeating each operation 1000 times to obtain a reasonable time difference.

From Fig. 8.5 it is obvious that compared to matrix multiplication and inversion, the other matrix operations require negligible computer

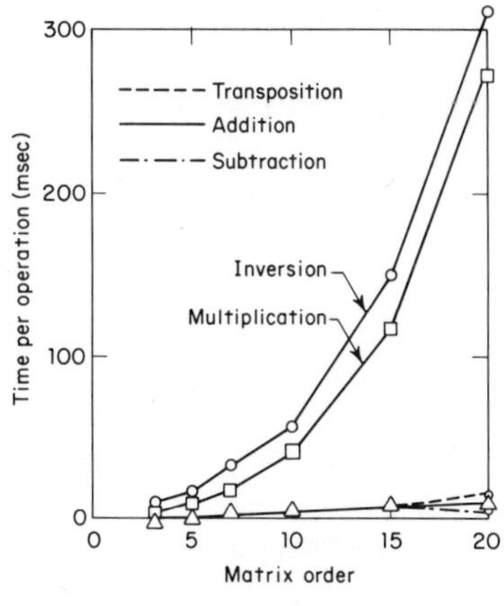

FIG. 8.5. Time per matrix operation versus matrix size. The matrix operations were performed with simple FORTRAN programs on a CDC 3600 computer.

time in the total solution process. A log-log plot of the time per operation versus the order of the matrix shows that, for multiplication, the time varies as the matrix order to the 2.7 power, and for matrix inversion as the matrix order to the 2.5 power. As an approximation, the inverse of a matrix takes about twice the time multiplication does up to matrix sizes of approximately ten by ten.

If we assume that, for a typical solution pass, the number of matrix operations is approximately the same for each operation, then the matrix multiplications and inversions will constitute the major part of the solution time. As the size of the network increases (i.e., contains a greater number of network branches and nodes), the size of the matrices increases. Consequently, the solution time will increase also and in approximately the same manner as the matrix inversion and multiplication time does. That is, the solution time should depend on the network size to approximately the same power as inversion and multiplication.

Two networks were used to verify the above statement. The first consisted of a common-emitter amplifier in which identical stages were successively added in cascade. Each stage added two nodes to the network. A plot of the program solution time versus the number of nodes for this circuit is shown in curve (a) of Fig. 8.6. For this curve, the following relationship holds:

$$\textit{Program solution time} \propto [\textit{Number of network nodes}]^{2.1} \qquad (8.1)$$

where α means 'proportional to'.

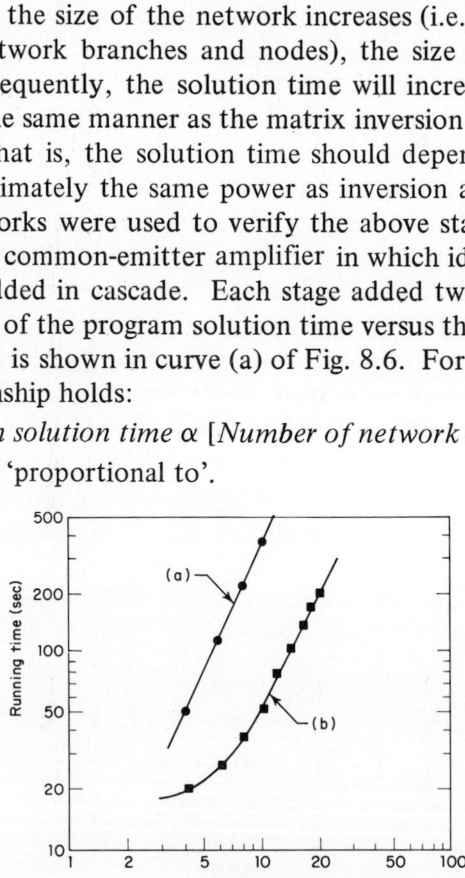

FIG. 8.6. Computer running time versus the number of network nodes. The PREDICT network analysis program was used for the analyses.

The second case consisted of cascading identical stages of a series-parallel RLC network. The solution time versus the number of network nodes is shown in Fig. 8.6, curve (b). Below solution times of approximately 50 sec, the influence of the faster matrix operations (addition, subtraction) and the programming transfer operations (calls to subroutines) cause a departure from (8.1). Solution times above 50 sec show a relationship that closely agrees with (8.1).

Interpreted simply, (8.1) means that doubling the circuit size (with other factors remaining constant) will increase the running time by a factor of approximately four.

Network Time Constants and Running Time – In addition to the solution time that arises when solving the matrix equations themselves, the network time constants and the integration routine utilized play a fundamental role in determining the solution time. Both CIRCUS and SCEPTRE make use of an exponential integration routine [15, 16] that provides integration step sizes of three to four times as large as the Runge-Kutta method [17, 18]. The machine time per step for the exponential method is about the same as for the Runge-Kutta. Consequently the exponential method can evaluate a given transient solution in one-fourth to one-third the time taken by a fourth-order Runge-Kutta method. The value of the shortest local time constant is computed in evaluating the matrix exponential and is used to maximize the time step.

Program running time, or solution time, is defined as the time required for the repeated solution of the network equations. It does not include the computer time used in formulating the network equations in matrix form or the reading out of the results by the computer. Problem time is defined in terms of the scaled time units the analysis is using and is the amount of time for which the solution is wanted.

Two examples show the nature of the variation of solution time to network time constants. The first is a simple two-stage RC circuit [curve (a) in Fig. 8.7] in which the analysis is performed for different values of

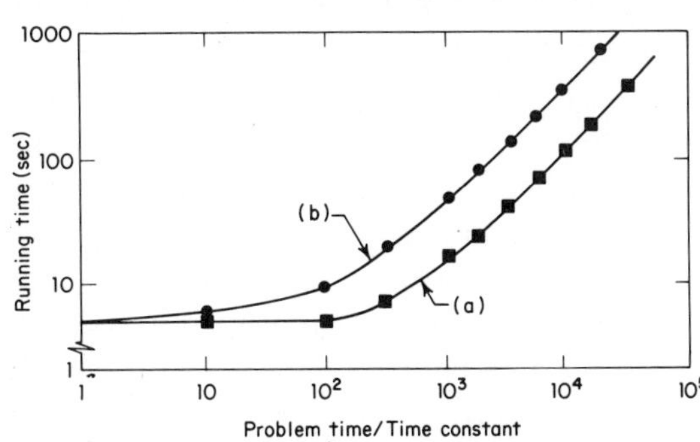

FIG. 8.7. Computer running time versus the problem time divided by the smallest circuit time constant. The problem time is four times the largest circuit time constant. The PREDICT program running on the IBM 7094 computer was used.

the predominant RC time constant. The solution time versus the problem time divided by the smallest time constant is shown in Fig. 8.4.

For solution times less than approximately 6 sec, most of the time is taken up by the mechanics of the computer program; for example, in the transfer of subroutine arguments in call and return statements. The solution time, for times greater than 6 sec, exhibits a near linear relationship to the abscissa. From curve (a) of Fig. 8.7 this relationship is

$$\textit{Program running time} \propto \left[\frac{\textit{Problem time}}{\textit{Smallest time constant}}\right]^{0.9} \tag{8.2}$$

In (8.2), a proportionality exists rather than an equality because, in general, the network being analyzed may contain any number of elements and nodes. In that event, the expression could be generalized to a form

$$\textit{Program running time} = k\left[\frac{\textit{Problem time}}{\textit{Smallest time constant}}\right]^{0.9} \tag{8.3}$$

where k is different for each network and depends on the complexity and size of the network. In (8.3), we assume the circuit contains a time constant that predominates over the other time constants (i.e., is much less than) and over other circuit natural frequencies. Two values of k are depicted by curves (a) and (b) in Fig. 8.7. Curve (b) represents the analysis results for a network containing seven more branches than that for the network of curve (a). Both networks, however, have a single time constant that governs the analysis time. Notice, however, that both curves have the same slope, indicating the same exponent for Eq. (8.3). The value of k is different though, for the two cases. In curve (a) $k = 0.023$ and in curve (b) $k = 0.097$.

Limitations of Existing Programs – In practice, most state-variable based circuit analysis programs share limitations that often severely restrict their usefulness. These limitations stem principally from two causes: 1) numerical errors including round-off errors and truncation errors, and 2) the time-constant problem generated by the network being analyzed. Often difficulties arising from round-off errors can be overcome by using better algorithms and double-precision arithmetic. The second limitation, the time-constant problem, arises when a time-domain response is being sought.

From a user's point of view, the significance of the time-constant problem is that many practical networks modeled for analysis, using the state-variable approach such as in CIRCUS or SCEPTRE, require excessive computer time. If the ratio of the largest to the smallest time constant is of the order of 10^4, then a solution time representing four times the largest time constant is appreciable. A preelimination of some network elements reduces the time-constant problem, but there are still many networks that cannot be evaluated by a CIRCUS-SCEPTRE-type program.

8.2.2 Program Features Available

Both CIRCUS and SCEPTRE were designed relying on several years of experience with general-purpose, network-analysis programs. SCEPTRE is a direct descendent of the PREDICT program [19] which found general dissemination throughout the country in 1964. CIRCUS was influenced by the earlier NET-1 program [20], became available in 1964. Because both the CIRCUS and SCEPTRE programs were designed to perform automatic transient analysis of electronic circuits made up of resistors, capacitors, inductors, semiconductor devices, and voltage and current sources in any arbitrary topology, they have several common features. In this section the principal program features are described separately at the risk of some repetition. The following section is more specific in regard to program input format and program usage.

CIRCUS Features

CIRCUS is a digital computer program which provides the electronic circuit designer with a means for obtaining both the dc initial conditions and transient solution for networks containing bipolar and field-effect transistors, conventional, tunnel and zener diodes, and 4-region devices. CIRCUS accepts as input any arbitrary network configuration made up of the above devices, resistors, capacitors, inductors, and sources.

Although developed specifically for studying nuclear radiation effects on electronic circuits, CIRCUS is applicable to the broad class of design problems that have a network analogue. CIRCUS has several highly useful features. Some of these features are [17]:

Stored Semiconductor Models and Device Parameter Library – One has only to specify the terminal connections of a given device and either (a) request the program to obtain the device data from a library or (b) supply the appropriate device parameters as part of the input. Unlike

SCEPTRE the models are programmed into CIRCUS and the user has little control over the model – only the model parameters. The advantage of the restriction of device models is the ability to increase the computation efficiency by using a knowledge of the model to simplify computation and speed up convergence.

Special Purpose Parameter Variations and Multiple Analysis – CIRCUS allows, in one computer run, multiple analyses of a given topology with arbitrary changes in any or all element values. CIRCUS permits a change in parameters at any predetermined time during an analysis. This feature is particularly useful in making changes in parameters affecting short time constants in low frequency circuitry after fast transients have passed, allowing for more rapid calculation.

Output Options – In addition to any or all node voltages and element currents, the user may request (1) internal device currents and voltages, (2) a monitor on voltage, current, or power dissipation in any element, (3) the sum or difference of any two node voltages, or (4) any variable minus its dc value. The variables selected to be printed will be printed only at intervals selected by the user. CIRCUS also has provisions for plotting any of the above variables versus time.

Computational Efficiency – CIRCUS has several programming features that make the program computationally fast. As pointed out above, CIRCUS takes advantage of a knowledge of the built-in models to provide efficiency. The exponential integration method was first used in large circuit analysis programs by CIRCUS and provides solution economy.

Machine Independence – Ninety-eight percent of CIRCUS is in FORTRAN IV and the program has been adapted to a variety of computers. For several years people claimed that some of the necessary operations in a large circuit analysis program could only be done in machine language. Both CIRCUS and SCEPTRE have shown that efficient programs can be implemented without resorting to assembly language.

Additional Features – As with most analysis programs today, CIRCUS has an input language that is problem oriented and natural to a circuit designer. Several special features incorporated in the program make it especially useful in design studies. Some of these features are: Multiple reruns with parameter changes, restart mode to continue analyses, stored models for convenient device descriptions, and relatively error free operation.

SCEPTRE Features

SCEPTRE is the most general of the available general-purpose, time-domain network analysis programs. The feature that distinguishes it from other transient-analysis programs is its flexibility in the representation of network elements. The solution procedure of SCEPTRE is made up of a two step process. The first step (which can be bypassed) solves for the initial conditions of a circuit. The second step (optional also) then uses the initial conditions and performs a transient analysis.

In performing an analysis, SCEPTRE first reads in the data and generates a FORTRAN program. This program is then compiled and its execution provides the initial conditions and transient solutions. On the IBM 7094 computer versions of the program, the first phase of writing the FORTRAN program uses approximately four minutes of computer time and represents a high overhead cost when analyzing simple circuits.

SCEPTRE has several extremely useful features which will be described briefly next.

Optional Stored Models – Equivalent circuits of semiconductor devices may be stored either permanently or temporarily by a user of the program. Further, any element or interconnected group of elements that can be described as a combination of passive elements, sources, and mutual inductances may be stored on tape and called into use at any point in a network. This feature is not limited to equivalent circuits of active devices but may include biasing networks and filter sections, for example.

Multiple Analyses – In studying circuit behavior for the purposes of optimization of performance, usually a designer is interested in the response changes when one or more components or forcing functions are varied. SCEPTRE, like CIRCUS, has provisions for repeated analyses for a given network topology in an efficient manner.

Selected Analyses and Automatic Termination – SCEPTRE can be operated in the dc mode (initial conditions), transient mode, or in a combination of the two. This permits the user to insert initial conditions for special studies. In addition, analyses may be terminated automatically depending on the behavior of specified network quantities.

Output Options – Although time is the usual independent variable in transient analyses, SCEPTRE allows considerable flexibility in both dependent and independent variables. For example, the current through a capacitor as a function of the voltage across it is allowed. In addition, the user is permitted to attach a unique label to the output quantities. Normally, SCEPTRE prints a maximum of 1000 spaced points for each output requested. However, this may be changed by the user for particular applications.

Computational Features – When the describing equations for the network response are being set up by SCEPTRE, the rows and columns of the incidence matrix are scanned and a reduced incidence matrix is then formed. This decreases the computation time during the solution process. Although the newer exponential integration technique is normally used, SCEPTRE also permits the user to specify either a Runge-Kutta or a modified trapezoidal method to be used.

Program Flexibility – The transient portion allows the description of linearly dependent voltage or current sources as functions of resistor voltages or currents thus encompassing applications involving small-signal equivalent circuits. FORTRAN subroutines may be inserted into SCEPTRE runs thus providing expanded program flexibility. A special section of the program enables the user to define and solve systems of linear or nonlinear first-order differential equations. Finally, SCEPTRE's feature of allowing elements and sources to be represented in the form of constants, tables, or analytic expressions, as well as expressions which combine these forms, yields a very powerful simulation program [18].

8.2.3 Program Description

Communication between a user and a circuit analysis program takes place through statements in a program-oriented language. Usually a general-purpose network simulator computer program requires several thousand instructions. Consequently, the program is stored on magnetic tape or disk. A user then uses a small starter deck (when operating in a batch processing mode) in conjunction with a problem describing data deck. The input formats for CIRCUS and SCEPTRE differ in several aspects and are described separately to avoid confusion. Many steps in the preparation are, however, common to both programs. For example, the following prerequisites must be met:

1. The network must be composed of resistors, capacitors, inductors (including mutual inductance), and voltage and current sources. All active devices, built-in or otherwise, must be represented by equivalent circuits made up of the above elements.
2. All the network nodes are labeled.
3. A consistent set of units must be used.

4. The network topology must be described.
5. The system control statements (what kind of analysis, how long the analysis is to be performed, etc.) must be given.

In conjunction with input requirements and description for the two programs, an example is used to illustrate the procedure required of a designer using the programs. The example is a two-stage, positive-feedback, emitter-coupled amplifier (Schmitt trigger circuit) generally used as an amplitude discriminator [21]. This circuit exhibits hysteresis when the loop gain is greater than one. The object of the analysis in this example is to investigate the static transfer characteristic (V_{output} versus V_{input}). The circuit is shown in Fig. 8.8.

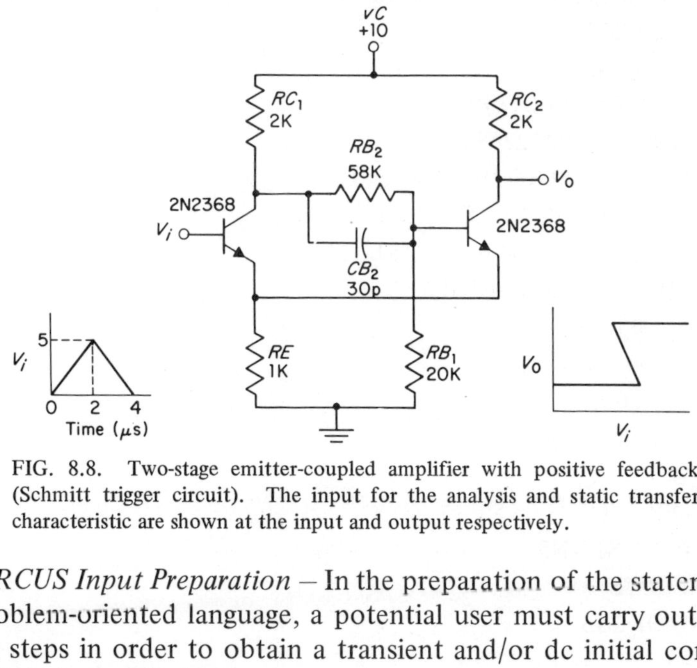

FIG. 8.8. Two-stage emitter-coupled amplifier with positive feedback (Schmitt trigger circuit). The input for the analysis and static transfer characteristic are shown at the input and output respectively.

CIRCUS Input Preparation – In the preparation of the statements in the problem-oriented language, a potential user must carry out the following steps in order to obtain a transient and/or dc initial conditions analysis (refer to Fig. 8.9):

1. *Network Description* – A set of statements (one per component) is written giving the label, node numbers, and value or model number for each component. Each component requires a series of comma-separated entries on the data statement. Blanks are ignored. Values can be positive or negative. Node numbers in CIRCUS must be numbered sequentially beginning with 0 which is designated as ground. Each component (resistor, inductor, transistor, etc.) is given a unique label. The current and voltage

```
'POSITIVE FEEDBACK AMPLIFIER (CIRCUS)'
USE MODEL LIBRARY
CB2,2,4,30.E-12
RE,3,0,1000.
RB2,2,4,58000.
RB1,4,0,20000.
RC1,6,2,2000.
RC2,6,5,2000.
VC,0,6,10.
PVI,0,1,0,5,0,2.E-6,0,2.E-6,5.E-6
T1,1,2,3,2T2368
T2,4,5,3,2N2368
DEVICE PARAMETERS
TRANSISTOR,2T2368,NPN,
   RB,62,RC,20,RE,.01,A1,4.9E-12,
   PHI1,1,N1,.29,A2,5.E-12,PH12,
   1,N2,.23,IES,5.E-16,ICS,2.9E-14,
   THETAN,39.6,THETAI,32.2
   BN,.001,30
   BI,.001,.05
   TCN,.001,3.E-10
   TCI,.001,.1E-6
END
INTERVALS,.1E06,5.E-6
PRINT,VN1,VN5
EXECUTE
END OF JOB
```

FIG. 8.9. Input statements representing the Schmitt trigger circuit for the CIRCUS analysis program. The parameters for T1 are supplied as input while those for T2 are extracted from the library.

convention assumes that when node n_2 is greater than n_1, positive current flow is from the higher numbered node to the lower. The set of units implied in CIRCUS is the volt, ampere, ohm, farad, henry, and second. Any consistent set of units will work equally well (however, the labeling of the output is based on the above set of units).

2. *Model Parameters* – Whenever the model library is not used in describing a semiconductor device, a set of statements defining the parameters needed for the built-in equivalent circuit must be supplied. In Fig. 8.9 T_2 uses the library and T_1 information is supplied.
3. *Display Parameters* – Three statements control the circuit variables to be printed and plotted and control the intervals the variables are to be output. The INTERVALS card also controls the duration of the analysis. If this statement is omitted, only the dc initial conditions solution is performed.
4. *Execution Control* – An EXECUTE statement signals that the data are complete and initiates processing. It is possible to change circuit parameters and re-execute without having to completely respecify the circuit by following the EXECUTE card with the data specifying element to be changed and its new value. Internal parameters of semiconductor devices may also be changed for reruns. Also, as mentioned earlier, in some cases it may be desirable to change a component value during the transient solution, for example in order to bypass a time-constant problem. This is achieved by including a HOLD FINAL CONDITIONS statement followed by parameter and interval changes. When CIRCUS encounters a data statement containing END OF JOB, it terminates the run.

A complete discussion of the problem preparation and coding procedure is given in the CIRCUS user's manual [4].

SCEPTRE Input Preparation – Figure 8.10 shows the required data statements to describe the Schmitt trigger circuit in the SCEPTRE format. Here again one of the transistors is described and the other extracted from the model library. In the SCEPTRE case, the transistor equivalent circuit must be supplied as well as the parameter data as only the diode equation is built-into program. The input consists of two groups; a model description and network description. The information in the problem deck has five header cards in the listing and under each heading the data are coded in the following manner:

1. *ELEMENTS* – The names of all circuit elements (resistors, capacitors, inductors, mutual inductances, voltage and current

```
MODEL DESCRIPTION
MODEL 2N2368T (TEMP) (B–E–C)
ELEMENTS
CE,1–E=EQUATION 1(3.7,1.1,VCE,.34,1.2521E–6,40.1,VCE)
CC,1–2=EQUATION 1(3.3,1.1,VCC,.1,1.7404E–2,29.4,VCC)
JE,1–E=DIODE EQUATION(2.6021E–6,40.1)
JC,1–2=DIODE EQUATION(1.6913E–5,29.4)
JI,E–1=0.23664*JC
JN,2–1=0.97778*JE
RB,B–1=.055
RC,C–2=.016
FUNCTIONS
EQUATION 1(A,B,C,D,E,F,G)=(A/(B–C)**D+E*EXP(F*G))
CIRCUIT DESCRIPTION
POSITIVE FEEDBACK AMPLIFIER (SCEPTRE)
ELEMENTS
CB2,2–4=30.
RE,3–0=1.
RB1,4–0=20.
RB2,2–4=58.
RC1,6–2=2.
RC2,6–5=2.
T1,1–3–2=MODEL 2N2368T
T2,4–3–5=MODEL 2N2368 (PERM)
EC,0–6=10.
EIN,0–1=TABLE 1
DEFINED PARAMETERS
PVN1=TABLE 1
PVN5=(EC–VRC2)
OUTPUTS
PVN1,PVN5,PLOT
PVN5(VOUT),PLOT(PVN1(VIN))
FUNCTIONS
TABLE 1
0,0,2.E3,5,4.E3,0,5.E3,0
RUN CONTROLS
STOPTIME=5.E3
INTEGRATION ROUTINE=TRAP
RUN INITIAL CONDITIONS
MINIMUM STEP SIZE=1.E–20
MAXIMUM INTEGRATION PASSES=1.E6
END
```

FIG. 8.10. Input statements representing the circuit of Fig. 8.8 for the SCEPTRE analysis program. Note that by using defined parameters, the apparent static transfer characteristic can be plotted directly.

sources), their connection nodes, and their values are coded in the format: (ELEMENT), (FROM NODE) – (TO NODE) = (ELEMENT VALUE). The positive current flow is specified by the form – to node ordering. If an element has a variable value specified by some function in either tabular or equation form the coded entry will be, for example, TABLE 1 or EQUATION 1 instead of a value. The table or equation must then be given under the FUNCTIONS header card.

2. *OUTPUTS* – All voltages, current, or other defined quantities the user wants printed or plotted are listed under this heading. Independent variables other than time may be printed and the labels changed as evidenced by the example.
3. *INITIAL CONDITIONS* – If the initial conditions are unknown, or are zero, then this card can be omitted and the program automatically computes them upon request.
4. *FUNCTIONS* – This section is used to define explicitly all tables and equations that were referenced under the ELEMENTS heading card. The general format for tables includes a statement giving the name and independent variable followed by statements giving data pairs separated by commas. Equations use dummy arguments with the variables being fed through in a subroutine fashion.
5. *RUN CONTROLS* – Although SCEPTRE requires only the problem time to be supplied, there are provisions for inserting integration control constants as well as specifying the integration routine to be used itself.

A complete discussion of the SCEPTRE input language is given in the user's manual [5].

As might be expected, the accuracy of a circuit simulation is governed not only by the numerical methods the program is based on but on how representative the solid-state equivalent circuits are. When typical values are used for parameters, the disparity between computed and laboratory results generally are less than five percent. If actual device measurements are used, then the agreement will be very good.

Both CIRCUS and SCEPTRE have many diagnostic messages built-in them. For the most part, the messages will be self-explanatory, but occasionally the user's manual must be consulted and sometimes, when no clue is given, the user must repeat the analysis.

8.3 COMPUTATIONAL ASPECTS OF NETWORK ANALYSIS PROGRAMS

Sections 8.1 and 8.2 have presented five network analysis programs which are available and in use today. The general features of these five programs are summarized in Table 8.1. This information alone however,

TABLE 8.1

Major features, capabilities, and program details for the network analysis programs described in this chapter.

	CALAHAN	CIRCUS	ECAP	POTTLE	SCEPTRE
Program input features					
Must use specific labels	No	Yes	Yes	Yes	Yes
Sequential node numbering	Yes	Yes	Yes	Yes	No
Voltage or current sources	Current sources only	Yes	Yes	Yes	Yes
Tabular input for branch description	No	No	No	No	Yes
Tabular input for signal sources	Yes	Yes	Yes	Yes	Yes
Analytic description of branch elements	No	No	No	No	Yes
Analytic description for signal sources	Yes (available from a modified subroutine)	No (has several built-in signal sources)	No	No	Yes
Automatic modifications of input (repeated runs)	Yes	No	Yes	No	Yes
Modeling capabilities					
Built-in models	No	Yes	No	No	Yes
Allows small-signal models	Yes	No	Yes	Yes	Yes
Allows large-signal models	No	Yes (built-in)	Yes (inconvenient)	No	Yes
Output options					
Transfer functions	Yes	No	No	Yes	No
Pole-zero locations	Yes	No	No	Yes	No
Symbolic expression for time response	Yes	No	No	Yes	No
Time-response output	Output node only – at fixed intervals	Node voltages and semiconductor voltages and currents	Selected variables at fixed intervals	Output node only at fixed intervals	Selected variables at fixed intervals
Steady-state solution	No	Yes	Separate analysis	No	Yes
Programming features					
Programming language	FORTRAN II, IV	FORTRAN IV	FORTRAN II, IV	FORTRAN II,63	FORTRAN IV
Memory capacity recommended	32 000	32 000 (uses overlays)	32 000	32 000	32 000 (uses overlays)
Network formulation	Topological	State-variable	Nodal	State-variable	State-variable
Primary integration routine	Runge-Kutta (and inverse Laplace)	Exponential	Implicit numerical integration	Runge-Kutta (and inverse Laplace)	Exponential

is not sufficient for choosing a network analysis program. The choice of an analysis program should be made after consideration of available computer support. Needless to say, if we have the world's largest computer facility and an unlimited supply of programming support, then no problems exist. However, if not, then we must decide on programs which will be compatible with our computational structure.

This section discusses five aspects of computational facilities which are important and then relates them to the programs discussed in sections 8.1 and 8.2. It is doubtful that a reader will find his particular situation outlined here, but he will become aware of the possible pitfalls and judgment errors that can be made.

Purpose of the analysis – From the discussions in sections 8.1, 8.2 and Table 8.1, a user should have a fair idea of which programs are best suited for his application and those which are not. For example, a person interested in the poles and zeros of a linear network would not be interested in ECAP, CIRCUS, or SCEPTRE, but would be interested in POTTLE and CALAHAN. Whereas, if time response were of primary interest, then any of the five programs described in sections 8.1 and 8.2 might be desirable. The considerations to be discussed next, will probably make the list of suitable programs a little smaller, or at least change their order of importance.

FORTRAN Compiler – All of the programs discussed, with a minor exception in CIRCUS, are written in FORTRAN. Since all major computer systems have a FORTRAN compiler supplied with the system, this usually presents little difficulty. However, it must be emphasized that the FORTRAN compiler should be a tested compiler and preferably not one that is under development. Even if the compiler has been well tested, it has been found that, under certain conditions, a compiler will generate incorrect coding for the FORTRAN program. This is particularly true in the compilers on newer computers which attempt to optimize the coding and in doing so, occasionally create incorrect coding. It is advisable to run sample problems for which the solutions are known and which exercise as many of the available features of the program as possible.

Memory and Computer Word Size – The memory size available to the user is important in determining which programs will or will not fit into the machine without resorting to programming techniques such as segmentation and overlay. Available memory size means the number of real words in central memory which are available to the user. This is necessarily always less than the total size of the memory. The reason is that system programs must also reside in core with the analysis program, thereby reducing the effective available memory size. Since the major computers differ in this measure, a clarification is necessary for meaningful comparisons. The number of real words is important because all of the programs discussed employ real arithmetic almost exclusively. When the measure real word is used it means a computer word of at least 32 bits. The CDC 6000 series (60 bit word) and the IBM 7000 series (36 bit word) have the number of real words equal to the number of memory words. Whereas the IBM system 360 uses a byte system in which 1 byte = 8 bits. And consequently one real word equals 4 bytes. In all computer programs in which repeated numerical calculations are performed, round-off error may be a contributing factor to numerical inaccuracies. This, in turn, can be related to the word size of the computer. It is found that greater precision can be obtained with the CDC-6600 computer (60 bit word) than with the IBM-7094 computer (36 bit word).

Experience and Programming Support – In many instances, the user's experience or available computational support are often deciding factors in the final choice of a program. Although the programs of section 8.1 and 8.2 are written in FORTRAN, this does not indicate the "degree of difficulty" for the program. By this is meant *how much computer programming and computer system experience will be necessary to intelligently use, adapt, and modify the program*? The programs discussed here run the gamut from requiring minimal experience in both areas (CALAHAN) to knowing a considerable amount about both areas (SCEPTRE). A rule of thumb is that nonlinear programs require more programming experience than linear programs; the more efficiently a program is written the more difficult it is to understand; the more general the program the more that must be known. A realistic appraisal of computer experience and support is vitally necessary for a realistic decision.

Mode of Operation – With the advent of time-sharing systems, the conventional batch processing scheme is slowly being displaced [22]. However many programs, expecially those discussed in section 8.1 and 8.2, were written primarily for the batch processing. Consequently, in their present form, the programs are not suited for time-sharing. Nevertheless, in some instances, the adaptation to time-sharing is straightforward, in others it is not [23]. It can be said that any adaptation must be made by those who know both the program and the time-sharing system.

The next section contains comments on the five programs of section 8.1 and 8.2. These comments will be made with reference to: machines used, user's experience, user's programming support, and mode of operation. It is assumed that the user has picked programs suitable for his application and needs only to determine which best fits his computational facility.

CALAHAN – The CALAHAN program is used primarily in applications requiring the knowledge of network poles and zeros. The computation time will become excessive for larger active networks (see Chapter 2). It will accept active devices described as voltage controlled current sources. However, once the function has been calculated by the program, it is exact because the topological technique suffers no round-off error.

From a programming viewpoint, CALAHAN is excellent. The entire program is written in a straightforward manner, requires little or no programming and system experience to use, and fits easily into a memory of 20 K decimal words. In addition, because of the manner in which it is written, the program can be decomposed into smaller segments which would require no more than 8 K decimal words. Such a decomposition would be of the form: a segment to calculate the transfer function, a segment to use the transfer function to generate the frequency response, and a segment to use the transfer function to generate the time response. This decomposition requires only a minimal experience with computer programming and systems.

The above memory requirements assume a maximum network size of 20 active devices, 100 passive elements, and 50 nodes. In most cases, this is an unrealistic network size (from the standpoint of computation time) and consequently, if the network size were to be reduced, the memory requirements would also be reduced.

The primary tradeoff with CALAHAN is between excessive calculation time and the user's experience and programming support. For those with little or no experience and support, the CALAHAN program might be the best choice.

The CALAHAN program, as it is today, contains little diagnostic help for the user, and consequently, if the program is to be used by a large user population, diagnostics must be put into the program. This task is relatively simple.

POTTLE – The POTTLE program is written entirely in FORTRAN II. Like CALAHAN, POTTLE is also used primarily by those who are interested in the poles and zeros of network functions. However, unlike CALAHAN, the POTTLE program is computationally fast, and active devices can be described by any of the four controlled sources. In its present form, the functional matrix inversion suffers from severe roundoff error, such that if the network poles are separated by 3 orders of magnitude or greater, the answer obtained is far from being correct. An interim solution is available to users who have a computer with at least 60 bit computer words (36 bit computer words have been found to be inadequate) which is to use double precision arithmetic. However the numerical problems still exist for networks with a large number of state variables. The eventual solution lies in the use of a functional matrix inversion approach which, although slower, will not suffer this roundoff problem.

From a programming point of view, POTTLE is well written, very efficient, and complete. The program requires approximately 32 K decimal words of central memory. Because of its efficiency, POTTLE is difficult to use, adapt, and modify. It requires a person sufficiently experienced in computer programming systems to intelligently use the program and make modifications.

The program itself can be segmented in a manner similar to CALAHAN, with one slight exception: the segment which calculates the function requires almost 32 K decimal words. Further reduction is possible but only after a thorough understanding of the program. Some improvement is possible by reducing the network size but again this requires advanced programming and system experience.

The primary tradeoff with POTTLE is between fast calculation versus extensive programming experience and support and its present numerical and system problems. Unless trained and experienced computer personnel are available, POTTLE may be more trouble than it's worth. However, if this personnel is available, the payoffs are substantial.

The program does offer some diagnostics for the user, however they must be expanded if the number of users is large. Again, this requires somewhat detailed knowledge of the program.

ECAP – ECAP is written entirely in FORTRAN, and is for the most part, the same program written for the 1620 computer with some expansion in network size and an updating of features compatible with the larger computers. ECAP is used primarily for any of three reasons: For its dc capabilities with worst case, standard deviation, and sensitivities; for its ac capabilities of frequency response of all branch and node variables; and for its transient capabilities which include the piecewise modeling of nonlinear elements. The ECAP user is interested in numerical rather than functional solutions (as in CALAHAN and POTTLE).

From a programming viewpoint, ECAP has several desirable features. Foremost is the degree to which the program can be easily segmented into three distinct programs. This point is important because ECAP, as it stands, requires more than 32 K decimal words. However, if the program is decomposed into three distinct programs then each would require no more than about 16 K decimal words. Additional space could be saved by reducing the maximum network size the program will accommodate. To go beyond this simple decomposition requires some detailed knowledge of the program and computer systems in order to further adapt and modify the program. In its present form, the program has an adequate number of diagnostics (user oriented).

ECAP has no major numerical problems with roundoff, especially if the word size of the computer is large. One might experience some slight difficulty with the internal tests for balance of currents in computers with a small word size.

CIRCUS – The CIRCUS program is primarily intended to perform a dc initial condition and transient analysis of discrete-component, semiconductor circuits. The program is computationally fast and straightforward to use. It is almost entirely in FORTRAN IV and had been adapted to a variety of computers (IBM 7090/7094, UNIVAC 1107/1108, GE 625/635, and the CDC 6600). The programming is efficiently written but, as with most circuit simulators which handle network nonlinearities, the program is large, complex, and any program modifications must be approached with caution.

The entire program requires about 42 K decimal words but it can easily be segmented into overlays. The outstanding feature of CIRCUS is its fast running time. The program input language is simple, natural, and easily learned. Although a user is limited to the built-in models for the active devices this limitation has been used to increase the program efficiency to provide the good running times.

The program does offer some diagnostics but, in many cases, the diagnostic messages do not give the full picture. It would be difficult to expand on diagnostics without a thorough knowledge of the program.

CIRCUS is currently one of the best large-signal network analysis programs available. This is due to its speed in analyzing networks (computer time is expensive) and convenience in describing a circuit for analysis.

SCEPTRE – The SCEPTRE program is the most general of the widely available time-domain network-analysis programs. The program provides the user with a greater flexibility in the representation of network elements then most other analysis programs currently available. In providing this flexibility, the program's efficiency, for many networks, suffers. For example, the Schmitt trigger circuit of Fig. 8.8 required over 40 minutes of IBM 7094 computer time (about one-half of the response was calculated), whereas CIRCUS completed the analysis in less than 43 seconds of CDC 6600 computer time. Even considering the speed difference in the two computers (a factor of 5 to 20), the above times are significant. Also significant are the running time differences for the tunnel diode-common base amplifier of Fig. 8.1. Table 8.2 shows a summary of the times each program used in analyzing the network. Again, a direct comparison cannot be made because ECAP and SCEPTRE were run on the IBM 7094 computer (IBSYS system) in a batch-processing mode whereas CALAHAN, POTTLE, and CIRCUS were run on the CDC 6600 computer in a time-sharing mode. The times are representative, however, of the amount of time a "typical" problem will require and do illustrate the price paid (in terms of computer time) for program generality.

The SCEPTRE program consists of approximately 15,000 FORTRAN statements and, when run on the IBM 7094 computer, utilizes the

TABLE 8.2

Solution times for each of the five analysis programs for the small-signal tunnel diode circuit of Fig. 8.1 (b). ECAP and SCEPTRE were run on the IBM 7094 computer using the IBSYS system.

Program	Computer (System)	Solution Time
CALAHAN	CDC 6600 (T-S)	1.175 sec.
POTTLE	CDC 6600 (T-S)	4.670 sec.
ECAP	IBM 7094 (Batch)	162.00 sec.
CIRCUS	CDC 6600 (T-S)	20.385 sec.
SCEPTRE	IBM 7094 (Batch)	237.60 sec.

IBSYS system. The program consists of two major phases. The first creates another FORTRAN IV program on magnetic tape and the second solves the equations described by this program. Prior to execution of the second phase the IBSYS system compiles the FORTRAN IV program written by phase one. Due to its size, the program is not particularly easy to maintain.

The diagnostics offered by the program, in most cases, are sufficient for a user to locate the trouble when the trouble is due to input format. However, not all violations of the SCEPTRE constraints are labelled clearly enough to enable an uninitiated user to find the cause of run termination. The fact that the program is as flexible as it is makes error diagnostics difficult at best.

SCEPTRE is currently the most powerful program of its type, and will provide a great deal of useful work and information for future program development.

REFERENCES

1. D. A. Calahan, "Linear network analysis and realization digital computer programs – an instruction manual," University of Illinois Bulletin, vol. 62, no 58, February 1965.

2. C. Pottle, "State-space techniques for general active network analysis," in *System Analysis by Digital Computer*. F. F. Kuo and J. F. Kaiser, Eds. New York: Wiley, 1966, Ch. 3.

3. "1620 electronic circuit analysis program [ECAP] [1620–EE–02X] user's manual," IBM Corporation, Data Processing Div., White Plains, N. Y., IBM Application Program File H20–0170–1, 1965.

4. L. D. Milliman, W. A. Massena, and R. H. Dickhaut, "CIRCUS – A digital computer program for transient analysis of electronic circuits – User's guide," The Boeing Company, Seattle, Wash., Harry Diamond Labs., Rept. AD–346–1, January 1967.

5. H. W. Mathers, S. R. Sedore, and J. R. Sents, "Automated digital computer program for determining responses of electronic circuits to transient nuclear radiation (SCEPTRE)," vol. I, IBM Electronics Systems Center, Owego, N. Y. IBM File 66-928-611, February, 1967.

6. S. J. Mason, "Topological analysis of linear non-reciprocal networks," *Proc. IRE*, vol. 45, pp. 829-838, June 1957.

7. S. Seshu and M. B. Reed, *Linear Graphs and Electrical Networks*. Reading, Mass.: Addison-Wesley, pp. 315-330, 1961.

8. L. Zadeh and C. Desoer, *Linear System Theory.* New York: McGraw–Hill, 1963.

9. E. S. Kuh and R. A. Rohrer, "The state variable approach to network analysis," *Proc. IEEE*, vol. 53, pp. 672-686, July 1965.

10. G. K. Pritchard, "A survey of transient circuit analysis computer programs," *NASA Computer-Aided Circuit Design Seminar (Seminar Record)*, Cambridge, Mass., pp. 97-104, April 1967.

11. H. Falk, "Computer programs for circuit design," *Electro-Technology*, Vol. 77, pp. 54-57, June 1966.

12. P. R. Bryant, "The order of complexity of electrical networks," *Proc. IEE*, vol. 106, pp. 174-188, 1959. (Part C, Monograph 335E.)

13. R. L. Wilson and W. A. Massena, "An extension of Bryant-Bashkow a matrix," *IEEE Tran. on Circuit Theory*, vol. CT–12, pp. 120-122, March 1965.

14. W. G. Magnuson, Jr., "Network characterization for efficient computer time-domain circuit analysis," *10th Midwest Symp. on Circuit Theory* (*Conf. Record*,) pp. I-4-1 to I-4-10, 1967.

15. D. A. Pope, "An exponential method of numerical integration of ordinary differential equations," *Comm. of the ACM*, vol. 6, no. 8, pp. 491-493, August 1963.

16. M. E. Fowler and R. M. Warten, "A numerical technique for ordinary differential equations with widely separated eigenvalues," *IBM J. Res. Dev.*, vol. 11, no. 5, pp. 537-543, September 1967.

17. R. H. Dickhaut, "CIRCUS, A digital computer program for transient analysis of electronic circuits," *NASA Computer-Aided Circuit Design Seminar* (*Seminar Record*,) Cambridge, Mass., pp. 41-43, April 1967.

18. S. R. Sedore, "SCEPTRE: A program for automatic network analysis," *IBM J. Res. Dev.*, vol 11, no. 6, pp. 627-637, November 1967.

19. "Automated digital computer program for determining responses of electronic systems to transient nuclear radiation (PREDICT)," vol. II, IBM Space Guidance Center, Owego, N. Y., IBM File 64-521-5, July 1964.

20. A. F. Malmberg, F. L. Cornwell, and F. N. Hofer, "NET–1 network analysis program," Los Alamos Scientific Laboratory, Los Alamos, N. Mex., Rept. LA-3119, 7090/7094 version, August 1964.

21. J. N. Harris, P. E. Gray, and C. L. Searle, *Digital Transistor Circuits*, New York: Wiley, pp. 68-81, 1966.

22. H. B. Lee, "The status of on-line circuit design," *IEEE Northeast Electronics Research and Engineering Meeting* (NEREM Record), vol. 9, pp. 200–201, 1967.

23. D. F. Dawson, F. F. Kuo, and W. G. Magnuson, Jr., "Computer-aided design of electronic circuits – a user's viewpoint," *Proc. IEEE*, vol. 55, pp. 1946–1954, November 1967.

CHAPTER 9

A SURVEY OF GRAPHIC DATA PROCESSING EQUIPMENT FOR COMPUTERS

G. A. MICHAEL, Lawrence Radiation Laboratory, Livermore and Haverford College
R. K. CRALLE, Lawrence Radiation Laboratory, Livermore

INTRODUCTION†

In the not too distant future, people are going to be working with computers that can execute hundreds of millions of instructions per second. This is about one hundred times faster than present computers.

† This survey first appeared in the proceedings of the AEC symposium, "Use of Computers in Analysis of Experimental Data and the Control of Nuclear Facilities" (CONF 660527), Argonne National Laboratory, May 1966. A first revision was published in the proceedings of the University of Waterloo (Ontario, Canada) International Conference on "Design and Planning," edited by M. Krampen and P. Seitz, Hastings House Publishers, New York 1967. The present revision now includes a section on graphic input devices. Other additions include some recently announced microfilm recorders and display terminals. Even so, a cautionary note is in order: The survey is less likely to be complete today than it was in 1966 because of the tremendous increase in activity in this field.

What will be done with them? How will we communicate with them? At present the usual mechanisms are printouts, punched cards, and an occasional graph. Clearly, these will be inadequate for almost everyone.

The idea of pictorial communication was used on the early (1950) computers, but most users were so taken up with merely getting the calculations to run that little time remained to struggle with the mechanics of picture generation. Today the situation is different. The reasons are clear:

1. The equipment is better, therefore the user is more inclined to try.
2. Users have more sophisticated programming tools available now.
3. The level of problem complexity has risen.
4. The number of new kinds of applications has increased.
5. There seems to be no other way to cope with the high volume of output.

Certain properties of pictures are particularly useful for presenting computer output. First, a picture is a very efficient way to compress information that would otherwise have to be printed out as huge lists of numbers. They easily show many interrelationships in the data that normally escape notice in printouts. Finally, computer produced pictures often allow us to see "things" we have never seen before. Such clues can significantly deepen and even foster new intuitions for the user.

In this article we will not explicitly discuss applications. Instead, our chief concern will be with graphic oriented equipment and methods.

Also, the terms *picture* and *graphic structure* are synonymous. While we might select from all sorts of flattering definitions for graphics, the clearest one simply says that graphic data processing is a very efficient input–output technique in which information is exchanged between the computer and the user, primarily in the form of pictures. When we look at a picture we perceive shapes, positional arrangements and other relationships. A long range goal of graphics is to make these perceptions somehow available to the computer also. This goal is a broader concept than our definition implies. What we have chosen however, is to emphasize the information compression gained by pictorial presentations that can be achieved now.

There are three other prejudices that undoubtedly influence the way things are going to be said, so it may be well to name them now.

1. The activities encompassed by graphical data processing may seem to be both diverse and numerous; this appearance is deceptive, or rather if our work is to have any relevance we had better make them unified.

2. Unless "graphics" is aimed at real problems it runs the risk of becoming inbred and inconsequential. This is not to say that toys are bad, but very expensive toys in this context, may be.

3. Due respect must be paid to what we call "The Parachute Folder's Syndrome." That is, anyone who wants to earn his livelihood by folding parachutes ought to be made to jump with his own handiwork on a reasonably regular basis. Old computer users can easily recognize the consequences of having system programs designed by people who don't use them for problem solving; the "company furnished software" includes assembly programs requiring twenty-six passes to compile a code, or the user ends up being crowded out of memory by the system executive and service programs.

The remainder of this article is divided into the following parts:

9.1 General features of displays and programs.
9.2 Microfilm recorders and peripheral equipments.
9.3 Visual displays (Monitors).
9.4 Graphic input devices
9.5 Future design criteria and summary.

9.1 GENERAL FEATURES OF DISPLAYS AND PROGRAMS

At present Cathode Ray Tubes (CRT) are the most usual way of producing pictures under computer control. It seems that the name "Cathode Ray" was first used by W. Hittorf (in 1868) [1,2] who was studying the phenomena of fluorescence in connection with the discharge of electricity in an evacuated glass bottle. W. Crookes showed (1879) that these rays were better thought of as negatively charged particles; he assumed that the particles were negatively charged atoms. Other researchers, H. Hertz (1892), J. Perrin (1895), J. J. Thomson (1897), succeeded in deducing that they were much smaller, and in fact today these particles are called "electrons" and are thought to be one of the several basic entities in the universe. (One of Thomson's collaborators, J. Stoney is credited with the first use of the word "electron"). Thus, while the cathode ray tube is more accurately called an electron beam tube, it is rather unlikely that the original name will ever be replaced.

A CRT is usually an elongated vacuum tube having a cylindrical section mated to a conical section (Fig. 9.1). The glass base of the cone is called a screen or face plate and is coated on the inside with a compound that emits light when struck by electrons. After the electrons have been

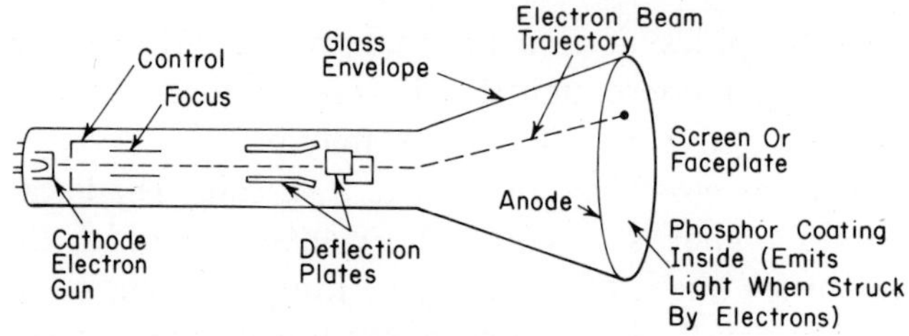

ELECTROSTATIC DEFLECTION

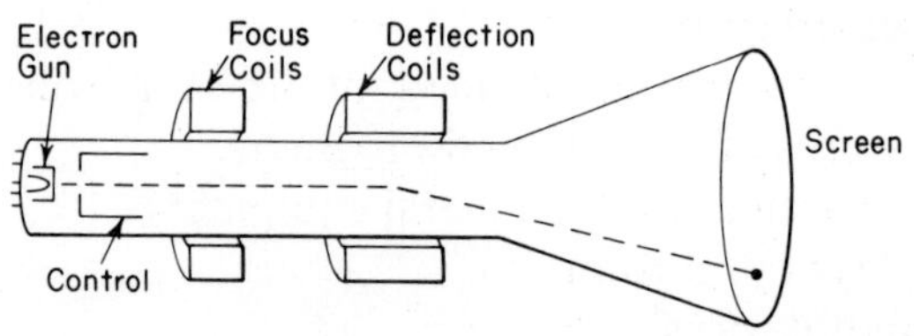

MAGNETIC DEFLECTION

FIG. 9.1. Simplified cathode ray tube construction.

ejected from the cathode or gun located at the other end of the vacuum tube, they are accelerated by establishing a difference in voltage levels between the cathode and the screen. Using other electrostatic or magnetic fields, electrons can be focused and deflected in a manner analogous to ordinary light beams. That is, there are electrical and magnetic equivalents to lenses, prisms, filters, shutters and so on. In any case, one reason for using CRT's under computer control is their ability to transduce electrical signals into light which we can then see or photograph. Furthermore, the electron beam can be valved on or off and deflected both very rapidly and precisely. Table 9.1 shows some elementary comparisons between light optics and electron optics. It should be emphasized, however, that this tabulation is analogical. We would not expect an electronics engineer or a photographer to agree with it.

The light-emitting compounds used as CRT screens are called "cathodoluminescent phosphors" or simply, "phosphors"; that is compositions of matter which radiate visible electromagnetic radiation (light) when stimulated by other, usually invisible, radiation (the electron beam). Chemically, the phosphors are composed of various salts of zinc, cadmium, magnesium silicates and sulfides or other alkali metal tungstates. To these a small amount of some other metal (e.g., silver,

TABLE 9.1

Comparison between Electron Optics and Light Optics

CRT components	Photographic components
Electron gun	Light source
Control elements for regulating electron current	Lens diaphragm, shutter
Accelerating elements	Filters
Focus elements	Focus ring
Deflection elements	Image-positioning prisms
Screen or faceplate	Photographic emulsion

copper, cerium) is added. These impurities help to determine spectral characteristics, and also function as "activators" to increase luminous efficiency, or as "killers" to reduce phosphorescence time [3, 4]. Some approximate characteristics for a number of phosphor types are given in Table 9.1a.

Within the CRT technology there are two general methods (Fig. 9.1), electrostatic and electromagnetic, for controlling the light-producing beam of electrons. The electromagnetic (or magnetic) deflection scheme is ordinarily held to be capable of higher positioning accuracy. Alternatively, electrostatic deflection is faster. Some CRT's use a combination of the two, but this is not now common. For this report we will simply note that both methods have their proper uses.

From a user's point of view we might be more properly interested in a list of elementary abilities that a display generator can have today. It is easy to see that whatever the display produces, must, in the final analysis, be combinations of its elementary functions. We have tabulated these in descending order of importance in Table 9.2. Thus, plotting a point seems to be the most basic in the sense that lines, arcs, and characters can be made with them. All the other features are time savers though admittedly of great importance.

Depending on factors such as frequency of use, execution times and costs, some of these functions will be expressed as computer programs, while others can be built into the display generator circuitry. In most cases the "wired in" functions are chosen on the basis of speed considerations, but these clearly make the equipment more complicated and expensive. A circle generator is an example of a built in function that might be so justified, although when time is not a serious problem, programming methods (algorithms) exist for producing equivalent results.

TABLE 9.1a

Approximate Characteristics for Selected Phosphors (1)

JEDEC Designation	(2) (6) Color of Fluorescence and Phosphorescence	(3) Spectral Peak (Å)	Spectral 10% Points Å	(4) Persistence Class	(5) Decay Time to 10% of Initial Brightness
P1	Yellowish-Green	5250	4900-5800	M	25 ms
P2	Yellowish-Green	5400	4400-6100	MS	70 μse
P4 sulfide	White (Blue/Greenish-Yellow)	4600/5600	4100-6900	MS/M	22 μs/60 μs
P4 sulfide-silicate	White (Blue/Greenish-Yellow)	4500/5400	3900-7000	MS/M	40 μs/12.5 ms
P4 silicate	White (Blue/Greenish-Yellow)	4100/5500	3200-7000	MS/M	40 μs/12.5 ms
P7	Blue White – Yellowish-Green	4400/5550	3900-6500	MS/L	60 μs/400 ms
P11	Blue	4600	4000-5500	MS	34 μs
P12	Orange	5900	5400-6800	L	210 ms
P14	Violet/Orange	4350/6000	3900-7100	MS/M	27 μs/5 ms
P15	Violet/Green	3900/5100	3700-6000	VS/S	.05 μs/2.8 μs
P16	Violet	3830	3400-4400	VS	.12 μs
P19	Orange	5950	5400-6700	L	220 ms
P20	Yellowish-Green	5600	4850-6700	M	200 ms
P21	Reddish-Orange	6000	5500-6500	M	––
P22	Violet/Green/Reddish-Orange	4500/5260/6350	4000-7200	MS	22/60/60 μs
P24	Green	5100	4300-6300	S	1.5 μs
P25	Orange	6100	5300-7100	M	45 ms
P26	Orange	5950	5400-6600	VL	∼10 sec.

Table 9.1a (cont.)

P28	Yellow Green	5500	4600-6300	L	500 ms
P31	Green	5200	4100-6000	MS	38 μs
P32	Purple Blue/Yellow	4700/5550	3800-6500	L	--
P33	Orange	5875	5400-6800	VL	∿ 10 sec.
P34	Blue Green/Yellow Green	4900	3900-6800	VL	--
P37	Blue	4700	4000-5400	VS	.15 μs
P40	White (modified P-4)	4400/5500	4100-6900	MS/L	150 μs/500 ms

Notes:

1. The contents of this table were abstracted from publications of the following sources.
   a) Dumont Div. of Fairchild Camera and Instrument Corporation, Clifton, New Jersey
   b) General Atronics Electron Tube Division, Philadelphia, Pennsylvania
   c) Thomas Electronics Inc., Wayne, New Jersey
   d) Radio Corporation of America, Harrison, New Jersey
   e) Eastman Kodak Corporation, Rochester, New York
   f) Tektronix Inc., Beaverton, Oregon
   g) JEDEC Electron Tube Council, Publication 16 A, Electronic Industries Association, Washington, D.C.
2. Color: Multiple components are separated by a slash eg. blue/yellow. This convention is used for all characteristics.
3. Spectral characteristics vary with chemical composition.
4. Persistence Class: Joint Electron Device Engineering Council persistence classifications are based on the time for the phosphorescence to decay to 10% of initial brightness.

VL	Very Long	1 second (sec) or more.	MS	Medium Short	10 microseconds (μs) to 1 ms.
L	Long	100 milliseconds (ms) to 1 sec.	S	Short	1 μs to 10 μs
M	Medium	1 ms to 100 ms	VS	Very Short	less than 1 μsec.

5. Decay Time: In practice, decay times depend on anode voltage and current density, thus the times given are for typical operating conditions as would be specified by the tube manufacturers.
6. Fluorescent light is produced by the interaction of the electron beam and the screen material (phosphor). After the exciting beam is removed (turned off) the phosphor continues to emit radiation. This light is called phosphorescence. The combination of fluorescence and phosphorescence is called luminescence.

TABLE 9-2
Basic Functions of a Display Generator

Elementary Function	Speed Range	Remarks
Plot a point	1-50 μsec	
Advance film	3-200 msec	Cameras are pulse operated, sprocketed, or claw pull down and pin registered
Draw a line	1-1000 μsec	Some generators have the writing rate independent of line length
Change the beam position (no unblanking)	1-50 μsec	To position the beam at the beginning of a line
Display a character	1-1000 μsec	Some characters are constructed in a dot matrix
Change brightness		
Change image size		
Change image aspect ratio		
Change effective raster size		

In any case a motion picture of a bouncing ball for example, is simply a sequence of plotted points, lines, film movements, and so on. For user convenience one can combine certain of these elementary functions into defined or derived functions of functions and do this in a fairly general way. Thus, the user need not think of, for example, three-line segments, but rather call for the "triangle" function. Further, one can think of combining the triangle function with others of these higher functions to any desired level. This is one of the properties of the artifice we have called a "glyph" [5]. A glyph is any combination of the elementary or higher functions to which is assigned an unique name. There can clearly be combinations of glyphs. We can also assign properties to them such as color, size, rotation and translation rates and so on. The total description is kept in a nesting of files and directories such that the programmer can, with a minimum of bookkeeping distraction, produce graphical presentations and alter them at any depth he desires. The suggestions listed in Table 9.3 are examples of properties that glyphs might be given. An important thing to notice is that such files imply a precise, reproducible description of a picture or an actual thing in a form suited to automatic processing; they are alterable. retrievable and communicable. Thus, they are clear statements of ideas or procedures in terms of a fixed, known set of primitives. The user guides the kind of processing done with these lists. For example, he may request a visual

TABLE 9-3
Examples of Operations and Properties Given to Glyphs

Function	Remarks
TURN	3-axis rotation
PAN, ZOOM-MOOZ	3-axis translation
PERSPECTIVE-PROJECTIVE	Geometric transformations
STEREO	2-views
BLINK	Attention direction
SHADE	Optical density variations; texturing
COLOR	Associate a color with the image (parts)
PHOTO	Take a picture
MOVIE	Take a sequence of pictures
NAME	Associate a name with the image(s)
ATTACH	Join line segments
CONNECT	Special glyph linking
DRAW	Follow the light pen
SMOOTH	Apply a relaxation operator to the data
MODIFY	Some change to a display list
DIMENSION	Assign lengths (areas, volumes, masses)
ERASE	Remove the image (parts)
WRITE	Prepare for text

display or hard copy or he may make modifications to the basic files or generate new subsets for them.

While Table 9.2 gives examples of things that a display might perform, we might pause a moment and consider the mechanics of display generation. When we wish to plot something we also have to say where to plot it. The addressing scheme normally chosen is obtained by imagining a two-dimensional Cartesian coordinate system superimposed on the screen of the CRT. In both the horizontal (X) and vertical (Y) directions we can assign integers for each point to which the electron beam may be digitally deflected. Most CRT's provide at least 1024 X and 1024 Y addresses, and any X may be associated with any Y value. The conventional way of denoting this is with powers of two since most computers use binary arithmetic. Thus, we have $2^{10} \times 2^{10}$ points in the example above. While distances spanned in each direction depend on the screen size, it is most useful for the purposes of graphics, to require that they be as equal as possible. The (square) array of points thus obtained is technically called a "lattice," but "raster" is more commonly used. In any case two numbers (*X* and *Y*) are needed to define the screen address. A number is needed to define the CRT function (point, line, character, etc.). Numbers would also be needed for intensity, for size control and whatever else there may be. The computer program must pack all these together in the appropriate "display word" format prior to sending them to the display where they will be decoded and acted upon. Whenever possible it is most efficient to make the display word size commensurable with the computer word [6]. Furthermore, the chosen positions of the various numbers within the word should be convenient relative to the instruction repertoire of the computer.

Some of the details about display-computer interfacing are best decided for each installation separately. A few particular issues are discussed in the other sections. There is one point however that is more pertinent for microfilm recorders. The problem of interfacing to a computer is logically subordinate to deciding whether to do it at all. An alternative is to write a tape in an agreed upon format and subsequently "play it" off-line on the display. Both approaches have advantages some of which are given below. From a user's point of view it is more satisfying to be able to plot and watch on-line.

Off-line operation advantages:

1. The tape record is permanent and can be produced into film at any later time.
2. Film wastage is minimized since unneeded plots can be skipped over on tape and recovered later if desired.
3. Picture quality can be higher because the display can be given a "special tune-up" prior to plotting.
4. Computer time is saved.

On-line operation advantages:

1. The time needed to plot a picture can be overlapped with other computer activities. If subsequent copies are needed the film can be duplicated.
2. The "average" display can accept output faster than standard tape writing takes place.
3. Very often errors can be caught (by observing the equivalent visual display) and quickly corrected, thereby saving a certain amount of computer time.
4. The more important time to save is that of people, not machines.
5. Off-line operation usually ties up a tape unit which is thereby unavailable to computer users.

9.2 MICROFILM RECORDERS AND PERIPHERAL EQUIPMENT

We have compiled a small listing of computer-controlled microfilm recorders, but equipment status changes so fast in the computing business that we feel it best to warn the unwary against assuming completeness in the survey. If a piece of existing equipment was omitted from the tabulation, it certainly was unintentional.

The information contained in Table 9.4 was obtained either from actual use or from certain more or less detailed descriptions supplied by company representatives. Table 9.5 is a synopsis in which the devices are listed according to speeds and costs. Of course this does not tell the whole story. That is, device timing depends heavily on the particular method used to interconnect the display equipment to the computer. For example, if a high speed CRT is connected to a tape channel then from the user's point of view it is only as fast as tape writing. There is another reason why the basis for speed rating is not uniform. The CDC 280, for example, can draw one continuous diagonal line while SC 4020 uses sixteen line segments for the same length. The question of costs can it seems, never have a common basis. In particular, the CDC equipment includes the cost of visual monitors, the others do not. At the low range, the cost estimates do not include the price of a tape transport (∼$30,000) which would be needed for off-line operation. Finally, the indicated price range does not reflect the cost of the various options listed in Table 9.4.

In microfilm recording the goal is to optimize the various parameters in order to achieve the highest resolution or the greatest contrast ratio and so on. The things that can be varied include beam dynamics, phosphors, lenses, films, developing machines, cameras, and so on. The lenses and films should be chosen so as to match the light output of the phosphor. If the phosphor emits light in the ultraviolet region for example, we should use quartz optics because ordinary glass transmits poorly in this range. For each of the quantities involved, it is useful to derive at least semi-quantitative data on the results produced when the parameters are systematically varied. A set of independent standards (e.g. [7]) must also be chosen. Since the ultimate goal is high quality microfilm, we might choose for example, several of the procedures used in scientific photography such as microdensitometry and sensitometry. It is wise to keep in mind that the optimization process involves adjustment of interrelated variables so that such things as record keeping and methodology

TABLE 9.4

Computer Controlled Microfilm Recorders

Equipment Name	Manufacturer	Number Delivered	Visual Monitor	Recording Unit & Deflection Method	Display Features and Times	Word Size	Camera	Options
CDC 280 (replaces DD80)	Control Data Corporation	20 (280) 8 (DD80)	19" CRT, P31, Independent gain and centering controls, intensity, focus, astigmatism.	Microfilm Recorder, Film output, 5" CRT, P11, Electrostatic, 1024 x 1024 points, Optional Hardcopy Recorder, 9" x 11" paper output, 7" CRT, P11, electrostatic, 1024 x 1024 points.	126 characters, 1 command code, 6 control codes, 4 sizes, 2 intensities, 2 fonts, 2 orientations, independent tube addressing 8 μsec/char. Random position or typewriter mode, Max 8192 char/page. Points: 5 μsec/pt. random position. Vectors: 2 intensities, any point to any point 5.5 to 50 μsec/line depending on length.	On line, 36 bits (12 bits at a time) Off line tape interface option: 200/556/800 b.p.i.	30 frame/sec., 4 perf. pull down, pin-registered 35 mm, 21.5 mm square exposing area, 1000 ft split supply and takeup magazines. Fault indicators: jammed, film supply low, supply exhausted	Hardcopy recorder on or off line processing, Hardcopy offline processor, film processor, 16 mm camera, extra monitor or recorder units, buttplot, buffer memory, A/N keyboard, light pen, trackball, switch modules, 137 char/line.
CDC 250	Control Data Corporation	14	19"CRT, P31 Independent gain and centering controls, intensity focus, astigmatism, A/N keyboard, function keyboard, light pen.	Same as CDC 280	64 characters, 4 sizes, 2 intensities, 2 fonts, 2 orientations, 3.3 μsec/char. type. mode, 6.6 μsec/char random position. Max 8192 char/page Points; 6.1 μsec/point. Vectors: 2 intensities, any point to any point 3.3 to 50 μsec/line depending on length. Recorder speeds are 3 times longer than above char and point speeds to assure proper exposure. Internal programming features allow off line text editing, light pen drawing, etc. and recording.	On line, 24 bits. 12 bit I/O	Same as CDC 280	Extra buffer memory, 3 visual and 2 recorder units, additional 64 characters, trackball, copy recorder on or off line processing, Hardcopy offline processor, film processor, camera.
BL 120*	Computer Industries	N/A	N/A	5" CRT Magnetic 1024 x 1024 raster	64 characters, 2 sizes, 50 μsec/char, 2 orientations. Vectors (lengths restricted to less than 1/16 raster diagonal), separate axes generation. Vector timing not available.	Off line tape formats for seven (6+1) channel tape drive provided.	10 frame/sec. 35 mm 400 ft. magazine.	Hardcopy forms projection, butting, 132 char/line, 64 additional characters, 9 channel tape, 16 mm camera.

*The BL is now off-line only.

TABLE 9.4.(Cont)

Equipment Name	Manufacturer	Number Delivered	Visual Monitor	Recording Unit and Deflection Method	Display Features and Times			
SC 4060	Stromberg-Carlson Div., General Dynamics	7	N/A	Same as SC 4020	96 characters 25 μsec/char. The SC 4060 is essentially a second-generation microfilm recorder. It has practically all of the main features of the SC 4020, at least the good ones. Three points are worth noting: It uses a small 16 bit computer (which looks like a DDP 116), pays great attention to the style of characters (using the CRAW MODERN font) and provides for on-line film development (2-3 sec/frame, MARK SYSTEMS model 1200 film processor) and hard copy production. Developed film can also be viewed on an 11 x 14 in. screen. There are 4096 x 3072 equi-spaced points addressable. Line generation is slower (in the order of 1000 μsec), random points plotting rate is 30 μsec/pt. The unit also provides for film retrieval coding (MIRACODE, KODAMATIC, etc.), has forms overlay and 2, 4, or 6 perforation film pull down (∿100 msec/frame). A "full" set of computer programs comes with the machine. The emphasis in this machine, which was designed in a sense by SC 4020 users, is very high quality recording.			
SC 4400	Stromberg-Carlson	N/A	N/A	Same as SC 4020	The SC 4400 in a sense is a cut down SC 4020, with the emphasis entirely on page printing. It can print 132 char/line with 64 line/frame and permits both forms projection overlay and information retrieval (KODAMATIC) coding of the film. The standard machine has no provision for line drawing.			
Series F	Minnesota Mining & Manufacturing Co.	N/A	Projection Screen	Direct electron beam recording	Off line device (uses 3M model 110 tape handler); character display only; features high resolution direct writing onto film by electron beam (no phosphor) and rapid access dry film processing ("the dry-silver process") using only heat. The dry silver process is also available in a hard copy unit directly onto paper. Several display manufacturers offer hard copy devices based on this method.			
					Display Features and Times	Word Size	Camera	Options
**IBM 2280 2282	International Business Machines	2 + (1 inventory)	2 projection screens. 23 inches sq.	5″ CRT electrostatic and magnetic 1024 x 1024 points	63 characters, average time 25 μsec/char, 3 sizes, 2 intensities. Characters generated by program, 6 strokes/char (avg.). Stroke time 1.5 μsec; space-between-character time, 15 μsec. Reposition time 150 μsec. Vectors: less than 1/4 max dist 102 μsec; between 1/4 and 1/2 204 μsec; greater than 1/2 408 μsec. Random points - same as vectors. Line stroke generation (length ∿1/100 max dist) 4 μsec/stroke.	Interfaced w/ system 360/x.	Unsprocketed film drive: Time .756″: 147 msec 1.008″: 182 msec 1.212″: 210 msec 2.004″: 321 msec	On line film processing and projection 40 in/min. Time to view first frame 50 sec.

**IBM 2280, 81, 82 only on special order

TABLE 9.4.(Cont)

Equipment Name	Manufacturer	Number Delivered	Visual Monitor	Recording Unit & Deflection Method	Display Features and Times	Word Size	Camera	Options
SC 4020	Stromberg-Carlson Div., General Dynamics	over 20	N/A	7″ charactron P-11, magnetic. Tube can be swung out for direct observation and also rotated about the beam axis $\pm 90^\circ$, 1024 x 1024 points	64 characters, 2 intensities, character formation by extruding the electron beam through a mask or template. Typewriter mode: 50 μsec/char. 128 char/line and 132 char/line (1403 mode) 64 lines/frame. Vectors point to point, max. length 1/16 raster diam, 50 μsec/line. Axes generator 220 μsec/axis. Points & characters, 50 μsec/pt. Program control for 2 raster sizes (butt plotting).	Off line 36 bits, 6 bits at a time from mag tape. Max block length is 1020 6 bit characters. 200/556/800 b.p.i.	16mm or 35mm 10 frames/sec friction drive or sprocketed drive, 400 ft. split supply and takeup magazines. 16mm and 35mm perforated and unperforated.	Hard copy camera option. Hard copy processor. Microfilm processor (Fulton). Forms projector
Model 1311	STRAZA Industries	N/A	16″CRT P-7 9″ square format	7″ CRT Magnetic 1024 x1024 tube can be rotated $-90^\circ, +90^\circ, +180^\circ$	63 characters 5 to 30 μsec/char. 2 intensities 4 sizes	On line, depends on computer.	10 frame/sec 16 or 35 mm 400 ft. magazine	Hard copy, forms projector, off line operation. Vector generator speeds increment 2 to 60 μsec 4 line types.
Model 1201	STRAZA Industries	N/A	N/A	6″ CRT Magnetic 1320 x 1360	Same as 1311	Off line, 556/800 bpi variable "word" length		Same as 1311
APD-5000	Link Group General Precision Systems Inc.	N/A	Option	5″ CRT P-11 Magnetic 4096 x 2730 (adjustable to 4096 x 4096)	Plots point only 10 μsec/pt variable spot diameter (4 sizes). Characters produced by subroutine.	On line, 32 bits for IBM 360 series (except model 20) Off line, link model 388 tape handler	10 frames/sec 35 mm 400 ft. split magazine	Visual monitor: Tektronix 611 storage tube.
CalComp 835	California Computer Products	25	11 x 17 projection screen	17″ CRT rectangular P-11 magnetic	The model 835 is logically equivalent to the CalComp pen and ink digital incremental plotters. That is any tape which would plot on a CalComp plotter can with one or two restrictions be plotted on the 835. (For example, a frame advance requires about 12″ of tape motion.) A 35mm sprocket driven camera is used (recording area 0.733 x 1.133 inches). In the sense that any conceivable plot (with .01″ increments) can be produced on the CalComp plotter, the 835 can do it also. When the exposed film is developed and viewed with a 15x enlargement the equivalent incremental step size is .01″. If the 835 is attached on line it is about 300 times as fast as the CalComp model 565 digital plotter. Off line it is about 100 times as fast. The tape drive is the same as is used for the CalComp 700 series plotters and moves at 60 in/sec in search mode. The following table shows the range of plotting rates.			

On line	Off line:	556	556/2	800/2	200	800/3	b.p.i.
100 KC (max)		33	16.5	24	12	16	kc

TABLE 9.5

Speed and Price Ranges of Recorders

Display Speed*		
Characters	Lines and Points	Approximate Purchase Price†
CDC 250	CDC 250	CalComp 835
	CDC 280	STRAZA model 1311
CDC 280		APD-5000
SC 4060	APD-5000**	CDC 250
STRAZA 1311	STRAZA 1311	CDC 280
SC 4020		SC 4020
	SC 4020	SC 4060
	SC 4060	
CalComp 835‡		

* Fastest to slowest.

† Least expensive to most expensive. Range is approximately $55,000 to $230,000. Price of options is not included.

‡ The CalComp 835 constructs everything out of incremental steps (the same way as the CalComp pen-and-ink plotters do).

** Points only.

will help to converge on the desired goals. Even more important, records will often point out when continued improvements are no longer worth the effort.

Beyond a preoccupation with making good recordings there are a number of things normally related to film use that we have included in Table 9.6. Some of these items are surprisingly unfamiliar to many users. However, it is hoped that the utility will be obvious simply from their names. Where possible, we have listed more than one example of each kind. In no case do we wish to imply that the listings are exhaustive.

The standard use of computer controlled microfilm recorders is in the production of graphs, listings, engineering drawings and so on. Some of this output is viewed with a variety of microfilm reader-printers or mounted on aperture cards. Various techniques have been devised for also marking the film with a machine readable code for automatic filing, sorting and retrieving. Other applications use the film as an intermediate material on the way to making prints, because no other present method is as fast, flexible and inexpensive. At our laboratory for example, we

TABLE 9.6
Microfilm Recording Accessories

Films For CRT Recording

Manufacturer	Type	Color Sensitivity	Recommended Phosphors	Remarks
Ansco Binghamton, N. Y.	Hyscan	Blue	P11, P24	
Eastman Kodak Co. Rochester, N. Y.	RAR 2498 RAR 2490, 91	Pan Blue (Ortho)	P11, P24 P5, P11, P16	Anti-hallation coating 2492 for P-31 2496 for P-7
Minnesota Mining and Manufacturing Inc. St. Paul, Minn.	Dry Silver	–	–	Direct Electron Beam Recording

Lenses For CRT Recording

Manufacturer	Type	Focal Length (mm)	Min f-stop	Magnification	Transmission Range Å	Resolution lines/mm
E. Kodak Co. Rochester, N. Y.	Formula M 236	50	f/2.0	6:1 to 2:1	3650 - 4350	∿128 ∿64
NIKON - Ehrenreich Photo Optical, Inc. Garden City, N. Y.	CRT lens	55	f/1.2	5.5:1 to 3.5:1	4600 - 5200	∿160 ∿100

Cameras For CRT Recording

Manufacturer	Type/Size	Rate frames/sec	Pin Registration	Remarks
Flight Research Div. of Gianini Sci. Richmond, Va.	M207-3A/ 35 mm	30	Yes	Used on CDC 280 - CDC 250
Trade Corp. Glendale, Calif.	Automax/35	10	No: (Sprocket Drive)	Used on SC 4020 APD-5000, S 1311, BL 120

TABLE 9.6 (Cont.)

Cameras For CRT Recording (cont.)

Manufacturer	Type/Size	Rate frames/sec	Pin Registration	Remarks
D. B. Millikan & Co. Arcadia, Calif.	DBM-5AP/16	8	Yes	
Vought Camera Co.	VDR-P4/16	15	No; (Sprocket Drive)	VDR-NP3 for unperforated film

35-mm film processors:	
Mark Systems 1200	Mark Systems, Santa Clara, Calif.
Kodak Prostar	E. Kodak Co., Rochester, N. Y.
Photomechanisms Datastat	Photomechanisms, Inc., Syosset, N. Y.
Fulton 35	Fulton Products. Tulare, Calif.
Houston Fearless (Model C)	Houston Pearless Corp., Los Angeles, Calif.
35-mm film viewers:	
Vanguard Instrument Corp. (motion analyzer)	Triad Corporation, Burbank, Calif.
Optical data analyzer	Richardson Camera Co., Phoenix, Ariz.
Optical effects printers (special effects):	
Research Products, Cinema Effects	716 North LaBrea, Hollywood, Calif.
Oxberry, Division of Berkey Photo, Inc.	New York, N. Y.
Producers' Sales, Inc.	North Hollywood, Calif.
16-mm flickerless variable-speed projectors:	
Bell & Howell (modified by Northridge Camera)	Marketed bv Triad Corp., Burbank, Calif.
Kodak Industrialist (modified by L & W Photo, Inc.)	Marketed by L&W Photo Inc., Van Nuys, Calif.
Stereo viewers:	
Stereo Realist	Realist, Inc., Menomonee Falls, Wis.
Abrams Stereoscope	Abrams Aerial Mapping, Lansing, Mich.
Polaroid glasses	Polaroid Corporation, Cambridge, Mass.
Stereobinokel, (Eletromedizinische Werkstatte Friedrich-Janus, Landau ILC)	Bush Electric Co., San Francisco, Calif. (importer)

Hard-Copy printers:
Complete information is contained in the "Guide to Microreproduction Equipment", pp. 343 ff., Hubbard W. Ballou, ed.; 1962, National Microfilm Association, Annapolis, Md.

use nearly 30,000 feet of film per month, and make hard copy for about 85% of it. About 75% of the printing is done one page per frame, but the rest consists of computer generated reports and many copies are made. The remaining 15% of the microfilm is devoted to motion picture production. A growing number of research centers, moreover, have begun using their equipment to produce these motion sequences. This use emphasizes the following features of microfilm recording.

1. Accurate film registration in the camera is quite necessary in order to minimize jitter effects when projecting the film.
2. The time for the CRT phosphor light to decay below the film sensitivity threshold should be short compared to the frame advance time.
3. The frame advance time should be as short as possible, consistent of course, with the two previous items. It is to be noted that the use of (mechanical) camera shutters is not standard practice for any of the recorders listed.
4. The CRT deflection and brightness control circuitry must possess as much repeatability and stability as is possible.

No amount of verbalizing has quite the power to convey information or stimulate our intuition as have motion sequences. Beyond the use of motion it is also possible to employ color and stereoscopic views to add even more content to the filmed output. Such techniques are relatively easy to employ using computer control, but at the same time they require some additional explanation. This is given in Appendix A at the end of this chapter.

When we consider motion it is in context of sensing changes in the position of one body relative to another at particular time-points. For example, a standard motion picture camera photographs the positional relationships every 1/24 of a second. This time taken as an independent variable, is called *real time*, or problem time, in distinction to the *computer time* necessary to actually do the calculation. Thus, a calculational model of say galactic evolution might require only a few hours to run on a computer, while the real time implied is several billion years. Again, it might take several hours to calculate the motion of a projectile fired from a gun barrel while the real time is only a few tenths of a second. Some of the space programs use computers to control the launch procedure in real time. In this case the problem (real) time is roughly the same as the computer time. The real time interval (Δt) between two consecutive frames is under computer control. Ordinarily the calculation "steps" from one time to the next by adding Δt. Thus,

$$t_{\text{now}} = t_{\text{previous}} + \Delta t$$

The Δt in the galactic evolution calculation might be as "short" as 10,000 years. Even so, there would have to be 100,000 time steps to represent a billion-year period.

In the example of the bouncing ball mentioned earlier, we might identify the following picture features:

1 An object or set of objects that are to be moved (or animated).
2. A set of procedures, algorithms or equations that describe each type of motion.
3. Special visual effects, cues and backgrounds.

Most likely we would use a circle (or sphere) to represent the ball, but we might for the sake of realism want to cause it to deform in a physically correct way as it bounces. From the computer program point of view a circle is first, a large set of points to identify the edge of the ball or, more ideally, a set of equations which are used to generate the points. The physical laws that describe the motion of falling bodies are, in this example, precisely the set of procedures that describe the motion of the ball. They simply "say" at a given time where the ball is located relative to some starting point. To these we could add descriptions (again mostly mathematical) of the elastic properties of the ball and the surface on which it is bouncing. The question of special effects has to be answered relative to a particular requirement. *Any* effect that can be described in terms of the elementary computer and display properties can be pictorialized. When modeling physical motions such as a bouncing ball we are often interested in seeing a "slow motion" representation. The real time interval at which the equations of motion are evaluated controls this type of visual effect.

Again, by changing the "size" of the ball we can give the illusion of motion toward or away from the viewer. A calibrated set of scales will permit an estimate of how high a "real" ball would bounce. The range of possibilities is predominantly limited by the designer's imagination. In programming rapid motions one caution is that they tend to produce a stroboscopic effect if successive positions are too dissimilar. This can be interpreted as setting a practical upper limit on Δt.

9.3 VISUAL DISPLAYS AND MONITORS (VCRT) [8, 9]

In the previous section we tried to warn about the problem of making relevant comparisons between the various microfilm recorders. The same is even more true for rating Visual Monitors. If it is not done in relation to well stated "bench mark" problems, then the results may be misleading, especially if they are used as a basis for selecting equipment. To do a thorough evaluation is not easy. On the one hand it is not particularly interesting when contrasted with actually using something, but the potential rewards are worth the effort. First, a clearer understanding of the application is almost always obtained. Another thing that quickly becomes rather obvious is the difference between actual equipment characteristics and their descriptions as obtained from sales brochures. Hence, we usually require the equivalent of a programmer's reference manual and an engineering description to find out how the device really works. For example, while a brochure will mention (in glowing terms) a line drawing feature it is only from the other manuals that we will learn that the maximum length permitted is some small fraction of the screen diameter. Thus, when a line drawing capability is needed, this limitation may be important.

The field of computer controlled display is relatively new as we have already noted, and is still undergoing basic changes. Consequently, at this time, there is hardly any agreement as to what a display device is. In particular, we see the name visual display used for everything from an illuminated instrument panel to devices which are so general as to even include a small general purpose computer. All the graphic consoles listed in Table 9.7 use a CRT to make visible images, but new display techniques are being uncovered quite regularly. Thus, we will not worry about definitions now, but simply say that the devices named in the table are examples of visual displays. Some designs only have the ability to display a string of characters. A number of these are given in Table 9.7, but since we are primarily interested here in graphical display, no further consideration is given to them in this chapter. Excepting the "advance film" function, all of the functions of Table 9.2 apply to visual display devices. However, since the measure of utility fundamentally depends on user acceptance, we need to further consider them, especially with respect to the speed.

TABLE 9.7

Visual Display Consoles

	A	B	C	D			E	
Identity								
Name	AGT-50	BR-90	CC-300	CDC 250	CDC 270	CDC 280	DEC 338	DEC 340
Tube/picture size	21" CRT/10" x 10"	13" sq.	Any TV receiver	19" CRT/12" sq.	24" CRT/14" sq.	19" CRT/12" sq.	16" CRT/9 3/8" sq.	16" CRT/9 3/8" sq.
Phosphor	P-31	P - 4	8 bit/char.	P-31	P-7	P-31	P-7	P-7
Input word size, bits	30	3(12-bit words)		24	12	36	18	18
Number of consoles per display generator	4	1		6	9	1 (more, optional)	1	5
Display functions								
Point generator	1024 x 1024	512 x 512	98 x 93	1024 x 1024	4096 x 4096	1024 x 1024	1024 x 1024	1024 x 1024
Line generator, † type	Analog	Analog		Analog	Analog incremental	analog	Digital incremental	Digital incremental
Longest line (raster count)		511		1023	7	1023	1023	127
Type of data	point-to-point and Δx,Δy	point-to-point		point-to-point and Δx,Δy	Δx,Δy	point-to-point	Δx,Δy ‡	Δx,Δy ‡
Brightness levels	Z-axis modulated intensity	1		2		2	8	8
Character generator								
Number of characters	64	48-BCD	ASCII §	64		126	64	64
Number of sizes	N/A	2		4		2	4	4
Construction	Stroke	Stroke		Stroke	Programmed	Stroke	5 x 7 Dot array programmed‡	Line segment
Fonts	N/A	1		2	Programmed	2		1
Typewriter mode	Yes	No		Yes	No	Yes	No	No
Speeds, μsec								
Points (random plotting)	N/A	35		6.6	Π	5	35	35
("almost" sequential)	N/A	35		6.6		5	15	35
Lines, short	3	35		3.3		5.5	1.5/pt‡	1.5/pt‡
long	38	35		50		50		
Characters	15	5		3.3		8	∼40	∼40
Special options	Hardware "windowing"							
Unless otherwise noted the following things are available either as standard or as optional features: keyboard light-pen, cursor, function switches, private buffer memory	Hard copy printer Extended Image Processor: (hardware implementation of certain of the programs) 3-D joystick or track ball	Blink control; circle generator; 64-character set; 1024 x 1024 raster††	Slave TV sets; typewriter hard copy	see CDC 250 Table 4	The CRT housing is tiltable from 15° to 75° with respect to the horizontal; 128-character set	See CDC 280 Table 4	128-character set DEC 338 has a complete PDP-8 computer built in to provide buffering and control	
Remarks								
	The AGT series has 3 models -10, 30, 50, with the 50, the most elaborate. A computer the AMBILOG-200 (8K - 30 bits/wd) controls the terminals. Analog components provide for (3 x 4) matrix multiplication (model 10 is 2 x 3). Thus the perspective mapping, rotation translation and scaling are all done in the Hybrid multiply-add array at very high speed.	Three smaller models ∷ with alphanumeric capability only are also available: BR-204, 211, 212, and 400					The DEC type 30 uses the same type of CRT but only plots points (50 μsec/pt) The DEC type 34 is a small (5") rack-mounted CRT for low-cost point plotting. Usually Tektronix RM-503 or RM-564 (storage) scopes are used	

TABLE 9.7 (cont.)

	F	G	H	I	J	K	L
Identity							
Name	IDIIOM	IBM 2250, I, II, III	MACC	WDA	READ	SEL 816A	SERIES 9000 (model 9100)
Tube/picture size	21EM10 P-31/13″sq.	21″ CRT/12″sq.	19″ CRT/12″sq.	17″ CRT/11″sq.	16″ CRT/9″sq.	16″/10″sq	23 Rectangular 13″x18″
Phosphor	P-31 (others available)	P-7	P-4	P-7	P-28	P-31	P-1 (approx.)
Input word size, bits	16	See remarks	See remarks	36	18	16	24
Number of consoles per display generator	8	4	1	1	15		8
Display functions							
Point generator	1024 x 1024	1024 x 1024	1024 x 1024	1024 x 1024	1024 x 1024	1024 x 1024	1400 x 2048
Line generator,† type	Analog	Analog	Analog and digital	Analog	Analog	Analog	Analog
Longest line (raster count)	1024	1023	1023	63	127	128/16 bit wd	2047
Type of data	Point-to-point and Δx, Δy	Point-to-point and Δx, Δy	Point-to-point	Point-to-point and Δx, Δy	Δx, Δy	Δx, Δy	Point-to-point (some models also use Δx, Δy)
Brightness control	4	1	1	16	3	4	4
Character generator							
Number of characters	62 (expandable to 128)	64	64	64	64	64	64 (128 option)
Number of sizes	4	2	2	3	7	2	4
Construction	Stroke	Stroke	Stroke	15 x 16 dot matrix	Stroke	16 x 8 dot matrix	Stroke
Fonts	1	1	1	1	2	1	1
Typewriter mode	Programmable	Yes	No	No	Yes	Yes	Yes
Speeds, μsec							
Points (random plotting)	20	120	17	10	35	14	48 μsec (max)
("almost" sequential)		17	17	5	5	14	4
Lines,† short	20 (<3″)	17	Analog line, 100	100	7	28	17 (.5″)
long	—	120	Digital line, 380	1600	10	14	54 (8″)
Characters	10	14-16	17	20	6		7 avg small ch.
Special options							
Unless otherwise noted the following things are available either as standard or as optional features: light-pen, cursor function switches, private buffer memory	Circle generator, microfilm recorder, hard copy,	No cursor; buffer memory is provided in the 2840 controller. The IBM 2260 is an alphanumeric display only	Blink; CRT is ported for rear projection overlay via a random-access (2-sec max.) slide projector ΠΠ	No cursor; 2048 x 2048 raster count; analog input via 15 high-precision potentiometers. Either film scanning or film recording can be done with an associated precision 7″ CRT (P-16 phosphor)	None	64 extra characters line texture control, light pen function switches	
Remarks	A small computer (eg DDP 116 16 bits/wd 4K wds) is an integral part of the display system.	There are 8 bits/byte, 1 byte per character, 4 bytes per point or vector. Model I is self contained, Model II requires 2848 controller for up to 8 displays, and Model III is about twice as fast as Model II. The times given are for Model III.	INPUT is via IBM 1410 tape channel (IBM 1414 TAU)	Deflection amplifiers permit three line-generating modes: TV raster scan, precision scan/record, and programmable length and direction (see Table 9 also)		Designed for SEL computer line	Model 9110 includes its own core memory for display refresh (50 cycle) service. Various other style CRT's can be chosen; also true for the raster counts.

TABLE 9.7.(Cont)

	B		D	G
Identity				
Name	Teleregister 212	Teleregister 403	CDC 210-211	IBM 2260
Tube/picture size	6″ CRT/4.5″sq.	10″ rect. CRT/7″ x 5″	14″ CRT/8″ x 6″	14″ CRT/9″ x 4″
Phosphor	P-31	P-31	P-4	P-12
Refresh memory/memory capacity /refresh rate	Delay line/768/48	Delay line/1536/48	Delay line/1000/50	Delay line/960/30
Number of terminals per display generator	216	16	12	24
Character repertoire	ASCII	ASCII	BCD	EBCDIC, ASCII
Character construction	5 x 7 dot matrix	5 x 7 dot matrix	5 x 7 dot matrix	Stroke
Characters/line	64	32	50	80
Lines/display	12	12	20	12
Remarks				

	M	N	O	P
Identity				
Name	DIDS 400	720 Communicator	SC 1100	UNIVAC 1551
Tube/picture size	12″ CRT/8.5″ x 6.5″	12″ rect./9.5″ x 7.5″	8″ CRT/6.5″ x 3.5″	17″ rect./19″ x 12″
Phosphor	P-31	P-4	P-12	P-31
Refresh memory/memory capacity/refresh rate	Delay line/1040/60	Delay line/1024/50	Delay line/500/25	Core/2000/50
Number of terminals per display controller	64	12	24	1
Character Repertoire	ASCII	ASCII	ASCII	ASCII
Character construction		Stroke	5 x 7 dot matrix	Stroke (8 max.)
Characters/line	80		50	80
Lines/display	13		10	25
Remarks	DIDS-1500 is a general-purpose display presently undergoing engineering development		SC 1200 is a digital display scanned by TV camera and broadcast on video channels, SC 1000 is a custom-designed general-purpose display system	

	AA		BB
Identity	HP 1300A	HP 6946	Type 611
Type	x-y oscilloscope	Television Picture Monitor 17 DWP-4	Storage (up to 15 min) CRT
Size	8″ x 10″	17″	11″ (∿8″ x 6″)
Phosphor	P-31	P4	∿ P1
Resolution		650 lines/full width	100 (stored) line pairs v. 125 (stored) line pairs h.
Linearity	∿10%	<2% full scale	∿5% (for ∿1 inch deflection)
Speed: (μsec)		Standard television	
Scattering time	∿.2	Frequencies (4.5 MHZ)	8.5
Writing time	∿.5		20.

TABLE 9.7 (cont.)

Erase time	N/A	N/A	.5 sec
Price	$1900	$950	$2500
Remarks	1. Very high speed electrostatic CRT	1. Unity interlace factor 2. Switchable display sizes 100% or 80% 3. Used with FPD Digital/Video disk memory Data Disk Inc. Palo Alto, Cal.	1. Non-store write feature provided 2. Type 601 5" storage CRT $1050 3. The longer a picture is stored the harder it is to erase 4. Computer Display Inc., Waltham, Mass. is using the 611 as the display part of a general purpose terminal (ARDS) designed to operate over phone lines with any time-shared computer service center.

* This table makes a number of compromises about important details in order to conform to editorial standards. Information about display-word formats, interfacing, and pricing was left out because of space limitations. Some display devices were not included in the table because they are still in an engineering prototype state. More-complete information can be obtained from the references. The manufacturers are indicated in the table according to the following key:

A	Adage	AGT-10, 30, 50
B	Bunker-Ramo Corp. Canoga Park, Calif.	BR-90, Teleregister 212, 403
C	Computer Communications, Inc., Inglewood, Calif.	CC-300
D	Control Data Corp., St. Paul, Minn.	CDC 250, 270, 280, 210
E	Digital Equipment Corp., Maynard, Mass.	DEC 338, 340, 30, 34
F	Information Displays, Inc., Mt. Vernon, N. Y.	M10000, M11000
G	International Business Machines Corp., New York, N.Y.	IBM 2250, I, II, III, IBM 2260
H	International Telephone and Telegraph Corporation, Fort Wayne, Ind.	MACC
	International Telephone and Telegraph Corporation, San Fernando, Calif.	LM 401, 411; 405, 906
I	General Precision, Inc., Link Group, Palo Alto, Calif.	WDA
J	Philco Corp., Willow Grove, Pa.	READ
K	Systems Engineering Laboratories	SEL 816A
L	Tasker Instruments Inc.	SERIES 9000
M	Raytheon Co., Wayland, Mass.	DIDS 400
N	Sanders Assoc. Nashua, N. H.	720 Communicator
0	Stromberg-Carlson Corp., Division of General Dynamics Corp., San Diego, Calif.	SC 1100, 1200
P	Univac, St. Paul, Minn.	UNIVAC 1551
AA	Hewlett-Packard, Palo Alto,Calif.	HP 1300A, 6946
BB	Tektronix, Beaverton, Ore.	TYPE 611

† Line-generation methods used by the various manufacturers fall into two major categories: analog and digital. The maximum length of line that can be requested by one display command is given in terms of the number of raster points that can be spanned. The type of data (point-to-point, or Δx, Δy) needed by the line generator is given.

‡ The DEC 338 and 340 both have a scale factor control on the line generator. Multiplications of 1, 2, 4, or 8 are allowed. The digital incremental nature of the line generator makes the output very similar to that of a CalComp digital plotter. The incremental steps each take 1.5 μsec. Memory cycle times are not included in these figures.

§ ASCII stands for American Standard Code for Information Interchange.

Π The CDC 270 is an incremental display generator. All beam movements (both invisible and visible) and all controls are supplied from a magnetic drum under computer control. The drum rotation rate thus supports the refresh rate. Each display command is processed in 1.67 μsec, but gross beam movements require more time. Therefore the synchronization between drum and display requires that varying numbers of "no-operation" words be written as spacers on the drum. Characters are constructed out of line segments by a suitable program. Some later models use core memory instead of disk memory.

†† The BR-90 has provisions for a computer-controlled random-access slide projector with a capacity of 100 slides. The projection, from the rear of the CRT, is "on-axis."

÷÷ The CC-300 can be any standard TV receiver. A display controller (CC-301) has buffer memory to hold up to 800 characters (ASCII format) and cycles the TV set 60 times per second. Graphic presentation is limited to a 99 x 93 dot matrix. A keyboard (CC-303) permits message composition prior to computer entry. The whole unit interfaces to any of the usual communications equipment or directly to a computer. One communications multiplexer (CC-70) can handle 32 receiving sets.

ΠΠ The ITT Industrial Products Div. in San Fernando, Calif., has a number of low-cost display sets or "do-it-yourself" applications. Among these are the LM 411, KM 405, and the KM 906 color set. Both a character generator CG 200 and several vector generators are available. Typical character plot times are quoted at 60 μsec. These sets have a resolution of about 20-25 lines/cm.

The human visual faculty is an extremely sensitive "data channel." Not enough is known yet about the relationships between it and the brain but if "very good visual data" is badly presented in part, the average viewer tends to reject the entire picture. For some, a flickering picture [9] is generally bad. For others, a lack of contrast is the annoying factor. In general, it is a matter of individual taste as to what is acceptable and what is not. The usual procedure is to provide a few manual adjustments to control such things as picture size, focus and brightness. The CDC Digigraphics consoles can even be tilted to suit individual tastes.

Beyond reducing the number of things to be displayed, a user cannot directly do very much to solve the problem of flicker. Most people do not like it although some of the early radar display operators often found that flicker actually helped them to notice new "blips" quicker. The rate of flicker is a function of the time needed to display the entire picture once. Other factors also having some influence are the type of phosphor used, the brightness and contrast of the picture parts, the level of background illumination and the sensitivity of the viewer's eyes. The average persistence time for human vision is usually given as 1/40 sec. This means simply that in order for our eyes to see some point of light on the CRT as a constant intensity source it must be refreshed at least once every 1/30 to 1/60 sec. Some phosphors have such short decay times that the illusion of constant brightness depends mostly on the persistence of vision. For example, the P-31 phosphor which emits strongly in the green portion of the visible spectrum, has a decay time of about 38 microseconds. Refresh cycles as low as 35/sec could be used with the P-31 phosphor, but the flicker is then obvious though not necessarily annoying. (Most manufacturers recommend a refresh rate of 40-60 per sec.)

Phosphor persistence times [Table 9.1a] are often given as the time needed for the brightness to be reduced to 10% of its initial value. Based on the persistence time, if the picture content is changing rapidly, a cleaner picture is obtained from a fast decay phosphor. However, many striking effects both visually and photographically can be produced when a long persistence phosphor such as P-7 (400 milliseconds) is used to show rapidly changing pictures. Some users have found that the effect of P-7 persistence can be reduced by placing a blue filter in front of the screen or by wearing sun glasses.

There are numerous ways to refresh a display. The most elementary of these simply requires the recomputation and retransmission of each picture part so many times per second. That portion of computer time used this way is obviously not available for other processing. Recomputation may be needed because the display list is too big to fit in the computer's memory. If enough computer memory is available, the usual procedure is to reserve a portion for the display lists which then can be buffered out to the CRT while the computer proceeds with other tasks. All new or large computers have as a standard feature, the ability to start an input/output process that proceeds more or less independently from central processor control. The idea of a buffered display list then evolves naturally.

Most of the display devices given in Table 9.7 can readily be equipped with their own private buffer memory. In this case the computer simply transfers the list to the private memory and starts the display running. Either way the display processes the list, command by command, into displayed picture parts. Points, line and characters are the "raw materials," but simple non-plotting actions are also useful. The last entry in a list might mean, for example, "start the list at the beginning if (say) 1/60 sec has gone by since the last time the list was started. If 1/60 sec has not elapsed, wait until it has and then start up again." The idea of waiting is mainly to provide for a uniform brightness that is independent of the shortness of the list.

The question of whether the display device should have a private buffer memory or should share the memory of the computer has several factors involved.

1. Private buffer memories tend to be rather more expensive then the same amount of computer memory.
2. If the buffer memory is computer memory, then it is always available for other computer needs when the display is not in use.
3. In a small but important number of cases the private buffer may be too small to contain the whole list.
4. If the display is attached to a small computer or if the computer does not have a buffered input/output capability, or if low speed communications lines are used, it may be necessary to equip the display with its own private buffer memory.

Figure 9.2a shows but two of the many ways that displays and computers could be interconnected. The second way alludes to the problem of long distance transmission of signals. As things presently stand, it is not usual to separate a high performance display from the control computer by more than a few hundred feet. In the future there may be a need arising for the ability to separate the two. The basic problem is the transmission at very high bit rates of at least digital data and at reason-

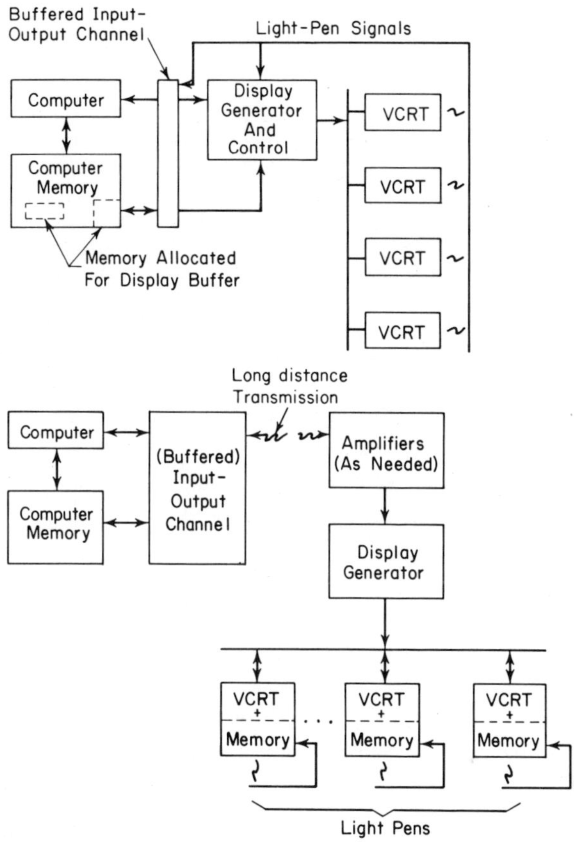

FIG. 9.2(a) Two possible ways to connect displays to a computer.

able costs. Aside from noting the existence of the problem no additional considerations of it need be made in this paper. It exists, people know about it; someday it will be solved unless suitable alternatives are developed first. Figure 9.2b shows a simplified block diagram of a display generator designed for multiple console control.

No discussion of visual displays would be complete without some consideration of the various input devices that have come to be associated with CRT use. Thus, while we attach some value to the visual presentation, it is often equally important to be able to conveniently and quickly interact with the computing process. There are several types of input devices that are commonly associated with this activity. An ordinary typewriter suitably connected to the computer is one example. Keyboards, arrays of toggle switches, pushbuttons, and tuning knobs are others.

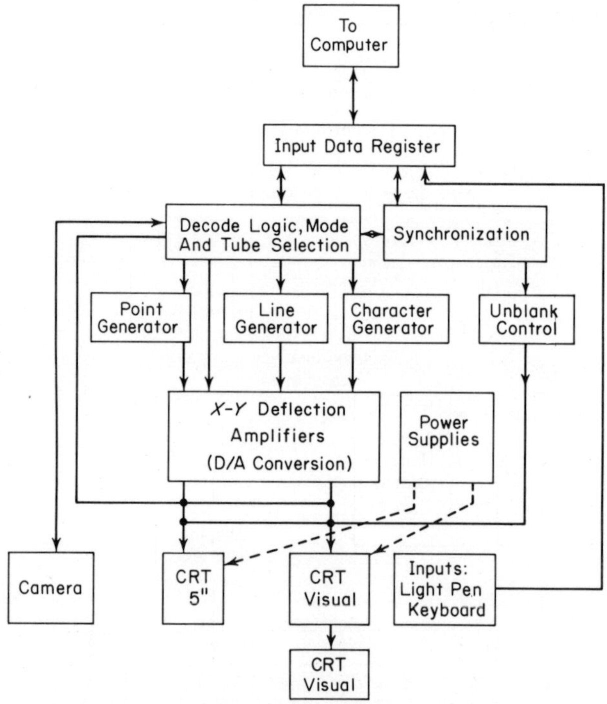

FIG. 9.2(b) Simplified display generator block diagram.

By program we can assign various special functions to any key or button. Striking the "R" key on a typewriter for instance, ordinarily causes an "R" to be typed. The program however, might be made to "change the perspective" of the image on the CRT when the "R" key is activated. As with any programmable device, how such things are used depends on the programmer, and their great utility derives precisely from the fact that they are programmable. At one instant we can be using the keyboard to type in a new program, and immediately thereafter use it to move an image around on the CRT screen.

Another kind of input device is epitomized by the so-called "light pen (or pencil)." As such, it is a hand-held thing that can sense the presence of a brightening spot of light when held at or near the CRT screen, and so signal the computer. The existence of the signal can be correlated with the display list to deduce (by program) which particular item was being displayed at the instant that the light pen "hit" was encountered. This is an example of a synchronous input. In other words, it is possible for the computer program to find out where or at what the user is pointing, and inferentially, what he wants to do. We can distinguish two programmed styles of using the light pen. First, we may wish to *follow* or track the light pen as the user moves it around on the screen thereby drawing lines or paths. Second, we may wish to *designate* or activate a new procedure or mark some particular picture part. One useful light pen application is given in Appendix B.

A common procedure is to program a series of bright points at, say, the lower portion of the screen. These can be treated as programmable push buttons. By selectively touching these buttons with the light pen the user can tell his program what he wishes to be done. Some common functions include "buttons" to control enlarging (zooming), rotating, erasing, and tracking. One light pen design (in use at LRL) is shown in Fig. 9.3. A large number of different designs are currently in use; each more or less appropriate for a given task.

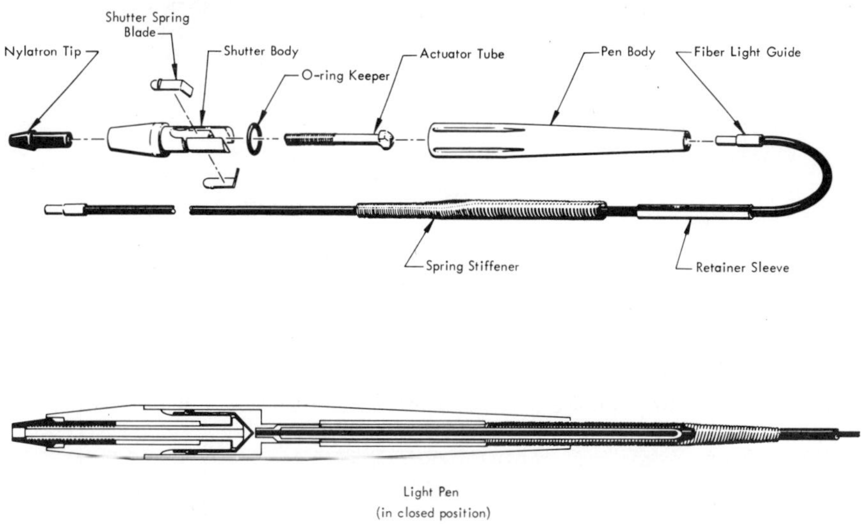

FIG. 9.3 The L. R. L. light-pen design.

Still other devices that provide functions similar to those of the light pen are available. This report would be unduly lengthened if each of these input devices were discussed separately. A more complete list of display associated input devices along with references to more detailed information is given in Table 9.8.

In some sense the light pen is the simplest and the most general or flexible input device. Even so, it is often found necessary [10] to supplement its function with one or more push buttons. One limitation is

TABLE 9.8a

Graphics Terminal Input Devices

Type	Remarks
Typewriter keyboard	Most useful for entry of alphanumerics
Track ball (see mfr. list) Joy stick S.R.I. mouse [33]	Control a visible (either hardware or program-generated) cursor on the CRT screen (see Ref. 9)
Push buttons Foot pedals Toggle switches	Procedural controls – usually a one-bit signal, often in conjunction with some other input device, e.g., the light pen (see Ref. 9)
Analog-to-digital converters (manually controlled) Potentiometers Shaft-angle encoders	Very useful for on-line input of quantities whose values are related to some qualitative aspect of the graphical display, i.e., for the cases where the precise digital value is less important than some given interrelation
Rand tablet-stylus [34, 35] Voltage-sensing pencil Light pen	See below See Ref. [20]

TABLE 9.8b

Tablet Input Devices

Name/Mfg	Size/Addressing	Stylus	Digitizing Speed	Resolution	Approx Price
GRAPHACON Bolt Beranek & Newman (Rand tablet)	10.24 x 10.24 1024 x 1024 pts digital	Ty's switch actuated by pressure can operate through conductive paper	4500 pts (220 μsec/pt)	.010″x.010″	$9500
DT-1 Sylvania Electronic Systems	11″ x 11″ 4096 x 4096 pts analog	functions within 1/2″ of tablet surface height is also quantized (4 levels)	200 pts/sec (5 ms/pt)	N/A	$6900

that it cannot be used to designate a place where no image is being displayed. That is, in order to know where the light pen is, the computer must "see" it. On the other hand devices such as the Track Ball, "volunteer" their position coordinates. The trouble is that the procedure for correlating this position with the display list often becomes more complicated because the input is not synchronized with the display list. It is not clear at this point which of the various schemes is most useful. In fact, the "best" things probably are not invented yet.

9.4 GRAPHIC INPUT†

So far in this chapter we have ignored the question of how the original data is supplied to the computer. It may be that graphic methods are indeed the best way to interact with a computer, but without data such a decision has no foundation. Most, if not all data is not in digital form when we first decide to use it. Hence the material, be it photographs, drawings or even lists of numbers, must first be digitized in order to feed it into the computer. Some applications require only a trivial amount of digital data and can therefore be serviced by ordinary computer input devices such as card or tape readers. In some cases we can get data in through a keyboard or with a light pen. There are, however, huge sets of potential data from sources that are completely outside of any computer environment. In order to be used these must somehow be gotten into a form that the computer can handle. This fact may be obvious, but very often it seems to have been overlooked. Establishing a data bank is not easy. For example modern sciences such as high energy physics annually produce millions of photographs often representing the only record of experiments, tests, and observations. Prior to 1960 most of this digitizing was done by hand. Since then several types of machines, and programs have been developed that attempt to automatically read (digitize) graphic material directly under computer control. In addition to these, there is also a variety of alphanumeric character reading machines for special applications. We will restrict our remarks to those general purpose graphical input devices that can be interfaced to a digital computer. We will require only enough description to see what limitations there may be in their use.

† This section is adapted from UCRL-70259, "Pictures, Computers and Input-Output", G. Michael, November 1966.

Basically these machines consist of

(a) a (more or less) controllable light source, e.g., CRT, LASER.
(b) appropriate optical components to use the light, e.g., lenses, mirrors.
(c) mechanical components to hold and perhaps move the graphical record during the digitizing procedure, e.g., film or paper transports.
(d) electronic and electro-optical components to measure the amount of light remaining after it has "interacted" with the material being digitized, e.g., signal processing electronics, photomultiplier tubes (PMT).

Figure 9.4 shows the functional relationships involved. There are two general ways to use light as an information probe; that is, to see. When the information appears on a transparent material such as film, the light is transmitted through it, while if the record is opaque, then reflected light is used. After transmission (or reflection) the light beam contains information about the object being digitized.

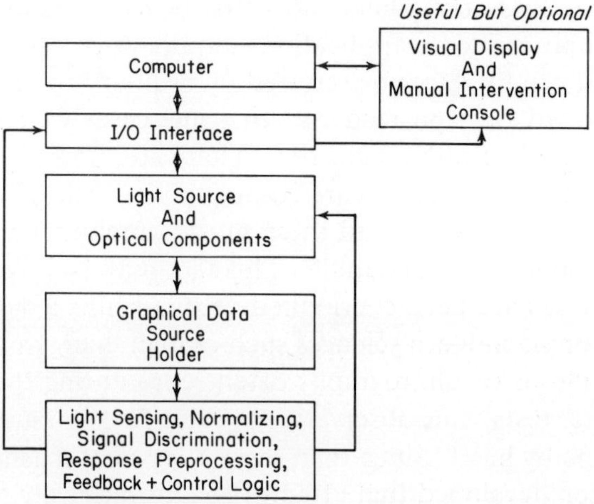

FIG. 9.4. Graphic input device - functional relationships.

The newer reading devices use a computer-controlled CRT as a light source, particularly if the input documents are transparent. There are two types of information we can obtain from any pictorial record. Qualitatively, we might wish to know what is in the picture, and this operation is called *scanning*.† Shape, for example, is a qualitative item. Quantitatively, certain of the images will be selected for *measuring*.

Position, distance and photographic density measurements are examples of quantitative data. Of the two, scanning (i.e., pattern recognition) is far more difficult to express as an automatic procedure.

In operation, the computer program tells the CRT to display a point of light at some designated location (X, Y). The lenses image this spot on the (film) sample. The amount of light that passes through the (transparent) film for example, depends on the clearness or blackness (photographic density) at that point. Whatever does get through is converted into an electrical signal by a photosensing device (PMT). This may then be compared with a suitable reference signal obtained by sampling the unattenuated CRT light. The comparisons are used to make the measurements independent of light source variations and for other control functions. Finally, some (or all) of the results are presented to the computer for appropriate use by the program. Figures 9.5, 9.6 show schematically the design of typical film readers for transparent and opaque materials.

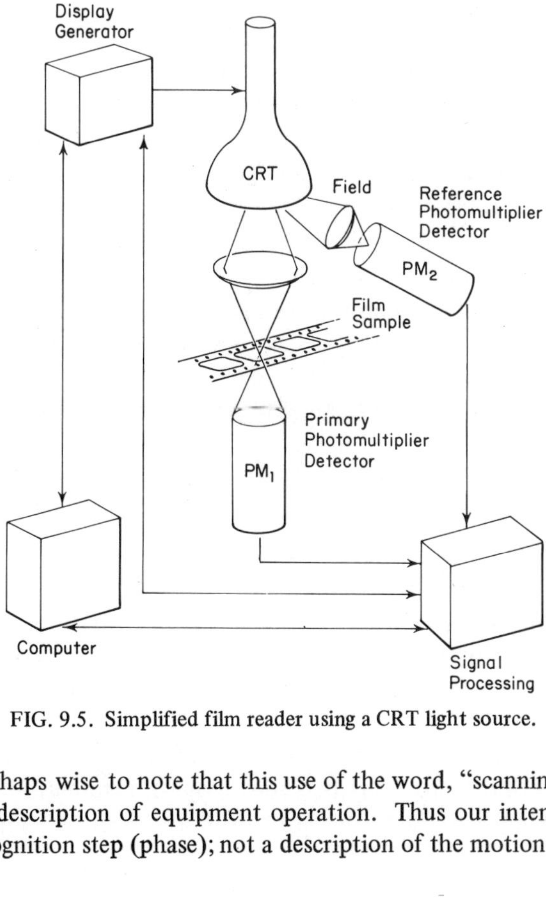

FIG. 9.5. Simplified film reader using a CRT light source.

† It is perhaps wise to note that this use of the word, "scanning," is more procedural than a description of equipment operation. Thus our intention here is to suggest the recognition step (phase); not a description of the motion of the electron beam.

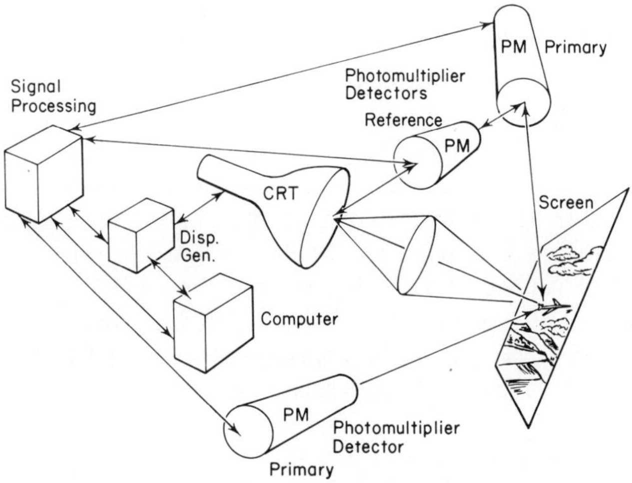

FIG. 9.6. The opaque reader using a CRT light source.

By repeated interrogation the digitizing program builds up various kinds of elementary "facts" about the input data. For example, we might be trying to deduce an ordered set of coordinates (x, y) that give the midpoint locus of some curved trace on the film. Or we might wish to find all the places where the photographic density is greater than some given value. There are some applications in which the (x, y) coordinates are precisely the total data needed from the graphic record. More often, however, these elementary facts serve only as input for a pattern recognition program. That is we are more interested in the logical orderings or sequences of the set of facts than we are in each fact individually.

When the graphic data is opaque it is generally better to use the method outlined in Fig. 9.7. In this case the light source does not have to be under computer control, but the light sensing part must be. We can regard this device as almost equivalent to a television camera. The principle difference is that tiny elements or portions of the visible scene are interrogated in the order that the computer program chooses while the "TV" camera is built to sequentially scan, line by line. In general the recently developed camera design is potentially more adaptable than a film reader though perhaps not as accurate. "Merely" by using the appropriate lenses it can function as anything from a microscope to a telescope.

Of the several different machines now in existence, each is optimized for different conditions. Most of these, by an interestingly large margin have been developed by the various users themselves. This is especially

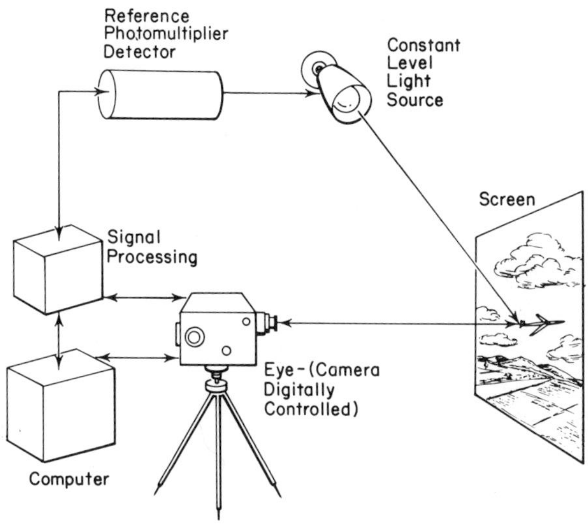

FIG. 9.7. The mobile eyeball.

true among the high-energy physics groups where, in the late 1950's, a pressing need arose to find faster ways to digitize the photographs of elementary particle events. These events, hidden so to speak amidst hundreds of thousands of pictures, needed first to be found and then measured with micron accuracy. In general the machines that are evolving in response to these and other needs are not yet fully automatic. Those machines that are now commercially available are listed in Table 9.9, but as usual we cannot vouch for its completeness [11].

Because the digitizing procedure is under computer control we could reasonably expect it to be dramatically faster than a human reader. Unfortunately this is not usually the case, the most persuasive reason being that humans are better than computers as general pattern recognizers. However, human readers make non-systematic errors whose rate depends on things like inexperience, fatigue, and even emotions. So we would like to use graphical input devices not only to handle large volumes of input data but also to minimize subjective biasing. The chief and only block to a full utilization of graphic input devices is the difficulty of programming them. Or, putting it another way, practically any of these machines have capabilities that have not yet been fully exhausted by programming.

It is easy to see, especially in the case of a CRT light source, that the graphic input document becomes an array of tiny picture elements ex-

TABLE 9.9
Computer-Controlled Graphic Input Devices

Name	Manufacturer	Control Computer	Light Source -- Phosphor Addressability - Size [1]	Document Format	On Order Or Delivered	Remarks[2]
WDA	Link Group: General Precision Inc.	IBM 7040 Others to order	7″ CRT - P-16 2048 x 2048 - 3″ sq.	70 mm rolls Other size optional	∿ 2	.15 high precision potentiometers for manual input. Film exposure included -16 levels of brightness 16 levels of density readout.[6]
PFR-3	Information International Inc. Los Angeles, Cal.	PDP-9 (PDP-10 option)	5″ CRT - Type A Ferranti 16392 x 16392 - 3″ sq.	16-35-70 mm film	∿ 7	Earlier PRF-2 (DDP-24) had a movable film holder 1″ format 20 steps in each direction 15:1 objective lens 64 levels of density readout
Computer Eye	Information International Inc. Los Angeles, Cal.	Optional or PDP-9	3″ (.002″) Image Dissector[3] S-20 4096 x 4096 - 1.5″ sq.	Anything[4]	∿ 5	Camera optics: Zeiss Synchro - Compur 80 mm 1:2.8 others per reqm't
CDC 272	Control Data Corp: Digigraphics Burlington, Mass.	CDC-3100	5″ CRT P-24 8196 x 8196 - 2″ sq.	2″ (max) cut film	N/A	Part of the CDC-270 Digigraphic System 64 levels of density readout
FIDAC	National Biomedical Research Foundation Silver Spring, Md.	IBM 7094	5″ CRT P-16 Fixed video scan[5] No interlace - 3″ sq.	35 mm film roll	∿ 4	NIKKON optics
Biological Scanner	Electronic Image Associates Boston, Mass.	IBM 360/40	5″ CRT P-24 2048 x 2048 - 3″ sq.	N/A	∿ 1	64 levels of density readout
DRC-150	Westinghouse: Astroelectronics Div. Newbury Park, Cal.	PDP-8S Others to order	1″ (.001″) Image Dissector[3] S-20 4096 x 4096 - .5″ sq.	Anything[4]	∿ 4	Camera optics: Adapted for NIKKON F series of lenses

NOTES:

1. No statements of stability, resolution and measuring accuracy are given only because no standard cross comparisons were available. Size refers to the light source since optical lenses can image this on to a variety of formats.
2. With the possible exception of FIDAC all systems provide for a visual CRT for manual control and intervention.
3. For a basic description of Image Dissectors see reference 10 section 5.5.
4. "Anything" implies that this device is a camera and as such, lenses determine what can be imaged (see section 9.4 this chapter).
5. FIDAC scan is a non-interlaced fixed raster scan. Two rasters are available (1000 x 800) points, or (700 x 500) points with 3 density bits/pt. Either scan, once started runs to completion loading 36 bits every 6 μsec into the IBM 7094 data channel.
6. A new special purpose scanning system will also soon be available from the Link Group. It is called AMACUS, an acronym for "Automated Microfilm Aperture Card Updating System". It is designed to do a full scan (spot matrix of 6144 x 4096 bits) storing the digitized data on a 3 x 10^7 bit drum, under control of a digital computer. Portions of the scan data can then be accessed by the computer, which transfers it to a smaller drum used to sustain a high intensity projection (26″ x 17″) CRT. The data can then be modified and a new aperture card produced.

actly corresponding to the raster points mentioned earlier. Furthermore, the only answers possible are "yes" or "no", although most machines are built so that an edited answer, "the optical density (at X, Y) is such and such," is also possible. Hence the successful use of these devices implies that we somehow must ask the right questions. The first problem then becomes one of deciding which points belong to the information we seek, given that we are limited in what we can ask. It also turns out that if we aren't careful about the way we ask, the machine quite often will "lie" to us. For example, probing the same point too frequently may give contradictory results. In any event we end up with a "mosaic" of yes-no responses and the computer program, by suitably combining these elementary responses, produces the desired input data.

The important point to emphasize is that the answers we get depend totally on the questions we ask, especially as these relate to the fundamental characteristics or properties of the information (light) probe. For (a somewhat strained) example, suppose a picture contains among other things, a round uniformly black blob. We cannot tell if it is hard or soft, spherical or flat because we have no way to ask such questions. We could, however, decide if it was really circular. It does seem that a broader truth is involved here: we ordinarily know things *only* in terms of their interactions with the particular measuring device we use.

Even though it is somewhat of an oversimplification, we might suspect that one basic task for pattern recognition is the production of computer-compatible descriptions of the objects we wish to identify [12]. That is, how *do* we describe what something "looks" like to a computer? (When we look at something we "recognize" it primarily by the amount of agreement between our visual image of it and our "memory" of things similar to it). We can either prepare these "property lists" or better yet make the computer build its own lists. Of course this begs the question a bit. As far as we know, there is no suitable theory of perception or cognition that can be drawn upon. Similarly, there is no concise set of heuristic procedures that can be made into a general computer program. One thing being done at present is to write codes that work for very limited applications. Another approach is to let the computer do the digitizing, but arrange for a human to do the pattern recognizing through a display terminal [13]. Conceivably he might even teach the computer.

We have already mentioned an instance when the machine might "lie" to us. This is a particular example of "noise" influencing the data. In general the digitizing will always contain extraneous data. Dirt becomes noise and so do scratches. As far as the program is concerned, anything about the graphic document that is *not* exactly the data *currently* being extracted is noise. Thus intermittent component failures show up as noise, as do scratches, dust, low contrast images, and so on. If two or more of the objects being examined touch or overlap they act as sources of noise for each other. The recognition procedures must be capable of deciding that some particular array of spots is an E even if it is smudged or touching a neighboring character. Noise not only complicates the problem, but to many people it is *the* problem.

Quite often, a certain amount of noise filtering can be safely applied to the data as it is produced by the reading machine, but to be acceptable, the pattern recognition scheme must at least be able to correctly function with noisy data. In addition it must be independent of data conditions like size, position, orientation, and perspective variations. That is recognition of an object, depends on the object not the apparent size, or where it is on the film or the direction it is facing. Finally a practical solution must be expressible in a comparatively small number of steps, require a finite amount of time and memory for execution and be iterative. That is, reapplication of the algorithm(s) should improve the results.

Despite the scope and number of difficulties confronting workers in the field of pattern recognition, it would be erroneous to conclude that little or no progress is being made. Quite the opposite is true. In the area of character recognition there are at present several commercially available machines to choose from, depending on the number of restrictions placed on the size of the font and document format. Similarly, a few general purpose designs are also on the market for research and production applications and more are coming. One such machine has now been in operation about four years. To date it has digitized oscilloscope traces at a rate of about one million pictures per year, becoming in the process a simple example of a research project that successfully evolved into a production application. So while not wanting to minimize the difficulties in this field, we can expect eventually some measure of success.

9.5 SOME FUTURES

In this last section we want to briefly discuss the following questions.

1. What are some of the features that future displays might have?
2. The multi-terminal problem?
3. How important is language development?

For the first item we have made a list; being at the same time aware that such things, while they may be useful, tend to be limited by current experiences. Thus, the features summary in Table 9.10 should be viewed as a partial list of things that may be useful in future developments. Also, the included comments are intended more as explanations than definitions. We have not mentioned such things as stability, repeatability, reliability, and addressing structure. Moreover, questions about timing were left out, but not because they are unimportant. Quite the opposite is true. Rather than launch into long discussions which are only slightly better than opinions, we have assumed that displays must have these properties in the same sense that we expect a camera to be "light tight." Continuing with the same analogy, we would demand that these properties be demonstrable.

The features listed in Table 9.10 are not necessarily aimed at all display equipment. That is, even if costs were not a factor, it is not at all clear that each display must embody all possible features.

There are two other things to be said about Table 9.10. First, it is entirely possible that some very important features are not in the list, while others that are may turn out to be unnecessary or not worth the effort to get them. The second point is that Table 9.10 is somewhat slanted toward CRT displays whereas there are a number of interesting substitutes now coming along. In fact, they have been "coming along" for some time now. The situation is somewhat like the computer memory business. When core memory was first introduced, it was so slow and expensive that people didn't expect it to last. However, it has continued to improve so that other more exotic memory mechanisms still have not really replaced it. The same is true for CRT's. So far their rate of improvement has tended to offset the expected gains from other display methods. Quite possibly some of the new technologies will (happily) someday supplant CRT usage. For those who would like to pursue the matter further, we have included a few references to new display techniques in Table 9.11 and Appendix A.

There is one final comment about future display devices. From an engineering point of view all the devices listed in Table 9.7 work as described. From a user's point of view, however, there are little, nagging "inconveniences" of one sort or another. Perhaps if the display designers could keep in mind the fact that their product is supposed to be used and maintained by people, things might be more thoughtfully done. Here are just a few examples. There is no good reason for not allowing some adjustment to the tilt of the display screen. Some people like to

TABLE 9.10

Summary of Some Useful Display Features

Feature	Description
Point plotting	
Line generation	Some of the current displays produce lines by sequential point plotting. This is not what we mean. Instead we intend that, given any two points on the screen, a "straight" line is drawn between them.
Edge fault detection	If the display screen is taken to be a window through which we are seeing a partial section of an arbitrarily large "drawing", then we want to know when the program that is producing the drawing tries to draw beyond the field of view.
Character-generation and rotation	Producing characters by dot plotting is time consuming at best; so the tendency has been to use the faster method of "wired-in" generators. These produce characters that are, say, oriented along a horizontal line. By rotation, we are saying that the programmer should be able to produce characters along any "slant" line. The range of selectability ought to be more than 128 characters (7 bits); 256 characters (8 bits) may be enough, especially if the user can insert special characters, possibly as many as 128, by program control and with no significant loss in the generation rate.
Multiple-tube addressing	One very-high-speed display generator can service several user terminals.
"Subroutine" features	If the display has buffer memory, it should be possible to send codes that cause it to do something other than display, such as set flags, sense flags, transfer control, etc.
Asynchronous deflection and unblank controls	There is no need to penalize every user by designing in a time delay based on a maximum deflection requirement. If the beam is moved only a small distance, the time needed is shorter, and the program ought to be able to benefit from this.
"Where am I" registers Computer-readable deflection registers	After a light-pen signal from the middle of a vector or while in the "typewriter" display mode, we normally do not know the beam position (x,y) explicitly. If we want to subscript or superscript a character, it is nice if the display tells where the beam is.
Multilevel and independent gain controls (raster size for each axis)	Usually CRT rasters are inscribed squares. This wastes portions of the screen since they are not addressable. Sometimes it would be nicer to circumscribe the raster even if some of the x,y addresses would ıen be illegal.
Color toning Three-dimensional viewing	Two-tube console and/or a view-isolator housing. The user should not be prevented from using a light pen, however.
Flat face tubes	Somewhat more convenient for light-pen work.
Variable-persistence (time) phosphors	For some applications the "storage" tube principle more or less solves the refresh problem.
Keyboard and other input-output devices	See text.
Light pens and other interaction devices	See text.
Adjustable display housings.	See text.

TABLE 9.11
New Technologies for Display Consoles [36]

High Resolution Storage Tube (Meshless Storage) [37]	(Uses Electroluminescent Panel but conduction (i.e., light emission) is stimulated by an electron beam; features: high resolution, long (∿ 5 sec) persistence, microsecond erase times, CRT addressing structure)
Plasma Display Panel [38]	(Low cost, high speed, *digitally controlled* display using the "Neon Sign" principle for producing light from a confined sample of gas. Some storage tube features, median resolution, long persistence, microsecond erase times. Digital addressing structures)
Multi-component CRT phosphors and multi-layered phosphors for producing color [8,9]	Various tube development laboratories and manufacturers. See Appendix A.
Color Image mixing by projection from ganged CRT's. [8,9]	See Appendix A.

work on tables, not walls. Some of the light pens presently being used are really more like light-logs; heavy, unwieldy and fatiguing. The console keyboard is often built into the counter top so that it cannot be moved to suit the user. The keys very often give neither audible nor tactual feedback when they are pressed. As a consequence, the user tends to "break his fingers" trying to push the keys down, and so on. The surprising thing about such complaints is that it would cost very little extra to correct them, provided they were considered before construction.

The progress of graphics research so far is almost exclusively epitomized in the union of one man using one limited CRT display and one limited computer, all being rather expensive and inadequate.

Considering the number of people who want to use such facilities, it is clear that simple duplication of existing equipment implies enormous costs. This alone is enough to stimulate the search for alternate methods that would provide adequate facilities at costs within the reach of a large number of people. So far no practical solutions have appeared. Beyond the issue of price there are technical problems that need to be solved. For example, the problem of refreshing a CRT display, responding to user interactions and doing calculations, all simultaneously, tends to fully saturate a computer both in speed and memory capacity. Thus, adding more CRT's would be pointless in such a configuration. Some of the newer computers are almost fast enough, but the costs are still far beyond a feasible range. Moreover, the technical requirements of high frequency, high accuracy data transmission between the computer and the CRT limit their separation so that the user is forced to work very near the computer. Where large numbers of people are involved this is not generally satisfactory. So, if we expect to provide a graphic terminal for everybody that wants one, it will be necessary to invent other devices and methods.

Often knowing who the potential users are and what, in general terms, it is they want to do will be almost as important as any new developments in electronic components are. For example, university faculties that wish to service the needs of thousands of students will be interested in something quite different from what an industrial design firm would require. Thus, while we recognize the "chicken in every pot" attitude, (everybody wants general purpose graphic terminals), the costs at least will force us to consider a spectrum of display features and qualities.

Most people feel that computer serviced graphic terminals are an excellent interface between men and machines. They might also agree that there should be a hierarchy of display devices. In other words, not all displays need to be of the highest quality. If we admit some usefulness to a low quality display then we might also agree that not all users need high speed display. That is, there are limits to the amount of information that a user needs to see at once and further that he may not need moving displays either. We might go on down the list; not everyone needs color, or line generators, or keyboards, and so on. There are several motives for making such compromises. First, we hope to get the cost per graphic terminal down into an acceptable range [14,15]. Second, we must make the total data requirement feasible for the computers, memories and transmission lines that are available now. Third, we expect to achieve better reliability by having fewer or simpler parts.

The lowest quality display so far considered is a commercial television (TV) receiver. Low cost, established electronic standards† and widespread familiarity with the technology make it worth the effort to try to use them with digital computers. One major difficulty that now arises, however, is the necessity of converting a user-oriented display list into a series of TV lines. The process known as "scan conversion" can be expressed as a computer program, but it would be a relatively slow-running affair. Alternatively, the electronic equivalent could be built,

† Internationally, there are at least four sets of standards [16], but this is not a major concern for our purposes.

again so that its cost is spread over many users. Figure 9.8 shows one of the many possible ways to implement a digital TV network for which all of the implied components presently exist.

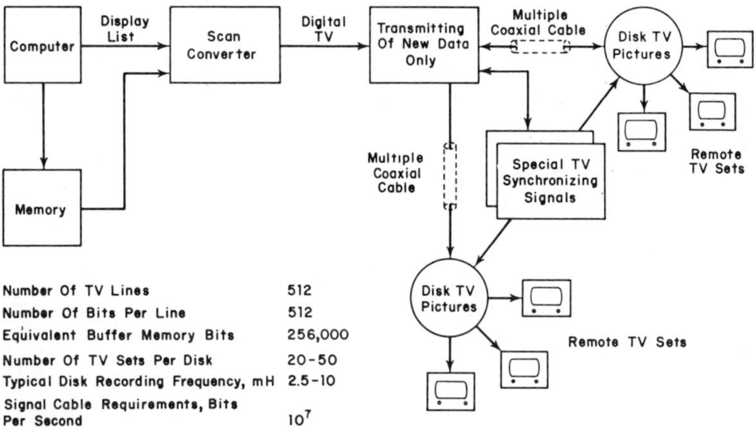

Number Of TV Lines	512
Number Of Bits Per Line	512
Equivalent Buffer Memory Bits	256,000
Number Of TV Sets Per Disk	20-50
Typical Disk Recording Frequency, mH	2.5-10
Signal Cable Requirements, Bits Per Second	10^7
Estimated Cost Per Terminal Excluding Cable Costs, Dollars	1000-6000

FIG. 9.8. Possible characteristics of a computer-driven TV display system.

There are two (at least) fundamental objectives of this or any comparable design. First, some method must be found for low cost picture storage. Second, the burden of picture regeneration must be removed from the computer. Periodic memories such as magnetic disks or drums are quite suitable on both points. User generated display lists are "passed through" a scan conversion procedure. The video equivalent of each list is then stored on the disk so that one rotation completely refreshes the TV screen. Thus, the disk needs to rotate thirty times per second. The amount of data to be sent to the disk for each TV set can be drastically cut by only transmitting those picture parts that have changed. This means that a number of pictures can be updated during each disk revolution. We can expect that a fast computer with a large overall memory capacity and operating in a time-shared mode will be able to service a reasonable number of such terminals. Light pen interaction can be provided, but would probably be limited because of the distance from the computer. Since the use of TV receivers is still being evaluated we will refrain from further comment.

The third and last question we wish to consider is the role of graphic languages. Of all the issues mentioned so far, the problems of language development are the most complicated and also the most important. A detailed treatment would not only go beyond the primary intention of this article, but would also be somewhat premature in the sense that right now there are more questions than answers; more problems than solutions. In the previous sections of this chapter a number of specific problems have already been mentioned. Hence, we will merely try to outline a few additional areas of immediate difficulty.

But first, to support the claims that these problems are the most important and complicated, let us make the following related assertions:

1. At least relative to the application of computers, the expression of concepts and methods is limited ultimately to the basic command or code set provided for the computer. In a sense there is now no way to easily program many properties that the user takes for granted in everyday conversation, that is, the degree of "expressiveness" fundamentally influences the (computer's) conception of the "universe of relationships between things."
2. In the sense of computability, no theory of perception or cognition has so far been successful. The language of ordinary discourse is related to human thought, but this may not be necessary for computability. In any case, there are very few guide lines at present that can be used to develop graphics as anything like a science.

This general view of the linguistic problem (if it is at all correct) suggests that solutions are a long way off. However, as we move in that dirrection there are a large number of very specific problems to solve. Of those we know about so far, many are in a sense not even exclusive to graphic data processing. The development of time (or resource) sharing is such an example, and itself contains other examples. For instance, we expect the executive system to maintain our files and protect them from other users. However, several geographically separated users who are jointly working on a certain design may wish to share specific files for some collaborations, but not for others. Or, suppose that someone is willing to allow others to use his program but does not want them to be able to modify it. If the user is working with a visual display, he should not be inconvenienced because his program is periodically interrupted. In the "old days" people were willing to wait many hours between computer runs, but under time shared operation even a two second delay is simply not acceptable.

Other problems arise from the necessity of getting around various equipment deficiencies, such as limited core memory, light pen noise, or low processing speeds. These are, in general, examples of questions that depend on overall system organization. Another type of problem comes from what we might recognize as pictorial realism or aesthetics. For example, the user may wish to suppress those lines not visible because of the given perspective.

One of the more basic questions that we continually face is, "How can or should pictures be described to a computer?" The answer (or answers) to this will fundamentally influence that part of graphic data processing known as *Pattern Recognition*. The overall scope of graphics covers more than simple displays. Indeed, a very large part of graphics is the *inputting* of pictorial data and here the role of Pattern Recognition is fundamental. Whether the data is an architectural photograph or an electronic circuit, we need to identify what the various shapes and lines and shadows mean. We need to be able to reconstruct shapes including the hidden lines from pictures or sketches. So, whichever language evolves for graphic output, ideally it ought to work for graphic input as well. Moreover, from the user's point of view there should be no logical difference whether the graphic data came from or is going to film or a sheet of paper or a CRT screen. We have already seen that ultimately a picture or graphic structure must be described in terms of those operations that a display generator can produce. Such a language, although precise, is not expressive enough for easy use by people. Thus, we need a higher level language [10, 17, 18, 19, 20] in order to express more than operations. Once there is enough structure we may find that we can describe certain kinds of interrelationships (e.g., A is part of B), qualities (e.g., A is green), and even concepts (e.g., find all the square parts). Fortran and Algol are good examples of this in that people actually communicate ideas through them. But, there are basic disagreements [21] about their utility as complete picture processing languages.

Some people feel that graphic data processing is not so different from "other kinds" of data processing that it needs separate language development. There are some special requirements, but these are not so numerous nor unique that they can not possibly be added to one of the already existing languages. So far, we have tended to follow this approach and one specification is discussed in the next chapter. If and when something better comes along it will be dropped with no regrets.

Others feel that because there is so much unknown it is premature to accept the restrictiveness of any existing language. Since there are very few good guide lines, and essentially no theories they prefer to depend on user feedback to influence the linguistic structure. This line of reasoning says that we should develop new languages that take their motivation from the "new problems." In this case it is especially important to be able to easily change the language; deleting, expanding and redefining very general procedures as the needs arise. In effect, they require a context-free compiler that compiles other compilers. According to this view Algol and Fortran are simply derived languages, each suited to a limited (albeit large) class of problems, but now with common source and respecification paths.

It is obviously too soon to insist very strongly on one point of view over another. This is the main reason for worrying about the "parachute folder's syndrome." If genuine users can really influence the development of graphic languages then the product, while not necessarily elegant, will work.

As things stand today the number of people working explicitly on graphic research is increasing rapidly and consequently so is the number of different approaches. Most of these (not surprisingly) seem to be quite sensitive, both to the type and configuration of equipment being used and the uncertainty about future equipment characteristics.

In the final analysis it is the presence of real equipment that particularizes and motivates any development.

ACKNOWLEDGEMENTS

The modest degree of progress made at the Lawrence Radiation Laboratory in graphics is the result of many people working together. The high quality of the color motion pictures is due first to M. Lehman and T. Rombough who keep the display equipment so well adjusted and second to D. Dixon, M. Milochek and F. Buys who are responsible for the quality photographic processing. D. Crawford, D. Dixon, and R. Krumhansl were responsible for the light pen design. In developing the graphics languages the following people have been fundamentally involved in the work, and their contributions are of the highest importance: C. Badger, D. Bailey, K. Bertran, G. Boer, A. Cecil, P. Crowley, E. Goodwin, N. Hardy, N. Ludkey, D. Mapes, G. Marshall and C. Springer. B. Saunders has been a principle source of information on visual processes. C. Haussmann, Associate Director of the Lawrence Radiation Laboratory, Livermore Site, and S. Fernbach, Director of the Computation Division, have fostered display research from the start. Their support of this type of work has been vital. It is appropriate finally to acknowledge the primary importance and the final authority of the long suffering USER.

APPENDIX A: TECHNIQUES FOR PRESENTING VISUAL DATA

We have already stated that graphic presentations permit very efficient information transfer between the computer and the user. Deciding how to present the data is generally a matter of individual tastes.

We will briefly outline four particular techniques that are often useful either separately or in various combinations.

1. Color Marking
2. Stereoscopic Projection
3. Perspective Mapping
4. Texturing

While the method we use to add color to a picture gives (so far) the best and most consistent quality, it is only appropriate to filmed displays. The total picture is organized in the computer as a number of separate picture parts according to the colors that we want the finished picture to have. When these picture parts are plotted (and therefore photographed) each frame of the black and white film contains only those picture parts that are to ultimately appear in one color. We call this procedure "Logical Color Separation." It is obvious that any reasonable number of separate black and white frames (and therefore hues) could be produced for each final frame of color film. In practice we have used eight separate colors and obtained satisfactory results. The separated picture parts are remerged through the appropriate color filters, onto color film using a "Special Effects Optical Reduction Step Printer." Figure 9.9 shows this schematically.

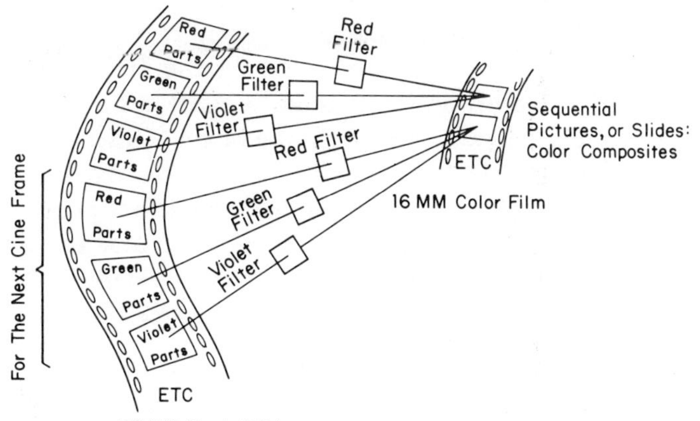

FIG. 9.9. Method for producing colored output on film.

Another method that also works only for film uses a white light phosphor (e.g., P-4) CRT screen. If then suitable color filters are sequentially inserted in the light path, a multi-colored image can be built-up directly on color film.

A number of other methods [8, 9] for more directly generating color displays have been, or are being, tried at various places. In one case three separate CRT's are arranged so that they image an exactly common area of projection screen. Each CRT is designed to produce just one of the primary colors so that by the use of appropriate dichroic filters and reflecting surfaces a color-blended composite image is produced at the viewing screen. The method, while workable, depends on great stability in the CRT electronics, and cleanliness for the optical components.

In another scheme,† a standard television color tube (either "shadow mask" or "chromatron") is used as the CRT. Suitable electronic connections are made to a computer interface so that the programmer can control each of the three electron (i.e., color) guns individually. One problem that arises is that for a digital deflection system the resolution is limited by the construction of the color tube, since it only has a finite number of centers for each color.

The illusion of three-dimensional projection requires [22, 23] that two related views be produced. One view is for the left and the other is for the right eye. It is a simple matter for the computer to generate these views.

One method was demonstrated* several years ago, using a spinning phosphor disk for the CRT screen. The disk was coated on both sides with a fast decay phosphor and from the point of view of the electron gun spun like a flipped coin. That is the axis of spin was orthogonal to the direction of the undeflected electron beam. It was spun fast enough to be essentially invisible to an observer. As the disk turned it passed through all angular positions and could be written on at any time except of course when it was edge on to the beam. Other than this single restriction when the display controller wanted to display a point of light at location X, Y, Z, for example, it would first position in X, Y and then wait until the disk had turned enough so that there was phosphor at the chosen Z coordinate. The beam then was unblanked. In this manner a "true" three-dimensional object could be displayed. All the

† C. Walter, Air Force Cambridge Research Laboratories, Lexington, Mass.

* R. D. Ketchpel, Hughes Advanced Research Laboratories, Malibu, Calif.

components were housed inside a clear glass vacuum jar so the viewer could actually move around the display. A small difficulty occurs whenever the disk is oriented nearly edge-on, because the beam then tends to just graze the screen and circular spots become very elongated ellipses.

A variation of the generated volume technique was recently developed at the Mitre Corporation in Bedford, Massachusetts.† In this case a standard CRT is imaged onto a vibrating, mirror-like membrane. Again, by knowing when the mirror will be able to reflect a chosen "space" point, the CRT can be unblanked at the proper instant. The amplitude of the vibrations need not be large to promote a stereoscopic illusion.

In the approximation we use, two separate projections (views) are generated, one for each eye. For one of the projections the object to be viewed is rotated relative to the other about a vertical axis through some angle which is normally set at 7.5° (Fig. 9.10a). This approximately, is the eye-separation angle subtended by a point in space when the interpupillary distance is 67 millimeters (mm) and the distance from the eyes to the point is 500 mm. This angle produces a satisfactory stereo effect for most people. Below 5° the view seems flat, and above 12° a pronounced eye muscle strain is experienced. Stereo viewing is achieved simply by permitting the left eye to see only the left eye view and similarly for the right eye. Both cross polarizing glasses or eye screening give satisfactory results (Fig. 9.10b). Usually stereo is not needed if motion is being used. Other than noting that the method works, (there are other methods, however,) no physiological conclusions are drawn from this model. It has been pointed out [24] that this technique does not yield orthostereoscopic* views. This may not be serious in certain applications, for example, the visualization of a mathematical function. Alternatively it may be very important to generate an accurate three-dimensional view of say, a building design. In this case the two views, one laterally shifted with respect to the other, are produced as shown in Fig. 9.11. [25, 29]

Very often the illusion of depth can be adequately conveyed by displaying only one perspective view [30]. Again Fig. 9.11 shows the interrelationships between the viewer's eyes, the CRT screen (film plane) and the viewed object. In this formalism each three-dimensional point is suit-

† A. C. Traub, Mitre Corporation, Bedford, Mass.

* Natural or normal vision (see Ref. 30).

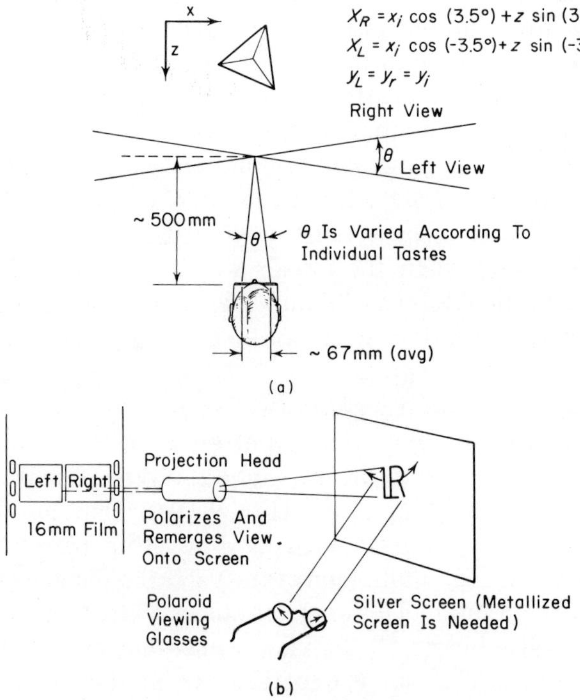

FIG. 9.10. Generation of stereoscopic views.

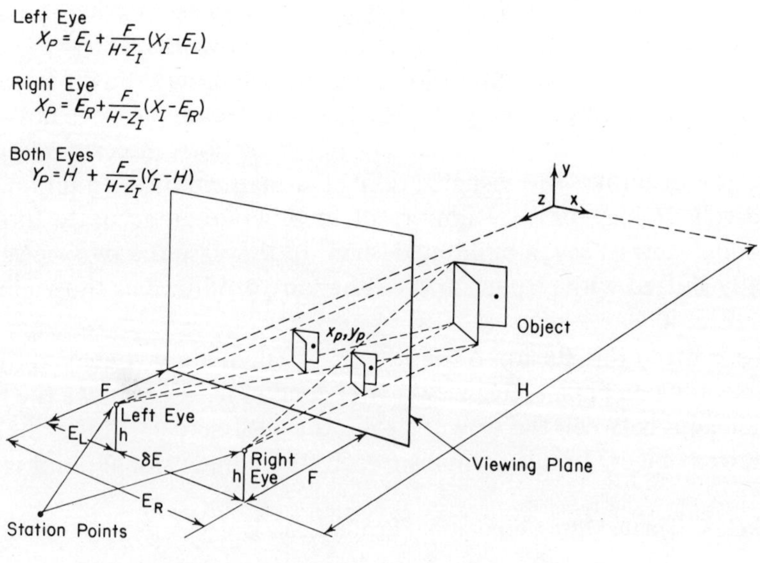

FIG. 9.11. Perspective relationships

ably projected in the viewing plane. The various parameters can take on desired values, and preferences tend to vary from person to person.

The fourth operation, texturing is useful to make the surfaces of an object visible. A planar solid is made visible in space by showing the lines of intersection of the defining planes. This obviously is not a very satisfactory method if the surfaces are not planar. Sometimes surfaces are regular enough so that they can be made visible by rulings, i.e., by drawing a number of parallel lines from one edge to another, or by drawing "latitude and longitude" lines. Another method somewhat more time consuming, is to randomly "spatter" dots on the surfaces. If the CRT has enough resolution, then controlling the number of dots per unit area will approximate the optical density variations obtained by ordinary photography. In many cases the spatter method has to be used with stereo in order to see the desired spatial distribution of spots. Figures 9.12-9.15† show examples of these various methods. The figures are given in stereo format to provide a basis for deciding when stereo is appropriate.

When we display solid three-dimensional objects, the degree of realism often depends on "hiding" those edges not now visible to the viewer. The general problem of hiding edges is not fully solved for arbitrary surfaces, but a number of particular solutions have been published. [12, 25, 31] The main limitation with most of these is the large amount of computing need to decide what can and what cannot be seen. For most applications it is not now possible to sustain a moving, hidden line display for real time viewing.* Although this is not a serious limitation, the problem is of sufficient interest to want to see it solved. These and any other techniques that come to mind are of value to scientific computations only to the extent that they convey useful in-

† Figure 9.15a was obtained through the courtesy of R. Conn, Lawrence Radiation Laboratory, Livermore, California. The digital input for the picture was prepared using the LRL Eyeball (Ref. 11 and 13).

* The General Electric Laboratories in Syracuse, N. Y. have recently (1967) delivered a full color real time display system NASAII to the NASA Manned Spacecraft Center, Houston, Texas, in which the hidden line problem is solved partly by using special purpose hardware to perform certain of the necessary calculations. Even more recently (1968) the Control Image Corporation in Denver, Colorado, has described a fully analog solution in which a VIDICON tube is used to detect electronically when edges or surfaces are about to overlap those picture parts already displayed.

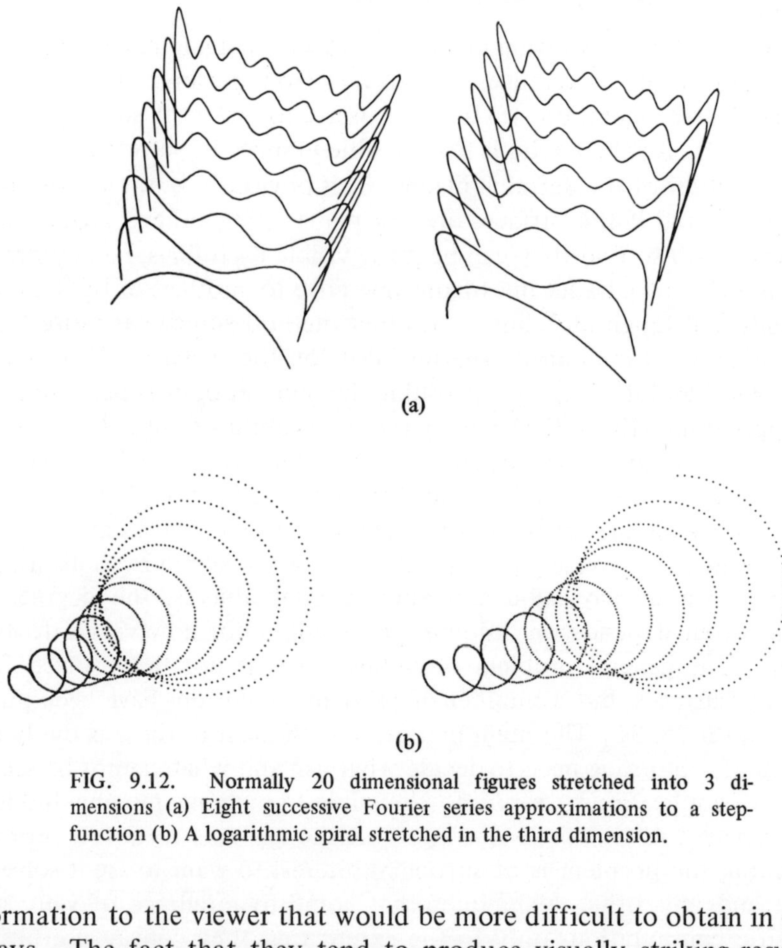

FIG. 9.12. Normally 20 dimensional figures stretched into 3 dimensions (a) Eight successive Fourier series approximations to a step-function (b) A logarithmic spiral stretched in the third dimension.

formation to the viewer that would be more difficult to obtain in other ways. The fact that they tend to produce visually striking results is accepted as a bonus and perhaps even as a stimulus. But the first importance has to be attached to the technical information content.

APPENDIX B: REAL TIME CHARACTER RECOGNIZER

Slate is a short program that can recognize sloppily constructed characters in real time applications. It is designed so that it can be "taught" to recognize a large number of variations in writing and printing styles and the recognition procedure is essentially independent of character size. At present it is implemented for an IBM 7094 that has a CDC 280 display and light pen. This version of the program is a modification of a method developed by W. Teitleman (M.I.T. and Bolt, Bara-

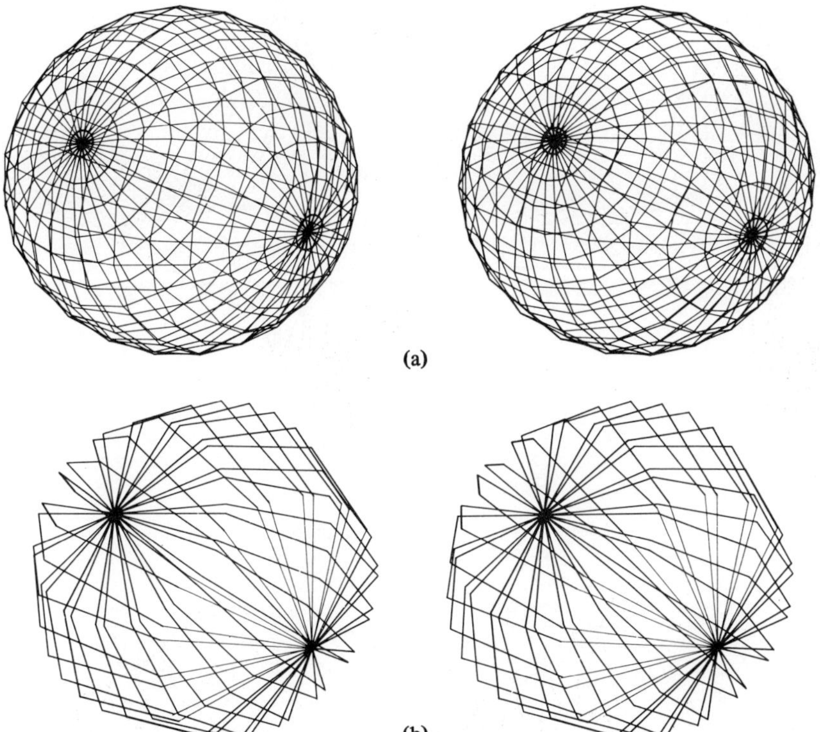

FIG. 9.13. Rulings on smooth closed surfaces (a) Latitude and Longitude lines on a sphere (b) "Longitude" lines on a Japanese Lantern."

nek and Newman, Cambridge, Mass.) [32] and later described also by J. C. Shaw (Rand Corporation, Santa Monica, Calif.).

The recognition is based on a 2 X 3 matrix (Fig. 9.16a). The drawing of a character is remembered in a time ordered way. In other words, the "2" depicted in Fig. 9.16a would be remembered as "365214" since that is a path some people take when drawing a "2." The "2" in Fig. 9.16b can be recognized also, as "365414." In this way a dictionary of characters is built up and in fact, new variants are added to the system as they are encountered in the individual styles of different users. Two problems arise in this method. The first problem is hand jitter. Looking at Fig. 9.16c, an exaggerated blowup of Fig. 9.16a at the interface of cell no. 1 and cell no. 2, we see that a non-smooth transition from cell 2 to cell 1 is possible. Thus, the original recognition of "2" in Fig. 9.16a as "365214" could have been "3652121214" where the underlined section illustrates the "jitter." Clearly, hand jitter would unnecessarily

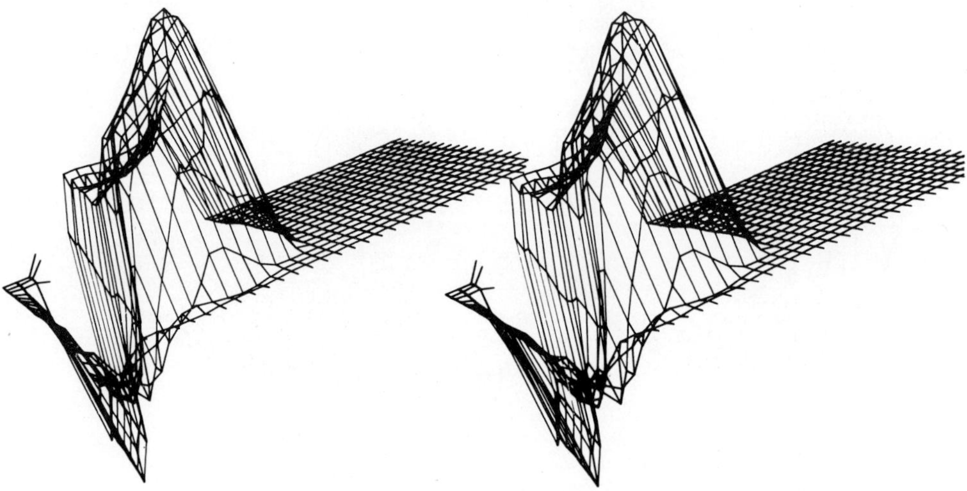

(a)

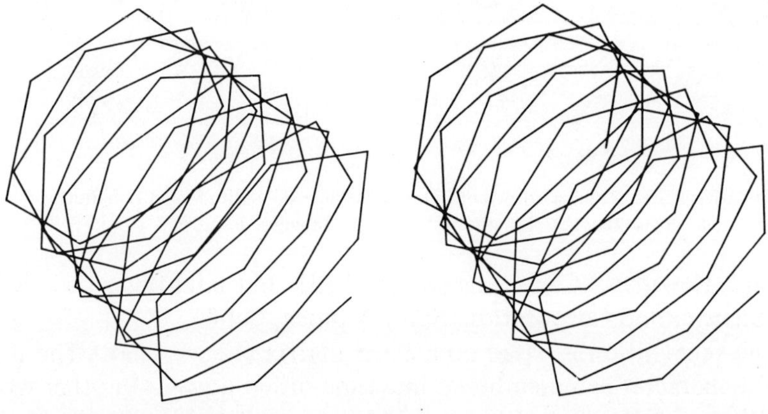

(b)

FIG. 9.14. Linings (paths) on regular surfaces (a) Rulings to show distortion out of the plane (b) A "segmented" wire wrapped around a cylinder.

complicate the number of different specimens (otherwise, essentially the same) to remember and compare against.

One solution of this problem is as follows: as a sequence of cell crossings is being recognized, a test is made to see if the current cell number is equal to the second to last cell number. If this is true we check to see if the last cell number is equal to the third to last cell

(a)

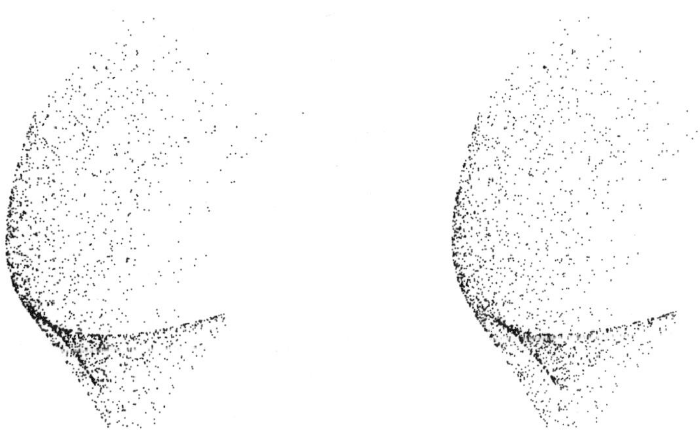

(b)

FIG. 9.15(a) Dot spatter digital simulation of a photograph. (b) Dot spatter method to show the "Peano" surface: $z = (x^2 - y^2)(x^2 - 2y^2)$.

number. If both tests are true, we throw away the current and last cell numbers. Otherwise, we keep the current cell reading as part of the time-ordered string. This test is sufficient to eliminate all hand jittering.

The second problem is illustrated in Fig. 9.16d where the time-ordered trace is "36214". Notice that going through one of the two interstices (6 to 2) requires some extra and unnecessary variants per charac-

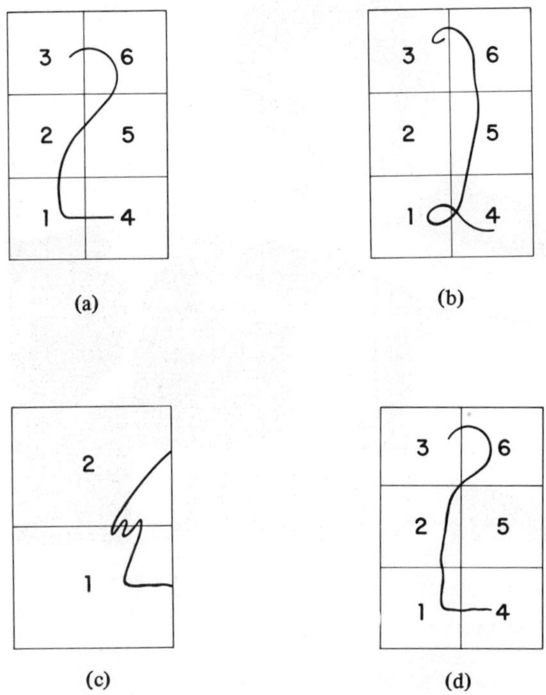

FIG. 9.16. Recognition matrix for the slate program.

ter. The solution to this problem is to check for an even to even (or odd to odd) cell crossing and force a choice (arbitrarily) of one of the odd (or even) cell neighbors. In this example "36214" would be taken as "365214" or "363214."

There are other special tests needed in Slate to make it work for a particular character set. In principle, at least, it is possible to derive a more elegant theoretical model so that special cases are not needed. A large number of particular cases leads to a turgid program. But, as the special cases were encountered, it seemed better to "glue and paste" and get a scheme that worked, than to reproduce a new, "neat and tidy" theoretical model that does not.

Slate has been interfaced to an interpretive algebraic equation evaluator called Scrawltran. It will calculate any symbolic arithmetic equation acceptable to, say Fortran, and display the answers. The attempt here was to provide an easy to use calculational facility. Literally, we want the user to be able to use the CRT screen as he would the back of an envelope.

APPENDIX C: LIST OF MANUFACTURERS

Manufacturer	Product
Bolt Beranek and Newman, Inc. Data Equipment Division 2126 South Lyon Street Santa Ana, California 92705	GRAFACON
California Computer Products 305 Muller Avenue Anaheim, California	M 835 Digital Incremental Plotters (Hard Copy)
Computer Industries - Graphic Systems Div. 14761 California Street Van Nuys, California	120
Control Data Corporation Data Display Div. 2401 N. Fairview St. Paul, Minnesota &	CDC 250, 280 CDC 210
Digigraphics Div. Northwest Industrial Park Burlington, Massachusetts	CDC 270
Data Disk Inc. Display Division 1275 California Avenue Palo Alto, California 94304	Digital/Video disk
Digital Equipment Corporation 146 Main Street Maynard, Massachusetts	DEC 30, 338, 340
General Precision Systems, Inc. Link Group 1077 E. Arques Ave. Sunnyvale, California 94086	APD 5000, AMACUS
Geospace Corporation 5803 Genmouth Drive Houston, Texas	DP-203 "TV" scan-organized, digital plotter on sheet film
Hewlett-Packard 1501 Page Mill Road Palo Alto, California 94304	HP 6946A PREC T.V. MONITOR HP 181A HP 1300A

Information Displays, Inc. 102 E. Sanford Blvd. Mt. Vernon, New York	Monoflec, Dualflec IDIIOM
Information International, Inc. 11161 W. Pico Blvd. Los Angeles, California	PFR-2,3 FR-80-microfilm recorder (new)
International Business Machines Data Processing Division Poughkeepsie, New York	2250, 2260
I. T. & T. Federal Laboratories 3700 E. Pontiac Street Fort Wayne, Indiana	MACC
I. T. & T. Industrial Products Div. 15191 Bledsoe Street	KM 402, 702, 105 KM 906 (color)
Minnesota Mining & Manufacturing Co. 2501 Hudson Road St. Paul, Minnesota 55119	Series F (EBR)
Raytheon Wayland, Massachusetts	DIDS 400
Saunders Associates, Inc. 95 Canal Street Nashua, New Hampshire	720 Communicator
Straza Industries 790 Garfield Dr. El Cajon, California 92021	Model 1311, 1201
Stromberg Carlson Div. of General Dynamics P. O. Box 2449 San Diego, California 92112	SC 4020, 4400, 4060 1100, 1200
Sylvania Electronic Systems 40 Sylvan Road Waltham, Massachusetts 02154	DT-1 data tablet CRT mfgr. (multi-color)
Tasker Instruments Corporation 7838 Orion Avenue Van Nuys, California 91409	Series 9000

Tektronix, Inc. P. O. Box 500 Beaverton, Oregon 97005	Type 601, 611 STORAGE CRT'S
Thomas Electronics 100 Riverview Drive Wayne, New Jersey 07470	CRT mfgr.
TRW, Inc. & Bunker Ramo Corporation 8433 Fallbrook Avenue Canoga Park, California	TRW 85 (1963) BR-90
UNIVAC Defense Systems UNIVAC Park St. Paul, Minnesota	UNIVAC 1551
Westinghouse Electric Corp. Astroelectronics Laboratory P. O. Box 245 Newbury Park, California 91320	DRC 150

REFERENCES AND BIBLIOGRAPHY

1. J. Soller, M. Starr, and G. Valley, *Cathode Ray Tube Displays*, New York, McGraw-Hill, 1948, (Dover Edition – reissue, 1965).

2. G. Parr and O. H. Davie, *The Cathode Ray Tube and its Applications*, New York, Reinhold Publishing Corp., 1959.

3. G. Cohn and H. Musal, "Transient Response of Phosphors," Proceedings of the National Electronics Conf., Chicago, Ill., 1957.

4. Optical Characteristics of Cathode Ray Tube Screens, J. E. D. C. Electron Tube Council, Publication no. 16, 1960.

5. R. K. Cralle, "A Language for Making Movies on a Computer," Talk presented at Annual Meeting of the American Association of Physics Teachers, New York, 1966.

6. A. Cecil and G. Michael, "DD80 Programmers' Manual," LRL Internal Report, (updated), 1966.

7. I. R. E. Standards on Electron Tubes, Methods of Testing, The Institute of Radio Engineers, New York, IRE 7.1, 1962.

8. J. H. Howard, *Electronic Information Displays*, Washington, D. C., Spartan Books, 1966.

9. H. H. Poole, *Fundamentals of Display Systems*, Washington, D. C., Spartan Books, 1963.

10. I. Sutherland, "Sketchpad: A Man-Machine Graphical Communication System (Thesis) M. I. T.," Cambridge, Mass., 1963.

11. G. Michael, "A Survey of Graphical Input Devices," UCRL, Lawrence Radiation Laboratory, Livermore, Calif., 1968.

12. L. G. Roberts, Machine Perception of Three-Dimensional Solids, in *Optical and Electro-Optical Information Processing*, Cambridge, Mass., M. I. T. Press, 1965.

13. L. Gales and G. Michael, "A Computer Controlled Film Reader with Manual Supervision Features," UCRL–70010, Lawrence Radiation Laboratory. Livermore, Calif., 1966.

14. H. S. McDonald, W. H. Ninke, and D. R. Weller, "A Direct View CRT for Remote Computing," 1967 International Solid State Circuits Conference, Philadelphia, Pennsylvania, Feb. 16, 1967.

15. R. H. Stotz and J. B. Cheek, "A Low-Cost Graphic Display for a Computer Time Sharing Console," Proc. 8th National Symposium on Information Display, San Francisco, Calif., p. 91, May 1967.

16. D. Fink, *Television Engineering Handbook*, New York, McGraw-Hill, 1957.

17. D. Englebart, "Augmenting Human Intellect: A Conceptual Framework," Stanford Research Institute Report, Project 3578, 1962.

18. K. C. Knowlton, "A Computer Technique for Producing Animated Movies," Proceedings – Spring Joint Computer Conference, Washington, D. C., 1964.

19. L. G. Roberts, Graphical Communication and Control Languages (private communication), 1966.

20. "A Laboratory for the Study of Graphical Man-Machine Communication," a series of papers by members of the staff of General Motors Research Laboratories, Proceedings – Fall Joint Computer Conference, San Francisco, 1964.

21. K. C. Knowlton, "A Programmer's Description of L^6," Communications of the ACM 9, 1966.

22. N. A. Valyus, *Stereoscopy*, London, The Focal Press, 1962.

23. B. Julesz and J. Miller, "Automatic Stereoscopic Presentation of Functions of Two Variables," Bell System Tech. Journal, pp. 663-676, 1962.

24. B. Saunders, "Stereoscopic Drawing by Computer – Is It Orthoscopic?", UCRL-70707, Lawrence Radiation Laboratory, Livermore, Calif., 1967.

25. D. J. Di Leonardo, "The Graphic Display of Numerical Data." Bettis Atomic Power Laboratory, WAPD-TM-611, 1966.

26. A. M. Noll, "Computer-generated Three-dimensional Movies," Computers and Automation, May 1965.

27. A. M. Noll, "A Computer Technique for Displaying n-Dimensional Hyperobjects," Communications of the ACM 10, 1967.

28. B. Saunders, "Stereoscopic Drawing by Computer," UCRL-70707, Lawrence Radiation Laboratory, Livermore, Calif., 1967.

29. B. Saunders, "Stereoscopic Charts," UCRL-70517, Lawrence Radiation Laboratory, Livermore, Calif., 1967.

30. E. E. Zajac, "Computer-Made Perspective Movies as a Scientific and Communication Tool," Communications of the ACM 7, 1964.

31. R. A. Weiss, BE VISION, Journal of the ACM 13, 1966.

32. W. Teitelman, "New Methods for the Real Time Recognition of Hand Drawn Characters," Bolt, Beranek, and Newman Report 1015, 1963.

33. W. English, D. Englebart and B. Huddart," Computer Aided Display Control," Stanford Research Institute Report, Project 5061, 1965.

34. M. R. Davis and T. Ellis, "The Rand Tablet: A Man Machine Graphical Communication Device," Proceedings – The Fall Joint Computer Conference, pp. 325-331, 1964.

35. M. R. Davis, "Description of Rand Graphic Input System," (private communication), 1963.

36. "Recent Advances In Display Media," Office of Technology Utilization, National Aeronautics and Space Administration, Washington, D. C., NASA SP-159, 1968.

37. R. Ketchpel and N. Lehrer, "Recent Progress on a High Resolution Meshless, Direct-view Storage Tube," Proceedings of the 1966 Fall Joint Computer Conference, pp. 531, Washington, D. C., Spartan Book Co., 1966.

38. B. Arora, D. Bitzer, H. Slottow and R. Wilson, "The Plasma Display Panel-A New Device for Information Display and Storage," Proceedings of the 8th National Symposium on Information Display, San Francisco, Calif., pp. 1, May 1967.

RELATED READING

R. Siders et al., *Computer Graphics*, New York, American Management Association, 1966.

A. van Dam, "A Survey of Pictorial Data Processing Techniques and Equipments," Defense Documentation Center, AD 626-155, 1965.

Periodicals:

Datamation	F. D. Thompson Publications 1830 West Olympic Blvd. Los Angeles, California 90006
Information Display	Journal of the Society for Information Display 647 N. Sepulveda Blvd. Bel Air, Los Angeles, California 90049
Computer Design	Computer Design Publication Corporation P. O. Box A Winchester, Massachusetts 01890
Communications of the A. C. M.	Association for Computer Machinery 211 East 43rd Street New York, New York 10017
S. P. I. E. Journal	Society of Photo-Optical Instrumentation Engineers P. O. Box 288 Redondo Beach, California 90277
Tech Bits	E. Kodak Company Professional, Commercial and Industrial Markets Division Rochester, New York
Computer Display Review	J. Gilmore ed., Chas. Adams, Inc. 575 Technology Square Cambridge, Massachusetts

CHAPTER 10

LANGUAGES FOR GRAPHIC DATA PROCESSING

K. R. BERTRAN, Lawrence Radiation Laboratory, Livermore
G. A. MICHAEL, Lawrence Radiation Laboratory, Livermore and Haverford College

INTRODUCTION

It is assumed from the furor around the country that the methodology of graphic data processing is gaining acceptance among more and more computer users. Possible reasons for this are suggested elsewhere.† One additional reason (and perhaps not as facetious as it appears) is that for the first time the non-specialist can tell what is going on, because he gets to look at pictures. Indeed, the picture is taken to be the visible proof that the computer is doing what it is told.

† See introduction to chapter 9.

In order to realize as much of the predicted benefits as possible, a great amount of very sophisticated system development must be and is being done. But for the most part things just are not ready yet, so in order to get started now, a number of perhaps unpleasant decisions have to be made.

Surely one of the most painful (especially for the "front office") is to face up to the fact that some graphic data processing jobs need to be done on a large computer. No one denies that tiny computers can control and recycle a display device. But, displaying a picture is normally the trivial part of a graphic data processing code. Actually using a large computer is not what bothers people; it is the sight of the users sitting in front of the display—*thinking*—while the "multi-million dollar servant" sits idle, eating up the company profits. We hope time sharing will help to resolve this dilemma.

From the user's point of view there are other annoyances. To get started now, he essentially will first have to get involved with programming details that he feels, rightly or wrongly, are non-productive relative to what he wants to do. On the other hand, if he tries to use one of the presently evolving "super systems" then it could be that he will have to contend with one or more of the following situations:

1. The software system has too many bugs in it, so his progress is slow to the point that he will not achieve what he set out to do.
2. The software system imposes what from his view are disabling constraints, or unreasonable work habits.
3. The software system is so large that he cannot really do the size problem he intended.
4. The software system is too slow.
5. The system is implemented on a computer that in one or more aspects is inadequate for the application.

There are no magic cures for these difficulties, but for those (and there are apparently, a large number) who are willing to make some effort, it turns out that a carefully chosen set of primitive display routines is often enough to allow really useful work to be done.

In this chapter we will first review the concept of (graphic) language levels, try to reveal some of their inner structuring, give an example of a lowest or most primitive level, and make a few direct comments on the interaction between the basic programs and the equipment characteristics.

The authors are aware of the non-trivial differences between the "level 0" language discussed in this article and some of the higher level language research projects now going on at various computer centers. Although "level 0" is the most hardware dependent part of the whole language structure, it is both surprising and gratifying to see just how much can be accomplished using just it together with FORTRAN. Microfilm recording, film reader (scanning) control, on-line (dynamic) control for electronic circuit and mechanical design have all been implemented.

The original development had three goals; first to aid in deciding what way to interface display devices to computers, second to particularize a data structure for possible use in a time shared graphics system, and third to get some operating experience with a minimum set of commands (functions) for exercising a display device.

10.1 THE NATURE OF THE PROBLEM

It is not really an issue that we want pictures as the result of program execution. The question of *how* to do this, of whether there is a "best" way, is now and will no doubt continue being explored for some time yet. We need to find the most efficient ways to use computers to help us achieve the goals we decide on.

One thing we can usually do is to develop a new program for each graphic data processing application as it arises. But it seems that there are so many of them that we cannot afford to write specialized programs every time. Hence we would like to generalize and "subroutine" wherever possible. This is done in effect, by analyzing various kinds of computer uses (e.g. circuit design, movie making, etc.) and looking for those procedures that are or, with a bit of adjustment, can be made common to all the applications. Trivial examples of this in general computer usage are such things as mathematical, editing, and sorting subroutines. Far less trivial are achievements such as LISP, FORTRAN and ALGOL.

When we go through similar considerations for graphic data processing the first and most obvious division is between computation and input-output (i-o) processes. Actually, analysis of the i-o activity tells us more about hardware requirements than about graphic languages. In fact as will be pointed out later (section 10.4), by trying to develop efficient graphic languages we are in a better position to say how the i-o facilities should be built.

Among the computation processes there exist many that are common to a variety of display applications. These would include procedures that prepare a given set of data for a display. Rotations, translations, or other mappings are examples.

As this "factoring" is carried further, commonality seems to depend less and less on the fact that the applications are graphics oriented. Indeed one begins to suspect that the major step was taken with the first separation (between computation and input-output). That is, the thing that primarily distinguishes graphics from other computer applications is the i-o usage because most of its other characteristics turn out to be common to uses not ordinarily thought of as graphics. For example, some of the internal manipulations we want to do on graphic data can be recognized as the same type of things done to list structures or matrices.

A possible difference (perhaps more in degree than in kind) is that we often need much higher efficiency in execution for graphics applications. That is, since the operations will result in a picture we can more easily tell when the computer is "dragging its heels." When we are waiting for our pictures to appear nothing is more obvious than a delay.†

The internal (computer) processes for graphics comprise then a large set of problems and are the significant area wherein the real research efforts must continue to go on. As an only slightly contrived example suppose we wish to display a house, with the windows and doors opening and closing and smoke rising from the chimney. It is not too hard to write a program that does this, but additional complications appear if we want to have "streets of houses" and "blocks of streets of houses" each with its own independent motion. Further suppose that our viewpoint is from an airplane flying around "overhead." The problem of keeping track of all of this and sustaining the display obviously adds complications. Aside from questions about the amount of computation needed, other problems could arise if we wanted to "erase" the windows of a particular house or "lock" a door or add another street of houses as the result of some user interaction.

This example, one of many we might have chosen, illustrates the two-fold aspects of graphic data processing, a heavy demand on i-o bandwidth in conjunction with highly iterative computation processes. Even in cases where picture motion is not needed, other requirements usually arise in effect maintaining a more or less constant computa-

† Although graphic data processing includes more activity than on-line interaction with a display, we will throughout this chapter use the on-line situation to exemplify many of the actions and decisions made. Thus the need for efficient program execution is most easily seen in the context of on-line display application.

tional burden. For example, the calculation of circuit response or of some kind of pictorial filter function can easily be equivalent to the same amount of computation needed for a motion sequence.

10.2 THE DATA STRUCTURE

In order to develop applications similar to the preceding example most of the evolving graphics systems have been based on the notion of list processing. For our purposes a list is simply a named and ordered sequence of items (operands) which are the information (data) to be processed. Thus if BOX is the name of a list of picture parts (glyphs) the statement ROTATE (BOX) requests an operation (ROTATE) on the list BOX. Needless to say, the actual (data) structure of the list and the allowed operators strongly influence both programming ease and efficiency. This chapter would be unnecessarily lengthened if an adequate treatment of list processing was attempted. The reader is directed to the included list of references [1 - 5] for a more thorough discussion. We will limit ourselves here to an enumeration of some of the features that a graphics oriented data structure might require.

1. An almost continuous problem is the lack of enough fast memory. Therefore the data structures must make reasonably efficient use of the storage resources.
2. The lists (that embody the data structure) must be capable of being linked (hooked) together or unlinked in totally general ways. That is, the logical sequence: a list, a previous list, and a next list, must be easy to obtain.
3. It must be easy to change both the state of the list as an entity and the number of elements in the list.
4. The lists should be homogeneous, that is each list looks like any other.

The totality of features such as these are not necessarily compatible with each other or with the requirements of the overall system development. Whatever results will obviously be some sort of compromise.

In addition to these general features, a data structure needs to have certain other characteristics. Since many of them may be linked together to form list chains, it is customary to include special information to characterize the conditions relevant to the list. The list *header* then contains links or pointers (both forward and backward)——these are essentially memory addresses that say which lists precede and follow the given list. There are also other pointers to:

contextual information———such things as the light pen value, the number of data elements in the list, flags and state vectors

and

procedural information———such as transformation rules.

Finally, the list *body* contains the actual data.

Before going on to consider the levels of graphic software, we will summarize what we have said so far:

1. We must try to make computers and their associated devices easier for people to use, especially in the interactive mode.
2. By system analysis we decide to try dividing the overall idea of graphic data processing into at least two parts: computation and input-output.
3. We identify many of the computation procedures as similar to list processing. Therefore we will try those techniques that have been useful in list processing. At the same time both program efficiency and human factors———ordinarily these are not fundamental constraints———will need to be emphasized.

10.3 LEVELS OF GRAPHIC SOFTWARE

The result of isolating common procedures and features in graphic data processing applications leads to considering them as various levels of language. The actual i-o programs are called "level 0". Our current implementation of the level 0 functions is called CALLIGRAPH and is discussed in more detail in the next section.

Procedures that generate or manipulate picture parts will be called "levels 1 to N." Except for level 0, the precise composition of each of the levels is largely up to the individual user. It should be emphasized moreover that the boundaries between levels are not always sharply defined and that the only level which might get to be complete is level 0. Furthermore the idea of levels is not intended to imply a rigid linear ordering. Thus if there is a level 3, we may wish to pass from it directly to level 0. Figure 10.1 indicates this and other possible sequences. Note that connections to level 0 are shown in one direction only. This may turn out to be wrong. The reasons for raising the question will be discussed in the next section.

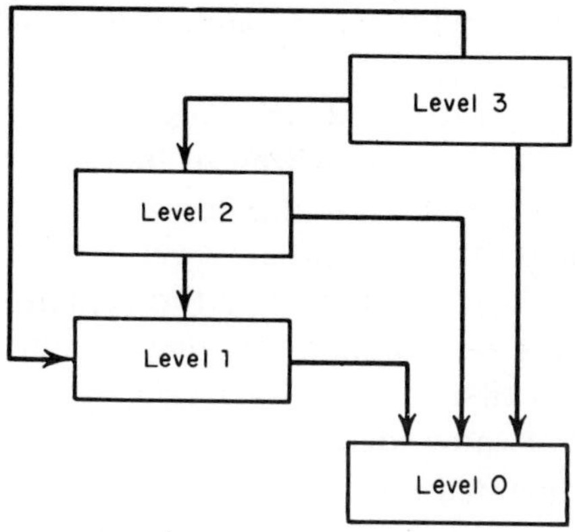

FIG. 10.1. Levels of graphic software showing communication paths.

Although it is up to the user to specify the operators he wants for each level, presumably a useful ordering is according to how close their function is to the actual display steps. Thus level 0 would include routines to operate the display as an i-o device and routines that build up lists of display commands. Similarly level 1 would have things (as already suggested) like rotation, translation, perspective and stereoscopic mapping operators. Table 1 shows one possible ordering.

TABLE 10.1

One possible categorization of graphic data processing subroutines.

LEVEL 0	LEVEL 1	LEVEL 2	LEVEL 3	• • • • • • •
Build display commands	Rotate	Mathematics	Specialized Process Languages (e.g. Circuit Design)	
	Translate	Computation		
Control the graphic i-o devices	Mapping	Movie		
	Chain lists			

10.4.1 CALLIGRAPH - A Level 0 Graphics Language†

Based on the foregoing considerations then, a level 0 language called CALLIGRAPH was developed. It is a set of i-o subroutines that can be invoked by a user as he needs them. The language was made explicitly compatible with FORTRAN II because of its broad user acceptance as a working language.

The subroutines that actually do the i-o steps need to be simple, efficient, small, flexible and yet specialized in the sense that they "know" the i-o devices extremely well. It is also desirable that the language does not rob the user of any of the potential power of the display equipment he has at his disposal. Clearly, they ought to be immune to user errors. That is, if the user does something wrong they should not just quit.

The language was first implemented using an IBM 7094 to which was attached a DD80 (now identified as a CDC 280) display device. It is also being implemented for a time shared computer system now under development at the Lawrence Radiation Laboratory, Livermore, California. Descriptions of those relevant hardware features will be left for section 10.5.

The format used to describe the CALLIGRAPH routines here is similar to that found in various kinds of programming and user manuals. We retain it for two reasons:

1. Since it is a somewhat standard format, the reader will find it easier to cross-compare the routines.
2. The format minimizes the amount of text needed to describe the routines.

The routines are as follows:

NAME	*PURPOSE*
CALL IGRAPH (*i, j*) or CALL IGRAPH (*i, j, k*)	Donate buffer memory
CALL NAME (*A, N*)	designate a glyph (picture part)
CALL CLEAR (*A*)	destroy a glyph

† This section is based on and includes most of an earlier, internal UCLRL report #104 which was a user manual for the CALLIGRAPH language.

NAME	*PURPOSE*
CALL IN (A)	enter the glyph into the display list
CALL OUT (A)	remove the glyph from the display list
CALL RUN (M)	turn on the display - i.e. send the glyph list to the display device
CALL STORE (ic, it, if, is, io, x, y)	construct one display word and add it to the list for the current glyph †
CALL STOREC ($it1, ic1, it2, ic2, it3, ic3, it4, ic4$)	construct one display word and add it to the list for the current glyph
CALL TRACK (X, Y, N)	"follow" the light pen
CALL NTRACK	disable light pen following
CALL LPEN (N, X, Y)	which glyph was seen by the light pen

IGRAPH

Form: call IGRAPH (i, j, k)
Arguments: i, j, k are FORTRAN II fixed point variables.
Description: This routine must be called *first* in any normal execution. Its purpose is to set up display buffer memory for use by the other routines. i is the beginning (first word address) of some array to be donated to CALLIGRAPH for buffer memory. j is the number of buffer words allocated. k is the buffer block size (SUBGLYPH), that is the minimum number of words that will be allocated for a glyph each time the NAME subroutine is called. Subsequently the user can interrogate i and j.

1. The display is inactive if $i = 0$
   The display is active if $i = 1$
2. The number of unused buffer words $= j$

† The actual structure of this function is determined by the hardware characteristics of the DD80 (cf. section 10.5).

NAME (*A*, *N*)

Form: CALL NAME (*A*, *N*)

Arguments: *A* is either a floating point or a fixed point variable, but formally it is a *name* or glyph;
N is a fixed point integer designating the light pen value for the following CALL STORE's.

Description: The NAME subroutine opens a (*k*-word) buffer block which is then associated with the name *A*, a glyph. The location *A* contains a pointer to the *k*-word block (see the remarks below). All display commands constructed by the STORE subroutine are associated with *A* until the next time NAME is called. There is no limitation on the length of the display list implied by the *k*-word block. If more space is required by the STORE routine, it will automatically call NAME with the arguments given in the most previous use of NAME. The *A* glyph can be closed (implicitly by other CALL NAME (*B*, *M*) sequences) and re-opened as often as necessary; limited only then by amount of unassigned memory. Each time *A* is re-opened a different "*N*" may be associated with it.

The light pen value *N* which the user assigns can be compared with a corresponding value returned by the light pen subroutines whenever the light pen "sees" a point of light. Thus the user can find out which glyph the light pen is pointing at.

Remarks: In an ordinary execution it is assumed that the *first* time a glyph name is used, the corresponding storage location contains zero. At any time thereafter this location contains a pointer. Being aware of this, a user then could use CALL IGRAPH more than once per execution provided that previously used glyph names are initialized to zero before using them again.

CLEAR

Form: CALL CLEAR (*A*)

Arguments: *A* is a glyph name

Description: The glyph *A* is destroyed and all storage assigned to it is returned to free storage.

IN

Form: CALL IN (*A*)
Argument: *A* is a glyph name
Description: The glyph *A* (i.e. all the display commands associated with the name (glyph) *A*) is added to the display loop.

OUT

Form: CALL OUT (*A*)
Argument: *A* is a glyph name
Description: The glyph *A* is removed from the display loop.

RUN

Form: CALL RUN (*M*)
Argument: *M* is a signed fixed point integer.
Description: The glyphs currently in the display list are shown; that is the display is "turned on."

If $M = 0$ the picture is regeneratively displayed; that is the DD80 is cycled continously. If $M > 0$ the display loop is started and control is passed immediately to the calling program. After *M* cycles the display goes inactive.

If $M < 0$ the display loop is cycled *M* times but control is *not* passed to the calling program until the *M* cycles are completed.

Remarks: The display list is processed by RUN as "last in - first out." Thus, if the user had sequentially "INed" glyphs *A, B, C, D*, the first one to be displayed is *D*, then *C*, then *B*, then *A*, then *D* and so on. Ordinarily this is not a concern for the user except for the case of frame advancing (cf example 4).

STORE

Form: CALL STORE (*iC, iT, iF, iS, iO, X, Y*)
Arguments: *iC, iT, iF, iS, iO* are fixed point integers.
X, Y are floating point numbers in the interval $0 \leqslant X, Y < 1$.

Description: iC is any one of the permitted characters (cf character codes Table 10.2 and also see the remarks below).

iT is the intensity control: 0 is normal
1 is high intensity

iF is the focus control: 0 is sharp focus
1 is fuzzy focus

iS is the character size control:
0 is miniature size: 128 ch/line - 64 lines/frame
1 is small size: 85 ch/line
2 is medium size: 64 ch/line
3 is large size: 45 ch/line

iO is the orientation control:
0 calls for erect characters
1 calls for characters rotated 90° counter-clockwise

X, Y are the coordinates, normalized in the interval (0, 1) at which the character is to be centered.

Remarks: If $0 \leqslant iC \leqslant 122$ the character is alpha-numeric and therefore displayable.

$iC = 123$ tells the display to merely move the beam to the point X, Y and set intensity, size etc. as given, but do not plot.

$iC = 124$ tells the display to enter the typewrite mode, (cf STOREC) set the size, focus and orientation and move the beam to the point X, Y.

$iC = 125$ tells the display to draw a vector from wherever it was to the point X, Y.

$iC = 126$ tells the display to advance the film one frame and *do nothing else*.

$iC = 127$ tells the display to plot a point at the location X, Y.

Clearly, the precise behavior of the STORE (and the STOREC) function depends on the display hardware characteristics. Comments that bear on this point are given in the next two sections.

STOREC

Form: CALL STOREC ($iT1$, $iC1$, $iT2$, $iC2$, $iT3$, $iC3$, $iT4$, $iC4$)
Arguments: All arguments are FORTRAN II fixed point integers
Description: Each pair of arguments iC, iT specifies a 7-bit character code and a 1-bit brightness control. In the typewriter mode, the DD80 simulates the typewriter sequence of displaying a character and automatically indexing to the next character position.
Remarks: The STOREC subroutine should be used only after the display has been placed in the typewriter mode via a -
CALL STORE (124, iT, iF, iS, iO, X, Y)

TRACK

Form: CALL TRACK (X, Y, N)
Arguments: X, Y are floating point variables in the interval $0 \leqslant x/y < 1$.
N is a fixed point integer.
Description: A "trackable" cross is established centered at the position X, Y. Then if during the last display cycle the light pen sees the point, N is set to 1; otherwise N is kept equal to 0. If $N = 1$, the values for X and Y are updated and are available for user computations (see example 2).
Remarks: As long as the run state 0 exists, tracking continues asynchronously.

NTRACK

Form: CALL NTRACK
Arguments: None.
Description: The tracking state ceases.

LPEN

Form: CALL LPEN (N, X, Y)
Arguments: N is a fixed point integer.
X, Y are floating point variables.
Description: The light pen value, N, of the first thing seen (after the most previous CALL LPEN, CALL IN, CALL OUT) is returned to the calling program. The coordinates X, Y (normalized to the interval $0. \leqslant x/y < 1.$) are also returned.

Programming Examples

```
1. DIMENSION i(3201)
   ..............
        j = 3201
        k = 32
   ..............
  CALLIGRAPH (i, j, k)
```

Note that here it is never appropriate to use literals for i or j.

```
  CALLIGRAPH (i, 3201, 32)
```

since the implied storage location will later not contain 3201 but the number of words remaining in the buffer.

```
2.          c This program will draw a vector from
            c the center of the screen to the current
            c position of the light-pen
            ..................
              CALLIGRAPH (N, J, K)
              X = .5
              Y = .5
              CALL TRACK (X, Y, L)
           10 CALL NAME (A, O)
              CALL STORE (123, 0, 0, 0, 0, .5, .5)
              CALL STORE (125, 0, 0, 0, 0, X, Y)
              CALL IN (A)
              CALL CLEAR (B)
              B = A
              A = 0
           20 CONTINUE
              IF (L) 10, 20, 10
```

At statement 10 a glyph is "named" and subsequent stores relate to this glyph. Next it is "placed in the display list" in the sense that it is made visible. A dummy glyph (B, in this example) is cleared although the first time nothing happens. The name B is then made equivalent to A so that it can be "cleared" on the next pass through the program. The glyph A is then "initialized" (by zeroing the contents of location A) and the program "idles" until the condition $L \neq 0$ "says" that the light pen sees the point x, y set up by the TRACK program. This illustrates one technique of "Double Buffering" in order to conserve memory space when a large number of temporary glyphs need to be constructed.

3.

```
c   AN EXAMPLE OF HOW TO MAKE ONE
c   PICTURE OF AN ALREADY RECIRCULATING
c   DISPLAY
    .
    .
    .
c   CREATE A FRAME ADVANCE "GLYPH"
    CALL NAME (FRAME, 0)
    CALL STORE (126, 0, 0, 0, 0, 0., 0.)
    .
    .
    .
c   SAVE THE PICTURE
c     A. STOP THE DISPLAY
    CALL RUN (-1)
c     B. INSERT A FRAME ADVANCE
    CALL IN (FRAME)
c   THE FRAME ADVANCE COMMAND IS NOW
c   AT THE BEGINNING OF THE DISPLAY LIST.
    CALL RUN (-2)
c   TWO CYCLES ARE NEEDED. THE
c   FIRST ADVANCES FILM AND PAINTS A PICTURE.
c   THE SECOND ADVANCES FILM TO PROJECT
c   THE PICTURE JUST PAINTED
c   RESTART THE DISPLAY.
    CALL OUT (FRAME)
    CALL RUN (0)
```

4.

```
c   THIS PROGRAM CREATES A SERIES OF
c   LIGHT BUTTONS (THE NUMERALS 1-9)
c   AND THEN TESTS THEM (A RUN (0) IS
c   ASSUMED) X = .05
c   REMEMBER THAT RUN STATE 0 NOW EXISTS
    DO 10 I = 1, 9
    CALL NAME (A, I)
    CALL STORE (I, 0, 0, 1, 0, X, .2)
10  X = X + .05
c   "A" NOW CONTAINS A SERIES OF LIGHT
    BUTTONS
c   WHICH WILL BE PUT ON THE SCREEN
    CALL IN (A)
```

```
20 CONTINUE
   PAUSE
c  A LIGHT PEN HIT RESTARTS THE COMPUTER
   CALL LPEN (L, X, Y)
   L = L + 1
   GO TO (20, 21, 22, 23, 29), L
21 (OPTION 1)
22 (OPTION 2)
   etc.
```

10.4.2 Specific Data Structure for Calligraph

Background. The sets of bits (instructions plus data) to be given to the display controller are stored in user-donated buffer memory (i.e. computer memory) as a chained list structure. The chaining procedure is realized through computer commands and input-output (data channel) commands. Although those actually used are specific to the IBM 7090-94 computer systems, they can be regarded (as is pointed out elsewhere in this chapter) as minimal requirements for any buffered channel operation. In order to understand how the chaining works it may be helpful to first consider the action of these commands and the data channel operation. We will therefore briefly describe these commands and channel operation before going on with the description of the glyph data structures.

The data channel is a device that can move data between the computer memory and various external devices such as tape recorders, printers, and in our case a display generator (controller). A computer instruction, "READ," conditions the data channel for data that will be transmitted into core memory. Similarly, "WRITE" tells the data channel that data will be copied from core out to the external device. After selection another computer instruction initializes the data channel and tells it where in core memory to look for its first instruction. Once started, given an adequate control description, the data channel can proceed autonomously to task completion, thereby leaving the computer free to continue processing its own instructions. We need describe only two data channel commands for purposes of this discussion.

TCH *Y* , , *C* (5) "Transfer in Channel"

This command tells the data channel to look in memory location *Y* for its next instruction. *C* is ignored.

IOCP Y , , C "Input (or Output) under count control and Proceed" This command to the data channel tells it to transmit C words of which the first is in location Y, the second in $Y + 1$, and so on to the last word in location $Y + C - 1$. After all the words have been transmitted the data channel will pick as its next instruction, the word immediately following the IOCP command in memory.

In what follows we have made use of a few notational conventions common in programming. Insofar as a computer word can be thought of as divided into fields, the comma (,) is used as a field delimiter. Thus TCH Y , , C implies four fields: The instruction TCH, the Y field, a blank field and the C field. The instruction then says (to the data channel), "get the next instruction from memory location Y." The C field does not influence the TCH instruction; it is merely a convenient place to store some data. In that an instruction is stored in the core memory, it is stored at some particular location. If the instruction needs to refer to its own location an asterisk (*) is used to signify "self." Thus, suppose location 104 contains TCH * + 2. This tells the data channel to get its next instruction as "self plus 2"; that is, at location 106.

Data Structure. It is the job of CALLIGRAPH (i, j, k) to allot and structure the dimensioned variable i which contains j available locations into k sized blocks to be assigned as free storage. The available buffer memory (i.e. the array i) is divided into as many k sized blocks as is possible. This forms the free storage area which will be divided up to form individual groups of blocks common to a given name which we will refer to as glyphs. Storage is allotted from this free storage area by the free storage allocator a block at a time and returned to free storage in the same manner. The storage is structured such that the first word of each free block points to the first word of the next free block. A forward pointer of zero indicates the end of the chain. The first block of the list is pointed to by a particular storage location, the free storage pointer (FSP), indicating the next free buffer to be assigned (See Fig. 10.2).

It is also the function of CALLIGRAPH to set up and initialize the interrupt structure needed by the display language. This structure consists of a light pen interrupt and a "channel end" interrupt for the display. The channel end interrupt occurs at the end of each display cycle.

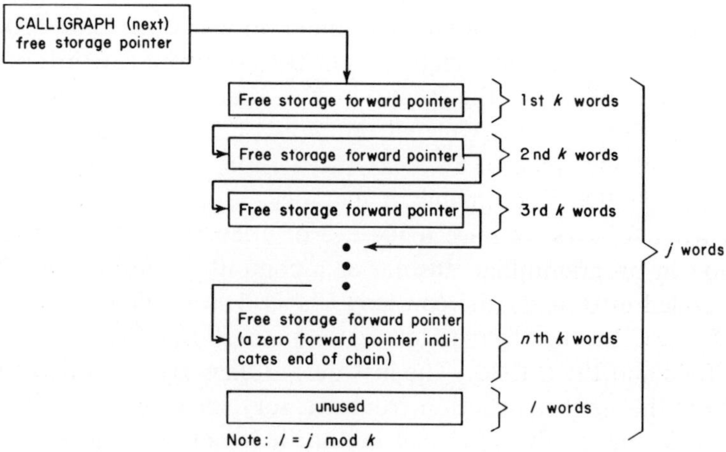

FIG. 10.2. User-donated buffer memory is divided into k-word blocks which are linked together.

10.4.3 The action of NAME

The first argument of NAME is examined to determine if it is zero. If the value is zero, this indicates that the name is unused and some initial buffer area must be assigned to it. A k-word block is removed from free storage and a value which points to this block is placed into the argument. The header which is to be used to link the storage areas designated by this name into the display list is placed into the first two words of this block.

The next three words, the sub-header, contain information about the particular block. This information consists of where the next block assigned to this glyph is to be found, what light pen value is assigned to this block, how many display commands are currently in this block. Calls to NAME with non-zero arguments indicate that some structure already exists. In these cases a new block is formed but the header is omitted and the new block is tied into the previously created structure by modifying the existing last sub-header, the header, and creating a new last sub-header in the new block.

Figure 10.3 shows how the linkage changes after the *first* use of NAME. Figure 10.4 shows in more detail, the composition of the k word block after NAME is used. If the actual display list becomes longer than the $(k - 5)$ words shown in Fig. 10.4, CALLIGRAPH automatically calls for another k-word block and constructs a linkage as shown in Fig. 10.5.

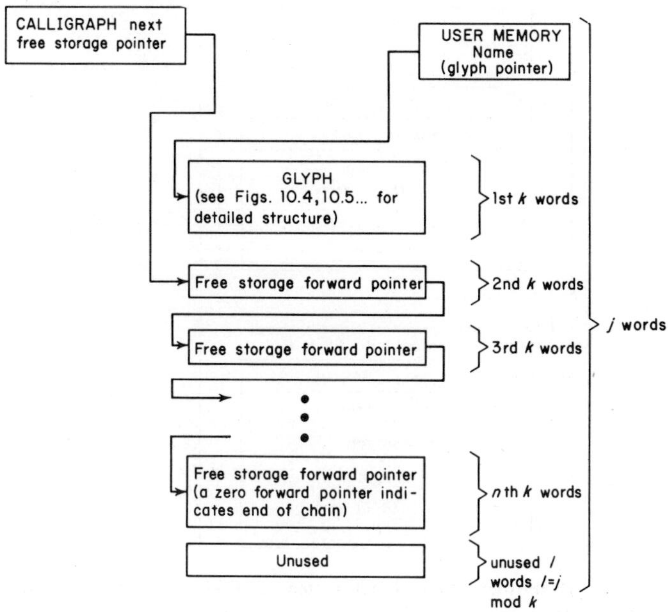

FIG. 10.3. After NAME is used a k-word block is assigned to the glyph name.

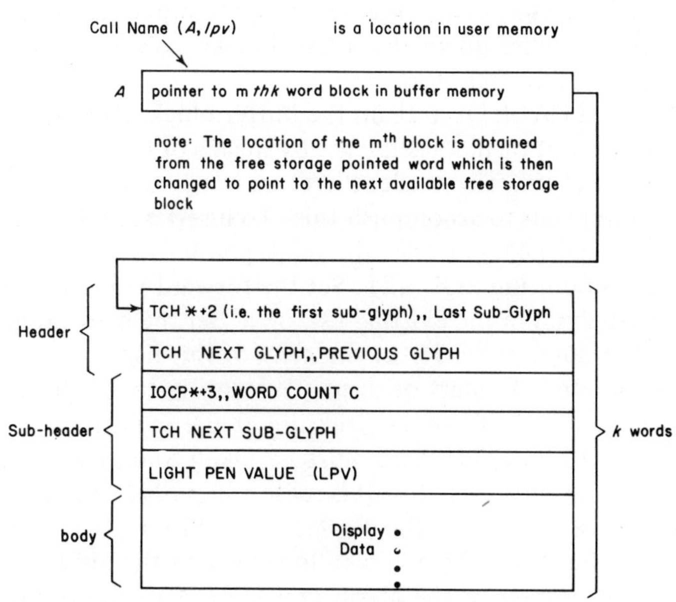

FIG. 10.4. More detailed illustration of the structure of a k-word block.

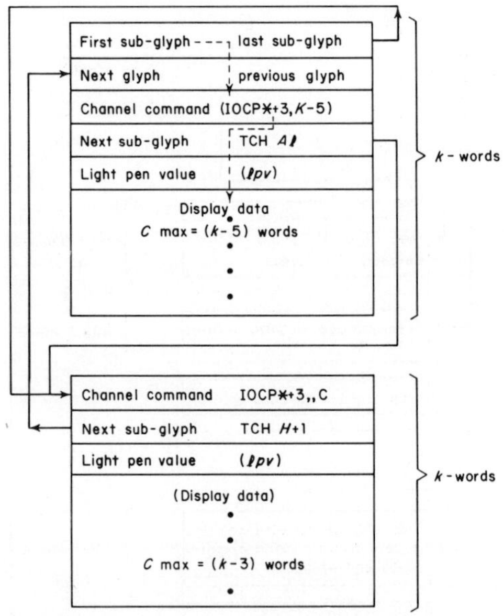

FIG. 10.5. If another k-word block is needed, CALLIGRAPH automatically calls for it.

After the user has been running awhile the linking of buffer memory becomes quite complicated. Figure 10.6 attempts to indicate this. CALL CLEAR searches down the list of blocks assigned to a glyph re-chaining each block to free storage.

CALL IN and CALL OUT chain the buffer blocks designated by the argument into or out of the display list by means of header found in the first block of the chain. The header therefore must contain the appropriate channel commands to accomplish this. To insert a glyph into the display list, CALL IN (A); (1). Set the back pointer of the existing first glyph in the list pointing to A. (2). Set the forward pointer of A pointing to the first glyph in the existing list. (3). Set the back pointer of A pointing to the dummy header at the start of the display list. (4). Set the forward pointer of "start of display" pointing to A. Since the display may be processing the data structure at the same time it is being modified by CALLIGRAPH, care must be taken to insure the integrity of the structure at all times. For this reason step (2) must precede step (4). To remove a name from the display list, CALL OUT (A). (1). Set the backward pointer of the glyph following A in the display list. (2). Set the forward pointer of the glyph preceding A to the glyph following A.

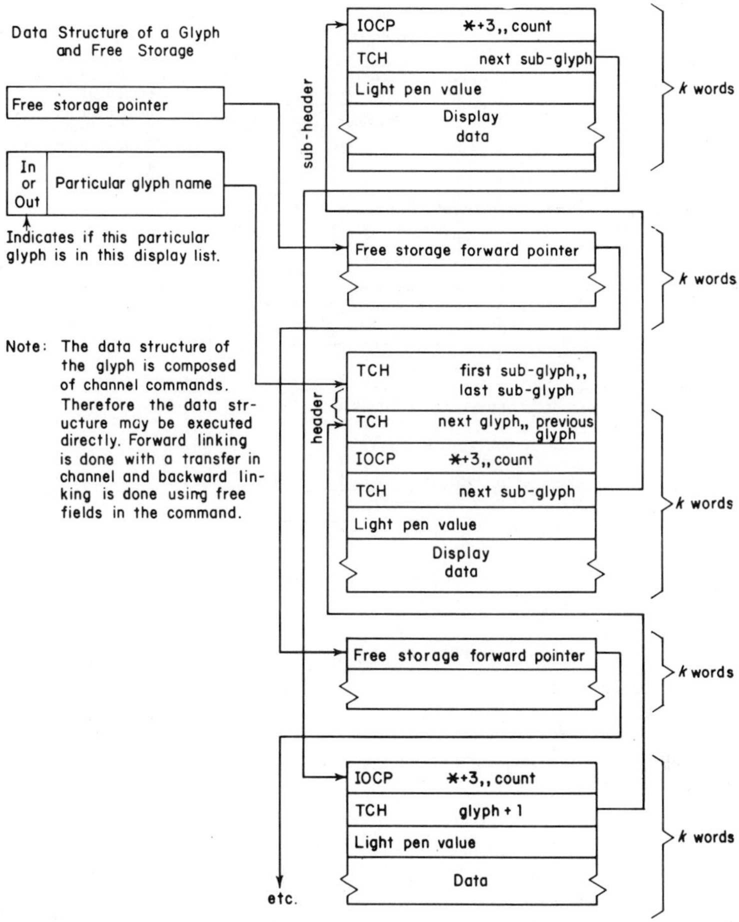

FIG. 10.6. The linkage becomes very complicated as a result of user actions.

As an example of how the user's actions cause the linkage to change let us assume that the lists are in the state shown in Fig. 10.7.† Note first that the CALLIGRAPH linking structure behaves like a push down list relative to the sequence in which glyphs are displayed. Thus if the user inserted them in the order name A, name D, name C, they are displayed in the order C, D, A. Now suppose that the user wishes to re-enter name B into the display list. Figure 10.9 shows the new linkage that results: name B, name C, name D, name A.

---

† Refer also to Fig. 10.8, which is a simplified version of Fig. 10.7.

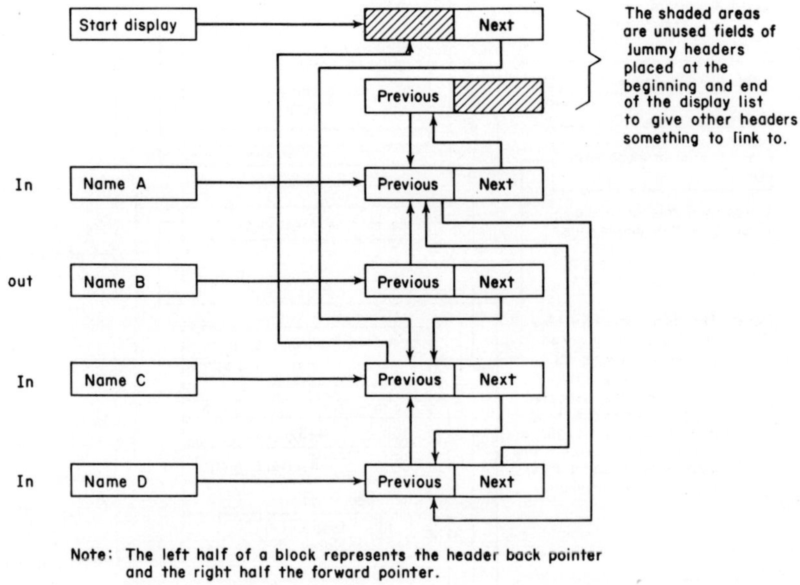

FIG. 10.7. The order in which the glyphs are processed is similar to that of a push down list.

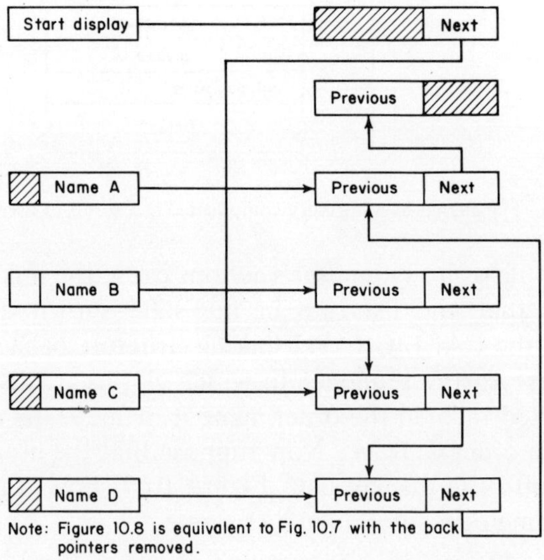

FIG. 10.8. A simplified version of Fig. 10.7 with the back pointers removed.

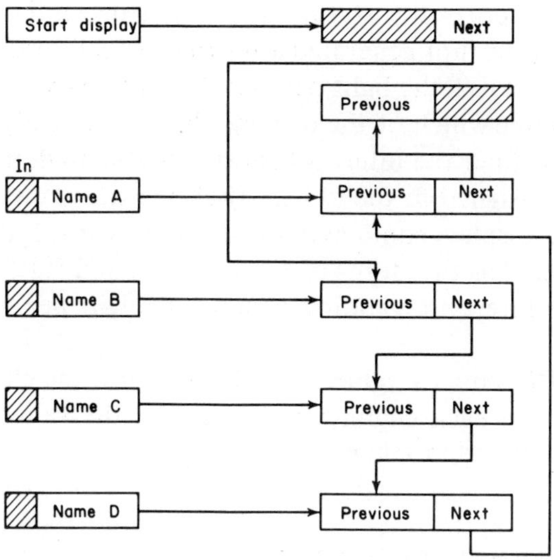

FIG. 10.9. The new linkage resulting from the reinsertion of a glyph

CALL STORE and CALL STOREC concatenate the argument string and place it into the next available location in the current block of the last designated glyph. If the block becomes full, STORE reinvokes the NAME subroutine with the most recent user-supplied arguments (name and light pen value), in order to assign a new block to that glyph. For this reason the last arguments in a call to NAME must be preserved for future reference. The arguments in a call to STORE form the actual command which goes to the display device. Additional comments about the STORE subroutine are given in the next section (concerning hardware details) because of its degree of dependence on the display word format.

A call to TRACK causes a tracking cross to be established at the location designated by the current values of the X, Y arguments of the call. Control is then returned to the user and tracking is continued asynchronously by means of the channel end interrupt routine. This routine uses the locations specified by the user in his original call to TRACK. In this manner the user may interrogate the position of the light pen at any time by examining the current value of the variable given in the argument string. The argument N indicates whether or not the light pen was seen during the last interrupt cycle.

CALL NTRACK removes the tracking cross by making the appropriate adjustments to the channel end interrupt service routine.

CALL LPEN reports the status of the light pen interrupt routine. On each light pen interrupt generated after the last call to this routine or CALL IN or CALL OUT the light pen service routine examines the data channel to determine which of the display blocks is actually being transmitted. In this manner the interrupt routine is able to determine which light pen value is assigned to the block. It also extracts the *x, y* portion of the particular display command word which caused the interrupt. These values are held as the status of the light pen and further interrupts beyond the first interrupt yielding a non-zero light pen value are not processed.

A CALL RUN causes a value to be loaded into the channel end interrupt routine. At each channel end interrupt the value is examined and the appropriate action taken.

10.5 EQUIPMENT CONSIDERATIONS

In this final section we will briefly consider three items: A description of the DD80 hardware characteristics, a general specification of the computer-display interface, and a few more comments about the STORE subroutine.

The DD80 is a display device consisting of [6]

1. A Display Controller (Generator)
2. A Microfilm Unit – 5″ CRT and Camera
3. A Visual Display Console – 19″ CRT and Light Pen

It is interfaced directly to an IBM 7094 via the Direct Data Entry option of the standard IBM 7607 data channel. This option allows for transmission of 36 bits as frequently as every 6 microseconds (μs).

The DD80 can plot points and characters, draw lines from any point to any point, and simulate a typewriter.

There are two display word formats:

I	*CH*	*X*	*S*	*O*	γ	*Y*
1	7	10	2	1	1	10

format #1

I	CH_1	I	CH_2	I	CH_3	I	CH_4
1	7	1	7	1	7	1	7

format #2

which have the following meanings

I	intensity control	1 bit	2 levels	bright normal
S	size control	2 bits	4 (character) sizes	
O	orientation	1 bit	2 orientations	erect 90° CCW rotated
F	focus† or font	1 bit	2 levels or 2 fonts	sharp focus fuzzy Roman Italic
X,Y	coordinates	10 bits each	$0 \leqslant X,Y \leqslant 1023$	
CH	character	7 bits	128 characters	$0 \leqslant CH \leqslant 127$

All character codes less than 123 are legal (displayable) relative to format #1

if CH = 123 the display will *move* the beam (without unblanking) to the position X, Y

= 124 the display enters the typewrite mode and and positions the beam at location X, Y; subsequent transmissions are then assumed, to be in format #2 until the first position CH = 123, 124, 125 or 127

= 125 draw a vector *to* the location X, Y

= 126 advance the film; do nothing else

= 127 plot a point at the location X, Y

The particular character code set used for our DD80 is given in Table 10.2.

† Controlled by a manual switch.

Low Order Digit

High Order Digit(s)	0	1	2	3	4	5	6	7	8	9
0	0	1	2	3	4	5	6	7	8	9
1	≡	=	∼	]	[	▽	+	*A*	*B*	*C*
2	*D*	*E*	*F*	*G*	*H*	*I*	π	.	)	→
3	λ	∧	—	*J*	*K*	*L*	*M*	*N*	*O*	*P*
4	*Q*	*R*	Σ	$	*	>	η	△		/
5	*S*	*T*	*U*	*V*	*W*	*X*	*Y*	*Z*	:	,
6	(	<	ρ	α	β	γ	≤	ε	θ	ν
7	τ	φ	χ	ψ	ω	δ	≥	μ	σ	∂
8	≠	*a*	*b*	*c*	*d*	*e*	*f*	*g*	*h*	*i*
9	∨	‘	{	}	’	⊂	⊃	*j*	*k*	*l*
10	*m*	*n*	*o*	*p*	*q*	*r*	|	\	↑	↓
11	←	∫	%	;	*s*	*t*	*u*	*v*	*w*	*x*
12	*y*	*z*	?	(move) *M*1	(type) *M*2	(vec-tor) *M*3	(frame) *M*4	(point) *M*5	▨	▨
	0	1	2	3	4	5	6	7	8	9

TABLE 10.2.

The character set used for the DD80 attached to the IBM 7094

The display cycle times are as follows†

points	4 μsec
characters	9 μsec
lines	6 – 42 μsec
frame advance	∼30 m sec

In reference to the second item it is convenient to restate the overall design philosophy.

† Assumes that the interface could supply the data fast enough. The actual times for the IBM 7094 interface will always be multiples of the channel cycle time, i.e. 6 μsec.

The design criteria underlying the CALLIGRAPH subprograms include the following considerations:

1. Based on our current understanding of what a user needs for on-line control of his program, elements of list processing needed to be included. Thus while he constructs a display list (commands) to produce a picture, he should be able to identify it by name, and chain it to other lists. These lists become *named picture parts,* glyphs, capable of being displayed with any chosen frequency.
2. Dynamic storage allocation while not absolutely necessary for the one-user-at-a-time mode of 7094 operation, makes the maximum amount of core available for display list storing.
3. The user must have intimate control of his own display cycle but must not be burdened with input output bookkeeping questions. Interference however inadvertent with other simultaneous users must also be prevented.
4. A light pen and a keyboard are connected to the display console. They represent the only proper mechanisms for providing on-line input to the computer. The user consequently needs to be able to use them in as unrestricted a fashion as possible while at the same time not being forced to contend with certain of their idiosyncrasies. We suppress multiple light pen responses for example, which arise mainly because of the difference in speeds between people and computers. At the same time the light pen interrupt must occur in such a way that the channel command and the data word which caused the interrupt can be quickly determined.
5. Within reason, the display package must exercise the equipment at the fastest possible rate, and allow the user as unrestricted as possible control over structure of the display lists. The i-o hardware should be capable of operating fully buffered.
6. The techniques of gather-scatter channel operation become necessary in view of 2.
7. The language has to be basic in the sense that a user can exploit any of the equipment features.
8. There are three general application areas that must be expressible in terms of this package:
    1. Man-computer (on line visual) communication.
    2. Microfilm recording
    3. Graphical Input Device (Eyeball) control.

9. The display command set must be sufficiently powerful to allow buffered channel operation without an excessively large number of memory accesses.

The last item to be discussed is the STORE subroutine.

As already noted, the routines described in this chapter are said to be machine dependent. We do not regard this as fatal on at least two counts. First, since they are basic they are short (presently less than 600 words) and hence rather easily recoded for other computers. Second, efficient graphic data processing is among other things, an input output problem demanding efficient equipment utilization, and this is generally easiest through machine dependence coding.

In addition the STORE subroutine is also dependent on the kind of display device features built into the hardware. For the DD80 as described here the sequence of arguments should then permit user access to these features: Thus

$$\text{CALL STORE } (iC, iT, iF, iS, iO, X, Y)$$

gives the user explicit control of all the programming features designed into the hardware. It is noted that simulation of these display features could be used to achieve upward compatability with programs written for less sophisticated devices.

REFERENCES AND RELATED READING

1. Saul Rosen, *Programming Systems and Languages*, New York, McGraw-Hill, 1967.

2. Wm. F. Miller and A. C. Shaw, "A Picture Calculus," SLAC-PUB-358 Stanford University, Stanford, Calif., Oct. 1967.

3. J. McCarthy, et al, *LISP 1.5 Programmer's Manual*, Cambridge, Mass., M.I.T. Press, 1962.

4. Kenneth Knowlton, "A Programmer's Description of L^6," Communication of the ACM 9, 1966.

5. Donald Knuth, *The Art of Computer Programming: Vol. 1, Fundamental Algorithms*, Reading Mass.: Addison-Wesley, 1968.

6. A. Cecil and G. Michael, "The DD80 Programmers Manual," Lawrence Radiation Laboratory, Internal Report, Livermore, Calif.

CHAPTER 11

GRAPHIC INPUT FOR NETWORK ANALYSIS

W. G. MAGNUSON, Jr., Lawrence Radiation Laboratory, Livermore
F. F. KUO, University of Hawaii
W. J. WALSH, University of California, Berkeley

INTRODUCTION

With the advent of on-line processing, the digital computer has become a real-time partner in the creative processes of engineering design. Today we are able to treat design problems previously too complex to put entirely on the computer. The interaction between man and machine in a real-time dialogue makes possible the successive testing of hypothetical solutions until a satisfactory design is achieved. The man controls the decision processes while the machine does the step-by-step solutions.

11.1 ON-LINE CIRCUIT DESIGN

Circuit Analysis Programs

In this chapter we will describe a system called GINA (Graphical Input for Network Analysis) for the on-line design of electronic circuits. The two major steps in the design process are *display* and *analysis. Display* involves the use of an on-line graphic input-output console, and *analysis* involves the use of electronic circuit analysis programs. Although we will concentrate on details of the display system, it is well to discuss briefly the capabilities of the analysis programs. Today there are powerful circuit analysis programs such as CIRCUS [1], NET-1 [2]. and SCEPTRE [3], (see Chapter 8), which can analyze sizeable electronic circuits (50 nodes and 200 branches). These circuits can contain up to 40 transistors and 75 diodes (NET-1 characteristics) and have nonlinear transistor models such as the Ebers-Moll [4], (see Chapter 7). The major factor that limits their use in electronic design is their unwieldy input-output formats. Under a batch-processing mode, these programs are difficult to use for interactive design because of the excessive turn-around time of an hour or more. With the advent of time-sharing by remote teletype, the turn-around times are reduced to seconds, but the user is still hampered by an inconvenient and bulky input language and large amounts of tabular output data. In circuit design the natural language of the engineer is the schematic. Thus the communications medium between the engineer and computer should be *graphic*. Ideally the designer should present the computer with a circuit schematic and the computer, in turn, should return a set of graphic response curves.

Graphic Console

Today, the principle means of on-line graphic communications is the computer-controlled cathode ray tube (CRT) display. The most popular graphic input device associated with the CRT is the *light-pen*, a photo sensor which produces a response when it detects a point of light in the CRT display. The graphic console used to develop GINA, the DD80, has a raster of 1024 by 1024 directly addressable locations. The display is tied to an IBM 7094, whose memory unit is shared with the display.

All graphic entities ("glyphs") are software generated. The display functions (point, line, character, etc.) and control functions (focus, intensity, size, position, etc.) are all software controlled. The language used to create the display list (which, when decoded, is the displayed graphics) is a simple, FORTRAN – compatible, set of 11 instructions written by K. Bertran and described in Chapter 10.

On-Line Circuit Design

The engineer, sitting at the console, (Fig. 11.1) can enter his schematic into the computer by using light-pen and a set of preprogrammed graphic subroutines called *light buttons*. The CRT face is divided into two areas: a *display area* at the center of the screen, and a *control area* at the bottom. On the control area are displayed the light buttons which may be sets of characters that allow the user to initiate certain actions such as ROTATE, COPY, or CONNECT, or may be problem-oriented glyphs – in our case, the circuit element – resistor, capacitor, inductor, diode, etc. (Fig. 11.2). The user initiates the light button by pointing the light-pen at the button, an action known as *picking*. Once a circuit element is picked, the designer can move this element to the display surface by means of a tracking cross and the light button, POSITION. The action buttons allow the designer to compose his schematic in a step-by-step manner. Along with the purely graphic information, the designer chooses a certain amount of nondisplay information such as element value, type of analysis desired, etc. When the circuit schematic is completed, the designer requests an analysis of the circuit by picking the ANALYZE button. The data representing the schematic is then stripped of its nondisplay part, which is converted into an internal form acceptable to the circuit analysis program. The circuit analysis program calculates the frequency response and produces data which is, in turn, translated into pictorial form. The engineer views the graphic output and makes changes in the original schematic. Then the process is repeated until a satisfactory design is achieved. The process can be likened to an engineer making actual changes on his breadboard circuit and viewing the results on an oscilloscope. By this procedure a circuit can be designed with a minimum amount of time and tedium and makes excellent use of the engineer's intuition and experience.

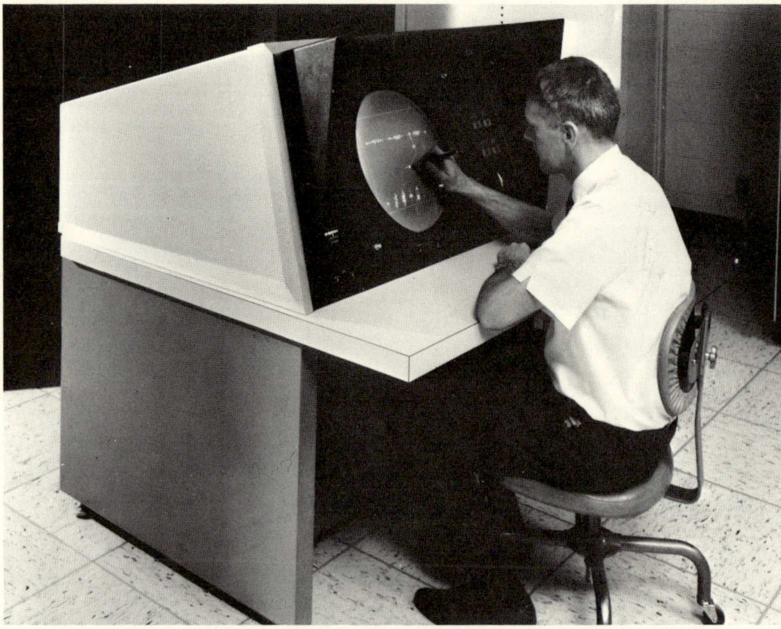

FIG. 11.1

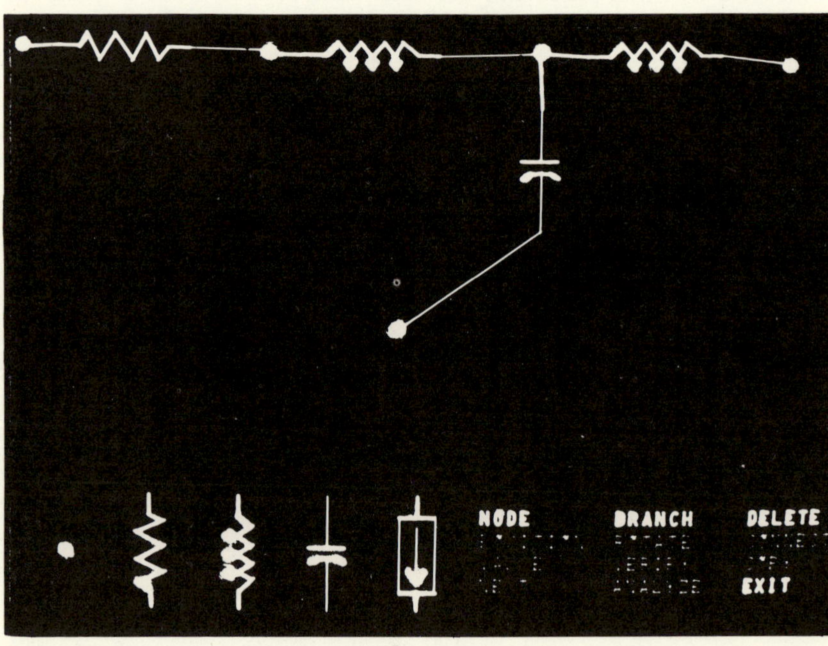

FIG. 11.2

Other Systems

GINA is by no means the only graphic input-output system for circuit analysis. Other programs that exist today are AEDNET [5.6] and CIRCAL [7], which operate on the Project MAC computer [8] with the Electronic Systems Laboratory display console [9]. The circuit elements are coded as glyphs and entered onto the display by a preprogrammed function keyboard. Another system developed at Bell Telephone Laboratories, is OLCA [10] which uses light buttons to draw the schematic. Still another system is the 2250 ECAP [11] system of IBM. The last system requires a function keyboard, alpha-numeric keyboard, the light pen, and foot pedal to enter in graphical information. Once the schematic is entered, it is translated into a set of card images which is the regular (nongraphic) input for ECAP. The card images are checked and edited and then sent to the analysis program. Of the five systems discussed, CIRCAL and AEDNET operate on a time-sharing system, OLCA operates on a local graphics computer (the GRAPHICS-I [12]), which generates and maintains the graphics but does not perform the circuit analysis. Graphic information is translated and sent to the central computer, an IBM 7094, in which resides the analysis program, HYBRID [13]. The 2250 ECAP system requires an IBM 2250 graphics console that shares the memory of an IBM 360 computer (a model 40 or larger). GINA uses an IBM 7094-II, but requires so little memory that an analysis program such as CALAHAN [14] or POTTLE [15] can easily fit into the memory also. Of the five systems, GINA is probably the fastest, although such comparisons are somewhat unfair because of different operating conditions. Finally, it should be said that only OLCA and GINA have hard copy provisions, and only GINA has a built-in movie-making function, as described in Chapter 9.

11.2 GRAPHIC ANALYSIS PROGRAM

Analysis Programs

The underlying analysis methods are of fundamental importance to successful on-line graphic electronic circuit design. Several network analysis programs exist today that perform analyses of both linear and nonlinear networks in both batch and time-sharing modes [1, 2, 3, 6, 7, 10, 11, 14, 15]. Each of these programs reflects many man-hours or

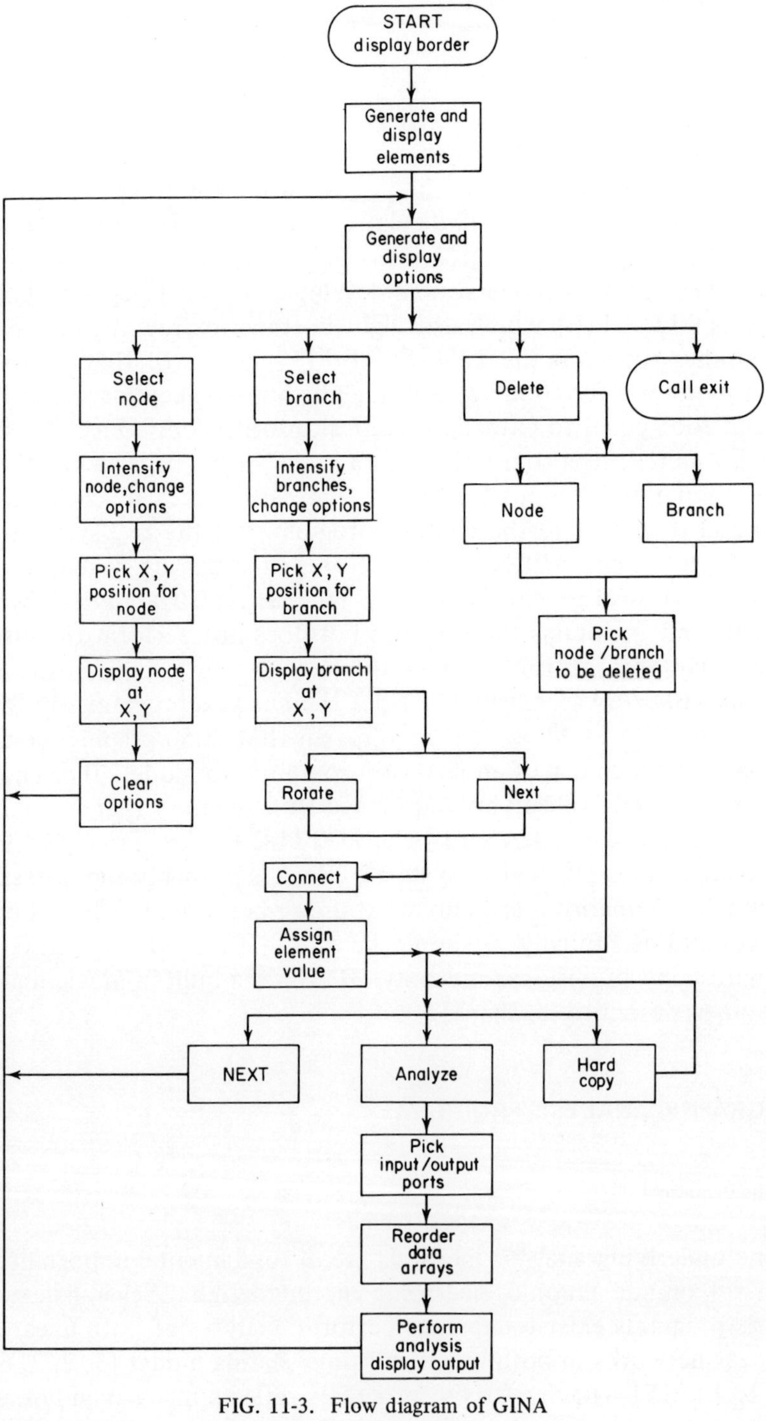

FIG. 11-3. Flow diagram of GINA

even years of development, and the ease in which they might be interfaced with a graphic console varies greatly. As mentioned in the introduction, several analysis programs have been developed concurrently with graphical communication to them. We chose two programs, initially, to link to a graphic console: POTTLE [15] and CALAHAN [14], (see also Chapter 8). Both of these programs perform a linear analysis of RLCM and controlled sources networks. POTTLE has distinct advantages for larger networks (greater than 10 nodes) due to its state variable formulation and consequently more efficient treatment of the solution phase. CALAHAN is based on a topological approach which involves finding all of the network trees. This may involve a considerable amount of time for larger networks and may not be suitable for use in an on-line environment. However, the programming in CALAHAN is very straightforward and is easily accommodated to GINA. For this reason, and also because CALAHAN requires slightly less computer core than POTTLE, it was chosen initially for coupling to GINA.

Graphical Generation Flow Diagram

The generation of the network schematic, when done on-line, should be accommodated to the user and not to the analysis program. Figure 11.3 shows the logical flow diagram for the complete GINA program.

At the start, the control area of the CRT display has two options active. The control area displays: (Fig. 11.4)

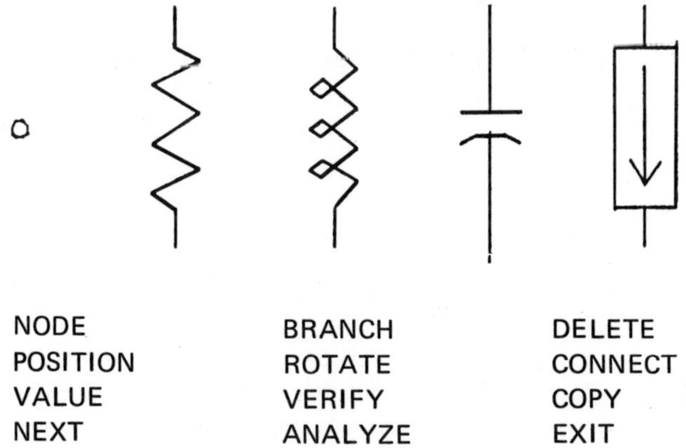

FIG. 11.4. Light buttons for GINA system.

The process of creating the network in the display area is as follows: Initially, only NODE and EXIT are displayed at high intensity while all other light buttons are at normal intensity. The light pen will cause a program response only when it detects a *hit* of NODE or EXIT. When the light pen is placed anywhere on the word NODE, the intensities of all options become normal, and the node at the left of the control area becomes bright (high intensity). Next we pick the bright node which causes both a tracking cross to appear in the center of the display area and the option POSITION to become bright in the control area. Then with the light pen, the tracking cross may be moved to the appropriate position on the working area of the CRT. Then the light pen picks POSITION and a node is positioned at the tracking cross and the cross disappears. The next available options are NODE, DELETE, and EXIT at high intensity. After two nodes are displayed, a BRANCH may then be selected.

Once the BRANCH option is active and is picked the flow goes:

1. Select an element type (RLC or Source).
2. POSITION it on the CRT by means of the tracking cross.
3. ROTATE the element (if desired).
4. CONNECT the element to two different nodes.
5. Assign a VALUE to the element.

The user is then given the option of making a hard copy, adding or deleting the next node or element, or performing the analysis.

The DELETE option becomes active only after a node has been displayed. The BRANCH option is activated only after two nodes are displayed. These conditions prevent the user from selecting DELETE when there is nothing to delete, or selecting BRANCH when there are no nodes to which to connect it. When the DELETE option is chosen the user picks a particular node or branch on the working surface. The selected item disappears from the screen, its memory space is released for use, and the data arrays are adjusted.

At each point in the development of a schematic, a number of methods are used in directing the user. One very effective method is intensity control. The options available are displayed at high intensity, thus directing the user not only to the proper choice, but also indicating what will follow. Each option displayed is actually a light button with a unique value associated with it (this point will be covered in more detail in the next section). Through programming, the value of the light button is repeatedly checked until the correct one is selected. This prevents the user from selecting an option from which a recovery cannot be achieved.

Data Structure: Display and Nondisplay

The purpose of the analysis is to gain insight into the circuit operation, and to verify design objectives. Once the output response has been observed the user can return to the representation of the network and add, delete, or change the topology and element values (or exit if finished).

The data structure is divided into two arrays: display and nondisplay. Both arrays are generated simultaneously. The display arrays are used to create, maintain, and change the graphic display. The nondisplay data structure is similar to the display data except that before entering the analysis phase the nondisplay files are reordered and edited before being transferred to the analysis program itself.

The data structure itself is very simple. Arrays are needed for the node numbers, for the branch numbers and types, for the element connections, and for the element values. The display data can only be changed by using the DELETE option. The DELETE option simply sets the proper member of the arrays equal to zero and removes the element or node from the display list.

Both display and analysis arrays have been arbitrarily dimensioned to accommodate 100 elements and 50 nodes which represent network sizes that will challenge most analysis programs.

When the VALUE option is picked, the entire option display is replaced by the display (Fig. 11.5).

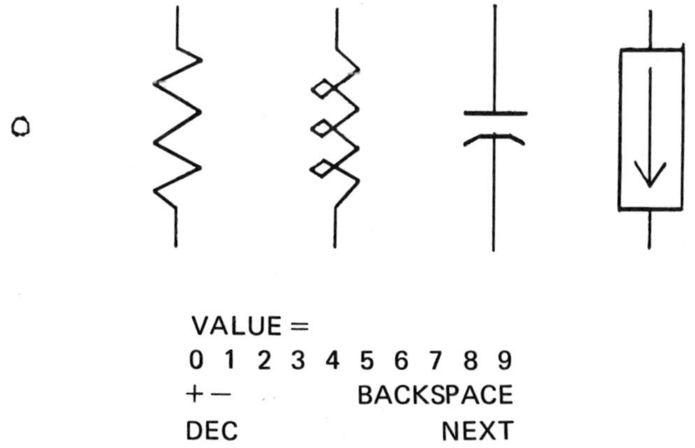

FIG. 11.5. Light Buttons for VALUE Subroutine.

The user then sequentially picks the numbers which are displayed to the right of the equal sign. The value must at present be in the range -10^8 to $10^{10} - 1$. Multiple light pen hits are prevented by a programmed lock-out feature consisting of a FORTRAN DO loop with 2000 iterations and by the requirement that the light pen must detect a light pen value of zero (not the number zero) before the succeeding numeral can be picked. The use of the light pen and programmed light pen buttons to assign values to elements has proved to be very convenient; we believe it is better for entering element values than the use of an adjacent keyboard because it is faster and the user's attention does not leave the display console at any time.

When the ANALYSIS option is selected, the user is next requested to pick the input and output ports. This is accomplished by replacing the options portion of the display by the display (Fig. 11.6):

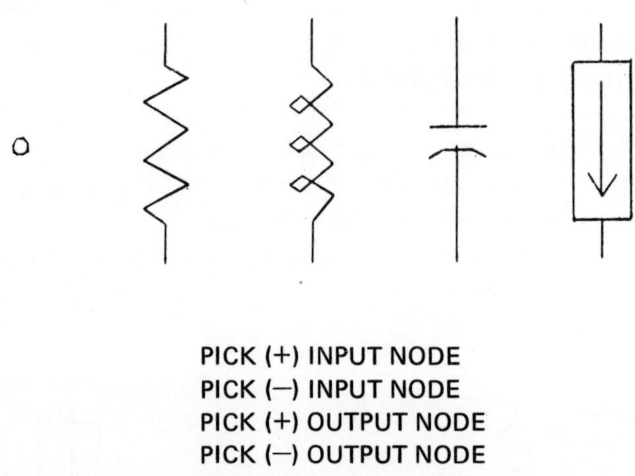

PICK (+) INPUT NODE
PICK (–) INPUT NODE
PICK (+) OUTPUT NODE
PICK (–) OUTPUT NODE

FIG. 11.6. Light Buttons for ANALYZE Subroutine.

Initially the top line is displayed at high intensity and the other lines at normal intensity. When the positive input node is picked, the second line becomes bright, and so forth.

After the input and output ports are picked, one set of the data (the nondisplay set) is edited by renumbering all nodes and removing all zeros. The analysis program (CALAHAN) is then entered and the analysis is performed. During this time and until the response appears, the options portion of the CRT displays RUNNING.

Analysis Flow Diagram

Figure 11.7 depicts the program flow after the analysis arrays have been reordered. From the network topological description, the CALAHAN program uses the Binet-Cauchy theorem to find the network transfer function. This involves finding the trees of the network graph and is the basic limitation of the topological approach, where in a practical problem a network may take a prohibitive amount of computer time. Thus, for on-line work the use of the CALAHAN program is limited to networks of less than approximately 12-15 nodes and 20-25 branches. The POTTLE program does not have the above difficulty due to its use of the state variable approach.

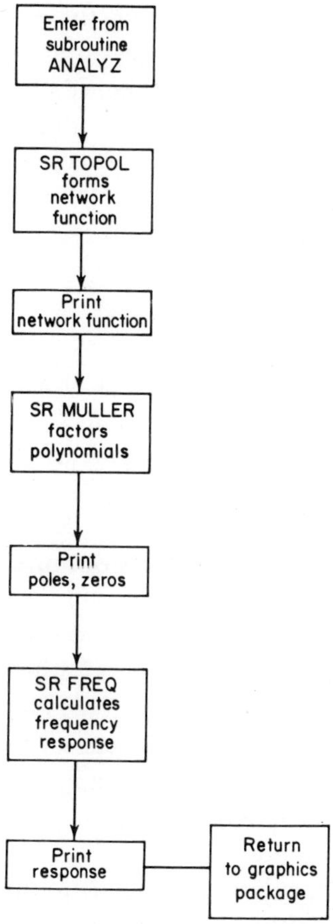

FIG. 11.7. Organizations of CALAHAN Programs.

Once the network transfer function is found, a polynomial factoring routine is used to find the network poles and zeros. At this point, either a time or frequency analysis could be made. In GINA, only the frequency response option has thus far been implemented. After the analysis is completed, the gain versus frequency response is displayed on the CRT. At the user's choice, successive displays of phase and delay versus frequency are displayed and then finally a return to the network topology display is made.

The entire programming is done in FORTRAN II. That portion of the program dealing with the generation of the graphic display and creation of both the display and analysis files requires 7144 words of memory. The CALAHAN analysis program requires 12876 words of memory (frequency analysis only).

Finally, the use of comments during the graphical construction are both specific and implied. They are implied, as mentioned earlier, through intensity control. They are specific in the case of assigning element values, picking the port nodes and during the actual execution phase by displaying RUNNING. The provisions for additional comments are in the program and can be easily implemented.

11.3 GRAPHICAL LANGUAGE

An ideal graphical language should be flexible to the user, require a minimal amount of computer memory, and be adaptable to any computer system. The language should be *multi-level*. The low-level part of the language should have as its basis, a set of subprograms to perform the most basic graphical functions (i.e., put a dot at the point x, y, display a character on the CRT, draw a vector from one point to another, etc.). The high-level part of the language should be FORTRAN compatible so that communication with the basic set would be by FORTRAN subroutine call. There are 11 basic subprograms in the low-level (FAP) graphics language. The main high-level graphics language (GINA) is FORTRAN II with FORTRAN subroutine calls to the low-level language. Thus the user is able to write the program in FORTRAN II, use subroutine calls to the basic language, and yet use only a portion of the main memory for the graphics.

The following sections describe the structure of the entire graphics language and specific machine configurations used in conjunction with the graphics language.

Low-Level Language

The low-level graphics language should perform the most elementary of operations – operations which are basic to the operation of the graphics system, rather than those which can be performed by the high-level graphics language. The low-level graphics language must perform such functions as:

(a) Given a picture or an object – display this on the CRT.
(b) Given an (x, y) coordinate pair – place the beam at that point.
(c) Given a picture being displayed – destroy this picture or remove it from the CRT but retain the picture information.
(d) Draw a line; create a letter; create strings of letters; etc.
(e) Give names to and allocate storage for picture parts.
(f) Enable and disable the beam.
(g) Track the light pen.
(h) Uniquely determine a picture part, when using the light pen.
(i) Be able to recover coordinate positions of any given picture part.

Specific Application of the Low-Level Graphics Language

The principal concern of the graphics language is one of beam (light pen) and picture part control. This section will illustrate how a portion of the low-level graphics language is used in the GINA program. Specifically the CONECT subroutine of GINA will be described in detail (see Fig. 11.8 for FORTRAN listing). This subroutine serves two functions:

```
      SUBROUTINE CONECT (N1,N2,NB,JB,XX,YY,IROT)
C     SUBROUTINE CONECT IS USED TO CONNECT AN ELEMENT TO THE
C     TWO DIFFERENT NODES PICKED BY THE LIGHT PEN.
C     N1 IS THE LPV OF THE FIRST NODE
C     N2 IS THE LPV OF THE SECOND NOBE
C     NB IS THE LPV OF THE BRANCH
C     JB IS THE BRANCH NUMBER
C     X IS THE X POSITION OF THE TOP OF THE BRANCH
C     Y IS THE Y POSITION OF THE TOP OF THE BRANCH
C     IROT IS THE ROTATE REFERENCE AND IS -1, 0, 1
      X=XX
      Y=YY
      XY = 0.15
      CALL NAME (JB,NB)
      CALL STORE(123,0,0,0,0,X,Y)
 10    CONTINUE
      PAUSE
      CALL LPEN(N1,X1,Y1)
      IF (N1)15,10,15
 15   IF(N1-50)20,20,10
 20    CALL STORE(125,0,0,0,0,X1,Y1)
      IF (IROT) 50,60,70
 50   X = X - XY
      GO TO 80
 60   Y = Y -XY
      GO TO 80
 70   X = X + XY
 80   CALL STORE (123,0,0,0,0,X,Y)
 25   CONTINUE
      PAUSE
      CALL LPEN (N2,X2,Y2)
      IF (N2)30,25,30
 30   IF(N1-N2)35,25,35
 35   IF(N2-50) 40,40,25
 40   CALL STORE (125,0,0,0,0,X2,Y2)
      RETURN
      END
```

FIG. 11.8. Listing of CONECT subroutine.

1) It is used to return to the main routine the two node numbers an element is being connected to, and 2) it adds to the display list the appropriate vectors connecting that element to the nodes.

The call to subroutine CONECT is:

CALL CONECT (N1, N2, NB, JB, *XX*, *YY*, IROT)

where

N1 is the light pen value (LPV) of the node which the top (left if the element has been rotated 90° CCW or right if rotated 90° CW) of the element is to be connected to.

N2 is the LPV of the node the other end of the element is to be connected to.

N1 and N2, node numbers, are both returned to the main routine and used to construct the arrays describing the network topology.

NB and JB are respectively the LPV of the branch being connected, and the branch number. Their use will be described later.

XX and *YY* are the coordinates about which the branch rotates.

IROT indicates whether or not the branch has been rotated and in which direction.

NB, JB, *XX*, *XY*, and IROT are all fed forward from the main routine.

Figure 11.9 shows the flow graph for subroutine CONECT and will be used to explain how information is added to the display loop and consequently appears on the CRT. The statements $X = XX$ and $Y = YY$ simply rename the variables so their values are not altered upon return to the main routine. The elements are 0.15 units long (an arbitrary length but one which yields the appropriate sized display) so $XY = 0.15$ is defined. This information is used when the element is connected to the second node.

Next subroutine NAME is called with the arguments JB and NB. The JB refers to a particular branch, in this case the most recent network element moved to the display area of the CRT. The call to NAME reopens the buffer block containing the picture part describing this element. Now more information can be added to the glyph or picture by subsequent CALL STORE's. We need to associate the same LPV of the element to the vectors connecting the element to the nodes and do this through the second argument NB. This enables us to remove both the element and the connecting lines simultaneously if the element is deleted at a later time. The CALL STORE (123, ..., *X*, *Y*) tells the display to merely move to the point (X, Y), (the top of the element) but do not plot. Next the (X, Y) location of the node the element is to be connected to must be picked with the light pen.

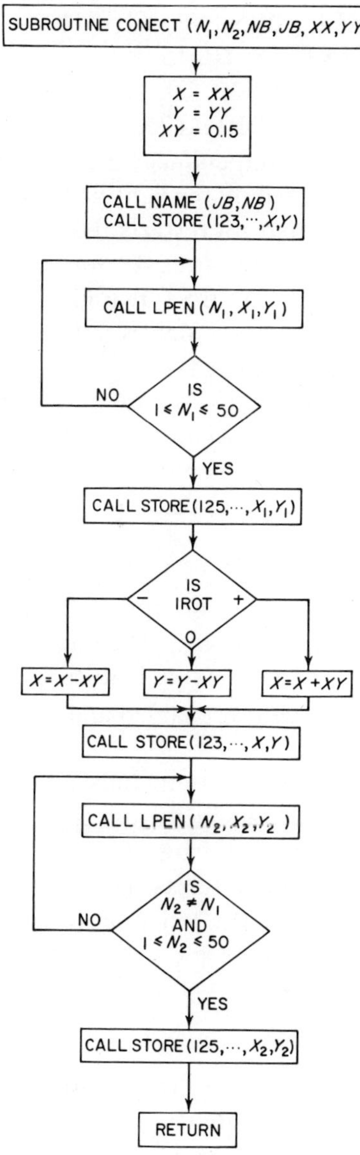

FIG. 11.9. Flow chart of CONECT subroutine.

Each node as it is displayed on the CRT (and entered in the data arrays) is given a specific number; the numbering is sequential from 1 to 51. The programming

```
10  CONTINUE
    PAUSE
    CALL LPEN (N1, X1, Y1)
    IF (N1) 15, 10, 15
15  IF (N1-50) 20, 20, 10
20  CALL STORE (125, ..., X1, Y1)
```

is used to pick the first node and connect the element to this node. The LPV, N1, of the first thing seen by the light pen is returned to the calling program (LPEN). A light pen hit restarts the computer after the pause statement. The LPV is then tested to make sure it is a node number (between zero and 51) and then CALL STORE (125, ..., X1, Y1) draws the vector from (X, Y) to (X1, Y1) which is automatically entered in the display loop and displayed on the CRT.

A new beginning value for (X, Y) is calculated if the element has been rotated and the above procedure repeated. The only additional protection in connecting to the second node is to insure the node number is different from the first node, a situation which would be meaningless electrically anyway.

This example has used the low-level graphic subroutines LPEN, NAME, and STORE. Two other groups of the low-level graphics language are extensively used by GINA. They are:

TRACK, NTRACK – The tracking feature (i.e., the electron beam following the light pen) is enabled and disabled with these commands.

IN, OUT, CLEAR – By these commands a picture part is: 1) added to the display list, 2) removed from the display list but still "remembered," 3) removed from the display list and all record of it destroyed (memory released).

A complete description of the ten subroutines, and the philosophy behind the low-level graphic language, is given in Chapter 10.

Machine Configurations

The low-level graphics language itself requires just 524 memory locations (real words). This permits us to use nearly the entire memory for the graphics program and the network analysis program. Let us consider machine requirements for a graphics analysis package such as GINA. First, let us consider the way in which the basic graphics language assigns memory space.

One of the commands in the low-level graphics language is the IGRAPH command. This enables the user to determine beforehand exactly where in memory the picture part information will be located (i.e., either in common or in a specific part of the main memory). In addition, the NAME command assigns specific relocatable segments of memory for the picture parts. When we reference (in the graphics program), the NAME command, a 1-word segment is established as a pointer to a 32-word block (this may be any number up to 32) which will contain the actual display (picture-part) information.

To give some specific numbers – for our graphics package, GINA on the IBM-7094 computer (36 bit words, 32K memory). Memory requirements are:

Low-level graphics – 524 words
GINA language – 7144 words
Analysis program – 12876 words.

The following comments can be made to the memory requirements of the system:

1. The IGRAPH command would allocate all of the unused memory for picture information.
2. The low-level language takes essentially no space.
3. The GINA language (which is designed to handle 50 nodes, 100 elements) really does not take a large amount of space.
4. The analysis package occupies about as much space as the graphics language plus the GINA language

If we are willing to allow problems of the size permitted by GINA (50 nodes, 100 elements), we do not need a very large system. In fact, as we will soon point out, the basic limitation will be the size of the network analysis program which is used in conjunction with the graphics package.

In general, a machine with 16K words will be more than adequate for use with a graphics language package such as GINA. However, we must then find an analysis program which will conveniently fit into 16K words. Most analysis programs are designed to be handled by a 32K machine. It may be possible to reduce this requirement by 30-40%, but in general, very few programs conveniently fit into 16K. Consequently, some form of overlay or linkage structure must be created which will usually involve the use of a tape system or a large disk.

However, with advanced programming techniques available on the modern computer, the problems of machine size are gradually disappearing and may not be a major concern in the future.

11.4 CONCLUSIONS

Two important ingredients for effective computer-aided circuit design are: (1) a versatile electronic circuit analysis program, and (2) a graphic display console tied on-line with a high-speed, large-memory computer. The problems of adapting (1) and (2) are not minimal. In this paper we have concentrated on the graphic input software package, GINA, but did not discuss in detail the interface between graphics input and the analysis program. Thus far we have interfaced only with the CALAHAN program. We envision in the near future, an interface with the POTTLE program which has some more general features than CALAHAN.

One of the major difficulties in the use of circuit analysis programs is that every program has a different input language for describing the topology and the element types and values of the circuit. There are two ways to handle the interface problem between the graphic I-O console and a large number of analysis programs. The first requires that the nondisplay data associated with the graphics input be tailored for each individual program. The second method involves leaving the graphic input system unchanged, and devising a common interface language for every analysis program in use. The common interface language seems inherently to be a more efficient approach, since this would enable an engineer to use a program on a batch-processing mode, also, without learning the details of the input language for the particular program. This common input language is the first step in establishing a *Design Supervisory Program*, as described in Chapter 1.

Finally, we should mention that we have deliberately underplayed the hardware features of the present graphics console at Lawrence Radiation Laboratory, Livermore. The reason for this is that the graphics input software only depends upon the low-level graphics language, which at present requires only 524 words of memory on the IBM 7094. Recently LRL installed and started using a Sigma 7 – Sigma 2 system as its main graphics computer. Very little effort is required to recode the graphics package for the Sigma 7 computer. Once the language is implemented in the Sigma 7, GINA will be available on the new computer.

Also, since CALAHAN, POTTLE, SCEPTRE, CIRCUS, etc., are written in FORTRAN IV, adapting these programs to the Sigma 7 could be achieved.

APPENDIX A

Example Of The Use Of GINA

This appendix contains a simple example of the use of GINA and the appearance of the hard copy output. Figure 11.10 is the network (a twin-tee) just prior to analysis. The figure is the hard copy output by means of the unit parallel to the visual display unit. Generally the hard copy is available in approximately two hours. Figure 11.11 is the frequency response output and Figure 11.12 shows a portion of the hard copy that is produced on the output tape during the analysis phase. This tape is later printed off-line, during the construction of the network and during the analysis the user is unaware of this operation.

APPENDIX B

Description of LRL DD80 System Hardware

The DD80 is a very fast and flexible computer-controlled display device. It consists of a 12-inch CRT for visual monitoring and a 5-inch CRT for recording on 35-mm microfilm. The camera has a framing rate of 30 frames per second and is specifically capable of accurately registering the film so that motion sequences can be more easily produced.

Generally, the flexibility includes the ability to generate and display *points, characters,* and *line segments* (*vectors*). One of two *intensities and* one of two *focus* conditions may be selected when displaying points or vectors. Additionally, one of four *sizes* and one of two *orientations* may be chosen when a character is to be displayed. There are 128 characters possible, of which 115 have been assigned.

The (x, y) coordinates of a point can take on any integral value in the range $0 \leqslant x$ or $y \leqslant 1023$. Any legitimate character (or point) can be displayed at any raster point. Characters are centered and line segments are constructed between any two points.

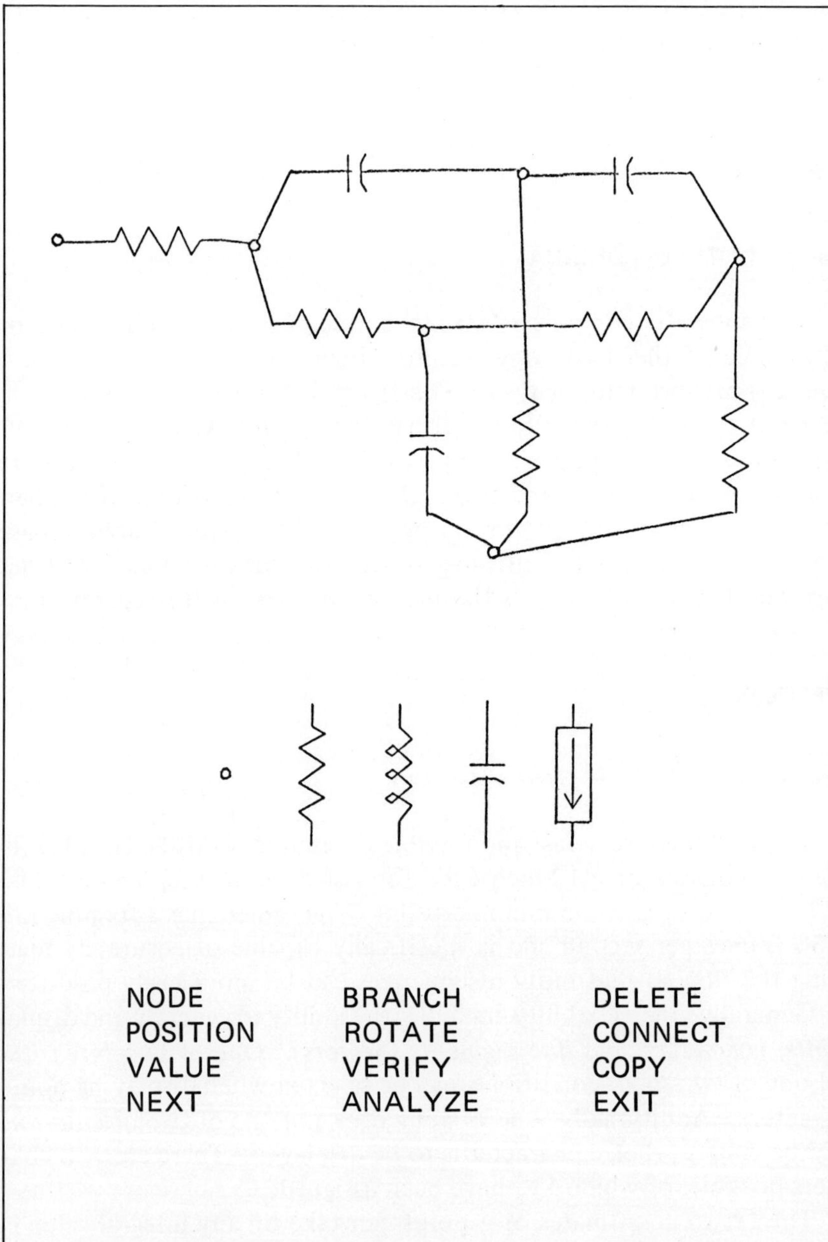

FIG. 11.10. Analysis of twin-tee circuit using GINA.

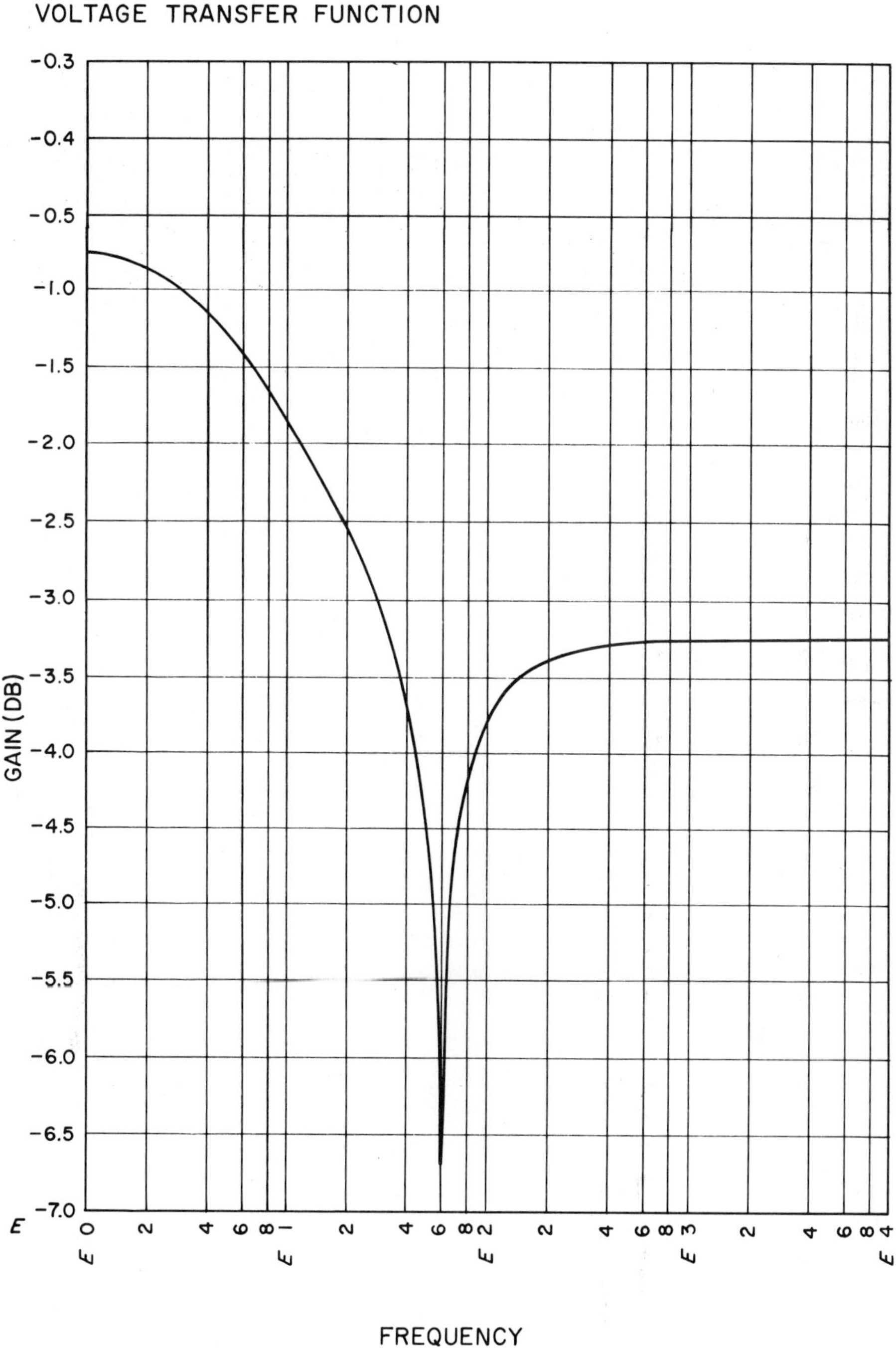

FIG. 11.11. Twin-tee circuit response from GINA.

```
CIRCUIT INPUT DATA

8  0  6  1  6  4  6  1  2  0  0
1  2 R       300.000000000
2  3 C         0.000002667
3  4 C         0.000002667
2  5 R      1000.000000000
5  4 R      1000.000000000
3  6 R       500.000000000
5  6 C         0.000005394
4  6 R       300.000000000


     NUMERATOR COEFFICIENTS,BEGINNING WITH THE CONSTANT TERM

0.66666663E-11 0.17799999E-13 0.47525497E-16 0.12832019E-18

     ZEROS

   REAL PART        IMAG PART            REAL PART         IMAG PART            REAL PART         IMAG PART
   -372.451073        0.000008     00001        1.040352        373.482758              1.040350      -373.4827

     DENOMINATOR COEFFICIENTS,BEGINNING WITH THE CONSTANT TERM

0.57777775E-10 0.43894663E-12 0.91495550E-15 0.41062462E-18

     POLES

   REAL PART        IMAG PART            REAL PART         IMAG PART            REAL PART         IMAG PART
   -232.762566        0.000001            -372.470512        0.000007            -294.990337        935.001472
```

FREQ (CPS)	GAIN (DB)	PHASE(DEG)	DELAY (SEC)
1.0000000	-18.7627282	-1.773	0.0049240
1.0964780	-18.7638783	-1.944	0.0049234
1.2022639	-18.7652609	-2.132	0.0049226
1.3182559	-18.7669230	-2.337	0.0049217
1.4454386	-18.7689211	-2.563	0.0049206
1.5848916	-18.7718232	-2.810	0.0049193
1.7377987	-18.7742107	-3.080	0.0049177
1.9054580	-18.7776818	-3.377	0.0049158
2.0892927	-18.7818542	-3.703	0.0049135
2.2908634	-18.7868690	-4.059	0.0049108
2.5118813	-18.7928967	-4.490	0.0049075
2.7542225	-18.8001411	-4.878	0.0049036
3.0199443	-18.8088477	-5.346	0.0048988
3.3113024	-18.8193079	-5.860	0.0048931
3.6307701	-18.8318818	-6.422	0.0048863
3.9810594	-18.8469861	-7.038	0.0048782
4.3651440	-18.8651307	-7.712	0.0048684
4.7862843	-18.8869245	-8.449	0.0048567
5.2480552	-18.9130964	-9.255	0.0048427
5.7543769	-18.9445188	-10.136	0.0048261

FREQ (CPS)	GAIN (DB)	PHASE(DEG)	DELAY (SEC)
1.0471284	-18.7632771	-1.857	0.0049237
1.1481533	-18.7645378	-2.036	0.0049230
1.2589248	-18.7660537	-2.232	0.0049222
1.3803833	-18.7678761	-2.448	0.0049212
1.5135598	-18.7700672	-2.683	0.0049200
1.6595850	-18.7727008	-2.942	0.0049186
1.8196984	-18.7758663	-3.225	0.0049168
1.9952592	-18.7796719	-3.536	0.0049147
2.1877578	-18.7842462	-3.877	0.0049122
2.3988282	-18.7897444	-4.250	0.0049092
2.6302623	-18.7963526	-4.659	0.0049055
2.8840247	-18.8042946	-5.107	0.0049013
3.1622695	-18.8138387	-5.597	0.0048961
3.4673589	-18.8253078	-6.135	0.0048899
3.8018827	-18.8390677	-6.723	0.0048824
4.1686805	-18.8556426	-7.367	0.0048735
4.5708664	-18.8755288	-8.072	0.0048628
5.0118543	-18.8994124	-8.843	0.0048500
5.4953879	-18.9280903	-9.686	0.0048349
6.0255717	-18.9625180	10.607	0.0048166

FIG. 11.12. Printout of twin-tee responses.

Our DD80 has been attached to the IBM 7094 computer system. Specifically, it is connected via I-O channel B through the IBM option known as Direct Data Connection (DDC). This option (IBM 7090 Programmer's Reference Manual A22-6528-1) permits the parallel transfer of 36 bits of information every 6 μsec. Only 32 bits are used in each transmission to the DD80. This set of 32 bits is called a DD80 word and consists of the following information:

Name	Bit Requirements
Intensity	1
j^{th} character	7
horizontal coordinate	10
character size	2
character orientation	1
vertical coordinate	10
focus	1
bit total	32

Upon receiving the 32-bit word the DD80 looks first at the character descriptor. This can be a normal character or a mode change character. Explicitly, the mode designators include: (1) Go into single-point mode, (2) plot the character, (3) go into typewriter mode, (4) plot a vector, (5) advance the film one frame, and (6) plot a point.

In addition there are 10 bits by which the DD80 can send information to the 7094. These are called incoming sense lines and may be sampled at any time. Presently six of the bits are being used in the following manner to indicate the status of the DD80: (1) Camera fault, (2) film supply low, (3) DD80 not ready, (4) illegal character code encountered, (5) camera motor off, and (6) a light pen signal has been sent.

Each point and character is plotted in eight microseconds or less. All vectors (regardless of length) are plotted in less than 32 microseconds. Every fourth point can be resolved as can every fourth vector regardless of the setting of intensity and focus. Generally, no difficulty is encountered in plotting either characters or vectors over the face of the tube. There is some character distortion near the corners but no irregularities occur due to lack of linearity, short-term instability, edge focusing, barrelling, or pincushioning.

ACKNOWLEDGMENT

The authors wish to thank the following people for their help and advice in the development of GINA: K. R. Bertran and G. A. Michael for answering many questions and offering suggestions regarding the use of the low-level graphics package; D. L. Braff and L. A. Loftis for making computer time available and the many operators for their assistance while we were on the computer; and the United States Atomic Energy Commission for not only funding but for creating a research laboratory having the necessary environment.

REFERENCES

1. L. D. Milliman, W. A. Massena, and R. H. Dickhaut, "CIRCUS -- A digital computer program for transient analysis of electronic circuits -- user's guide," The Boeing Company, Seattle, Wash., Harry Diamond Laboratories Rept. AD-346-1, January 1967.

2. A. F. Malmberg, F. L. Cornwell, and F. N. Hofer, "NET-1 network analysis program." Los Alamos Scientific Laboratory, Los Alamos, N. M., Rept. LA-3119, August 1964, 7090/94 version.

3. H. W. Mathers, S. R. Sedore, and J. R. Sents, "Automated digital computer program for determining responses of electronic circuits to transient nuclear radiation (SCEPTRE)," vol. I, IBM Space Guidance Center, Owego, N. Y., IBM File 66-928-611, February 1967.

4. J. J. Ebers and J. L. Moll, "Large signal behavior of junction transistors," *Proc. IRE.*, vol. 42, pp. 1761-78, December 1954.

5. D. S. Evans and J. Katzenelson, "Data structure and man-machine communication for network problems," *Proc. IEEE*, vol. 55, no. 7, pp. 1135-1144, July 1967.

6. J. Katzenelson, "AEDNET: A simulator for nonlinear networks," *Proc. IEEE* vol. 54, pp. 1536-1552, November 1966.

7. M. L. Dertouzos, "CIRCAL: On-line circuit design," *Proc. IEEE*, vol. 55, pp. 637-654, May 1967.

8. R. M. Fano, "The MAC system: The computer utility approach," *IEEE Spectrum*, vol. 2, No. 1, pp. 56-64, January 1965.

9. J. E. Ward and R. H. Stotz, "Operating manual for the ESL display console," Electronic Systems Laboratory memo 9442-M-139, Project MAC memo MAC-M-217, Massachusetts Institute of Technology, Cambridge, Mass., March 1965.

10. H. C. So, "Analysis and iterative design of networks using on-line simulation," in *System Analysis by Digital Computer,* F. F. Kuo and J. F. Kaiser, Eds., New York: Wiley, 1966, ch. 2.

11. G. R. Hogsett, D. A. Nisewanger, and A. C. O'Hara, Jr., "An application experiment with on-line graphics-aided ECAP," in Conf. Digest *1967 International Solid-State Circuits Conference,* pp. 72-73.

12. W. H. Ninke, "GRAPHIC-1-A remote graphical display console system," in *Proceedings of the FJCC,* vol. 27, Washington, D. C.: Spartan Books, pp. 834-846, 1965.

13. H. C. So, "Analysis and design of linear networks with adjustable parameters using on-line simulation," in *Proc. Third Allerton Conference in Circuit and System Theory*, October 1965.

14. D. A. Calahan, "Linear network analysis and realization digital computer programs - An instruction manual," *University of Illinois Bulletin* vol.62,No.58, February 1965.

15. C. Pottle, "State-space techniques for general active network analysis," in *System Analysis by Digital Computer,* F. F. Kuo and J. F. Kaiser, Eds., New York: Wiley, 1966, ch. 3.

CHAPTER 12

COMPUTER-AIDED DESIGN OF PRINTED CIRCUIT ARTWORK

R. ZANE, Lawrence Radiation Laboratory, Berkeley
D. A. WILBER,† Lawrence Radiation Laboratory, Berkeley

12.1 PRINTED CIRCUIT CARD DEVELOPMENT

The starting point in the design of printed circuit cards is the schematic diagram, logic diagram, or Boolean algebraic expression for the circuit. The circuit is usually a part of a larger system so that a common first step is deciding which portions of the overall system should be confined to a particular printed circuit card. This step establishes the *boundaries* of a particular circuit, Fig. 12.1.

The printed circuit card *board outline* chosen for a particular circuit greatly affects the layout of the circuit. The dimensions of the card and the type of connector used are usually dictated by systems

†Editor's note: Miss D. A. Wilber = Mrs. D. A. Harrell

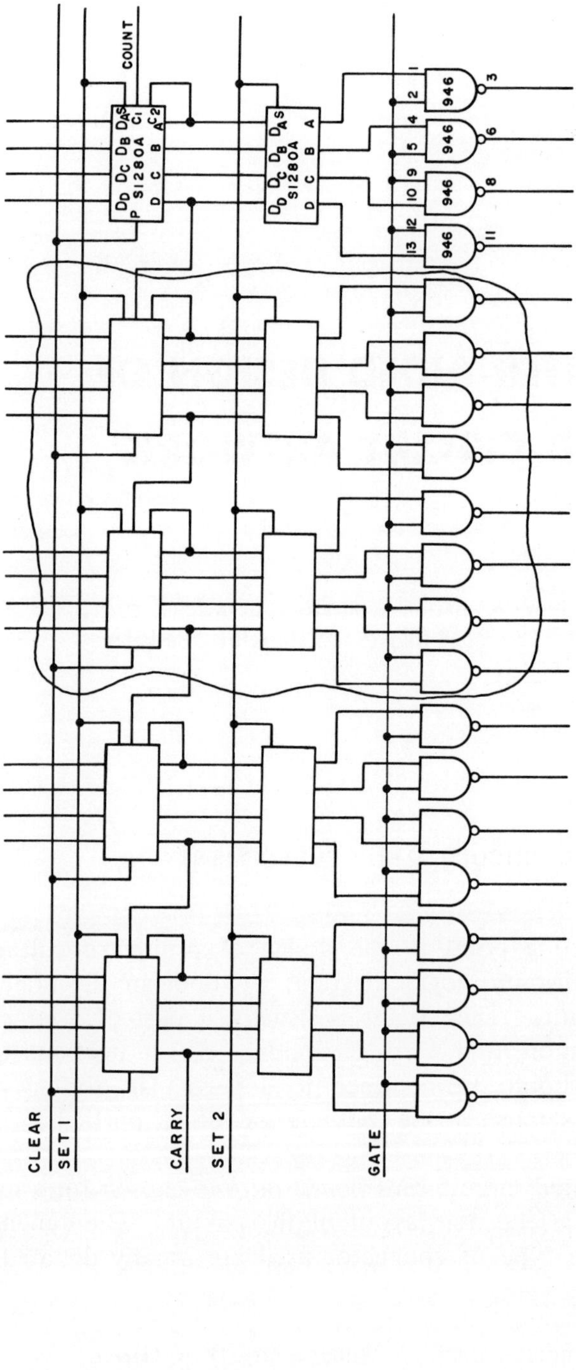

FIG. 12.1. The outlined portion of the Scaler and Register Circuit would be allocated to one printed circuit card. All times crossing the outlined boundary would constitute connections to the printed circuit connector.

packaging considerations, but the effect of the outline on the topology of the circuit must not be neglected. The more important considerations are:

1. Pin spacing and number of pins in the connector.
2. Board dimensions and aspect ratio relative to the connector.
3. Number of surfaces on the board, i.e., single sided, double sided, or multi-layer.
4. Power distribution and filtering techniques as well as the presence or absence of the need for a ground plane.

Numerous techniques are available for the etching of printed circuit cards. The choice of technique ranges from the rather crude prototype technique of direct application of an etch-resistant tape or paint to printed circuit laminated stock, through the use of a photo-resist such as KPR (Kodak Photo Resist) for small production quantity of good quality, to silk screening techniques for large production quantities.

The loading of boards with components is considerably eased if a systematic placement of components is used. To this end it may be desirable to align resistors and capacitors in neat arrays. Integrated circuits may usually be placed in systematic patterns, Fig. 12.2. On the other hand, particularly in high frequency linear circuits, design considerations may preclude the use of systematic placement and instead require the placement of the components in positions and orientations determined by performance characteristics of the functional circuit.

12.2 THE ROLE OF THE COMPUTER

The highly variable nature of the considerations briefly touched upon in the preceding section suggest that the construction of a general algorithm to deal with all the known variables in printed circuit card design is a formidable task. In addition, the construction of such an algorithm is complicated by the fact that technological innovations in circuitry and packaging are prone to introduce new sets of variables in the problem. The complexity

the manual and automated approaches are dissimilar in many respects, they have some common characteristics. Once a board size is chosen, it is necessary to position the necessary components. This placement will be influenced by such considerations as placing interconnecting components close together, placing components connecting to the connector tabs near the respective edges, spacing components in order to evenly distribute board density, etc. Either approach may permit modification of placement during actual routing of interconnections. However, the time and amount of modification depend on the actual routing philosophy.

In addition, electrical and practical considerations impose certain restrictions on layout. One of the most obvious concerns is to avoid short circuits. Therefore lines that cross each other must appear on opposite sides of the board or be rerouted unless they are part of the same branch of the circuit. Parallel lines on opposite sides of the board, which are superimposed upon one another and separated only by the dielectric material of the printed circuit board, should in general be avoided because they would have a relatively large capacitance between them. It is desirable to minimize overall wire length, in order to minimize the inductance and to optimize use of board space. The number of holes should be minimized in order to reduce production cost. All of these factors must be considered whether the printed circuit artwork is manually or automatically produced.

12.3.1 Manual Design

When a man sets out to design a printed circuit card, he has several distinct advantages over a computer program. The primary advantage being that a human designer may employ a more flexible variety of procedure, whereas a computer is rigidly confined to a predetermined set of rules. A man would first establish a rough layout of the components, His next step would probably be to route the obvious connections, keeping lead length as short as possible and attempt to avoid crossing through the board, yet all the while planning ahead in order to leave area for the other paths. As he proceeds to the more difficult connections he may begin to cheat a bit. For example he may shift components or related lines. When the most difficult connections are to be made, he may take advantage of the idiosyncracies of the particular circuit. For example, the capacitance effect may be a relatively minor concern for the particular pair of lines involved. Or it may be feasible to extend the leads on a particular resistor in such a way that its terminals are 0.5 in. apart rather than 0.4 in., and consequently allow an additional path to cross the line of the resistor. He is not restricted to a particular grid, hence curved lines and diagonals are readily available to him. In general, he has a free space to work with and the capability to keep an accurate and overall picture, and this visual image can constantly cause him to alter his ground rules.

There are disadvantages to the manual system, however. Two practical concerns are *time* and *cost.* A man may spend many hours trying to route a connection, and later have to reroute portions of the path or entire connection. He may also have to shift large blocks of routing, which is a slow and tedious procedure. This can be costly. Also there is the unavoidable problem of human error. No system can be completely independent of this factor, but with automated systems it can be minimized. The human designer may also tire and resort to jumpers sooner than necessary. The above reasons are some of the factors which are causing the development of automated systems that are faster and less costly.

12.3.2 Automated Systems

Basically two different design philosophies are presently being used in the automated systems for printed circuit design. A three-dimensional model is used by the topological approach. Other systems maintain a two-dimensional model [1].

The first step in the topological approach is to set up a three-dimensional model with peaks, slopes, plains, and valleys based on the component pin locations.† So again the first step is to locate components. A grid* is then set with weighting factors to represent the board's *terrain*. These weighting factors may also be influenced by voltage considerations. As paths are routed, the topography of the model changes, with new paths creating ridges. The valleys are chosen as preferred paths and the ridges and peaks are the unavailable area. An incrementing process is then used to work the path through from origin to destination [2, 3].

† This is the system currently being used by Sandia Corporation and the Thomas Bede Foundation in their system, ACCEL (see Chapter 1).

* The establishment of a fixed grid is one of the disadvantages of the automated systems. However, because wires cannot be forced to be too close together, if the grid is made fine enough it may very closely approximate the usable working space.

Interested people have played maze-solving games for centuries. With the advent of the computer, automating chess and similar games has yielded many algorithms to approximate human effort. As the result of similar experimentation, the Lee Algorithm [4] for maze solving was developed. The procedure Lee used was to set up two series of concentric wavefronts, radiating from the origin and from the destination within a maze. These wavefronts were set up originally as circles which are forced to yield to obstacles. The path from any point on a *circle* is then begun by the line segment forming the perpendicular to the tangent of the circle. Line segments are added by successively taking the perpendicular to the tangent of each inner circle at the point of intersection of the previous line segment. The building of wavefronts ceases as soon as the outer wavefronts intersect. Then the line segments are determined. Lee's Algorithm can assure a solution if one exists and can guarantee the shortest path. The implementation of Lee's Algorithm per se requires an unreasonably long computer run, therefore modifications have been made by concerns using this approach.† However, such systems *may* still be expensive in terms of computer costs.

At Lawrence Radiation Laboratory we are using a simplified stepping procedure. In some ways it resembles both the topological and the Lee approach; however, its basic precepts are markedly different. The logic very closely resembles that of a rat in a maze. The only knowledge our hypothetical rat has is the general direction in which he must go. Motion that reduces the distance to the destination is defined as *positive*, and motion that increases this distance is defined as *negative*.* The rat may move in only a horizontal or a vertical direction, and as he completes each new path, it is blocked to prevent his reuse of it. The rat must be assumed to have *tunnel vision*, that is, he can see only in the direction he is moving. When he runs into a wall, he backs away from it and turns 90 degrees and starts in that direction. He then moves until he hits another wall, then rotates again, and so on until he reaches his destination. He has been trained sufficiently that if he is blocked when he changes direction, he goes back to his original direction, backs up one

† Modified forms of Lee's Algorithm are being used by Friden, San Leandro and by Norden Division of United Aircraft.

* At the present time the only motion allowed is orthogonal, i.e., along grid lines. At a future time diagonals will be incorporated allowing eight possible direction choices rather than four.

more space, and again tries to turn. One additional trick has been installed in the rat's maze. That is. his maze is a *two level* maze, with all vertical paths in one level and all horizontal paths in the second level. (In the real situation these two levels correspond to the two sides of a board). Therefore, in order to change direction the rat must change levels. It is when this level change is impossible that the rat must move backwards.

The rat has been given a bit of a helping hand. When he works himself into an impossible situation, he is put at the beginning again and given a new starting direction. He is given a total of four tries to start out. If the fourth try fails, his destination and starting point are reversed and he is given four more tries.

The sophistication necessary in the routing techniques will vary with the individual needs of an electronic production organization. The greatest advantages seem to be gained, however, by keeping the logic for the machine simple and employing some form of man-machine interaction to achieve the more sophisticated logic.

12.4 IMPLEMENTATION OF PUZZLE

The problem of computerizing various phases of printed circuit card development has been handled in several different ways. [1, 2, 3, 4, 5, 6, 7, 8] Some of these methods involve highly sophisticated algorithms and consequently require several hours of machine time [2, 3]. These are the programs based on the terrain methods or involved maze solving techniques discussed in the previous section. Other systems depend on some type of man-machine interaction and hence allot the user more control over the finished board [1, 4, 5, 6, 7, 8]. These systems have the added advantage of minimizing the costly computer time necessary for the completion of a circuit card. Such a system is used at Lawrence Radiation Laboratory.

The basis of LRL's system is a program called PUZZLE [5, 6, 7, 8]. PUZZLE is currently running on a CDC 6600, using a core of approximately 125000 (octal), and spending approximately 10 to 15 seconds of execution time.†

† The current Software system at LRL is CDC's *Chippewa*. PUZZLE is written entirely in FORTRAN IV and it would be reasonably convenient to convert the program for use on other machines.

12.4.1 Basic Assumptions

Before any actual coding can be done for a system, a working framework must be established. The ground rules for PUZZLE were determined after considerable discussion and experimentation. These rules centered around such factors as component placement, board size and dimensions, line spacing, and scaling factors for the graphic output.

Most of the rules came from considerations outside the computer itself. For example, what size boards were actually needed? What components were available and currently being used? If graphs were to be drawn twice scale, what effect would photographic reduction have on pad size and line width? What line width would be necessary with current fabrication techniques to insure reliable circuitry? In addition there are limitations imposed by both hardware and software. These include the size of the plotting surface,† the grid nature of the pen motion on the CalComp plotter, pen widths available with the plotter, and increments used in plotting. Choices were also influenced by programming difficulty.

It was decided that it would be convenient to use 0.1 in. spacing for lines and pads, because component leads and pins are usually fabricated for use on 0.1 in. centers. A double sided board would be produced with essentially all horizontals on one side and all verticals on the other.* In addition, to simplify input, the working area is blocked into 1-in. squares. The present version of PUZZLE can handle up to fifty of these squares in a 5 X 10 array, [Fig. 12.3]; hence the working area maximum is 5 X 10 in. Any rectangle within this array may be used. Also the circular configuration for eight and ten pin micro-logic presented a problem. Therefore, it was decided that in order to fit the grid pattern, it would be convenient and quite workable to use a square configuration for the eight pin and a rectangular configuration for the ten pin micro-logic, (Fig. 12.2). Then a centered position was chosen for these packages with rotation provided for the ten pin can. At a later date, the dual-in-line packages were incorporated, allowing two positions for each. The choice of standard positioning is a convenience for the user, not a restriction. Facility is also provided for specifying discrete points within the grid area. This facility allows the user to use variable length resistors, inductors, etc., to place transistors,

---

† LRL is using a 565 CalComp drum plotter. The drum is 11 in. wide.

* This has been modified within the logic of the program, but all plotting is begun with this assumption.

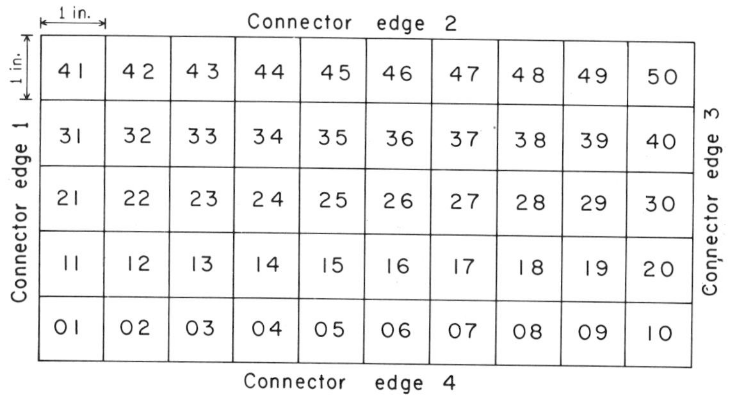

FIG. 12.3. PUZZLE's orientation is based on a 5 in. x 10 in. rectangle divided into fifty 1-inch squares.

and to shift the larger packages in other than the standard position. These various restrictions were then used as the framework to build a working system.

12.4.2 Data Transmission

A unique coordinate system was devised to facilitate easy definition of the points to be connected. Each point of the board is specified by a combination of three codes: The *type* of component, the *square* the component falls in, and the *pin* being specified with the component. The points that are to be interconnected are listed together on a single data card. The program will accept two to eleven of these points per card. There may be as many of these cards as are necessary to describe the circuit. There are two additional blocks of information which may be necessary to complete a board. The first of these is optional. It is a list of variables describing the board dimensions and location of connector tabs. The other is an identifier which includes such information as the user's name, date, job number, and the scale to be used for the graphic output. PUZZLE has been designed to recognize any such identifiers as the beginning of a new data group. This allows the user to sequentially run several boards through the computer in a single run.

With some working experience we found that unnecessary failures were occurring because data points were improperly coded. Consequently, provisions have been added to PUZZLE to examine the data as thoroughly as possible and delete any data that cannot be interpreted. A message is then printed to indicate to the user what data has been deleted. The remainder of the data is processed normally.

12.4.3 Board Outline

Earlier versions of PUZZLE yielded graphic output which called for some manual tape laying. That is, the user was responsible for adjusting the layout which PUZZLE produced within an appropriate board outline and then using a tape overlay to bring connections to the connector tabs. However, the current version of PUZZLE allows the user to code his board dimensions (with variable spacing on the connector tabs). Hence PUZZLE now yields a complete representation of the finished board. Any or all edges may have connector tabs on them, and within the physical limits any number of connectors may appear on the tab. It is also possible to produce a board with no connector tabs.

If the specifications for the board outline are omitted or found faulty or incomplete, the outline will be omitted and the program will revert to the previously used method. That is, a scan is made of the data to determine the maximum and minimum in each direction. A working area is then described which includes the necessary maxima and minima. This area will be x by y inches where x and y are integers. Hence during the data scan it is necessary to check for the square-inch a component is located in, not the actual physical coordinates of the point.

12.4.4 Sequencing

Very little effort has been made in the routing logic to optimize path length. It is evident that, in general, by decreasing path length, the remaining working area is increased, and hence more interconnections may be completed. Therefore two levels of data sequencing have been introduced to indirectly reduce path length.

The first step is to order the individual data strings. They are often not listed in the most convenient order. Therefore a procedure is used to obtain the shortest total path length for the string. The string is broken into pairs and no restriction is made in their configuration other than optimizing length. That is, if the points A, B, C, and D are to be connected, they might all be in a string A to B, B to C, and C to D, but if the total path length would be shorter, they could be paired in such a way that B, C, and D are each connected to A.

A system of matrix operations is used to perform the necessary ordering. One matrix, **OM**, is used to hold the path length for any possible pair. (Because path length does not depend on direction, only the triangular matrix is necessary). A second matrix, **BM**, is used to record what connections are yielded as choices are made among the possible pairs. This Boolean matrix is initialized with a truth value of false.

The first step is to fill the first matrix, **OM**, with the indicated path lengths. Because of the grid restrictions that routing move either horizontally or vertically rather than diagonally, path length is based on the rectilinear length and any obstacles are ignored. The matrix is then scanned for the shortest path length. The appropriate data pair is chosen as the first leg of the branch, and its path length is set at a maximum so it will not be selected again. The appropriate location in the second matrix is assigned a truth value of true. A second pair of points is then chosen and the appropriate value is set in **BM**. To obtain the full extent of the connections now yielded, it is necessary to square **BM**. (Using the Boolean "and" operator is equivalent to algebraic multiplication). For example if AB were the first choice, and AC the second, squaring **BM** fills the location corresponding to BC. The choosing process continues until **BM** has a full truth value of true. At each step **BM** receives one new value of true and is operated on until the new value of $\mathbf{BM}=(\mathbf{BM})^N$, where N is the number of choices that have been made.

The second level of sequencing concerns the entire data block. It is based on the premise that by routing a long and involved path prior to a short one, unnecessary obstacles may completely block the simpler path. In order to prevent this problem, after the data has been broken into connecting pairs, the data is scanned and ordered so that the shortest paths will be routed first.

12.4.5 Routing in PUZZLE

The routing procedure in PUZZLE depends on successful interrelation between several somewhat simple phases. The basic mapping is an implementation of the rat-in-the-maze technique described in Section 12.3.2. Initially, a positive vertical and positive horizontal direction are chosen. These are the orientations necessary to yield the most direct path to the end point. Hence if A is going to B, and A is to the left and above B, right and down are the positive directions. The path is then begun using increments in the positive vertical direction. This path continues either until it reaches an obstacle, such as a predefined path or a pad necessary for the placement of a component, or until it reaches the vertical coordinate of the end point. This basic pattern is followed until a connection is completed. Of course, only the simpler routes will be completed by the direct stepping process. Therefore there are various provisions for backing up on a line, extending a line beyond the endpoint and dropping back into the destination, restarting a line, or causing the line to clear a component. Arbitrary limits have been set to prevent a line from backing up too far (such as beyond the origin), or its extending the line too far beyond the endpoint. A line may be restarted in several ways. A total of eight tries is possible. The first try calls for starting in the positive vertical direction, the next in the positive horizontal, the third in the negative vertical, and the fourth in a negative horizontal direction. The next four tries parallel these, except that the origin and the destination are reversed. Because of the greater difficulty encountered when a path passes through a component various segments cause a line to avoid moving into the center of a component and being blocked. As a path is being traced it may develop unnecessary kinks or bends; hence a *smoothing process* has been built into the routing algorithm. Each time the program causes the path to change direction, the two previous line segments are examined. If it is possible, the line segments are redetermined in order to remove a bend. In Fig. 12.4, if *ABC* comprise the two completed line segments and a direction change is desired at *C*, then if no obstacles block the path *ADC, ADC* would eliminate at least one and possibly two bends in the overall path. (If *A* or *C* is a terminal, no bend would appear at that point, therefore the smoothing process

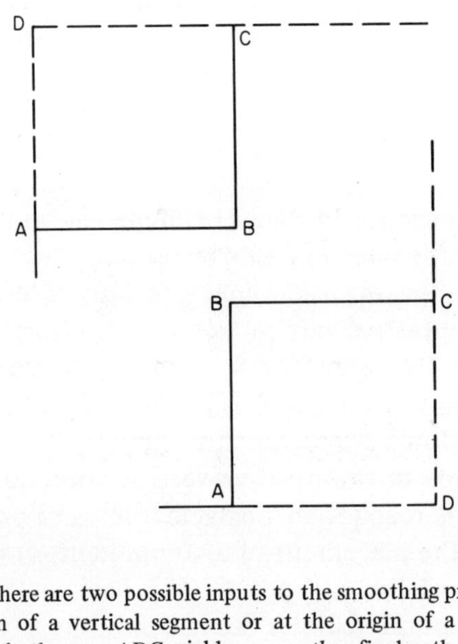

FIG. 12.4. There are two possible inputs to the smoothing procedure, i.e., at the initiation of a vertical segment or at the origin of a horizontal line segment. In both cases ADC yields a *smoother* final path than does ABC.

would eliminate only one bend). At first glance this smoothing seems unnecessary due to the stepping procedures. Direction changes usually result when an obstacle is encountered and hence the smoothing process would achieve nothing. However, the facilities for retracing a line, for moving around a unit of obstacles, for going beyond the end point, etc. make the smoothing necessary.

Each of these principles has been used to increase the number of completed paths. However there may be some connections that resist all efforts to be routed. Should such problems occur the program searches for other connections to correctly complete the circuit. That is, if *A*, *B*, and *C* are to be connected and in that order and *B* cannot be connected to *C*; the program attempts to connect *A* to *C* in order to complete the circuit.

12.4.6 Graphing Procedure

The graphing procedure which consists of two portions, begins with the mapping of a composite picture of both sides of the board. This graph is sufficient for debugging purposes and becomes a convenient reference for board loading. The composite graph will always be drawn.

The second phase of the graphing is optional, as it consists of graphing the individual sides of the board, and it is a waste of both computer and plotter time to plot these graphs on a run which does not complete all paths. Hence, when a graph is incomplete due to either bad data or missing connections, PUZZLE branches around this section.† It is at the beginning of this stage that the various line segments are assigned to a particular side of the board. A complete path is examined and any line segment that crosses a line perpendicular to it is assigned first, all vertical lines being assigned to the first drawing and all horizontals to the second. Any remaining segments are then assigned in such a way as to avoid unnecessary holes through the board. Should an entire path be unrestricted, it will appear on the second graph. The second graph is a mirror image thus allowing the user to place the two graphs back to back to duplicate the composite.

Figures 12.5 through 12.8 show the composite graph, graphs of each side of the board, the artwork with tape overlay to the connector (the border outline option was not used), and the finished board. A simplified flow chart of PUZZLE is shown in Fig. 12.9.

† There is an exception to this rule. Provisions have been added to PUZZLE to force the graphing if the user so chooses.

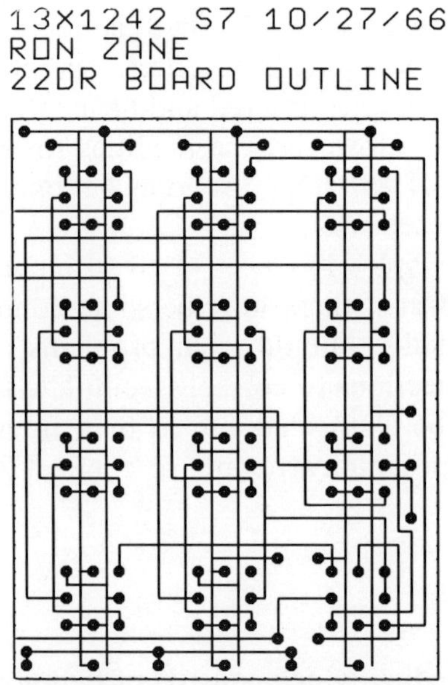

FIG. 12.5. The first graph drawn by PUZZLE is a composite picture of both sides of the board as viewed from the parts side.

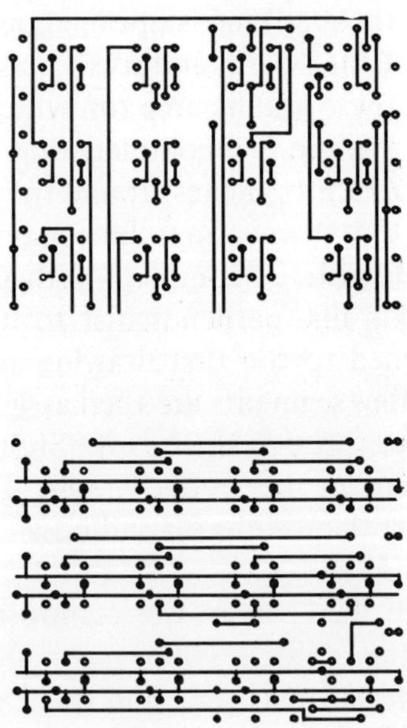

FIG. 12.6. The second and third graphs are pictures of the component side and wiring side.

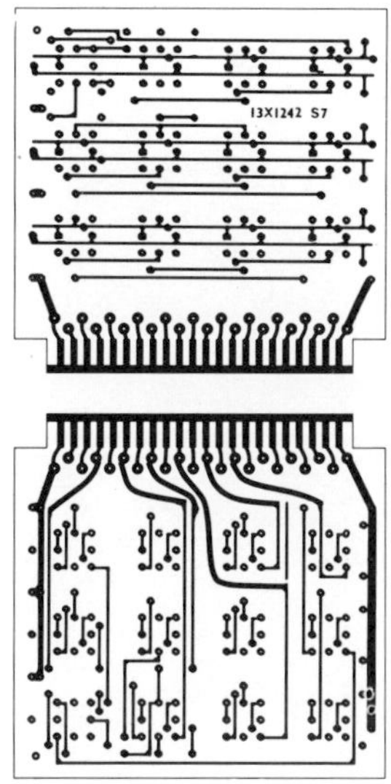

FIG. 12.7. When no border outline is specified, it is necessary to place the graphs produced by the CalComp onto a board outline and lay tape to the connector.

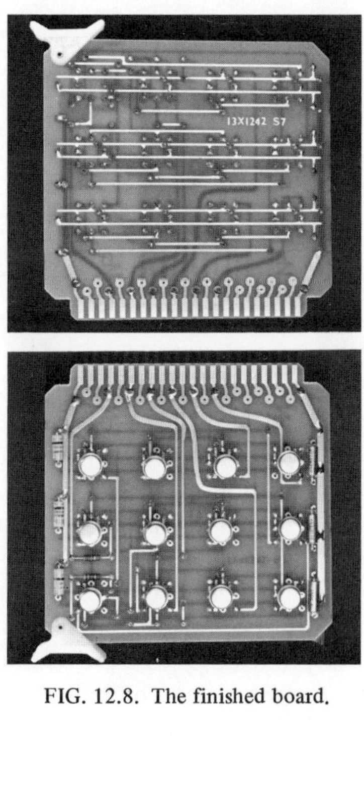

FIG. 12.8. The finished board.

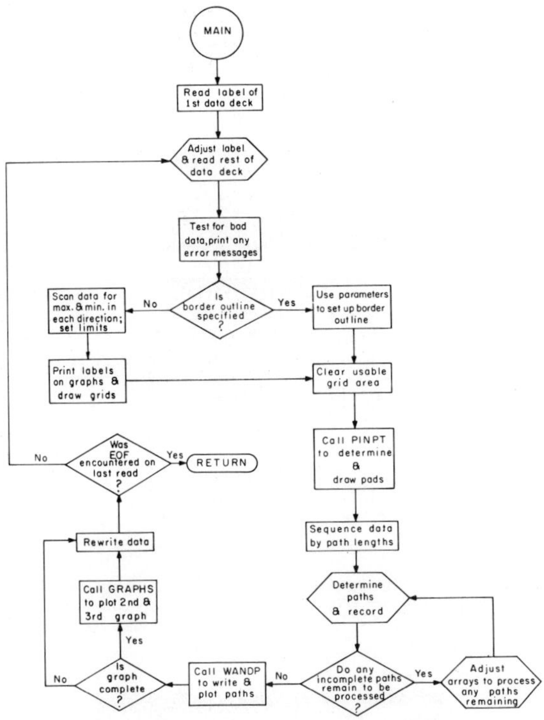

FIG. 12.9. The flow chart for PUZZLE.

12.5 LIMITATIONS AND EXTENSIONS OF PUZZLE

PUZZLE has been designed to provide *a* solution rather than the optimum solution to the printed circuit card design problem. As a result there are several limitations and inefficiencies in PUZZLE's solutions:

1. The path lengths are not minimized. The routing logic tends to produce short paths but no effort is made to assure that they are the shortest paths available.
2. Through holes are not minimized. The smoothing routine tends to reduce the number of holes carrying lines back and forth through the surface of the boards but the algorithm does not minimize them.
3. Solutions to portions of the topological routing may be incomplete, requiring unnecessary jumpers or touch-up of the artwork.

In line with the basic philosophy of the design of PUZZLE, projected improvements in its performance are predicated upon efforts to improve its solutions rather than try to optimize them. Some of the modifications that will possibly be incorporated in the near future are:

1. Enable the program to utilize the 45 degree diagonal paths. Geometrical and topological considerations indicate that this procedure will insure adequate line separation while increasing the number of available path routings by about 40%.
2. Require that the program test all available route solutions and use the shortest one. The present logic stops seeking paths as soon as a trial completes the path.
3. Relocate the lines which block a path when all attempts to complete a given path have met with failure. The offending lines could be eradicated to permit completion of the blocked path and then the eradicated line could be retried to find an alternative path.

The CalComp pen and ink plotters have proven themselves to be of limited value in the production of the high quality graphic output needed in the production of printed circuit artwork. The pen and ink drawings may be used to produce prototype boards by direct reproduction but if the high quality needed for a production run is to be produced an optical plotting system is indicated.†

REFERENCES

1. James M. Spitze, "Printed Circuit Card Design Automation System," Friden (unpublished), November 14, 1967.

2. L. E. West and D. L. Caskey, "Topographic simulation as an aid to printed circuit board design," Sandia Laboratory, Tech. Rept. SC-RR-66-424, August 1966.

3. C. J. Fisk and D. L. Isett, " 'ACCEL' Automated Circuit Card Etching Layout," Sandia Laboratory, Technical Memorandum, SC-TM-65-544, October 1965.

4. M. J. Goldberg and R. H. Schroeder, et. al, *Development of On-line System for Computer Aided Design of Integral Circuits,* vol. 1, Norden Division of

† The Lawrence Radiation Laboratory, Livermore, is using a Gerber optical plotting system to plot PUZZLE output. It produces very high quality film positives directly.

United Aircraft Corp., Air Force Materials Laboratory, Research and Technology Division, Wright-Patterson AFB, Ohio, October 1967.

5. D. A. Wilber and R. Zane, "Development of 'PUZZLE': A program to aid in printed circuit layout design," Lawrence Radiation Laboratory, EE1215, UCID 3079, July 28, 1967.

6. R. Zane, "PUZZLE/Version 4: Operational description," Lawrence Radiation Laboratory, EE1219, UCID 3081, August 9, 1967.

7. Ronald Zane and Deanna A. Wilber, "PUZZLE: A program for computer aided design of printed circuit artwork," Lawrence Radiation Laboratory, UCRL 17814, September 12, 1967.

8. Deanna A. Wilber and Ronald Zane, "Computer-Aided Design of Printed Circuit Artwork," Lawrence Radiation Laboratory, UCRL 18172, April 9, 1968.

APPENDIX 1

PUZZLE IV Input Procedure†

PUZZLE's input has been designed to give the user as much flexibility as possible. The data deck may have enough information to do several boards or only enough for one. The number of interconnections to be completed may vary from *1* to *550*.* And the number of interconnected points in a single branch may vary from two to eleven. The user may consider the information necessary to yield one board a data block. The deck which contains all data blocks for a given turn will be referred to as the *run deck.*

The data blocks consist of three types of cards.

1. The first card of the data block may be considered the label card.
   (a) Columns 1-60 must contain 60 characters to be used as an identifier on both printed and graphic output. It will appear as a single line in the printed output and 3 twenty character lines on the graphic output. This label may include such information as user's name, identification number, a date, or the particular application. At least one non-numeric character must be included in the label.

† It is possible to increase the number of allowable interconnections by a very minor program modification.

* A Copy of the listing of PUZZLE is available through Mrs. D. A. Harrell of LRL, Berkeley.

(b) Column 61 must contain the scaling factor for the graphic output. At present actual size or twice scale drawings are the only sizes permitted. The program then expects a *1* or *2* in column 61. If any other character appears, actual size drawings will be produced and the print out will flag the defective data.

(c) Column 62 is used to control the graphing of the individual sides of the board. A zero or blank will allow the program to control the graphing. That is, if there is any bad data or any missing connections only the composite will be graphed. If the graph is complete, however, all three graphs will be produced. If column 62 contains a *1* all three graphs will be produced regardless of the completeness of the mapping.

(d) Two additional arguments are necessary if and only if the user chooses to use PUZZLE's options of placing the interconnecting components in an appropriate outline which will include connector tabs. Components are placed within a 5 by 10 inch rectangle. (See section on seven-digit codes.) Columns 64-70 should contain the horizontal distance† in inches from the left-hand lower corner of the board to the point (0.1, 0.1), the first usable point in the rectangle. Columns 71-77 should contain the vertical distance to the same point.

2. The second data card is optional. If used it should carry the parameters necessary to describe the board outline. If any value was placed in columns 64-77 the program will expect this second card to exist.

   (a) Columns 1-7 should contain the width of the board in inches. The number may be placed anywhere in the seven columns, but a decimal point is required to convey magnitude. Examples of acceptable value are 3.5, 5.0, 7.0, or .96.

   (b) Columns 8-14 should contain the height of the board in inches. The acceptable forms of the number are the same as for the width.

   (c) Columns 15-21 should specify the number of connector pins for edge 1, (Fig. 12.3). The number must be right adjusted within the seven columns.*

---

† The number is a decimal written in field-free format.

* If the number of pins and the location of the connector tab, on a given edge, are unspecified, (or left blank) no connector will be drawn for that edge.

(d) Columns 22-28 should specify the vertical distance from the lower left hand corner. The number should be in inches and the format is the same as that used for the height and width.

(e) Columns 29-35 should contain the number of connector pins on edge 2.

(f) Columns 36-42 should specify the horizontal distance in inches from the upper left hand corner to the edge of the connector tab on edge 2.

(g) Columns 43-49 should specify the number of connector pins on edge 3.

(h) Columns 50-56 should specify the vertical distance in inches from the lower right hand corner to the connector on edge 3.

(i) Columns 57-63 should specify the number of connector pins on edge 4.

(j) Columns 64-70 should specify the horizontal distances in inches from the lower right hand corner to the connector on edge 4.

(k) Columns 71-77 should contain a number to convey the pin spacing on the connector tabs. This number should be in inches and of the same format as the width, height, and distance measures.

3. The remainder of the data deck constitutes the wiring list. Each individual card will contain a string of interconnected points. The program will continue to look for these cards until it encounters either a new label card or a blank card. Each card may contain up to eleven points, each point being represented by a seven-digit code. The composition of these seven-digit codes is as follows: *Ommnnpp*.

   (a) The first digit is always zero or blank.

   (b) *mm* is determined in the following manner:

      (i) If the unit is an eight-pin micro-logic can

      $mm = 00$ (or two blanks)

      the can will be centered in the one-inch square, Fig. 12.3.

      (ii) If the unit is a ten-pin micro-logic can

      $mm = 01$ (or a blank and a 1)

      centers the can in the square, (Fig. 12.3);

      $mm = 02$

shifts the can 0.1 in. to the right of center, (Fig. 12.2);

$$mm = 03$$

shifts the can 0.1 to the left of center;

$$mm = 04$$

sets the can in the vertical position with pins 4 and 10 on the vertical center;

$$mm = 05$$

places the can in the vertical center.

(iii) If the unit is a discrete component

$$mm = 1x$$

where x specifies the distance in tenths of an inch from the center of the square, (Fig. 12.2).

(iv) If the point is being used for forced routing and pad is not desired at this point

$$mm = 2x$$

where again x is the distance in tenths of an inch from the center.

(v) If the point being specified is on the connector edge

$$mm = 3x$$

where x is the edge referred to.

(c) The next two digits (nn) refer to the square in which the unit is placed. Figure 12.3 shows a 5 by 10 inch rectangle. The user may choose to use any portion of this by assigning nn appropriately.

(d) The last two digits are the pin number. Figure 12.2 shows the location of the pins of the assigned component. For example, to specify the upper left hand corner of an eight-pin can $pp = 07$. For the discrete components pp is the point in the ring, (Fig. 12.2). If the point is on the connector edge pp equals the connector number. The connectors are numbered from right to left if the connector is facing down.

APPENDIX 2

Interpretation of Output

There are two forms of output produced by PUZZLE, *graphic* and *printed.* Each serves a valuable function.

There are several parts to the printed output. The data will be listed twice, once at the beginning and once at the end of the listing. The first listing is produced as soon as the data is read. It will therefore list any bad data. The final listing will be the list of data actually operated on.

The second portion of the printed output is a listing of any appropriate error messages. These may warn of errors in the board outline, in the scale factor, or in the interconnection list. If the bad data appears in the interconnection list, that one data will be deleted and the remainder of the deck will be processed normally. When this occurs, an additional line will be printed (item 7) giving the user an approximate location of the bad data within the data deck. The error messages include:

1. BAD DATA-FIRST COLUMN OF SEVEN DIGIT CODE NOT ZERO.
2. BAD DATA-COMPONENT CODE IS NOT VALID.
3. BAD DATA-PIN CODE DOES NOT AGREE WITH COMPONENT CODE.
4. BAD DATA-CONNECTION CALLED FROM NON-EXISTENT SQUARE.
5. BAD DATA-CONNECTOR CANNOT BE MAPPED FOR THIS BOARD.
6. BAD DATA-THIS CONNECTOR NUMBER GREATER THAN THAT ALLOWED FOR THIS BOARD.
7. xxxxxxx WAS DELETED FROM THE DATA LIST, IT IS THE xxxx PIECE OF DATA AND IS THE xxx ITEM ON THE CARD:
8. MISSING DATA-SIZE SPECIFICATION FOR GRAPHS 2 AND 3 IS MISSING OR IS INCORRECT. IT WAS SET TO ONE.
9. AN ERROR WAS FOUND IN YOUR BOARD SPECIFICATIONS-OUTLINE CANNOT BE DRAWN.
10. AT LEAST ONE DATA POINT EXCEEDS THE BOARD DIMENSIONS-BOARD DIMENSIONS WILL BE IGNORED.

The third portion of the printed output gives the user the results of PUZZLE's processing. The majority of this will be a listing of completed routing: PATH FOR PIN PAIR xxxxxxx yyyyyyy WAS COMPLETED. At the end of this listing will be a message, PATHS COULD NOT BE FOUND FOR THE FOLLOWING INTERCONNECTIONS. followed by the codes for any missing connections.

The most important part of the output produced by PUZZLE are the representations of the board produced by the CalComp plotter. There are three graphs produced when the routing is completely successful. If there are any incomplete paths, only the first or composite picture will be mapped. The first graph pictures the board from the parts side but shows both sides. This composite picture is very useful for checking for bad placement of components or for bad data unobserved by the program's error checks and for the general density of the board or of given segments of the board, and as a guide to the actual builder of the board as to placement of components. If the program determines that for any reason this board is incomplete, that is, if there are any incomplete paths or rejected data, this composite picture is the only one produced, and may be used to adjust the input data to correct the insufficiencies.

The second graph shows all lines which are to appear on the *parts side* of the board. These will be predominantly vertical lines. The third graph shows the *wiring side* of the board, and will be essentially all of the horizontal lines (this includes the buses).

Experimentation was done to determine the best material to use for this graphical output. The standard CalComp paper is thin and not so stable as is desired for best results. Therefore the final graphs are plotted on vellum, which is produced by the makers of the CalComp. This is much more stable than the paper, and the lines are sharper. In addition, a pen tip wider than that normally used on the plotter is used to produce lines wide enough for direct production of the board from the CalComp graphs. These should be requested by the user when the job is submitted.

CHAPTER 13

AUTOMATIC ARTWORK GENERATION FOR INTEGRATED CIRCUITS

LESTER H. HAZLETT, Motorola Semiconductor Products Division
GERALD V. BEEMILLER, Motorola Semiconductor Products Division

The fabrication of integrated circuits involves processing steps which demand the preparation of precise artwork. Procedure improvements have been made in the generation of integrated circuit artwork through the application of automatic drafting equipment and computer-aided techniques. In this chapter we describe an evolving computer-aided design system for integrated circuit layout and mask master generation. The evolution is from current manual procedure toward automation. We review the history of mask making and discuss the motivation for automating the procedure. Then we present a specific system, CAMP, to demonstrate the requirements for such a system. We conclude with a discussion of the role of computer-aided design in *LSI* technology and the ramifications of this technology in the requirements of the computer-aided design system.

13.1 HISTORY OF MASK MAKING

The processing of integrated circuits involves selective *diffusion* into a semiconductor wafer and the *etching* of metal patterns upon the wafer. These processing steps are controlled by *lithographic* techniques using photographic masks. A *mask* is a high resolution glass photographic plate used for direct contact exposure on a semiconductor wafer. This determines the size and location of patterns diffused into or etched on the wafer. Many integrated circuits are fabricated in each semiconductor wafer (Fig. 13.1). Each wafer mask is a matrix of identical chip size patterns generated by repetitively stepping the appropriate mask master. Mask stepping is commonly handled by automated equipment, but until recently, the generation of mask masters has been by manual procedure. Manual mask making dates back to transistor masks. These first masks were hand cut (patterns were cut out of an opaque material with a straight edge and a knife) and then reduced to actual size. With the advent of the integrated circuit, this technique was no longer satisfactory; increased complexity and higher precision were required to get satisfactory results. Today, manual procedures go something like this:

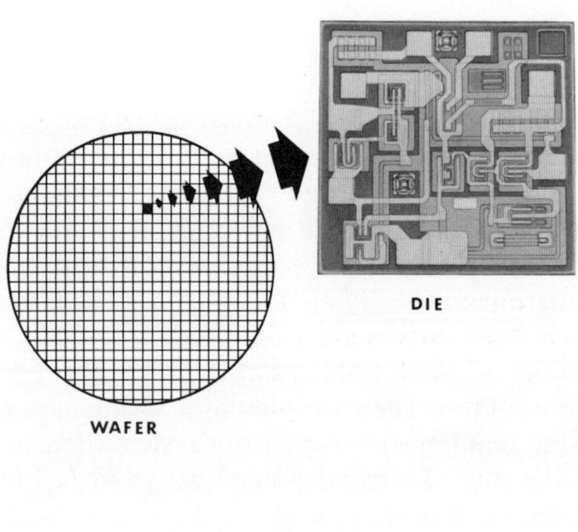

FIG. 13.1. Many Identical Circuit Die are Processed Together in One Wafer.

The layout is done by a layout draftsman. He starts with a set of layout rules and a circuit schematic. He then creates a dimensioned, detailed drawing for each required device type. He makes geometrical models of the devices and tries various configurations to determine optimum *placements* and metal *interconnection* patterns. The entire circuit is then drawn to scale (Fig. 13.2(a), a composite drawing). Each layer of the circuit is shown in the composite using a pre-established line code. When the drawing is complete, it is checked for possible omitted parts. The composite is then taken to a *coordinatograph* operator where the mask masters for the various layers are cut on pieces of ruby-cut and strip film† (Fig. 13.2(b)). Each layer must be cut and stripped (Fig. 13.2(c)) and checked before being reduced.

The number of required masks (5-15) depends upon the number of processing steps. There is usually one mask per diffusion and two or more per metal layer. The entire process from circuit schematic to mask set takes from 80 to 320 man hours depending on the complexity of the circuit. Because in each step of preparation, it is necessary to wait for the preceding, the turn-around time from conception to mask set is not shortened by training more personnel; this just helps to eliminate backlog. A process had to be developed to reduce preparation time by automating portions of the manual design process. The availability of automatic drafting equipment made such a development feasible. Figure 13.3 shows a time exposure of an automatic drafting machine in motion. This equipment can be used to make mask masters by directly exposing photographic material with a moving light source.

13.2 WHY AUTOMATED MASK GENERATION FOR INTEGRATED CIRCUITS

The pervasive goals of *cost* and *performance* are keenly felt in the semiconductor industry. It is the quest for better integrated circuit products at a lower cost that motivates the development of improved procedures through automation. The prime motive behind mask making

---

† Ruby-cut and strip film is a material composed of two parts; a hard stabilene base and a soft rubylith overcoat. When a mask is cut, only enough pressure is applied to cut the soft overcoat. Then, portions of the opaque material are removed and only the clear stabilene remains. This process is known as stripping. Thus, we end up with a solid contrast master of opaque and clear areas, or ultimately diffused and non-diffused areas.

(a)

(b)

(c)

FIG. 13.2 (a) (b) (c). The Manual Method for Making Integrated Circuit Mask Masters.

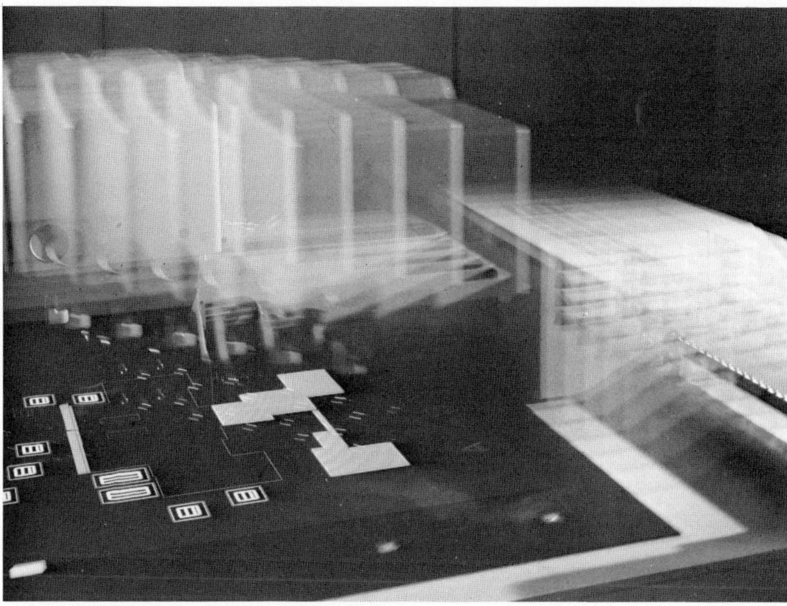

FIG. 13.3. Time Exposure of Automatic Drafting Machine in Motion.

automation is the reduction of design time and cost for complex parts. Especially beneficial is any reduction of the turn-around time required to modify a design. Automated drafting equipment exceeds manual accuracies and allows drafting at smaller scale factors. This results in less time, lower cost, and less distortion due to photographic reduction. Through automatic drafting, many manual errors and omissions are avoided; for example, if only one emitter contact is inadvertently omitted from one transistor, an entire complex design could be incorrect and the error defies visual detection. The speed of automated drafting increases the number of custom designs which can be done, and the improved accuracy makes feasible designs which were previously too big or complex.

In the attempt to automate any design procedures, there is this basic question to be answered: how much can really be automatic and where must the designer manually intervene to achieve the required results? Algorithmic techniques for *metal routing, element placement* and *assignment*, and *area allocation* have been developed and tested and found to be very useful. Unfortunately, designs generated by these algorithms are seldom entirely complete. As a result, it is a practical necessity to have some form of designer interaction so that these 99 percent solutions can be completed. Manual intervention techniques are particularly efficient and powerful for design modifications. Simple modification can be made without requiring the computational costs of re-executing the algorithmic procedures for the whole design. The type of man-machine interaction employed by the system is a major factor in establishing equipment requirements and system objectives.

The objective of the CAMP system described below is to provide improved design procedures through the use of automated and computer-aided techniques. This objective leads to the following goals.

1. To construct a *library* as a central store of information about device and circuit configurations.
2. To create a *layout procedure* which extends the productivity of a layout designer.
3. To automate the *drafting* of mask masters.
4. To initially build up a computer-aided technology which will lend itself to future incorporation of algorithmic procedures.

13.3 COMPUTER-AIDED MASK PREPARATION

13.3.1 CAMP Hardware

The CAMP system is a system used to generate complicated integrated circuit mask masters. There are two hardware systems used in this process; the first is a computer and CRT design console and the second is an automatic drafting machine.

The CDC 1700 Computer and CRT. The Control Data 1700 is a small general purpose digital computer. Some of its features are sixteen thousand 16-bit words of memory, 1.1 μsec. memory cycle time, 16 external interrupt lines, mass memory with 1,536,000 words of interchangeable disk storage, teletype, paper tape reader, and paper tape punch (Fig. 13.4).

The CRT design console is a Control Data 274 display system. Its cathode ray tube is 22″ in diameter and has magnetic deflection. The entire surface of the CRT is available as a working surface and positions are addressed on a 4096 X 4096 matrix. It is a graphic display whose principle of operation is based upon incremental vectors. Maximum line drawing speed is 1.2″ per 1.67 μsec. Beam size is .015″ and when moved, yields uniform lines of this width. The controller for the display system has a 4096 word core memory buffer. It has a *subroutining* cap-

FIG. 13.4. CAMP System Design Computer.

ability which makes possible repeated drawings of shapes without using a lot of additional storage. In other words, the amount of picture that can be displayed depends on the uniqueness of shapes in the design rather than the number of shapes. The refresh cycle of the controller is set at 25 millisecond intervals for flicker free operation. The display system has a *light pen* for use by the designer in selecting and pointing to elements displayed on the screen. This light pen consists of a tube which channels light into a fiber optics bundle, where it is transmitted to a photo-multiplier. A signal from the photo-multiplier causes a program interrupt in the computer so that the program can interrogate the display system to determine to what element or position the designer is pointing.

The Gerber 2032 Automatic Drafting Machine. The Gerber 2032 is a high resolution plotting table with a DDP 116 computer as a controller (Fig. 13.5). The computer reads numerical controlled (*NC*) tapes and steps the optical head in the X and Y directions. The step size is .0001 inches in either X or Y. The optical head or light source has 24 program selectable apertures which range in diameter from .002″ to .187″. There is also a variable density filter which changes light filtration proportionally with the speed of the head. This allows for uniform exposure of the film during drafting. The resident program in the DDP 116 controls the velocity, acceleration, aperture selection and light filtration of the drafting head. The machine is capable of linear, circular, and parabolic interpolation. Other features include symbol generation and independent axis scaling. The previously described hardware configuration was chosen for its relatively inexpensive cost and for its adaptability to computer-aided mask design. However, the heart of the CAMP system is the *software* (programs) that makes the hardware work.

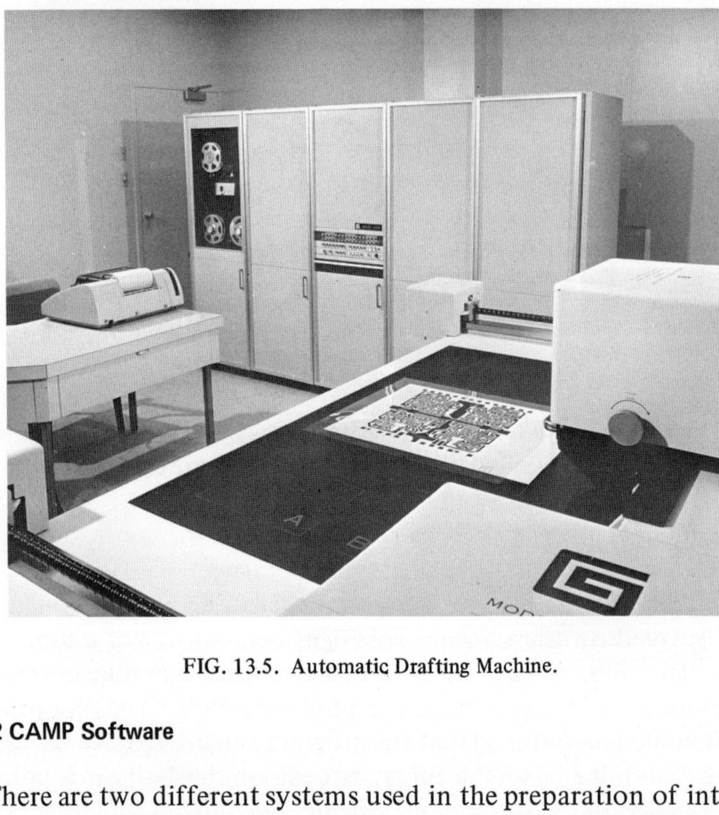

FIG. 13.5. Automatic Drafting Machine.

13.3.2 CAMP Software

There are two different systems used in the preparation of integrated circuit master artwork. The first, also the first to be developed, is known as the *Off-Line CAMP* system. Essentially, it is a *graphic compiler*, written in FORTRAN, and can run on any 16K small scale computer with FORTRAN capabilities. The second system, *On-Line CAMP*, involves the use of the CRT as an input media and requires the previously described 1700 hardware system to run. The designer has at his disposal both On-Line and Off-Line CAMP systems; the choice of which to use depends on the type of artwork to be generated and the designers own personal preference.

13.3.3 Off-Line CAMP

The Off-Line system is a card input system. The cards contain commands and graphical data for the generation of integrated circuit masks masters. There are six main commands for artwork generation; STORE, PLACE, ROUTE, DATA, DRAFT and HALT. With these commands and the appropriate subcommands, any mask set can be efficiently created (Fig. 13.6 shows a flow chart of the Off-Line system).

The Off-Line compiler is divided into three parts; (1) resident I/O drivers and system monitor program, (2) CAMP software programs, (3) COMMON data storage. The creation of this COMMON list structure is done through the use of the six previously mentioned commands. Each command either builds or modifies one of the COMMON lists. The system takes advantage of repetitive shape descriptions as well as repetitive device usage in an integrated circuit.

STORE. The STORE main command is used to create the initial *data base* for the layout. Upon detection of the STORE main command, the program reads the next card to determine if a device or a polycell is being stored in the data base.

Device Being Stored. If a device is being stored into the system, next card must contain the word *DEVI* and a four letter name of the device. The following cards contain the description of the device and all its shapes on the various layers that they appear. This is done by first naming the layer and then the shape and position.

There are two shape types that can be understood by the system. The first, and most general, is the *line-arc* (LIAR). A line-arc is any closed figure whose edges are defined by straight line and arc segments. The second shape was put into the system because of its frequent appearance in device geometries; this shape is RECT, or rectangle. A rectangle is defined by its width and height; this is simpler than using the LIAR command to describe rectangles.

All shapes are referenced from the *base point* of the device. The base point can be any point on the device, however, the lower left hand corner has usually been found to be most satisfactory. All shapes, no matter what layer they are on, are defined relative to the base point; therefore, when the device is moved, every shape of every layer moves with it.

There are no constraints placed on the number of shapes in the device or on the number of layers. Consequently a complex gate could be described as a device to the system. However, a better way to describe such a gate is to identify it as a *polycell* using the instructions outlined

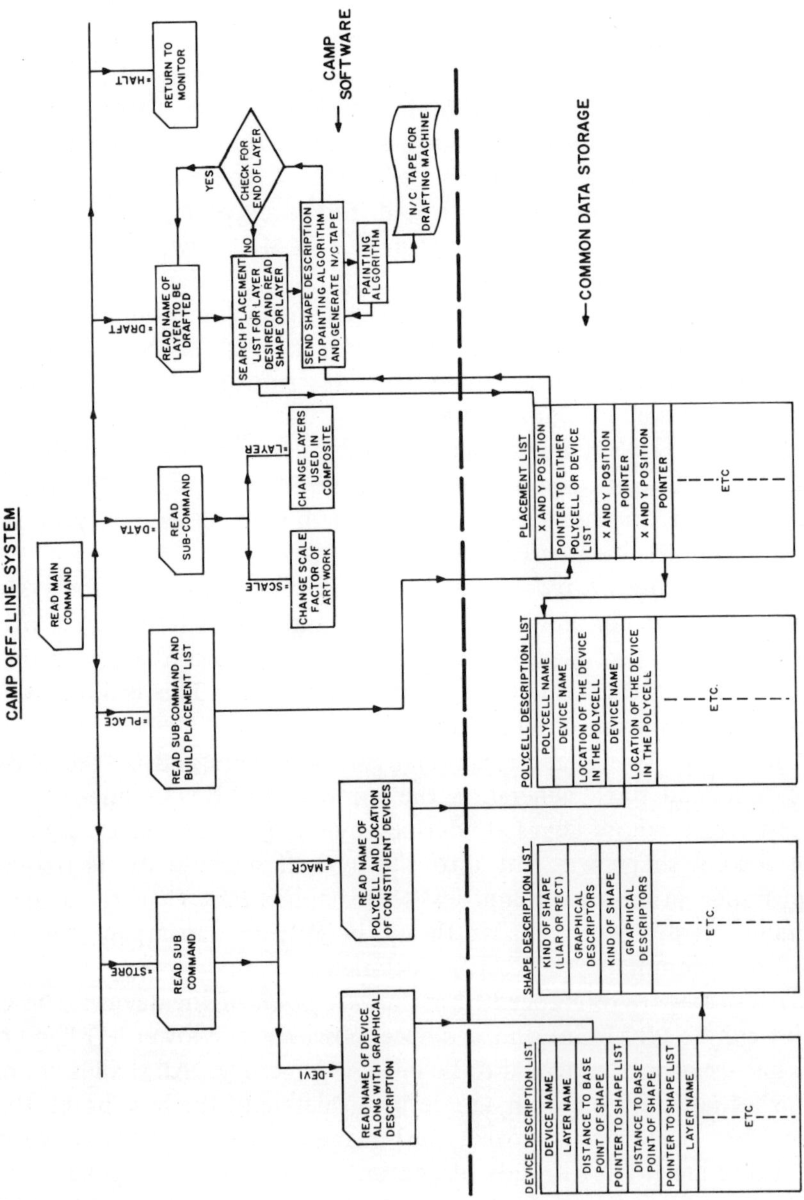

Fig. 13.6. Off-Line CAMP Flow Chart.

below. A polycell is considered to be a logic entity. It can vary in size from a single gate to a collection of circuits which realize a complex logic function.

Polycell Being Stored. If after reading the STORE main command, the next card contains the command MACR, the system prepares for the description of a polycell. The polycell must be named and then its component parts must be defined. The named polycell is then given a base point. All component parts are specified relative to this base point; therefore, when the polycell is moved, all the entities within it are also moved.

All device and polycell geometries are defined by STORE. These descriptions are referred to later when doing component and polycell placement.

PLACE. Components and polycells are placed at various positions on the integrated circuit by the command PLACE. After the command PLACE is read, the program reads the next card which contains (1) the name of the entity to be placed, (2) the *X* and *Y* position of the base point of the entity, (3) number of degrees of rotation (blank means no rotation), and (4) whether or not the entity is mirror imaged. Mirror imaging plus rotation allows all possible planar orientations.

Only polycells and devices are placed; all their constituent parts are defined by the placement and need not be re-defined.

ROUTE. This is the command that specifies where the metal runs go on the integrated circuit. There are two subcommands provided to specify metal, each with a specific type of metal in mind. TRACE is for the large power busses or irregular shapes of metal. This metal is coded in exactly the same manner as LIAR shapes in the STORE main command. It has been found that TRACE is seldom used in coding large amounts of metal. The majority of metal on an integrated circuit consists of narrow metal paths that interconnect components as required to perform various logic functions. This metal is coded efficiently by using STRIP. The designer merely specifies the width of the metal desired and the points to which it runs and CAMP takes care of the rest. After the STRIP subcommand is read, the next card contains an *X-Y* position and a width. Continuous *X-Y* data is read until another main or subcommand is encountered. All inputs are *modal*; therefore, if a fixed width metal path is being coded, the width only has to be given with the first *X-Y* point.

The STRIP technique for metal description is extremely versatile. For example, it allows metal to change width between points and to curve smoothly. The metal can even be made to change width while curving.

DATA. The specific mask layers that are desired on the composite drawing can be specified by cards following the DATA main command. These cards contain the subcommand LAYER, and the names of the layers desired. If all layers are desired, the layer name is ALL.

DRAFT. After the entire data base has been specified, the system is ready to create the *NC* tape for the drafting machine. The main command DRAFT is followed by a subcommand which specifies the layer to be drafted. This is done by the subcommand LAYER and the name of layer desired. If all layers are desired, the layer name is ALL. If a composite drawing is needed, the layer name is COMP. The *NC* tapes for the layer descriptions are generated by interrogating the entire placement list and checking to see if each device or polycell appears on the specified layer. If the device has a shape on the layer, the shape description is read and sent to the painting program (the painting program will be described later). After the entire placement list has been checked for the desired layer, the next layer to be drafted is read and the placement list is searched again. Although very exhaustive, the DRAFT routine executes quickly. Usually the limiting factor is the speed of the output device used.

HALT. This command signifies that the job is completed.

The ease and efficiency of coding masks of the Off-Line manner can be demonstrated by the fact that the device in Fig. 13.7 was coded in less than 50 graphic CAMP statements.

13.3.4 On-Line CAMP Capabilities

The Off-Line system can be used whenever it is impossible to use the facilities of the On-Line station. However, maximum speed of design and optimum system usage can be accomplished only in the On-Line mode.

The On-Line designer has at his disposal all the advantages of the Off-Line system plus the computer power of a *dedicated* small scale computer. On-Line graphics are done on the CDC 274 system with the light pen augmenting the normal means of communication (Fig. 13.8).

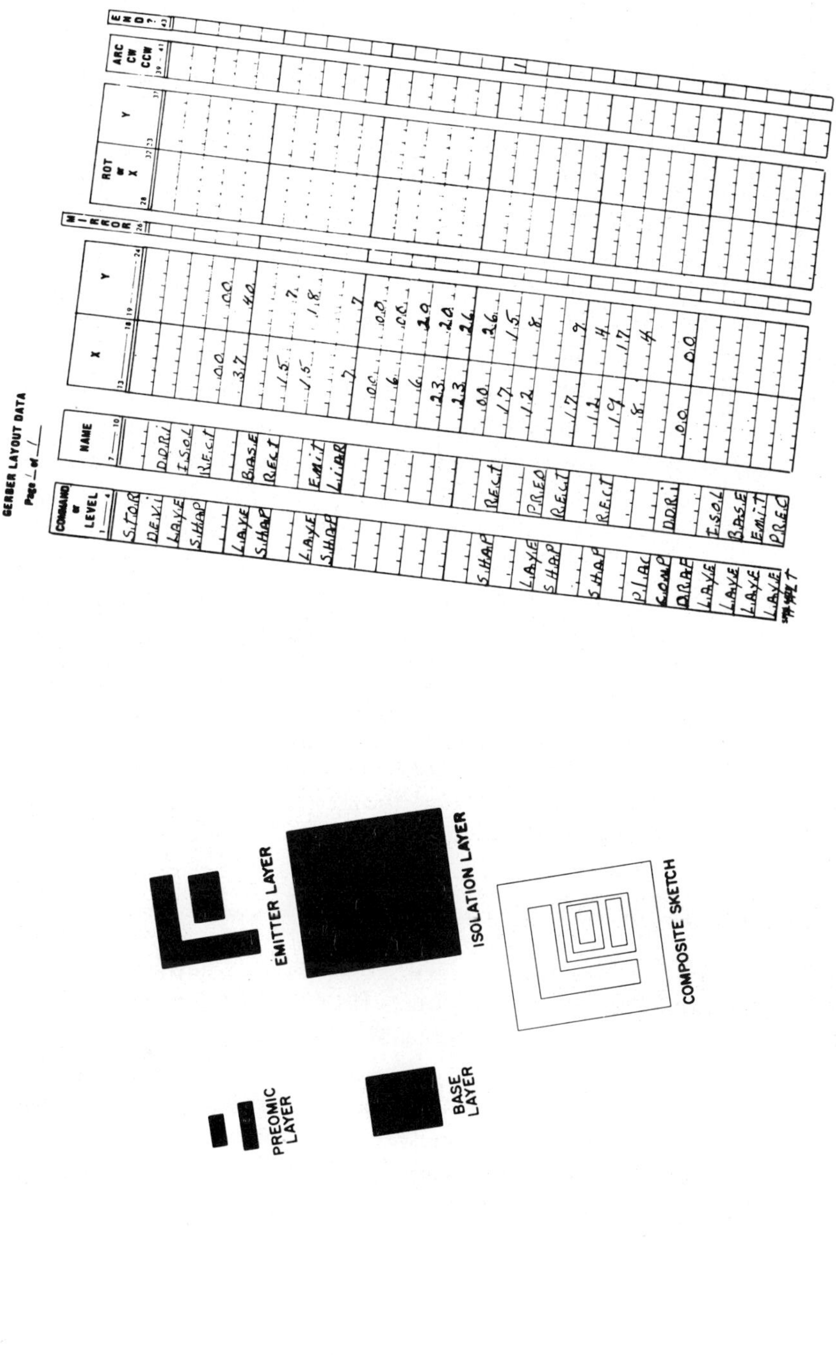

FIG. 13.7. A Simple Example of the Use of Off-Line CAMP.

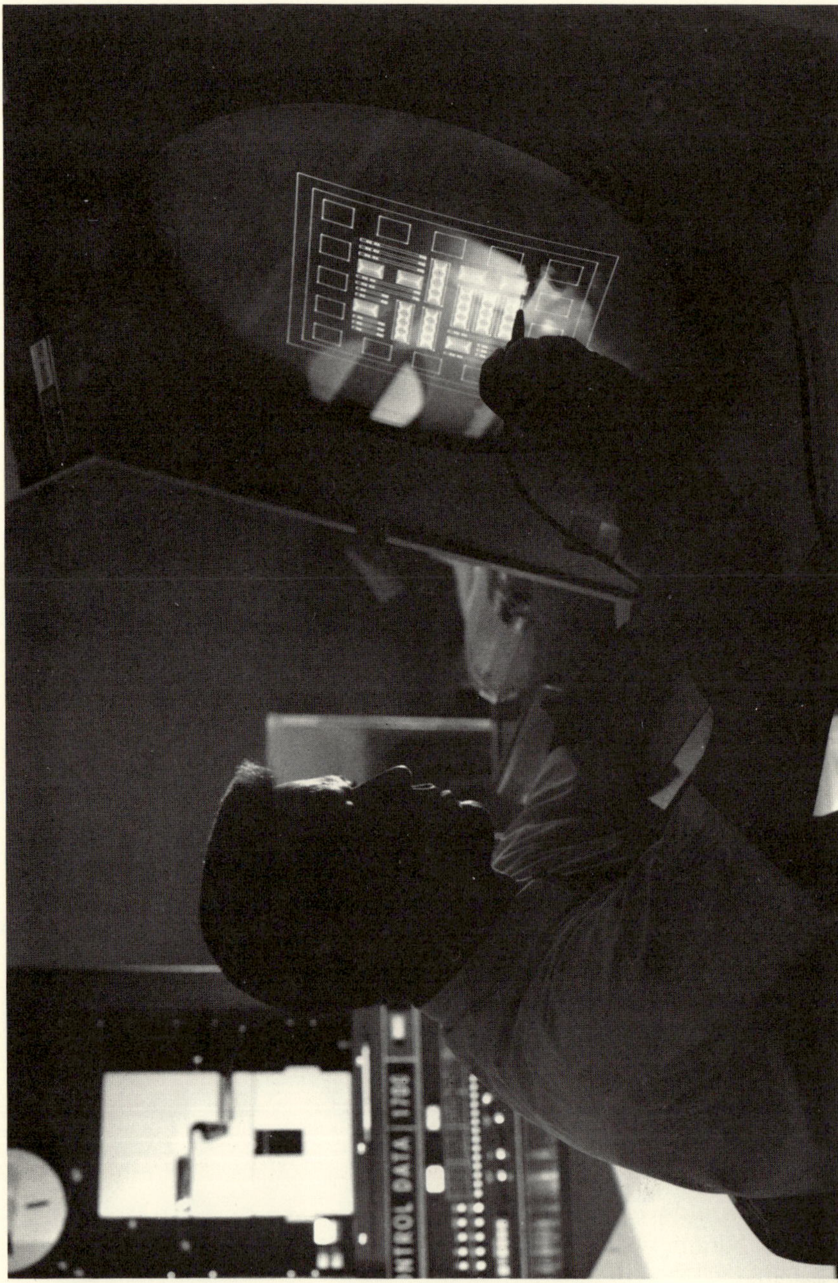

FIG. 13.8. Graphic Communication in Integrated Circuit Layout.

To design a complex integrated circuit, the designer needs only a circuit or logic schematic, and a list of suitable elements.

In the Off-Line system, each device was read from cards to create the desired data base. The On-Line system differs from this in that it uses a mass memory resident library to supply devices and polycells to the designer.

Device Library. The device library is built by a program called LIBRAR. Input to LIBRAR is divided into two parts: (1) electrical and physical data (Fig. 13.9), and (2) graphical data.

The electrical and physical data is information about the performance and processing characteristics for each entry in the library. This information is checked by the On-Line system to determine if different devices used in a layout are compatible. For example, if one part requires base diffusions of 1.7 microns to function properly while another is designed for 1 micron diffusions, these two elements cannot be used together without one or the other being modified. The graphical descriptions for the devices are the *X-Y* coordinates which define the boundaries of the various shapes in the device. These shapes are defined in the same manner as in the Off-Line system. However, in the On-Line system, the shapes do not have to be core resident; they can be kept on the disk and accessed when needed.

The layout design session begins with the designer specifying the scale factor at which he wishes to work. He then chooses, from the device library catalog, the transistors, diodes, and polycells to be used in the design. At this point, he can either begin the layout (specify the placement) of the elements or request the *Resistor Design* program.

Resistor Design Program. There is an endless number of resistor values and configurations used in the layout design of integrated circuits. Because of this fact, it is impossible to try to catalog resistors in the same manner as other components. Therefore, a program has been written to create resistors as they are needed and store them in the temporary section of the device library. Each layout uses this temporary section for the resistors in its mask set. When a new design is started, the previously defined resistors are wiped out and a new set is specified. Inputs to the program are such variables as *junction depth, end effect constant, preohmic width and length, and resistance* of the device. When sufficient parameters have been entered to define the device, the computer calculates the overall length of the resistor and displays it at an appropriate scale factor. At this point the designer can modify the resistor by bending or folding it to achieve a desired shape. As he does so, the com-

```
                              DEVICE DATA

DEVICE NAME- _ _ _ _ _ _               BASE RESISTIVITY=     _ _ _ _     OHM-CM
DESIGNER- _._._ _ _ _ _ _ _ _ _ _ _ _  EPI THICKNESS=            _ _   MICRONS
DATE OF DESIGN- _ _/_ _/_ _            GOLD DOPING=              _ _
FTM (FREQUENCY)=_ _ _ _X10+_ _HZ       TS (STORAGE TIME)= _ _ _ _ _X10- _ _SEC
CIRCUIT DELAY/BW=_ _ _ _X10_ _SEC/HZ   BVCEO=                _ _ _._     VOLTS
MAXIMUM CURRENT=_ _ _ _X10-_ _AMPS     MAXIMUM POWER=        _ _ _ _ _X10-_ _WATTS
                                       NUMBER OF LOTS=       _ _ _

BIPOLAR ELECTRICAL PARAMETERS-

TEMP=_ _ _DEGREES C.     ME=     _._ _              BVCEO=_ _ _._ VOLTS
A11=    _._ _X10-_ _AMPS  MC=     _._ _              BVCES=_ _ _._ VOLTS
A12=    _._ _X10-_ _AMPS  MS=     _._ _              BVCBO=_ _ _._ VOLTS
A21=    _._ _X10-_ _AMPS  BM=_ _ _ _._ _             BVCSO=  _ _._ VOLTS
A22=    _._ _X10-_ _AMPS  ICM=   _ _._ _ _ _AMPS     BVEBO=  _ _._ VOLTS
RSE=  _ _._ _ OHMS        ICMM=  _ _._ _ _ _AMPS     FTM=  _ _ _ _ MHZ
RSC= _ _ _._  OHMS        BIM=   _ _._ _ _           ITM=      _ _._ _ _ _   AMPS
RBF= _ _ _._  OHMS        IEM=   _ _._ _ _ _AMPS     RBCC=      _._ _X10-_ _SEC
RBI= _ _ _._  OHMS        IEMM=  _ _._ _ _ _AMPS     T11=       _._ _X10-_ _SEC
RD=  _ _ _._  OHMS        CE=    _ _._ _ _ PF        T12=       _._ _X10-_ _SEC
                          CC=    _ _._ _ _ PF        T21=       _._ _X10-_ _SEC
                          CS=    _ _._ _ _ PF        T22=       _._ _X10-_ _SEC

BIPOLAR PHYSICAL DESCRIPTION-

STARTING WAFER-                               DOPING-_     ORIENTATION- _ _ _
                THICKNESS= _ _._-_ _._ MILS  RESISTIVITY= _ _._ _-_ _._ _ OHM-CM
EPI LAYER 1-    THICKNESS= _ _._ U      RESISTIVITY= _ _._ _ OHM-CM   DOPING- _
EPI LAYER 2-    THICKNESS= _ _._ U      RESISTIVITY= _ _._ _ OHM-CM   DOPING- _
BURIED LAYER-                   PS=  _ _._   OHMS/SQ      XJ=  _ _._ _ MICRONS
BURIED CHANNEL-                 PS=  _ _._   OHMS/SQ      XJ=  _ _._ _ MICRONS
CHANNEL-                        PS= _ _ _._  OHMS/SQ      XJ=  _ _._ _ MICRONS
DEEP N+ -                       PS=  _ _._ _ OHMS/SQ      XJ=  _ _._ _ MICRONS
TRANSISTOR BASE-                PS=_ _ _ _   OHMS/SQ      XJ=  _ _._ _ MICRONS
RESISTOR BASE-                  PS=_ _ _ _   OHMS/SQ      XJ=  _ _._ _ MICRONS
EMITTER-                        PS=  _ _._ _ OHMS/SQ      XJ=  _ _._ _ MICRONS
GOLD CYCLE-                   TIME=  _ _._ _ MINUTES   TEMP=_ _ _      DEG. C.
BASE-           INTRINSIC BASE PS=   _ _._ _ KOHMS/SQ  WIDTH=  _ _._ _ MICRONS
MAXIMUM METAL THICKNESS=             _ _     K ANG.

HISTORY-        CIRCUIT LAYOUT NAME       DATE         DESIGNER

                _ _ _ _ _ _ _ _ _ _ _ _ _ _  _ _/_ _/_ _   _._._ _ _ _ _ _ _ _ _ _ _ _
                _ _ _ _ _ _ _ _ _ _ _ _ _ _  _ _/_ _/_ _   _._._ _ _ _ _ _ _ _ _ _ _ _
                _ _ _ _ _ _ _ _ _ _ _ _ _ _  _ _/_ _/_ _   _._._ _ _ _ _ _ _ _ _ _ _ _
                _ _ _ _ _ _ _ _ _ _ _ _ _ _  _ _/_ _/_ _   _._._ _ _ _ _ _ _ _ _ _ _ _
```

FIG. 13.9. Format of Device Data in Library.

puter recomputes segment lengths to maintain desired resistance. When the designer is *visually* satisfied with the design he either accepts or rejects it for use in the layout. The design of the first resistor takes the most time because each succeeding resistor has many of the same characteristics (such as junction depth and ohms per square), and therefore, input specification time is reduced.

When the preliminary device specification has been done, the designer is ready to lay out the chip. He starts by specifying the scale factor at which to work. If the layout, at the selected scale factor, exceeds the screen size, he can elect to *zoom in* on any portion of the layout. With the light pen, he picks devices and moves them about the CRT screen and finally places them in any orientation he desires. CAMP furnishes the designer with On-Line commands to do *scaling*, *rotation*, *mirroring*, *metallization* of components, and the creation and *storage* of polycells. The ability to create polycells relieves the designer of the monotonous burden of having to redescribe geometries to the system. He merely has to name the polycell, define the boundary and specify the constituent parts. From this point on, the polycell is recognized as an entity and can be manipulated as if it were a device. Polycell structures are completely *hierarchic*. Consequently, polycells can be made of polycells.

The culmination of the design session is the creation of the *NC* tape for the drafting machine. The designer can specify a documentation drawing or the mask of any layer that is part of the integrated circuit. At this time, a magnetic tape record of the results of the design session is created and filed for later use. If, at a later time, the designer needs to update or modify the previously created circuit, he can do so by reading in the tape and starting where he left off.

The Painting Algorithm. One of the greatest disadvantages in using ruby on a coordinatograph is that a very large scale factor is needed to be able to accurately strip and check the ruby masters (typically 500 to 1). With component size decreasing and chip size increasing, the trend is toward increasing rather than decreasing the scale factor of ruby artwork. In fact, as complexity continues to increase, it is quite likely that the required ruby artwork will be of such excessive size that it will be completely unmanageable.

The scale factor at which most of our mask masters are prepared on the drafting machine is 100x. This minimizes drafting time and eliminates the previously required first reduction. Masks of reasonable complexity take about 30 to 50 minutes drafting time as opposed to 2 to 5 days ruby cutting and stripping time.

All masters that are generated by the drafting machine are complete; all shapes are opaque and require no further manual operations. The technique used to direct the drafting machine to create solid images is called *painting*. The ability to paint shapes is one of the features that makes CAMP so successful. The painting program that has been developed allows for any shape at any orientation and at any scale factor. These are the basic steps taken to paint in a device shape:

1. The closed figure is analyzed and *in-lined*† with a very fine aperture. This gives a good sharp edge.
2. The shape is in-lined with a medium size aperture (40 mils) to opaque the corners.
3. The shape is now ready to be filled in solidly. This is done by painting parallel to the *longest* side until the entire shape is opaque.

The painting algorithm does not check for acute angles; consequently, the corners of a shape with acute angles are not entirely filled and have to be completed manually. Thus far, this has not caused any problems as there are no acute angles in any mask set made to date.

Although painting is time consuming, it does allow for a smaller scale factor than can be used in cutting and stripping (the smallest shape that can be cut reliably on ruby is about 60 mils at the smallest point). The program is capable of painting in any shapes wider than 10 mils at the narrowest point. Small shapes (less than 40 mils) are painted using smaller apertures and the procedure outlined above.

13.3.5 Accomplishments and Goals

The CAMP system language has been developed to expedite mask engineering while minimizing cost. All programs were written in FORTRAN and can be run with several overlays on a 16K computer.

The future goals for CAMP are to continually interact with designer needs and to keep pace with product specifications. Generality of system design is a key to the usefulness of CAMP. The system has already been extended to the design of complex multi-layer printed circuit boards. Algorithmic design techniques are constantly being evaluated and tested in the CAMP system. However, the importance of ultimate designer control and responsibility cannot be overemphasized. Today, CAMP's main usage is in the area of complex chips and unit cell arrays (Fig. 13.10). The complexity of the design implemented with CAMP is limited only by state-of-the-art processing techniques. Limits on chip size and number of components and gates are not relevant to the CAMP

† In-Line – A line of finite width drawn such that its outer edge (not its center defines the shape boundaries.

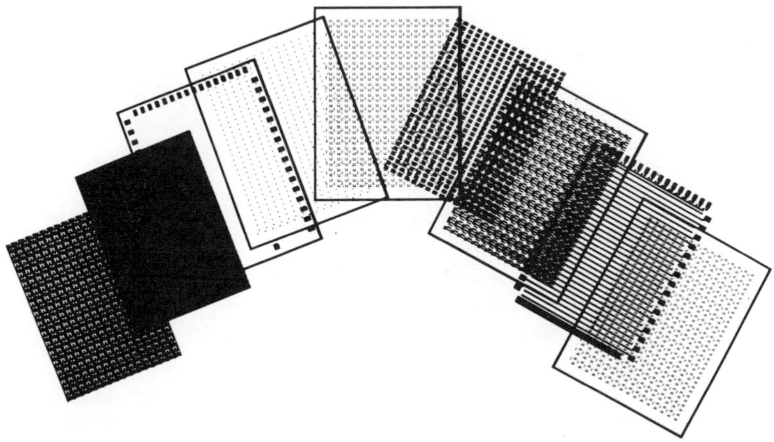

FIG. 13.10. Complete Mask Set for an LSI 256 Bit Bipolar Read Write Memory on One Chip. This Mask Set Was Made Using the CAMP System.

system. It is intended that through the use of the On-Line CAMP system, customers will lay out their own custom integrated circuits and have their mask masters available on the same day.

13.4 DESIGN AND IMPLEMENTATION FOR LSI

The incentive to automate integrated circuit layout and mask making is amplified by the advent of *Large Scale Integration (LSI). LSI* is a new semiconductor technology which promises to bloom into a new generation of semiconductor products. The history of the semiconductor industry can be traced through three generations of product. Single components, simple integrated circuits, and multi-function integrated circuits have each been milestones in the short history of increasing semiconductor sophistication. Across the electronics industry, products from each generation are now available. *LSI* subsystems will be the next milestone.

LSI represents almost the same quantum jump over conventional integrated circuits that the first integrated circuits showed over discrete components. The *LSI* concept proposes to place on a *single monolithic chip* – not an individual circuit or not even a small number of such circuits – but a large number of circuits all interconnected to form a complete electronics system or subsystem. Figure 13.11 shows a photomicrograph of an 8-bit adder.

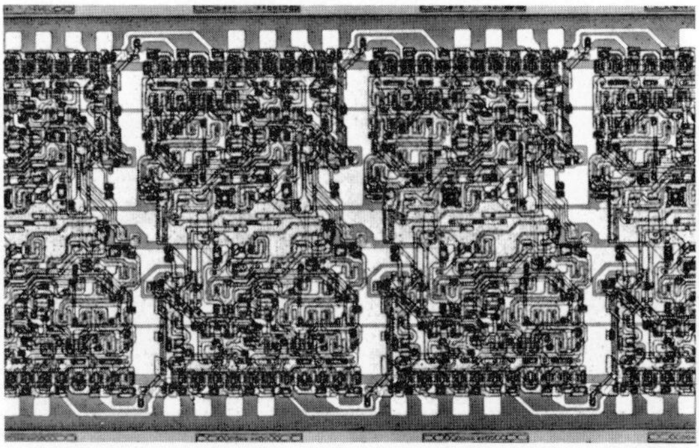

FIG. 13.11. Photomicrograph of an Eight Stage Full-Adder Chip. There are 448 Devices and 3 Layers of Metal Interconnection. Operable Parts Average Less Than 20 Nanoseconds for an Addition of Two Eight-Bit Operands.

Just as conventional integrated circuits have overcome the cost of assembling circuits from discrete components with more efficient semiconductor batch fabrication techniques, *LSI* promises to utilize these same batch techniques to overcome system assembly costs.

Performance in advanced digital systems requires high speed, and high speed is attainable with *LSI*. The elimination of the loading effect of individual circuit packages and the reduction in interconnection lengths from inches to mils, opens the doorway to attainment of heretofore unaccomplished system speeds.

Currently, the *degree of impact* and the *time* at which *LSI* will become available are very controversial. The controversy will undoubtedly continue for some time but the trend is clear. Circuit complexity is on the increase. There exists both the processing ability and the demand for integrated circuits with more components, greater density, and bigger packages with more input - output pins. As circuits get larger and more complex, each circuit has a more specific application and more different types are required. This trend not only cements the need for automation in layout and mask making, but clearly indicates the need for computer aids in other design and implementation functions. As stated before, CAMP is an evolving computer-aided design system. Thus, it is consistent to survey discernible problem areas in *LSI* technology and to discuss currently proposed computer-aided design techniques for *LSI* implementation.

The overall goal of any approach to computer-aided *LSI* implementation is to realize the promise of low cost. Set-up costs include the costs of product specifications, design and engineering. Production costs include the many phases of material processing and testing. Through the use of simplified computer-assisted procedures for design, considerably lower set-up costs can be achieved. Since set-up costs are distributed across an entire production run, their significance depends upon production volume. The incremental costs of fabrication and testing are incurred for the duration of the products' life and are of prime significance to low product cost.

In the past, the formula for low semiconductor costs has been the batch fabrication of identical items. It was the economics of batch fabrication which brought the cost of individual devices below that of the least expensive vacuum tubes. It was the economics of batch fabrication that brought the cost of integrated circuits below the price of circuits made of interconnected discrete devices. The most likely vehicle to reduçe the cost of integrated systems (*LSI*) below that of interconnecting individual circuits is the batch fabrication of *identical* integrated subsystems. This means that truly low cost results from high volume, as it always has. This is not to be interpreted to mean that nothing can be done through automation to reduce the cost of low-volume integrated systems, but that low-volume items will never be priced competitively, per logic function, with high-volume items. Production costs are incurred in proportion to quantity of material processed. Thus, the attainment of satisfactory processing yield is essential for *LSI*.

13.4.1 Yield

Semiconductor fabrication involves a high number of *sequential* operations. The semiconductor material must repetitively pass through the processing steps of photo-resist development, etching, diffusion, passivation, metal deposition, etc. Although each of these operations is conducted with great care in a tightly controlled environment, there can be imperfections resulting from each processing step. Processing losses are measured by the percentage of starting material which is still good following each operation (yield). If there are 50 total operations, the yield of the whole process is the product of all 50 yields. For example, if the average yield per operation were 98 percent, the resultant yield for the whole process would be 36 percent. What yield level is satisfactory depends upon total processing cost and the value of the good parts.

The process imperfections which lead to yield loss are nearly uniformly distributed throughout an entire semiconductor wafer. Given a number of imperfections, the chances of one occurring in an individual circuit chip depends upon the area of the chip. When the number of chips per wafer exceeds the total number of imperfections, the yield expectations are better than if the number of imperfections exceeds the number of chips. *LSI* involves the processing of more complex chips. There are two ways of increasing complexity. One is to increase chip size and the other is to decrease the size of components. Achievement of high circuit speed is also accomplished through the decreasing of component size. In practice, *LSI* designs include both smaller components and larger chip sizes. Although increasing chip size hurts yield, satisfactory yields for larger chips potentially can be achieved through improved control on all processing steps.

Total yield is one of the major problems in implementing an *LSI* technology. Any significant *LSI* approach involves techniques for the minimization of yield loss. There are two basic things which can be done to maximize yield. Improvements in control of processing steps and careful adherence to area minimization in design result in better yield. Another much more sophisticated technique which has become known as *discretionary wiring* also promises to enhance yields.

13.4.2 Discretionary Wiring

The rationale underlying the discretionary wiring approach is that an array will not function unless all of its circuits are good and that this requires an unrealistically high yield if a fixed mask pattern is used. This translates into two arguments. First, the absolute yield of arrays fabricated with a fixed interconnection pattern will be uneconomically low. Second, discretionary wiring offers substantially lower final test costs. The testing cost per array is the same regardless of how it is made, but the number of arrays subjected to final test in order to fill a given requirement may be lower with discretionary wiring. While the ratio of final test yields cannot now be accurately estimated, the yield of discretionarily wired arrays will be higher since all circuits used were known to be good before metallization.

The starting point for discretionary wiring is a silicon wafer containing a large number of diffused components connected into simple circuits. These circuits are the type required in the final array, but the number of each type is considerably in excess of the array requirements. Each circuit on the wafer is then tested individually and the results retained as a record of the particular wafer. On the basis of the test results, interconnection patterns are generated including only those circuits shown to be good. The interconnection patterns are then used to apply multilayer metallization to the wafer. After metallization, the array is ready for packaging and final testing.

Several limitations of the discretionary approach are apparent. First, there is the problem that each interconnection mask is *unique* and so is each wafer. Since the interconnection of a complex array is difficult and represents a substantial portion of the total design effort, this unique mask is extremely expensive. Even if the design can be fully automated, there will be a considerable amount of expensive computer time tied up in a mask that can be used on only one unique wafer. Like the other steps in integrated circuit fabrication, metallization yields are not 100 percent. This means that a substantial probability exists that this expensive design effort will be wasted.

There is a second inherent drawback. The economics of batch fabrication are directly affected by the *efficiency* of converting the surface of the silicon wafer to saleable product. It is not that the silicon itself is so expensive, but that a great deal of value is added in the processing. Only a certain number of wafers can be processed in a given furnace at one time and mask registration is performed on a wafer-by-wafer basis. This means that costs are incurred on a "wafer" basis. Discretionary wiring assumes that a surplus of circuits will be formed on a wafer. Therefore, a significant portion of the wafer will be wasted. This limits the maximum complexity of an array (or the number of arrays on a wafer) below that possible with a given level of technology. As an additional drawback, it also reduces ultimate operating speed by introducing unpredictable and longer than necessary interconnections. These drawbacks are aggravated by the need for test pads on each circuit. Integrated circuit fabrication technology has already advanced to the point where a moderately complex circuit can be constructed in less area than required by the contact pads needed to test it.

An even more basic drawback to discretionary wiring is the loss of the batch fabrication process. From the time individual circuits are tested, each wafer has a unique identity and must be handled individually (Fig. 13.12).

13.4.3 Alternatives for Yield Enhancement

If discretionary wiring is rejected as a practical *LSI* technique, what can be done to enhance yield? The best solution is found by analyzing the yield problem itself. With a better understanding of semiconductor failure mechanisms, a better job of design can be done. With an adequate

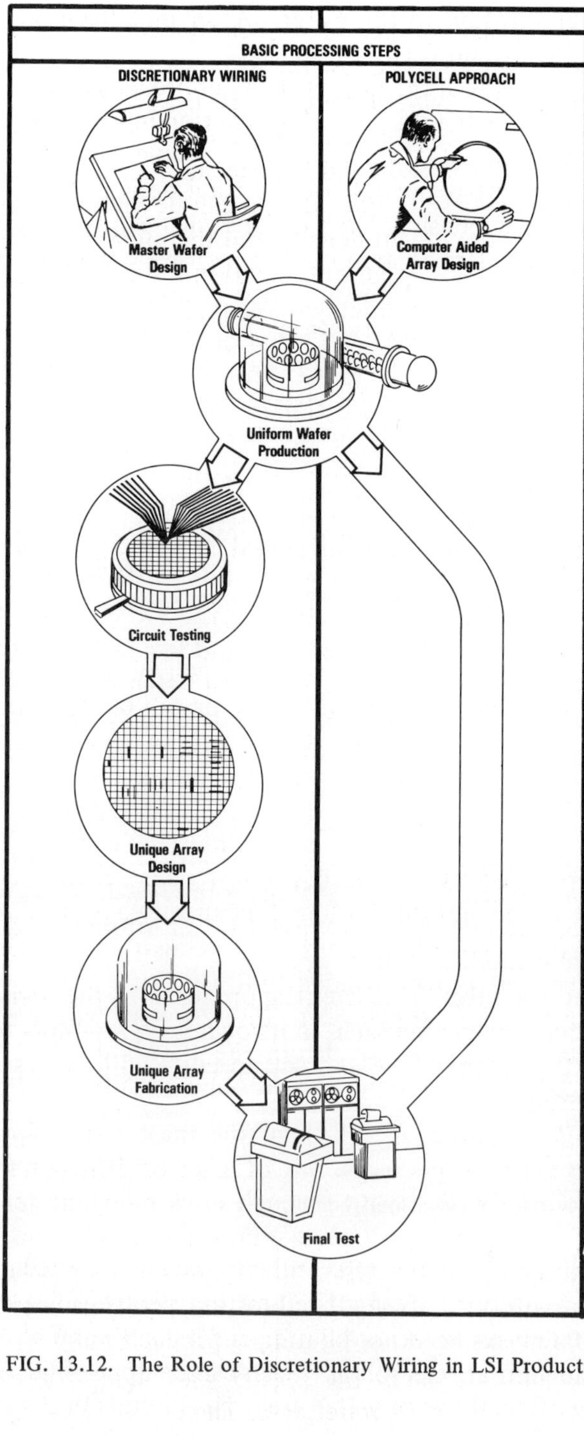

FIG. 13.12. The Role of Discretionary Wiring in LSI Production.

computer-aided design facility, design optimization becomes practical. Computer-aided procedures allow the ultimate in design flexibility. Exactly the number and type of components needed for the design can be used. Different circuit types can be used to allow optimization of both speed and power. Both circuit and logic techniques can be applied to implement the desired result. Each design can be optimized as to the best trade off between component density and the number of layers of interconnecting metal. The design optimization resulting from the available flexibility can yield a product with maximum producibility through batch processing techniques. Evidence of the historic results of design optimization is shown in Fig. 13.13. Note how the number of components per unit has increased with time. Taking advantage of the greater component density made possible through design optimization permits a large number of circuits to occupy a space that promises to be relatively free of surface imperfections and other flaws. Hence, the chip yield can be expected to be high despite increased complexity. Thus, today, reasonably complex subsystems are producible, with comparable yield on the same size chip as individual logic gates used to be. The combined effects of increasing component densities and increasing processing yields has demonstrated the feasibility of low cost batch fabricated *LSI* arrays.

Widespread acceptance of *LSI* depends equally upon very low cost and very fast vendor response time. Drastically reduced response time for both initial production parts and early samples of modified designs, is required if *LSI* arrays are to be designed into large (computer) systems. Due to the increasing complexity of *LSI* arrays, often some arrays require alteration when the total system is modified. What today is considered as excellent response time, will be an intolerable delay to wait for each system modification.

There are a number of existent approaches to the logistics for *LSI* implementation. Each approach, in its own way, attempts to attack the problem of turn-around time and set up costs without resorting to discretionary wiring.

Master Slice Approach. Through the master slice approach, it is attempted to get multiple usage out of a set of diffusion masks. Minimum turn-around time is sought through stock piling master slice wafers and applying custom metal to create unique parts. The master slice concept is also the basis of the discretionary wiring approach. The differences are that without individual cell testing, probe pads are not needed and that metal masks need not be unique for each wafer.

The basic limitations to the master slice approach are its inflexibility and inefficient use of wafer area. The circuits in the master wafer

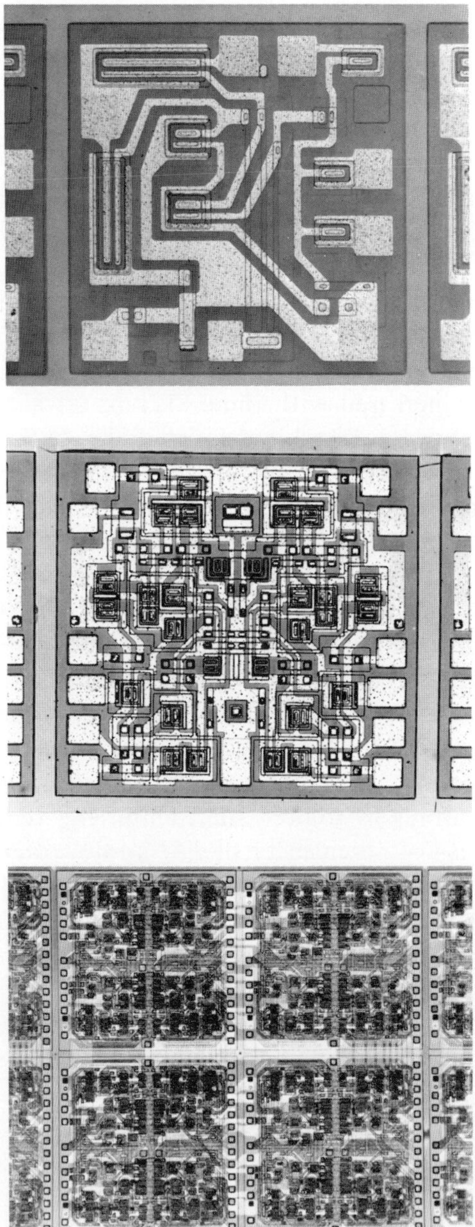

(a)

(b)

(c)

FIG. 13.13. A Historical Sequence of Increasing Complexity on Same Size Chip. (a) 1962, 14 Components, (b) 1965. 54 Components, (c) 1967, Over 500 Components.

are all that are available to the designer and there is seldom an appropriate mix of circuit types if any mix at all. The size of the chip is bigger than actually essential due to the number of unused elements. The attainment of rapid turn-around time is the most significant advantage to the master slice approach. Use of a computer-aided design system such as CAMP for generating or modifying the custom metal masks amplifies this important advantage.

The Polycell Approach. The CAMP system does not use the discretionary wiring technique to attempt to enhance yield. What the CAMP system provides is the capability for expediting *LSI* design through computer-aided techniques. This allows a high quality of design, optimized for minimum chip size and maximum processing yield. It is the concept of the polycell that makes rapid but efficient design by computer practical. A polycell is a logic entity such as a gate, flip-flop or complex function. Data structures representing these polycells are stored in a computer library. For the design of a particular array, the designer simply specifies the desired configuration of polycells using the CAMP procedures discussed in Sec. 13.3. The computer translates the specific configuration into instructions for the automatic drafting machine which prepares a complete set of mask masters for the required array. Although the polycell approach cannot match the turn-around time of the master slice approach, the resultant design makes for lower costs through higher yields. Another advantage of the polycell approach is the low cost and ease of modifying an existent design.

It appears that both the master slice and polycell approach have their place in the emerging *LSI* technology and that they both depend upon useful computer-aided design systems such as CAMP. There are additional significant problems ahead to be solved in the development of an *LSI* technology. The following problem areas have not been discussed to this point because although solutions to these problems are required, the form of solution is irrelevant to the above arguments between design technology and implementation logistics.

13.4.4 Specifications

LSI breaks down the conventional boundary between device and system design and requires the involvement of both disciplines. *LSI* design encompasses both circuit and logic design. With the complexity achievable for an *LSI* part, the logic structure is inseparably linked to the system structure. This changes the manner in which *LSI* circuits are specified. Specifications must still include circuit parameters but logical structure must also be completely defined. For example, an ordinary integrated circuit often contains as little as two interconnected NAND circuits where an *LSI* circuit might contain 50 to 500 interconnected circuits. To specify the function for such a unit requires detailed system structure data, most likely in the form of logic schematics and also logic equations. Much of this data is often proprietary. Techniques and procedures for respecting and protecting intercompany proprietary rights must be worked out. The use of a computer-aided design system in the device system interface can both help and complicate the requirement for protecting proprietary rights. Because of the increased complexity possible in an *LSI* part, its desired system function cannot be clearly implied by ordinary specification data. To assure that the *LSI* parts will function as desired in their intended system, it is important that the specification include certain functional tests to augment any standard testing procedures.

13.4.5 Design Verification

Along with increasing complexity goes an increasing concern that although the end product satisfies specification criteria, it may fail to function properly in the intended system. When such problems arise, the cost and time required for redesign and reprocessing are great. What is needed is a comprehensive design verification system. Such a system would include the capability for circuit analysis of various circuits in the design. From the results of the circuit analysis, parameters for a macro model of each circuit could be generated. The macro parameters and the logic structure of the design would then be used in a system (logic) simulation. Through simulation, the behavior of the design can be predicted and compared against functional requirements. For digital subsystems, functional simulation in its simplest form is merely logical simulation for combinatorial networks. This is the simplest form of simulation and is the one most often discussed. Simulation of sequential digital logic is also quite feasible and requires slightly more sophisticated techniques. A dynamic model of the actual logic elements must be established. Then, with initial conditions fixed, the sets of logic equations for all internal functions can be iteratively solved until the network becomes stable. Then, input transitions are considered in sequence. For this type of digital simulation to be successful, the time increment needs to be as small or smaller than the typical transition time for the functional model.

13.4.6 Testing

Testing of *LSI* presents a number of problems not encountered with ordinary integrated circuits. The requirement for AC and DC parameter testing is complicated by more input and output pins. Functional testing of digital *LSI* arrays requires the application of techniques heretofore, unnecessary in testing integrated circuits. Proper functional operation of simple digital integrated circuits can be exhaustively tested by monitoring outputs while sequencing through all possible input and internal state combinations. In many *LSI* designs the number of inputs and internal feedbacks is so large that the number of possible states is astronomical. Even for very fast automatic testers, exhaustive testing is impractical and reasonable testing sequences must be generated. There are two distinct needs for different testing sequences. Until a design is in full production status, it is highly desirable to diagnose failures. By doing so, design imperfections can be eliminated and process procedures optimized. For diagnosis, tests which isolate individual failures are sought. When in production, efficient testing sequences are needed to detect all faults with a minimum number of tests. Minimizing final testing complexity is necessary because testing cost adds directly to the cost of the product.

There are a variety of evolving philosophies for *LSI* testing. Most of these philosophies attack the problem of production testing, that is to assure a certain percentage probability of a part being good with a minimum amount of testing. One interesting approach is comparative testing. Once a good *LSI* unit has been found, or an appropriate physical model constructed, this good unit is exercised in parallel with a unit under test and the outputs compared. The sequence of input changes during comparison testing could be either random or generated by a variety of algorithms from the desired testing criteria. Test by comparison appears to have merits, but cannot be used until a good unit has been found or built. There is a variety of types of *LSI* test equipment being developed that do not depend on comparison. The common feature of all such testers is that they require test sequences to be generated for the control of their operation. Techniques for generating these test sequences need a lot of developmental work. The most successful procedures today have involved pseudo languages. These languages, much like a program compiler, translate statements of testing criteria into testing sequences in the correct data format for controlling test equipment. Promising steps toward automatic test sequence generation are being made by computer simulated comparison techniques. A number of hypothetical circuits (one which is good and the remainder which differ from each other and from the good circuit by the presence of specified faults) are simulated while being subjected to the same input conditions. Simulation continues until a difference is detected between the good circuit and the faulty copies. This technique promises potential utility in generating test sequences for either rapid fault detection or selective fault isolation.

SUMMARY

There is no doubt that *LSI* is just around the corner; however, the distance to the corner still remains nebulous. The problems of yield, design verification and testing are all under investigation by every major semiconductor manufacturer. Many new techniques are being tested and evaluated to enhance the yield as well as reliability of large complex integrated circuit chips. The CAMP system is being used to prepare the mask masters for these current tests. CAMP provides the design engineer with the ability to update or modify a design layout in the same time it previously took to describe the change.

ACKNOWLEDGEMENT

The work reported on in this chapter represents the results of a cooperative effort by many people in the Computer-Aided Design department at Motorola Semiconductor Products Division. While it is not possible to list all who have contributed, we would like to cite the contributions made by Kenneth Hutto for his work in the development of the CAMP language, Ronald Leatherman for his contributions in successfully achieving automatic drafting, Carole Martin for her work on the Device Library program, and Ronald Pfau for his work on the Resistor Design program.

REFERENCES

1. M. Ahrens, "Computer design of microelectronic circuits," *Electronic Design*, vol. 13., pp. 26-29, Jan. 4, 1965.

2. T. R. Allen, and J. E. Foote, "Input/output software capability for a man-machine communication and image processing system," *Proceedings of the F. J. C. C.*, vol. 26, pp. 387-396, 1964.

3. P. W. Case, et al., "Solid logic design automation," *IBM Journal*, vol. 8, no. 2, pp. 127-140, April, 1964.

4. S. H. Chasen, "The introduction of man-computer graphics into the aerospace industry," *Proceedings of the F. J. C. C.*, vol. 27, pp. 883-892, 1965.

5. S. A. Coons, "An outline of the requirements for a computer-aided design system," *Proceedings of the S. J. C. C.*, vol. 23, pp. 299-304, 1963.

6. H. S. Corbin, "A survey of CRT display consoles," *Control Engineering*, vol. 12, pp. 77-83, Dec. 1965.

7. M. R. Davis, and T. O. Ellis, "The RAND tablet: a man-machine graphical communication device," *Proceedings of the F. J. C. C.*, vol. 26, pp. 325-331, 1964.

8. Ruth M. Davis, "A history of automated displays," *Datamation*, vol. 9, pp. 24-28, Jan. 1965.

9. G. S. Devere, B. Hargreaves, and D. M. Walker, "The dac-I system," *Datamation*, pp. 37-48, June 1966.

10. H. Freitag, "Automatic mask generation for integrated circuits," *NEREM Record,* 1966.

11. H. Freitag, "Design automation for large-scale integration," *International Solid State Circuits Conference Digest*, Session 10, Aug. 23-26, 1966.

12. L. H. Hazlett, D. K. Lynn, and J. A. Narud, "Design and automatic artwork generation tradeoffs," *NEREM Record*, 1967.

13. L. Hazlett, "The polycell approach to large scale integration," *Electronics,* Feb. 20, 1967.

14. E. L. Jacks, "A laboratory for the study of graphical man-machine communication," *Proceedings of the F. J. C. C.*, vol. 26, pp. 343-350, 1964.

15. J. S. Koford, et al., "Using a graphic data processing system to design artwork for manufacturing hybrid integrated circuits," *Proceedings of the F. J. C. C.,* pp. 229-246, 1966.

16. J. W. Lathrop, "Discretionary wiring approach to large-scale integration," *IEEE Technical Papers,* Session 10, Aug. 23-26, 1966.

17. C. Y. Lee, "An algorithm for path connections and its applications," *IRE Transactions on Electronic Computers,* Sept. 1961.

18. D. K. Lynn, C. S. Meyer, and D. J. Hamilton (eds.), *Analysis and Design of Integrated Circuits*, New York: McGraw-Hill, 1967.

19. R. W. Mann, and S. A. Coons, "Computer-aided design," *McGraw-Hill Yearbook of Science and Technology*, McGraw-Hill, New York, N. Y., pp. 1-9 1965.

20. C. Hugh Mays, "Computer-aided design for large-scale integration," *International Solid State Circuits Conference Digest*, Feb. 16, 1967.

21. W. H. Ninke, "Graphic 1 - a remote graphical display console system," *Proceedings of the F. J. C. C.*, vol. 27, pp. 834-846, 1965.

22. W. H. Ninke, "Man-computer graphical communication," *System Analysis by Digital Computer*, Edited by Franklin F. Kuo and James F. Kaiser, John Wiley & Sons, 1966.

23. Donn B. Parker, "Solving design problems in graphical dialog," *On-line Computer Systems*, Edited by Walter J. Karplus, McGraw-Hill, Inc., 1966.

24. C. D. Phillips, W. C. Seelbach, J. A. Narud, D. K. Lynn, "Complex monolithic arrays: some aspects of design and fabrication," *IEEE Journal of Solid-State Circuits*, vol. SC-2, no. 4, Dec. 1967.

25. W. Raisanen, "Wally Raisanen of Motorola speaks out on the real world of integrated circuits," *EEE*, vol. 15, no. 12, Dec. 1967.

26. D. T. Ross, and J. E. Rodriques, "Theoretical foundations for the computer-aided design system," *Proceedings of the S. J. C. C.,* vol. 23, pp. 305-322, 1963.

27. W. C. Seelbach, "Computer subsystems on a chip of silicon," IFAC/IFIP *Symp.*, sec., 6.3, pp. 609-629.

28. D. M. Sheppard, et al., "A computer-aided method for checking and making monolithic integrated circuit masks," *International Solid State Circuits Conference Digest*, Session 10, Aug. 23-26, 1966.

29. I. E. Sutherland, "Sketchpad, a man-machine communication system," *Proceedings of the S. J. C. C.*, Detroit, Mich., pp. 329-346, May 21-23, 1963.

30. S. Triebwasser, "Programmable interconnection techniques," *Proceedings Solid State Circuits Conference*, pp. 124-126, Feb. 11, 1967.

31. Interim Engineering Progress Report, "Development of on-line system for computer-aided design of integral circuits," Norden Division, Project No. 9-521, United Aircraft Corporation, Norwalk, Conn., Mar. 1, 1966, May 31, 1966.

INDEX